AF778236

Catalytic and Noncatalytic Upgrading of Oils

ACS SYMPOSIUM SERIES **1379**

Catalytic and Noncatalytic Upgrading of Oils

Ajay K. Dalai, Editor
Department of Chemical and Biological Engineering, University of Saskatchewan
Saskatoon, Saskatchewan, Canada

Dady B. Dadyburjor, Editor
Department of Chemical Engineering, West Virginia University
Morgantown, West Virginia, United States

Ying Zheng, Editor
Department of Chemical and Biochemical Engineering, Western University, London, Ontario, Canada

Aijun Duan, Editor
College of Chemical Engineering and Environment, China University of Petroleum
Changping-qu, Beijing, China

William L. Roberts, Editor
Mechanical Engineering Program, King Abdullah University of Science and Technology
Thuwal, Mekkah, Saudi Arabia

Sonil Nanda, Editor
Department of Chemical and Biological Engineering, University of Saskatchewan
Saskatoon, Saskatchewan, Canada

Sponsored by the
ACS Division of Energy and Fuels

American Chemical Society, Washington, DC

Library of Congress Cataloging-in-Publication Data

Library of Congress Cataloging-in-Publication Data

Names: Dalai, Ajay Kumar, 1959- editor.
Title: Catalytic and noncatalytic upgrading of oils / Ajay K. Dalai, editor, Department of Chemical and Biological Engineering, University of Saskatchewan, Saskatoon, Saskatchewan, Canada, Dady B. Dadyburjor, editor, Department of Chemical Engineering, West Virginia University, Morgantown, West Virginia, United States, Ying Zheng, editor, Department of Chemical and Biochemical Engineering, Western University, London, Ontario, Canada,Aijun Duan, editor, College of Chemical Engineering and Environment, China University of Petroleum, Changping-qu, Beijing, China, William L. Roberts, editor, Mechanical Engineering Program, King Abdullah University of Science and Technology, Thuwal, Mekkah, Saudi Arabia, Sonil Nanda, editor, Department of Chemical and Biological Engineering, University of Saskatchewan, Saskatoon, Saskatchewan, Canada ; sponsored by the ACS Division of Energy and Fuels.
Description: Washington, DC : American Chemical Society, 2021. | Series: ACS symposium series ; 1379 | Includes bibliographical references and index.
Identifiers: LCCN 2021017493 (print) | LCCN 2021017494 (ebook) | ISBN 9780841298422 (permanent paper) | ISBN 9780841298415 (ebook)
Subjects: LCSH: Petroleum--Refining. | Catalytic reforming. | Chemical reactors. | Heavy oil.
Classification: LCC TP690.3 .C38 2021 (print) | LCC TP690.3 (ebook) | DDC 665.5/3--dc23
LC record available at https://lccn.loc.gov/2021017493
LC ebook record available at https://lccn.loc.gov/2021017494

The paper used in this publication meets the minimum requirements of American National Standard for Information Sciences—Permanence of Paper for Printed Library Materials, ANSI Z39.48n1984.

Copyright © 2021 American Chemical Society

All Rights Reserved. Reprographic copying beyond that permitted by Sections 107 or 108 of the U.S. Copyright Act is allowed for internal use only, provided that a per-chapter fee of $40.25 plus $0.75 per page is paid to the Copyright Clearance Center, Inc., 222 Rosewood Drive, Danvers, MA 01923, USA. Republication or reproduction for sale of pages in this book is permitted only under license from ACS. Direct these and other permission requests to ACS Copyright Office, Publications Division, 1155 16th Street, N.W., Washington, DC 20036.

The citation of trade names and/or names of manufacturers in this publication is not to be construed as an endorsement or as approval by ACS of the commercial products or services referenced herein; nor should the mere reference herein to any drawing, specification, chemical process, or other data be regarded as a license or as a conveyance of any right or permission to the holder, reader, or any other person or corporation, to manufacture, reproduce, use, or sell any patented invention or copyrighted work that may in any way be related thereto. Registered names, trademarks, etc., used in this publication, even without specific indication thereof, are not to be considered unprotected by law.

Foreword

The ACS Symposium Series is an established program that publishes high-quality volumes of thematic manuscripts. For over 40 years, the ACS Symposium Series has been delivering essential research from world leading scientists, including 36 Chemistry Nobel Laureates, to audiences spanning disciplines and applications.

Books are developed from successful symposia sponsored by the ACS or other organizations. Topics span the entirety of chemistry, including applications, basic research, and interdisciplinary reviews.

Before agreeing to publish a book, prospective editors submit a proposal, including a table of contents. The proposal is reviewed for originality, coverage, and interest to the audience. Some manuscripts may be excluded to better focus the book; others may be added to aid comprehensiveness. All chapters are peer reviewed prior to final acceptance or rejection.

As a rule, only original research papers and original review papers are included in the volumes. Verbatim reproductions of previous published papers are not accepted.

ACS Books

Contents

Indexes

Chapter 1

Catalytic and Noncatalytic Upgrading of Bio-Oil to Synthetic Fuels: An Introductory Review

Sonil Nanda,[1] Falguni Pattnaik,[2] Venu Babu Borugadda,[1] Ajay K. Dalai,[*,1] Janusz A. Kozinski,[3] and Satyanarayan Naik[2]

[1]Department of Chemical and Biological Engineering, University of Saskatchewan, Saskatoon, Saskatchewan S7N 5A9, Canada

[2]Center for Rural Development and Technology, Indian Institute of Technology Delhi, New Delhi 110016, India

[3]Faculty of Engineering, Lakehead University, Thunder Bay, Ontario P7B 5E1, Canada

[*]Email: ajay.dalai@usask.ca

Biofuels can potentially address greenhouse gas emissions and related environmental issues caused by fossil fuels. Fossil fuels such as gasoline and diesel have been the preferred fuels for the automotive sector. Although promising, crude bio-oil derived from pyrolysis and liquefaction of waste biomass does not meet the fuel standards for direct use in combustion engines and power plants. Bio-oil has a considerable amount of water as well as components containing oxygen, nitrogen, sulfur, metals, and aromatic compounds. Such components add many undesired properties to bio-oil such as high viscosity, low fluidity, low heating value, greater acidity, and thermal instability. This chapter is an introductory review of some notable catalytic and noncatalytic bio-oil upgrading technologies that make them compatible with transportation fuels. The catalytic upgrading technologies reviewed include hydrogenation, hydrocracking, esterification, and transesterification. The noncatalytic upgrading techniques reviewed are emulsification, solvent addition, supercritical fluids, and electrochemical stabilization. The strengths, weaknesses, opportunities, and threats for each of these bio-oil upgrading technologies are comprehensively discussed along with their operational mechanisms and challenges.

© 2021 American Chemical Society

Introduction

The transportation sector, aviation, marine, as well as combined heat and power (CHP) facilities will continue to rely on liquid fuels in the future. However, CO, CO_2, NO_x, SO_x, and other greenhouse gases emitted from the burning of liquid fossil fuels represent a significant contribution to global warming. According to the International Energy Agency, to keep the global CO_2 emissions below 450 ppm, 27% of fossil fuel consumption should be replaced by biofuels by 2050 (*1*). Therefore, to overcome the challenges posed by fossil fuels, producing sustainable fuels from renewable resources has gained considerable interest. Renewable resources (e.g., forestry residues, agricultural crop biomass, invasive crops, dedicated energy crops, algae, food waste, municipal solid waste, cattle manure, waste fats, and discarded cooking oil) appear to be the most promising feedstock to produce synthetic liquid transportation fuels (*2, 3*).

Biofuels could play an important role in partially or completely replacing fossil partially, thereby reducing greenhouse gas emissions. Examples of carbon-neutral biofuels include bio-oil, biodiesel, bio-jet fuel, bioethanol, biobutanol, biohydrogen, biomethane, and syngas (*3*). The technologies used to produce biofuels can be categorized as biomass-to-liquid (e.g., liquefaction, pyrolysis, transesterification, fermentation, solvent extraction), biomass-to-gas (e.g., gasification, steam reforming, and anaerobic digestion), and gas-to-liquid (e.g., Fischer–Tropsch catalysis and syngas fermentation) processes (*4, 5*). Among all the waste-to-energy conversion technologies, pyrolysis and liquefaction have gained the most attention in the production of synthetic liquid fuels and chemicals. Pyrolysis operates under an inert atmosphere to transform biomass into bio-oil, producer gas, and biochar at variable temperatures, heating rates, and vapor residence times (*6*). In contrast, hydrothermal liquefaction operates under subcritical water to liquefy biomass without the requirement of latent heat to remove its in-situ moisture (*7*).

Bio-oil generated from pyrolysis and liquefaction contains higher levels of water and oxygen that result in a mediocre heating value and poor fuel properties (*8*). The feedstock requires drying before pyrolysis to reduce the presence of water and oxygenated compounds in the resulting bio-oil, which could be costly and inefficient at industrial scales. In contrast, the use of subcritical water as a reaction medium in hydrothermal liquefaction could lower the cost of the process by eliminating the biomass drying step and utilizing a wide variety of high-moisture feedstock. Bio-oil derived from pyrolysis and hydrothermal liquefaction requires further processing to produce synthetic transportation fuels. This involves the removal of unwanted elements (i.e., oxygen, nitrogen, and sulfur) through upgrading techniques. Through upgrading, the inherent physicochemical properties of bio-oil (e.g., high acidity, moisture and oxygen content, viscosity, and low heating value) can be improved. Upon addressing these challenges, bio-oils can be used for corefining at the refineries with conventional crude oil or as a direct drop-in biofuel in the existing liquid transportation infrastructures (*9*).

Presently, many catalytic and noncatalytic processes are being used to obtain high-quality liquid transportation fuels, which include hydrotreating, hydrocracking, blending, emulsification, and steam reforming, to name a few. Catalytic upgrading processes are classified based on the type of catalyst used such as homogeneous (e.g., soluble organic acids, inorganic acids, alkalis, and carbonates) and heterogeneous (e.g., insoluble inorganic salts, metals, and molecular sieves) (*10*). Upgrading of bio-oil is extremely crucial to determine its utilization in combustion engines, power plants, automobiles, and biochemical fractionation. Bio-oil upgrading technologies have recently gained wide interest in research and development. However, the current literature on catalytic and noncatalytic upgrading technologies appears scattered. This chapter aims to resolve the scattered literature by systematically reviewing the status, benefits, and limitations of some notable catalytic and noncatalytic bio-oil upgrading techniques. Furthermore, the performance of each catalyst relative to the upgrading technique and the physicochemical properties of the raw and upgraded bio-oils are discussed. The characteristics of a single- and multi-stage upgrading process are compared. The type of suitable catalysts along with the process conditions optimal to produce synthetic liquid transportation fuels is described. The chapter closes with the challenges and perspectives of catalytic and noncatalytic upgrading techniques and recommendations for future research.

Methods for Bio-Oil Production

Pyrolysis

Pyrolysis is performed under an oxygen-free atmosphere, which is achieved by the purging of an inert gas such as nitrogen or argon into the reactor. During pyrolysis, the biopolymers present in lignocellulosic biomasses (e.g., cellulose, hemicellulose, and lignin undergo) thermal cracking to generate various lower molecular weight products that recondense to produce bio-oil and biochar along with noncondensable gases such as CO_2, CO, H_2, and CH_4 (*6*). Bio-oil can be used as a precursor to producing synthetic transportation fuels and biochemicals. Alternately, biochar is a carbon-rich solid product that has applications in agronomy, carbon sequestration, catalyst support, activated carbon, absorbent, graphene production, biocomposites, gas storage, engineered carbon materials, cosmeceuticals, and pharmaceuticals (*11*). The producer gas contains some combustible fractions (e.g., CO, H_2, and CH_4) with applications for energy generation in CHP plants or gas-to-liquid conversion technologies. Figure 1 illustrates the underlying mechanisms of pyrolysis and liquefaction of biomass.

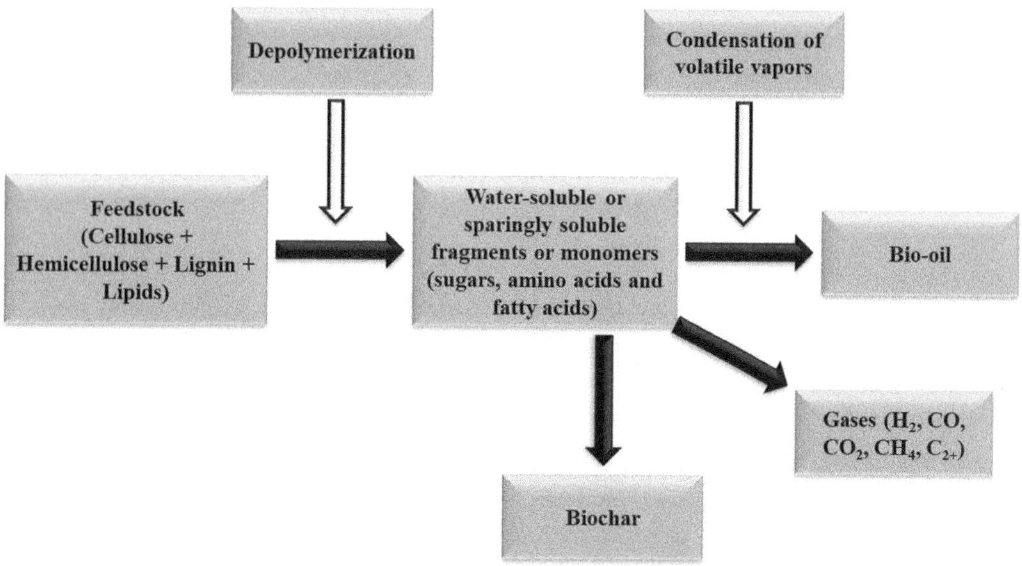

Figure 1. Basic reaction pathways for thermochemical conversion of biomass through pyrolysis and liquefaction.

Pyrolysis can be broadly categorized into slow, fast, and flash pyrolysis, based on the temperature, heating rate, and vapor residence time. Slow pyrolysis is implemented for maximizing biochar production because of its slow heating rate (i.e., 0.1–1 °C/s), moderate process temperatures (i.e., 300–600 °C), and longer vapor residence times (>30 min) (*12, 13*). The longer residence times allow the volatile vapors generated from the thermal cracking of biomass to undergo further heterogeneous secondary reactions (e.g., repolymerization, oxidation, and carbon deposition) leading to higher yields of biochar (*14*). In contrast, fast pyrolysis produces higher yields of bio-oil because of its characteristic fast heating rates (i.e., 10–1000 °C/s), higher process temperatures (i.e., 300–700 °C), and short vapor residence times (<2 s) (*15*). Lastly, flash pyrolysis is operated at higher temperatures (i.e., 900–1300 °C), rapid heating rates (>1000 °C/s), and shorter vapor residence time (<0.5 s). In flash and fast pyrolysis processes, the hydrocarbon vapors and volatile species are rapidly quenched and condensed to produce higher yields of bio-oil due to shorter vapor residence times (*16*).

Pyrolysis-derived bio-oil contains several types of oxygenated aromatic and aliphatic organic compounds, which subsequently increases the oxygen to carbon ratio and lowers the heating value (*17*). Catalytic pyrolysis can help decrease the content of oxygenated organic compounds and increase the hydrogen to carbon ratio in the resulting bio-oils. Several homogeneous and heterogeneous catalysts (e.g., zeolites, mesoporous aluminosilicate, MgO, NaOH, KOH, Na_2CO_3, Al_2O_3, K_2CO_3, $Ca(OH)_2$, and other acid/base catalysts) are used for catalytic pyrolysis (*16*, *18*). Catalytic pyrolysis improves thermal cracking of organics, increases the aromatic content, and simultaneously enhances the heating values of bio-oil. Besides the conventionally powdered zeolites, several metals supported zeolites (i.e., Y-zeolite with Fe, Co, Ni, Ga, Mo, and Ru) have proven efficient catalytic materials to enhance the bio-oil quality (*19*).

Zainan et al. performed the catalytic pyrolysis of microalgae *Chlorella vulgaris* using nickel-loaded zeolites (*20*). The authors reported obtaining a high-grade bio-oil with decreased oxygenated and nitrogenous compounds. Also, the bio-oil had a higher H/C ratio, which is dependent upon the catalyst/feedstock ratio (i.e., the higher the catalyst/feedstock ratio, the higher is the hydrocarbon contents of the bio-oil). Dai et al. performed the microwave-assisted catalytic pyrolysis of soapstock using HZSM-5. At optimal temperature (600 °C) and catalyst/feedstock ratio (1:1), hydrocarbon-rich pyrolysis oil was obtained (*21*). At these conditions, the hydrocarbon content of bio-oil increased from 65% to 88%, whereas the aromatics contents escalated from 6% to 29%.

Hydrothermal Liquefaction

Hydrothermal liquefaction is a kind of hydrated pyrolysis involving thermochemical depolymerization to the biomass under hot and compressed aqueous conditions to produce biocrude oil, gases, and hydrochar. Subcritical water provided the aqueous medium for hydrothermal liquefaction acting as a green solvent and catalyst. Subcritical water occurs at temperatures and pressures below the water's critical points. The critical temperature (T_c) and critical pressure (P_c) of water are 374 °C and 22.1 MPa, respectively (*22*). The advantages of using subcritical water in hydrothermal liquefaction are its nontoxicity, nonflammability, eco-friendliness, increased diffusivity, solvation ability, self-ionization, reduced viscosity, surface tension, as well as less polarity (*23*). Hydrothermal liquefaction also enhances the solubility of organic compounds in subcritical water owing to the decreased water density (*24*). As the reaction state reaches the critical point of water, many thermophysical properties are dramatically altered to create a homogeneous medium for hydrolysis and cracking reactions.

Hydrothermal liquefaction is operated at a temperature between 250 and 350 °C under high pressures (i.e., 5–20 MPa) (*25*). Hydrothermal liquefaction is largely independent of process heating rate and biomass particle size since subcritical water has a leading role in heat transfer, thus providing a homogeneous extraction medium (*26*). The yield and physicochemical characteristics of the products from hydrothermal liquefaction are mainly influenced by feedstock type, processing conditions (e.g., reaction temperature and time) and catalyst type (*26*).

The primary comparison between hydrothermal liquefaction-derived bio-oil and pyrolysis-derived bio-oil relies on their yields, composition, and heating value. Hognon et al. conducted a comparative study on bio-oil quality obtained from pyrolysis and hydrothermal liquefaction of green algae *Chlamydomonas reinhardtii* (*27*). This comparative analysis determined that the bio-oil obtained from hydrothermal liquefaction at 310 °C had higher yields, greater aromatic components, as well as lower moisture, nitrogen, and oxygen contents than the pyrolysis oil. Bach et al. performed hydrothermal liquefaction of Norwegian brown algae *Laminaria saccharina* and reported bio-oil yield of 79% with a high heating value of 36 MJ/kg at the optimal temperature, reaction time, and biomass/water ratio of 350 °C, 15 min, and 1/10 w/w, respectively (*28*).

Catalytic and microwave-assisted hydrothermal liquefaction are also used to enhance the quality of biocrude oil by altering its compositions through dehydration, dehydrogenation, isomerization, deoxygenation, and other chemical reactions (*29, 30*). Hydrothermal liquefaction involves hydrolysis and depolymerization of cellulose, hemicellulose, lignin, protein, and other biopolymers into various water-soluble or sparingly soluble monomers, which subsequently react with each other to form several aromatic and aliphatic compounds with higher heating values and fuel properties (*31*). Upon upgrading by catalytic and noncatalytic methods these organic compounds can be converted into several hydrocarbons through decarboxylation, decarbonylation, cracking, hydrogenation, and deoxygenation reactions (*32*).

Requirements for Bio-Oil Upgrading

Bio-oil and biocrude generated from pyrolysis or liquefaction contain considerable amounts of moisture, heteroatoms (i.e., oxygen, nitrogen, sulfur), and heavy metals. Moreover, several oxygenated compounds (i.e., carboxylic acids and alcohols) are produced during pyrolysis and liquefaction, which increases the acidity of the bio-oil. The presence of carboxylic acids leads to condensation and polymerization reactions causing an increase in the acidity and viscosity of the oil. While bio-oil acidity leads to the corrosion of reactors and storage units, a greater viscosity creates problems during transportation (*33*). These oxygenated compounds subsequently decrease the heating value of the bio-oil to approximately 50% than that of fossil fuels (*34*). Unprocessed bio-oil may consist of 15–30% of moisture decreases its heating values and delays the ignition process (*33*).

Besides carboxylic acids and alcohols, more than 400 types of oxygenated compounds are present in bio-oil (e.g., phenolics, furans, polyols, aldehydes, and ketones) (*35*). The presence of such highly reactive organic compounds makes the bio-oil thermally and chemically unstable under ambient conditions. The high-ash contents (e.g., alkali and alkaline earth metals like Na, K, Ca, and Mg) can be deposited in the reactor and pipelines causing severe wear. Wearing of metallic components of engine, refineries, fuel pumps, meters, and piping systems could also be due to the poor lubricity of crude oil (*36*). However, the lubricity of bio-oil could be optimized to match with fuel standards by upgrading and blending with conventional fuel oils. Unprocessed bio-oil contains a larger fraction of nitrogen and sulfur-containing compounds retained from the feedstock, which could release sulfur oxides (SO_x), nitrogen oxides (NO_x), and particulate matter upon combustion (*37*).

To enhance the fuel characteristics of bio-oil, some specific catalytic and noncatalytic methods are implemented, which include esterification, deoxygenation, denitrogenation, desulfurization, dehydrogenation, hydrodeoxygenation, cracking, condensation (e.g., aldol condensation), and other organic reactions (*34*). Figure 2 illustrates some commonly used bio-oil upgrading technologies. Considering the adverse effects of oxygenated compounds on the bio-oil properties, hydrodeoxygenation plays a crucial role in eliminating oxygen by forming H_2O, CO_2, and CO. A major disadvantage of this upgrading process is the loss of carbon in the form of CO_2 and CO along with the undesired formation of coke to compromise catalyst selectivity and cause catalyst poisoning (*38*). Another way of upgrading crude bio-oil is through catalytic cracking using zeolites, which eliminates the excessive oxygen contents through the cracking reactions (*39*). The process is affected by the negative effects of the catalyst poisoning by the deposition of heavy tar formed during the condensation reactions. Using appropriate reaction parameters and catalysts, oxygenated compounds from the bio-oil can be removed in the form of water to lower its viscosity and enhance the fluidity through the breaking of aromatic or polyaromatic rings (*5*).

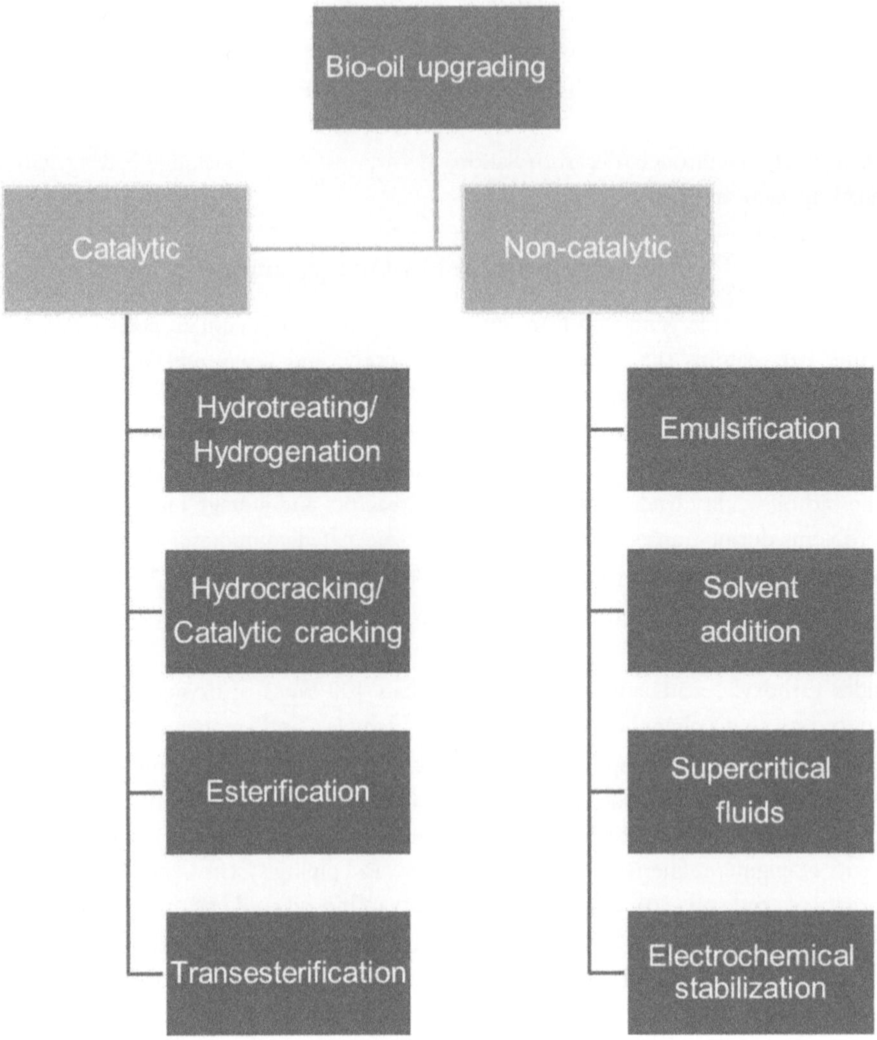

Figure 2. Catalytic and noncatalytic techniques used for bio-oil upgrading.

The overall quality of bio-oil can be improved by the removal of low-molecular-weight compounds to enhance its flashpoint and stability. This can be achieved by the solvent evaporation method at a lower temperature (*40*). Another method for bio-oil upgrading, steam reforming, produces syngas (H_2 and CO) that can be used as a direct gaseous fuel or as a feedstock for gas-to-liquid technologies (e.g., Fischer–Tropsch synthesis and syngas fermentation) (*41*, *42*). Through the Fischer–Tropsch process, syngas can be catalytically converted to higher alcohols and green diesel, whereas through syngas fermentation, *Clostridium* spp. can ferment syngas to produce ethanol (*22*, *43*). Moreover, hydrogen can also be separated from the syngas for use in fuel cells, hydrotreating, energy vector, energy carrier, and as a precursor for chemicals synthesis (*44*).

Catalytic Upgrading of Bio-Oil

Hydrotreating and Hydrogenation

Hydrotreating is a mature catalytic upgrading technique operating typically at high-temperatures (345–500°C) and high-pressures (5–15 MPa) in the presence of heterogeneous acidic catalysts used to reduce oxygen, sulfur, nitrogen, metals, aromatics, and unsaturated hydrocarbons from bio-oil (*45*). Under the spectrum of hydrotreating, several other processes take place such as hydrodeoxygenation, hydrodenitrogenation, hydrodesulfurization, hydrodearomatization, hydrodemetallization, and conversion of unsaturated compounds into saturated components (*46*). As the names suggest, hydrodeoxygenation, hydrodenitrogenation, hydrodesulfurization, hydrodearomatization, and hydrodemetallization serve to eliminate oxygenates, nitrogenous compounds, sulfurous compounds, aromatics, and metals from bio-oil.

The hydrodeoxygenation ability of different catalysts on several feedstocks and process conditions are summarized in Table 1. Sulfided CoMo and NiMo supported on γ-Al_2O_3 are traditionally used for hydrodeoxygenation of bio-oil (*52–54*). However, these catalysts deactivate at a faster rate due to sulfur leaching and clog the reactor by forming char, coke, and tar (*54*). In contrast, noble metal catalysts (sulfur-free) are used for hydrogenation with higher catalytic activity and product selectivity (*55*). Such catalysts are expensive for large-scale operations. Inexpensive and novel amorphous catalysts have been developed for hydrodeoxygenation of bio-oil without compromising the catalytic activity and product selectivity. Wang et al. have demonstrated the preparation of amorphous, sulfur-free, and non-noble metal catalysts (i.e., Ni(Co)–W–B and Co–Mo–B) (*56*, *57*). The catalysts demonstrated superior bio-oil hydrodeoxygenation with higher thermal stability and catalytic performance. Furthermore, metal phosphides and metal carbide catalysts are also developed for hydrodeoxygenation of bio-oil (*58*, *59*).

Hydrodenitrogenation and hydrodesulfurization are common processes used to remove nitrogen and sulfur compounds from conventional crude oil. Compared to conventional crude oil, biocrudes have relatively less nitrogen and sulfur contents. These heteroatoms can be removed via hydroprocessing in the presence of a suitable catalyst, high hydrogen pressure and high-temperature conditions. Biocrude produced from algal feedstocks is high in nitrogen (*60*). As such, algal-derived biocrude oil is subjected to hydrodenitrogenation to remove nitrogen and prevent NO_x emissions upon combustion in diesel or gasoline engines. Catalytic hydroprocessing can remove unwanted oxygen, sulfur, and nitrogen in the form of H_2O, H_2S, and NH_3 over commercial NiMo or CoMo on γ-Al_2O_3 catalysts (*52–54*).

Table 1. Recent Studies on Catalytic Hydrodeoxygenation of Different Feedstocks with their Selectivity and Hydrocarbons Yield

Feedstock	*Catalyst*	*Hydrocarbons*	*Process conditions*	*Reference*
Beechwood	10% Ni - 15% WO_3-ZrO_2	• Alkylcycloalkanes • Alkyl-benzenes	Pyrolysis (120–210 °C, 3 MPa H_2)	Zerva et al. (*47*)
Cornstalk	Ru–$NbOPO_4$	• Alkanes • Methylcyclohexane • Ethyl cyclohexane • Propyl cyclohexane	Pyrolysis (170 °C, 1 MPa H_2, 1 h, 0.1 g catalyst)	Zhao et al. (*48*)
Pine sawdust	Mo_2C/CNF	• Phenols (56–78 wt%) • Cyclic ketones (7–30 wt%) • Carboxylic acids (2–8 wt%) • Esters (0–9 wt%) • Aromatic compounds (0–20 wt%)	Fast pyrolysis (350 °C, 4 MPa H_2, 1 h, 0.19 g catalyst/g bio-oil)	Remón et al. (*49*)
Straw and residual wood	NiCu/SBA-15, NiMn/SBA-15, NiZn/SBA-15	• Char (8.2–22.3 wt%) • Light oil (3.7–11.7 wt%) • Heavy oil (40.8–60.6 wt%) • Gases (16.3–43.7 wt%)	250–350 °C, 45 min, 3 MPa H_2	Oh et al. (*50*)
Woody biomass	Ni/C, Ni/SBA-15, Ni/Al-SBA-15, NiMn/SBA-15	• Char (2.3–11 wt%) • Light oil (10.7–35 wt%) • Heavy oil (29.6–42.3 wt%) • Gases (26.3–37.9 wt%)	Fast pyrolysis (550 °C, 3 MPa, 3 h, 3 g catalyst)	Oh et al. (*51*)

Hydrocracking and Catalytic Cracking

As the name suggests, hydrocracking and catalytic cracking of bio-oil are processes that break down larger hydrocarbon molecules into smaller range hydrocarbon molecules followed by hydrogenation (for hydrocracking) in the presence of a heterogeneous catalyst at temperatures >350°C and a hydrogen pressure of 0.7–21 MPa (*61*). Catalysts play a dual role in hydrocracking: (1) the removal of oxygenated compounds like water and (2) cleavage of the long-chain hydrocarbons into smaller hydrocarbons (Eq. 1 and Eq. 2). Most hydrocracking catalysts are designed with active metal loading (e.g., Ni, Mo, Co, etc.) on acidic supports such as zeolites or aluminosilicates that aid in cracking.

$$\text{Bio-oil } (CH_{1.4}O_{0.4}) + 0.7H_2 \rightarrow 1CH_2(\text{hydrocarbons}) + 0.4H_2O \qquad (1)$$

$$R_1\text{-}CH_2\text{-}CH_2\text{-}R_2 + H_2 \rightarrow R_1\text{-}CH_3 + CH_3\text{-}R_2 \quad (\text{where } R_1, R_2 - \text{Alkyl groups}) \qquad (2)$$

The hydrocracking and catalytic cracking of different catalysts on several feedstocks and process conditions are summarized in Table 2. Hansen et al. reported that the upgraded bio-oil via catalytic cracking could meet the ASTM specification for heating value, pour point, and viscosity (*67*). Though the catalytic cracking approach is effective to reduce oxygen and improve the quality of bio-oil, it is not feasible because of high-temperature and high-pressure requirements. Catalytic pyrolysis can be integrated to produce bio-oil with low oxygen content followed by catalytic cracking to upgrade the bio-oil over $Fe/\gamma\text{-}Al_2O_3$ (*68*). Furthermore, Xu et al. concluded that, compared to traditional catalytic cracking, the integrated approach performed better in promoting the liquid hydrocarbons yield and quality of the bio-oil (*68*).

Esterification and Transesterification

Esterification converts organic acids (e.g., formic, acetic, and propionic acid) present in bio-oil into their respective esters in the presence of catalyst using alcohol-based solvents (Eq. 3 and Eq. 4). The addition of alcohols to sugars, aldehydes, and ketones present in the bio-oil produces acetals that lower the acid value, water content, viscosity, flow properties, and increases the heating value of the bio-oil by reducing oxygen and water contents (*67*).

$$\text{Bio-oil} + \text{R-OH} \leftrightarrow \text{Upgraded Bio-oil} + H_2O \qquad (3)$$

$$R_1\text{-CHO} + 2(\text{R-OH}) \leftrightarrow R_1\text{-CH(OR)}_2 + H_2O \qquad (4)$$

Schulzke et al. esterified fast pyrolysis bio-oil with butanol in the presence of zinc oxide and zeolites (*69*). The authors reported the esterified bio-oil as the potential candidate for boiler fuels (heating oil) and gas turbines for CHP generation. During esterification, water is produced as a by-product that affects the catalytic activity. Furthermore, phenolic compounds were found to be abundant followed by ketones and high molecular weight hydrocarbons in the esterified bio-oil, limiting its direct application as a liquid transportation fuel (*70*).

Table 2. Recent Studies on Hydrocracking and Catalytic Cracking of Different Feedstocks with their Selectivity and Hydrocarbons Yield

Feedstock	*Catalyst*	*Process conditions*	*Products*	*Reference*
10 wt% algae bio-oil blended with heavy vacuum gas oil	Commercial fluid catalytic cracking	Hydrothermal liquefaction (520 °C, 12 s)	Decant oil, gasoline, light cycle oil	Santillan-Jimenez et al. (*62*)
Bio-oil ($CH_2.12O_{0.89}$)	La/HZSM-5	500–750 °C, 6 wt% La/HZSM-5 catalyst, 0.4 h^{-1} WHSV	C2–C4 light olefins	Gong et al. (*63*)
Blend of pine sawdust and kitchen waste oil (100:0, 50:50, 0:100 wt/wt)	HZSM-5	Fast pyrolysis (550 °C, 0.5 mL/min, 90 min)	Olefins and aromatic hydrocarbons	Ma et al. (*64*)
Pine sawdust	HY, HZSM-5, and HBeta	Fast pyrolysis (550 °C, 6 s, 1.8–7.3 g catalyst/g bio-oil)	Gasoline for HY, liquefied petroleum gas for HZSM-5; Hydrocarbon yield 56% for HBeta	Ibarra et al. (*65*)
Surrogates of long-chain paraffins	NiAg/SAPO-11	320–380 °C, 3 MPa, 2 h^{-1} WHSV, 13.6 H_2/alkane molar ratio	Jet-fuel range products	Chen et al. (*66*)

Note: Weight hourly space velocity (WHSV).

Transesterification substitutes long-chain esters in bio-oil using alcohols (Eq. 5) with solid acid catalysts, which separates water and oxygen as well as reduces the viscosity. The major difference between esterification and transesterification is that the former produces an ester and water as product and by-products, whereas the latter utilizes ester as a reactant and produces alcohol and ester with varying chain lengths (*67*). Solid acid catalysts are found appropriate for esterification and transesterification of bio-oil to stabilize to be blended at the refinery (*71*). Esterification and transesterification cannot remove unwanted sulfur, nitrogen, and metals from the bio-oil. Unlike hydrotreating and hydrocracking, esterification and transesterification are less popular bio-oil upgrading technologies. The esterification and transesterification of several feedstocks and process conditions are summarized in Table 3.

$$2\text{R-COOR}^1 + 2\,(\text{R}^2\text{-OH}) \leftrightarrow 2\text{R-COOR}^2 + 2(\text{R}^1\text{-OH}) \quad (\text{where } \text{R}^1, \text{R}^2 - \text{Alkyl groups}) \qquad (5)$$

Noncatalytic Upgrading of Bio-Oil

Emulsification

Emulsification is a process of dispersing one liquid into another liquid with low miscibility to form an emulsion by implementing certain surfactants. For instance, the mixing of two immiscible liquids like water and oil results in an emulsion. In an emulsion, there are two phases, namely the dispersed phase and the continuous phase. As an emulsion fuel, bio-oil is considered as the *dispersed phase*, whereas fossil fuel or conventional hydrocarbon fuel is the *continuous phase*. Unlike other upgrading techniques, emulsification is considered a comparatively instantaneous method that can upgrade bio-oil by improving its heating value, cetane number, and reducing the viscosity (*76*). However, emulsion fuels encounter poor thermal stability and phase separation that create issues for long-term storage (*77*). To obtain a stable emulsion oil, an appropriate surfactant (i.e., emulsifying agent) and an efficient method of mixing the two immiscible liquids are required.

During the preparation of the emulsion, the bio-oil should be mixed properly or dispersed into the fossil oil by vigorous stirring. In comparison with conventional stirring, for effective mixing, ultrasonic emulsification is adopted by several researchers (*78, 79*). Farooq et al. upgraded ether-extracted bio-oil by enhancing its stability by an emulsion with diesel using two different surfactants (Span 80 and Tween 60) (*80*). Both the surfactants were used individually and in combination. The authors reported the bio-oil/diesel emulsion to be highly stable for 40 days with a high heating value of 44 MJ/kg, which was comparable with that of diesel fuel (i.e., 45 MJ/kg).

By using these two emulsifying agents (i.e., Span 80 and Tween 60), Du et al. prepared an emulsion by ultrasonic method that resulted in an extremely stable emulsion of bio-oil and diesel with an optimum hydrophilic-lipophilic ratio (*81*). The researchers reported an emulsion oil with more enhanced fuel properties than bio-oil and that can be implemented in several engines. Moreover, for higher emulsification, Li et al. used 1-butanol as a coemulsifier with Tween-85 (*82*). An emulsion oil with a stability time for 20 days was obtained with 20–40 °C and 5 min of reaction time.

Besides using diesel as the continuous phase, for enhanced sustainability, bio-oil can be emulsified with biodiesel. This could improve the efficiency of the fuel and its combustion properties, thus reducing engine emissions (*83*). A reduction in NO_x emissions from emulsified bio-oil has also been reported (*67*).

Table 3. Feedstocks, Catalysts, and Alcohols Used to Esterify and Transesterify Different Bio-Oil Samples

Feedstock	*Catalyst*	*Alcohol*	*Process conditions*	*Reference*
Beechwood and cereal straw	ZnO, Amberlyst, zeolites	Butanol, ethanol, hexanol	Fast pyrolysis (80–120 °C, 6 h, 100 mL alcohol, 5 g catalyst)	Schulzke et al. (69)
Pine sawdust	$MoNi/\gamma-Al_2O_3$	Methanol	Fast pyrolysis (100 °C, 3 MPa, 2 h, catalyst 0.50\~g per 10 mL bio-oil)	Xu et al. (72)
Rice husk	$NaHSO_4$	Butanol	Pyrolysis (116 °C, 4 h, 50 g alcohol, 4 g catalyst)	Xu et al. (73)
Rice husk	SO_4^{2-}/ZrO_2-TiO_2, SO_4^{2-}/SiO_2-TiO_2, and SO_4^{2-}/TiO_2	Ethanol	Pyrolysis (80 °C, 2 h, 10% catalyst with bio-oil)	Song et al. (74)
Rice husk	Solid acids ($SiO_2/TiO_2-SO_4^{2-}$) and solid bases (K_2CO_3/Al_2O_3-NaOH)	Ethanol	Pyrolysis (50 °C, 5 h, 2:5 v/v ethanol/bio-oil ratio, 5 wt% catalyst)	Zhang et al. (75)

Solvent Addition

The high viscosity of bio-oil can create several problems like transportation difficulties, lower heating value and corrosion to the engine, machinery, and refinery. Accordingly, the addition of several polar solvents like methanol, ethanol, and furfural can amalgamate with bio-oil while reducing its viscosity, enhancing flowability, and promoting its heating value (*61*, *84*, *85*). The added solvents also contribute to an increment in the heating value of the upgraded bio-oil. More, the reduction of the viscosity depends on the addition of solvents into the bio-oil such as physical dilution, molecular dilution, esterification, and acetalization reaction between the solvent and the bio-oil. Such reactions prevent further growth of long or short-chain organic molecules present in the bio-oil (*61*).

Solvents can be homogenized with bio-oil after the completion of the pyrolysis process, which enhances its stability by preventing polymerization reactions. Besides diluting the bio-oil, acetalization, and esterification convert oxygenated components like aldehydes or ketones and carboxylic acids into acetals and esters (Figure 3). Further chain growth can be prevented because of the conversion of oxygenated compounds into acetals and esters (*86*). Such reactions decrease the viscosity, acidity, and oxygenated components in the bio-oil while improving its volatility, heating value, and miscibility with fossil fuels to prepare blended fuels (*87*).

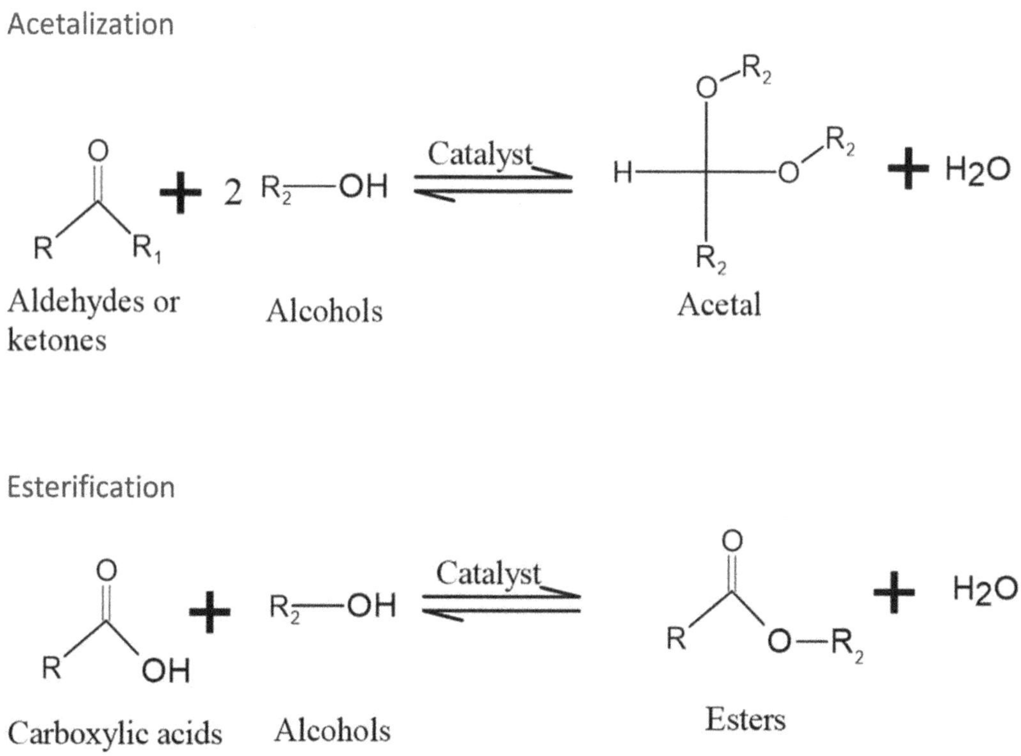

R, R_1 = Alkyl groups or –H, R_2 = Alkyl groups

Figure 3. Acetalization and esterification reactions.

The additions of alcohol-based solvents (e.g., ethanol and butanol) can be associated with several acid catalysts to speed up the reaction (*88*). Considering the recyclability of homogeneous acid or base catalysts, various solid acid or base catalysts, aluminosilicates, and ionic liquids can be used for esterification using heterogeneous catalysts in the presence of alcohol-based solvents (*89–92*). Due to the low cost of some solvents, this approach of bio-oil upgrading is considered as one of the most economically feasible techniques for enhancing the fuel properties, physicochemical stability, long-term storage, and transportation of bio-oil.

Supercritical Alcohols

Any substance above its critical temperature and critical pressure turns into a supercritical fluid with no phase distinction between gases and liquids. At the supercritical state, a solvent has its density, viscosity, and diffusivity intermediated between its liquid and gaseous state (*93*). With higher diffusivity and comparatively lower density, the penetration power of the solvents increases, acting as an efficient reaction and extraction medium to organic matter (*94*). In addition to the physical properties, the chemical properties like lower dielectric constant play a crucial role in tuning the reaction environment and subsequently optimizes the rate of the reactions involved in the bio-oil upgrading with a higher level of selectivity towards the specific reactions (*93*). Supercritical fluids such as supercritical methanol (*93, 95, 96*), supercritical ethanol (*97, 98*), supercritical butanol (*99*), supercritical isopropanol (*100, 101*), and supercritical water (*102–105*) are considered noncatalytic upgrading media for bio-oil.

Table 4 summarizes recent studies on the supercritical fluid upgrading of bio-oil. Omar et al. studied the noncatalytic upgrading of fast pyrolysis bio-oil using supercritical methanol ($T_c > 240$ °C and $P_c > 7.9$ MPa), supercritical ethanol ($T_c > 241$ °C and $P_c > 6.3$ MPa), and supercritical isopropanol ($T_c > 235.6$ °C and $P_c > 5.4$ MPa) (*101*). The results showed that supercritical treatment of crude bio-oil with methanol resulted in complete acid removal and a substantial increase in esters. Consequently, after the supercritical methanol reaction, the pH of the crude bio-oil increased from 2.4 to 4. The deacidification for reducing bio-oil might be due to ester formation.

In another study, Prajitno et al. performed noncatalytic upgrading of bio-oil in supercritical ethanol (*97*). Various reactions facilitated by supercritical ethanol are hydrogen donation, esterification, breaking of alcohol bonds or linkages, alkylation, and cracking. These reactions decreased the water, oxygen, and nitrogen contents while simultaneously increasing the yield, calorific value as well as hydrocarbon content of bio-oil. The bio-oil upgraded at 400 °C exhibited corresponding performance to the heavy fuel oil, which can be considered as a boiler combustion fuel. Supercritical ethanol also reacts with carboxylic acids in the bio-oil to generate esters through esterification (*107*). Consumption of carboxylic acids reduced the acidity of the bio-oil as well as its aqueous contents.

Besides the noncatalytic supercritical fluid upgrading, catalyst-assisted supercritical fluid upgrading is also a promising technique to enhance the fuel properties of bio-oil. The catalysts used along with supercritical fluids include acids, bases, zeolites, mesoporous aluminosilicates, bimetallic catalysts, metal oxides, and carbon-based catalysts (*97, 107–109*). The catalysts along with supercritical fluids influence a wide variety of reactions occurring during bio-oil upgrading including esterification, alkylation, hydrogen donation, alcoholysis, and aldolization. (*93*).

Table 4. Supercritical Fluid Upgrading of Bio-Oil

Unprocessed bio-oil	*Supercritical fluid*	*Catalysts*	*Key outcomes*	*References*
Bio-oil from pine sawdust	Methanol	Co–Zn/HZSM-5	Esterification removed carboxylic acid from bio-oil.	Cheng et al. (95)
Pyrolysis oil from palm fruit bunch	Ethanol	-	Moisture, oxygen, nitrogen, and the total acid number of the bio-oil decreased. The calorific value of the bio-oil increased. The upgraded bio-oil had fuel properties equivalent to heavy oil.	Prajitno et al. (*97*)
Pyrolysis oil from pine sawdust	1-butanol	Ru/C	Oxygen content was decreased. The calorific value was increased.	Xu et al. (99)
Pyrolysis bio-oil	Isopropanol	NiCu	Increased hydrogen content. Dehydrogenation removed *p*-cresol and furfural.	Kannapu et al. (*100*)
Pyrolysis bio-oil	Methanol	-	pH increased from 2.4 to 4. Carboxylic acids were converted to ester.	Omar et al. (*101*)
Biocrude from hydrothermal liquefaction of microalgae	Water	Pt/Al_2O_3 and HZSM-5	Deoxygenation removed oxygen from bio-oil.	Barreiro et al. (*102*)
Pyrolysis bio-oil	Water	-	Lower oxygen content Minimal char formation. 28 wt% heavy oil and 1314 wt% light oil were obtained.	Isa et al. (*103*)
Pyrolysis bio-oil	Water	Activated charcoal	Lower oxygen, nitrogen, and sulfur content. The calorific value reached 44 MJ/kg.	Duan et al. (*104*)
Gumweed derived bio-oil	Water	Pd/C	Oxygen content decreased from 24.1 to 4.8 wt%. Calorific values ranged from 41 to 42.8 MJ/kg.	Saha et al. (*105*)
Pyrolysis bio-oil from corn stalk	Carbon dioxide	-	pH increased from 3.4 to 5.1. Volatile contents were unaffected due to low operational temperature.	Cui et al. (*106*)

Xu et al. performed Ru/C catalyst-assisted supercritical 1-butanol (T_c > 289 °C and P_c > 4.5 MPa) upgrading of bio-oil obtained from fast pyrolysis of sawdust (*110*). Compared to catalytic upgrading, Ru/C catalyst in supercritical 1-butanol medium accelerated the process of upgrading through reactions such as hydrogenation, acetalization, hydrodeoxygenation, etherification, and esterification. Supercritical 1-butanol played a vital role in the upgrading process by providing a suitable or selective reaction medium, thus enhancing the dissolution of H_2, and protecting the solid catalyst from being gradually consumed. The catalyst-assisted supercritical fluid improved the properties of the crude pyrolysis oil by increasing the calorific value from 17.9 to 32 MJ/kg while decreasing moisture content from 21.2 to 6.7 wt%. The viscosity of the bio-oil also decreased from 22.3 to 5.7 cSt along with reduction in oxygen content from 40 to 14.5 wt%. Besides, because of the constraints exerted by the supercritical reaction medium, there was a minimal coke formation of 0.2 wt%.

Supercritical Water

Supercritical water results when the critical temperature and critical pressure of water exceeds its critical point (T_c > 374 °C and P_c > 22.1 MPa). The higher acidic environment, comparatively lower dielectric constant, lower density, and higher diffusivity of supercritical water enhance several reactions involved in the upgrading of bio-oil. Among various parameters, a lower dielectric constant provides a lower polarity to the reaction medium, which expedites the dissolution of organic compounds present in the bio-oil and subsequently accelerates the reactions involved in the upgrading process (*103*).

Isa et al. reported nearly 90% conversion of bio-oil during supercritical water upgrading at 410 °C with a bio-oil/water ratio of 1:9.3 in 1 h (*103*). Besides, 28 wt% of heavy and 13–14 wt% of light oil fractions were recovered with minimal char formation and lower oxygen content. However, at a higher temperature of 450 °C, a larger fraction of gas phase was obtained accompanying a greater yield of char due to secondary reactions. Duan and Savage (*111*) and Duan et al. (*112*) reported that noncatalytic supercritical water upgraded bio-oil had comparatively lower oxygen and nitrogen content and no sulfur contents.

Although supercritical water is considered a noncatalytic upgrading medium for bio-oil, certain catalysts can also be implemented in the process for improving efficiency, product yield, and quality. Xu et al. performed a dual catalytic upgrading of algal bio-oil in supercritical water (*113*). In this study, dual catalytic mixtures such as Ru/C + Rh/γ-Al_2O_3, Ru/C + Mo_2C, Ru/C + Pt/γ-Al_2O_3, and Ru/C + Pt/C were used for the upgrading of bio-oil at 400 °C for 4 h under 24 MPa pressure. The catalysts were tremendously efficient for deoxygenation, denitrogenation, and desulfurization to remove the heteroatoms like oxygen, nitrogen, and sulfur from the bio-oil. Ru/C + Mo_2C was found to be the most versatile catalyst mixture to obtain the maximum yield (77.2 wt%) of the upgraded bio-oil with the utmost retention of the heating value compared to the crude algal bio-oil. Moreover, the upgraded bio-oil obtained from the catalytic supercritical water upgrading resulted in a similar fuel quality as the diesel fuel. In another study, Saha et al. performed the Pd/C-assisted supercritical water upgrading of gumweed-derived bio-oil at 400 °C for 30 min under 29 MPa pressure (*105*). From this study, it was concluded that the oxygen content of the upgraded bio-oil decreased from 24.1 to 4.8 wt% while the caloric value increased from 41 to 43 MJ/kg.

Through catalytic supercritical water upgrading of bio-oil, activated charcoal can replace metal catalysts (i.e., Ni, Mo, Ru, Rh, Pd, Pt, etc.) to decrease the metal contamination in the upgraded bio-oil. Duan et al. reported an upgraded bio-oil with 76% conversion and calorific value of 44 MJ/

kg at 400 °C for 1 h in presence of 10 wt% of bamboo-derived activated charcoal from subcritical water pretreated bio-oil (*112*). Furthermore, the upgraded bio-oil contained less ash and acidic compounds along with lower levels of oxygen, nitrogen, and sulfur through deoxygenation, denitrogenation, and desulfurization reactions. Therefore, both catalytic and noncatalytic supercritical water upgrading can provide clean media for the upgrading of bio-oil.

Supercritical Carbon Dioxide

Supercritical CO_2 ($T_c > 31$ °C and $P_c > 7.4$ MPa) acts as a green solvent in various organic reactions and extraction of organic molecules. Supercritical CO_2 is widely implemented as an extraction medium for several natural products, nutraceuticals, cosmeceuticals, pharmaceuticals, and biochemicals due to being nontoxic and nonflammable, as well as having high product selectivity and low-cost. Unlike other supercritical fluids, CO_2 possesses relatively lower critical temperature (31 °C) and critical pressure (7.4 MPa) and do not require expensive infrastructures and high-energy input (*114*). Furthermore, the lower operating temperature provides retention of several volatile species and prevents the degradation of various useful compounds (*115*). Supercritical CO_2 provides appealing extraction characteristics in addition to the minimal temperature factor, desirable diffusiveness, viscosity, and other physical properties. Moreover, with an increment in pressure, the abovementioned solvent properties of supercritical CO_2 can be tailored to meet specific operating conditions of different processes. Therefore, it is possible to selectively extract various kinds of bioactive compounds from crude bio-oil at specific CO_2 pressures (*114*, *116*). In addition to the esterification reaction, supercritical CO_2 efficiently removes moisture, carboxylic acids, and alcohols from bio-oil due to the varied polarity at certain pressures (*106*).

The higher moisture content in pyrolysis-derived bio-oil (around 25–40%) is a major drawback for its processing. Such high levels of moisture create an azeotropic mixture with various polar compounds that subsequently interfere with the distillation of bio-oil. Supercritical CO_2 is used as a medium to fractionate the bio-oil by extracting the nonpolar organic compounds like hydrocarbons, while simultaneously removing water, alcohols, carboxylic acids, and other polar compounds (*117*).

Cui et al. reported supercritical CO_2 upgrading of bio-oil at 80 °C and 28 MPa in 3 h with facilitated esterification reaction (*106*). The total acid conversion reached up to 86.8% under such operating conditions. The pH value of the upgraded bio-oil increased from 3.8 to 5.1 and reduced the acidity by retaining the volatile compounds. Moreover, the lower temperature requirement for supercritical CO_2 retains most of the volatile species in the bio-oil without degradation. Bio-oil's stability and heating value are improved with lower contents of water, volatiles, and acids.

Naik et al. performed an upgrading of biocrude oil from wheat hemlock biomass in supercritical CO_2 at 40 °C for 2 h under variable pressures (10–30 MPa) (*114*). Bio-oil fractionation at 10 and 25 MPa showed comparatively better results than that at 30 MPa. Conclusively, at these pressures, the moisture content of the crude bio-oil decreased from 45 to 0.5 wt%. Similarly, the nitrogen, sulfur, and oxygen contents reduced significantly. Most importantly, the caloric value of the bio-oil increased to 41, 40.2, and 43.2 MJ/kg at 10, 25, and 30 MPa, respectively. Therefore, supercritical CO_2 seems to be an advanced clean technique for bio-oil upgrading and fractionation.

Cheng et al. performed a three-step fractionation of bio-oil (*118*). The first step was conducted with 100% supercritical CO_2 at 10 MPa for 5 min. The second step was conducted with 90% supercritical CO_2 with 10% methanol as cosolvent at 20 MPa for 25 min. In the final step, methanol concentration was raised to 25% under 30 MPa pressure for 40 min. This fractionation process decreased the concentration of nitrogenous compounds (e.g., amines) and oxygenated compounds (e.g., carboxylic acid and alcohols) in the upgraded bio-oil with other chemical properties on par with transportation fuels.

Electrochemical Stabilization

Electrochemical techniques are used for upgrading bio-oil without the requirement of external hydrogen gas. Beyond the omission of external hydrogen, another advantage of this process is its wide range of operational parameters (i.e., temperatures and pressures) (*119*). In the electrochemical method for bio-oil upgrading, there is a reduction of carboxylic acids, aldehydes, and ketones into corresponding alcohols along with the formation of hydrogen gas (*119*). This process is performed in an electrochemical cell in the presence of a membrane between the anode and cathode. With the application of electricity, hydrogen ion (H^+) is generated at the anode and subsequently moves towards the cathode by penetrating the membrane (Eq. 6). At the cathode, in presence of the electrons, the undesired oxygenated compounds (e.g., carboxylic acid, aldehydes, and ketones) are reduced or hydrogenated to the corresponding alcohols (Eq. 7) (*67*). The acidity of bio-oil is decreased with the conversion of carboxylic acids.

$$H_2O \rightarrow 0.5O_2 + 2H^+ + 2e^- \quad (6)$$

$$\text{R-CHO or R-CO-R or R-COOH} + 2H^+ + 2e^- \rightarrow \text{R-CH}_2\text{-OH} \quad (7)$$

Electrochemical stabilization establishes a low-energy hydrodeoxygenation of bio-oil. This technique is advantageous over other catalytic hydrodeoxygenation procedures because of tar deposition in the latter techniques owing to higher process temperatures and pressures (*120*). Electrochemical stabilization can also be operated using polymer electrolyte membrane with low-energy consumption and without any external electrolytes unlike the conventional electrolyzers used for bio-oil stabilization (*121*). The absence of the external electrolytes is privileged for bio-oil to avoid phase separation and challenges associated with the recovery of electrolytes after the process.

Studies have reported the use of polymer electrolyte membrane , especially cation-exchange membranes (CEM) (e.g., Nafion) for the dehydrogenation of unsaturated C=C in bio-oil, unsaturated compounds (i.e., decene), and reduction of various carbonyl compounds (*121–124*). Lister et al. used both CEM and anion-exchange membrane (AEM) bound to an anode and cathode, respectively (*121*). In such dual electrochemical cells, AEM transfers the small acid anions rather than blocking the hydrogen ion (H^+). Therefore, at the AEM bonded cathode, two specific reactions occur that can be captured and used for other hydrotreating reactions [e.g., hydrogenation of unsaturated compounds present in the bio-oil and generation of molecular hydrogen (H_2)]. The generation of molecular hydrogen is an efficient value addition to the process. The researchers have also reported the removal of small acid compounds from the upgraded bio-oil, increasing its pH value and decreasing the concentration of carbonyl compounds.

In addition to the reduction reaction of carboxylic acids like benzoic acid into benzyl alcohol, hydrogenation of unsaturated C=C occurs during the electrochemical transformation. Li et al. conducted Ru-assisted electrochemical upgrading of water-soluble phenolic compounds present in bio-oil (*125*). A certain ionic membrane serves as the conducting electrolyte for proton transfer. Weak bio-oil conductivity requires the use of membranes to improve performance (*126*). More studies are required for largescale use of electrochemical stabilization for bio-oil upgrading.

Strengths, Weaknesses, Opportunities, and Threats of Bio-Oil Upgrading

Several thermochemical technologies have been developed to produce biocrude oil followed by its upgrading into liquid transportation fuels. Nevertheless, maximizing the efficiency, productivity, and profitability of biofuels are important to remain competitive in the marketplace. When compared to conventional fossil fuels, thermochemical technologies for biofuel production (e.g., pyrolysis, gasification, and liquefaction) strive to be energy-efficient and cost-effective (*127*).

The integration of pyrolysis and liquefaction with selective bio-oil upgrading systems could be a promising strategy to efficiently and economically produce biofuels (*128*). An integrated approach could reduce capital and operational costs, simplify requirements for processing, and utilizing most by-products. Further research and development are needed in the determination of an efficient integrated conversion pathway.

The other approach is corefining, coprocessing or drop-in biocrude (via blending) with refinery distillates to produce diesel, gasoline, and jet-fuels (*129, 130*). Coprocessing of hydrotreated biocrude with conventional crude oil could generate hydrocarbon fractions like that of gasoline and diesel (*131*). This approach was found to be economically favorable for commercialization, but it still demands biocrude pretreatment to remove oxygenated compounds before blending or coprocessing. Thus, the optimization of the current process using concurrent monitoring and product characterization could lead to achieving affordability and sustainability of corefining technologies (*67*).

Despite several advantages, the weaknesses of upgrading bio-oil should also be accessed. As discussed in this chapter, catalytic hydrogenation, cracking, and esterification play vital roles in the upgrading process. However, in these catalyst-assisted upgrading processes, several drawbacks can be encountered due to the high process temperatures and expensive catalysts used. Some common challenges associated with catalytic upgrading techniques are catalyst deactivation due to coking, catalyst poisoning and sintering, as well as blocking the catalytic sites (*132*). Furthermore, the formation of tar decreases the fuel quality and performance (*133*). Therefore, the formation of coke and tar in catalytic upgrading processes increase the process expenditure associated with their removal and equipment maintenance. While retaining recyclability for catalysts, the recovery and regeneration could also increase the expenditure and energy input to the catalytic upgrading processes (*22*). Along with complex catalytic reactions, the impurities present in the bio-oil can undesirably impede the upgrading process leading to low-carbon conversion and increased processing cost (*134*).

In the case of noncatalytic bio-oil upgrading techniques, especially supercritical fluids, the fuel quality of upgraded bio-oil is significantly enhanced due to the removal of oxygenated compounds and lowering of viscosity. However, the costs of the solvents and equipment are expensive for the pilot-scale implementation (*133*). Moreover, in the presence of certain catalysts, supercritical water at high-temperatures and high-pressures could be corrosive to stainless steel reactors due to salt formation and precipitation (*22*).

From the aforementioned approaches, there is clearly a significant need to develop efficient biocrude upgrading pathways in order to become a viable solution to overcome the challenges to produce liquid transportation fuels. Therefore, further research to explore integrated energy systems for biofuel production followed by upgrading, blending, and corefining, as well as process modeling and simulation, could advance the commercialization of liquid transportation biofuels.

Conclusions

The fuel qualities of bio-oil produced via thermochemical technologies (i.e., pyrolysis and hydrothermal liquefaction) are inferior to that of conventional fuels. Therefore, the direct application of bio-oil is limited because of the presence of unwanted heteroatoms (e.g., oxygen, sulfur, nitrogen, and metals), which directly render high oxygen content, acidity, high viscosity, and thermo-oxidative instability to bio-oil. Thus, there is a need to upgrade bio-oil using appropriate techniques before they can be employed as liquid transportation fuels in gasoline and diesel engines. The most common bio-oil upgrading technologies operate using heat, pressure, and controlled conditions to separate different fractions of bio-oil. Though catalytic upgrading technologies for bio-oil are identical to refinery methods; bio-oil is incompatible with conventional crude oils due to the presence of high oxygen and water contents. Bio-oils need to be stabilized for coprocessing at the existed refineries to produce various liquid transportation fuels that can be used in direct gasoline and diesel engines.

Catalytic upgrading of bio-oil through hydrotreating, catalytic cracking, and hydrocracking are found favorable because of the faster process, product selectivity, and high-purity of products. A corefining approach is the most appropriate approach to commercialize bio-oils to produce diesel, gasoline, and jet-fuels. Further, when the bio-oils are coprocessed at the refinery, unwanted heteroatoms could be removed via hydrotreating. High molecular weight hydrocarbons could be reduced to low-molecular weight hydrocarbon fuels via fluid catalytic cracking and hydrocracking. On the other hand, the noncatalytic upgrading techniques are relatively environmentally friendly because of the absence of metal catalysts and associated concerns of their reusability or sequestration after the completion of the upgrading process. Noncatalytic upgrading techniques are also considered as pretreatment methods to reduce the oxygen content, blend bio-oil, and refinery crude via emulsification, lower viscosity, and extract specialty and value-added chemicals from bio-oils. More research is underway to make both catalytic and noncatalytic upgrading process cost-effective and energy-efficient by integrating with the pyrolysis and liquefaction systems.

Acknowledgments

The financial support from the Natural Sciences and Engineering Research Council of Canada (NSERC), Canada Research Chairs (CRC) Program, and MITACS Canada are greatly acknowledged.

References

1. *International Energy Agency (IEA). Global Energy & CO_2 Status Report 2019*, IEA, Paris. https://www.iea.org/reports/global-energy-co2-status-report-2019 (accessed 17 Nov 2020).
2. Okolie, J. A.; Nanda, S.; Dalai, A. K.; Berruti, F.; Kozinski, J. A. A review on subcritical and supercritical water gasification of biogenic, polymeric and petroleum wastes to hydrogen-rich synthesis gas. *Renew. Sustain. Energy Rev.* **2020,** *119*, 109546.
3. Nanda, S.; Berruti, F. A technical review of bioenergy and resource recovery from municipal solid waste. *J. Hazard. Mater.* **2021,** *403*, 123970.
4. Parakh, P. D.; Nanda, S.; Kozinski, J. A. Eco-friendly transformation of waste biomass to biofuels. *Curr. Biochem. Eng.* **2020,** *6*, 120–134.
5. Nanda, S.; Mohammad, J.; Reddy, S. N.; Kozinski, J. A.; Dalai, A. K. Pathways of lignocellulosic biomass conversion to renewable fuels. *Biomass Conv. Bioref.* **2014,** *4*, 157–191.

6. Fahmy, T. Y. A.; Fahmy, Y.; Mobarak, F.; El-Sakhawy, M.; Abou-Zeid, R. E. Biomass pyrolysis: past, present, and future. *Environ. Develop. Sustain.* **2020**, *22*, 17–32.
7. Taghipour, A.; Ramirez, J. A.; Brown, R. J.; Rainey, T. J. A review of fractional distillation to improve hydrothermal liquefaction biocrude characteristics: future outlook and prospects. *Renew. Sustain. Energy Rev.* **2019**, *115*, 109355.
8. Eke, J.; Onwudili, J. A.; Bridgwater, A. V. Influence of moisture contents on the fast pyrolysis of trommel fines in a bubbling fluidized bed reactor. *Waste Biomass Valor.* **2020**, *11*, 3711–3722.
9. Chand, R.; Borugadda, V. B.; Qiu, M.; Dalai, A. K. Evaluating the potential for bio-fuel upgrading: A comprehensive analysis of bio-crude and bio-residue from hydrothermal liquefaction of agricultural biomass. *Appl. Energy* **2019**, *254*, 113679.
10. Xu, D.; Lin, G.; Guo, S.; Wang, S.; Guo, Y.; Jing, Z. Catalytic hydrothermal liquefaction of algae and upgrading of biocrude: A critical review. *Renew. Sustain. Energy Rev.* **2018**, *97*, 103–118.
11. Nanda, S.; Dalai, A. K.; Berruti, F.; Kozinski, J. A. Biochar as an exceptional bioresource for energy, agronomy, carbon sequestration, activated carbon and specialty materials. *Waste Biomass Valor.* **2016**, *7*, 201–235.
12. Senneca, O.; Cerciello, F.; Heuer, S.; Ammendola, P. Slow pyrolysis of walnut shells in nitrogen and carbon dioxide. *Fuel* **2018**, *225*, 419–425.
13. Safdari, M. S.; Amini, E.; Weise, D. R.; Fletcher, T. H. 2019. Heating rate and temperature effects on pyrolysis products from live wildland fuels. *Fuel* **2019**, *242*, 295–304.
14. Anca-Couce, A.; Dieguez-Alonso, A.; Zobel, N.; Berger, A.; Kienz, N.; Behrendt, F. Influence of heterogeneous secondary reactions during slow pyrolysis on char oxidation reactivity of woody biomass. *Energy Fuels* **2017**, *31*, 2335–2344.
15. Roy, P.; Dias, G. Prospects for pyrolysis technologies in the bioenergy sector: A review. *Renew. Sustain. Energy Rev.* **2017**, *77*, 59–69.
16. Nzihou, A.; Stanmore, B.; Lyczko, N.; Minh, D. P. The catalytic effect of inherent and adsorbed metals on the fast/flash pyrolysis of biomass: A review. *Energy* **2019**, *170*, 326–337.
17. Kabir, G.; Hameed, B. H. Recent progress on catalytic pyrolysis of lignocellulosic biomass to high-grade bio-oil and bio-chemicals. *Renew. Sustain. Energy Rev.* **2017**, *70*, 945–967.
18. Miandad, R.; Barakat, M. A.; Aburiazaiza, A. S.; Rehan, M.; Nizami, A. S. Catalytic pyrolysis of plastic waste: A review. *Process Saf. Environ. Protect.* **2016**, *102*, 822–838.
19. Akubo, K.; Nahil, M. A; Williams, P. T. Aromatic fuel oils produced from the pyrolysis-catalysis of polyethylene plastic with metal-impregnated zeolite catalysts. *J. Energy Inst.* **2019**, *92*, 195–202.
20. Zainan, N. H.; Srivatsa, S. C.; Li, F.; Bhattacharya, S. Quality of bio-oil from catalytic pyrolysis of microalgae *Chlorella vulgaris*. *Fuel* **2018**, *223*, 12–19.
21. Dai, L.; Fan, L.; Duan, D.; Ruan, R.; Wang, Y.; Liu, Y.; Jiang, L. Production of hydrocarbon-rich bio-oil from soapstock via fast microwave-assisted catalytic pyrolysis. *J. Anal. Appl. Pyrolysis* **2017**, *125*, 356–362.
22. Okolie, J. A.; Rana, R.; Nanda, S.; Dalai, A. K.; Kozinski, J. A. Supercritical water gasification of biomass: A state-of-the-art review of process parameters, reaction mechanisms and catalysis. *Sustain. Energy Fuels* **2019**, *3*, 578–598.

23. Möller, M.; Nilges, P.; Harnisch, F.; Schröder, U. Subcritical water as reaction environment: fundamentals of hydrothermal biomass transformation. *ChemSusChem* **2011**, *4*, 566–579.
24. Toor, S. S.; Rosendahl, L.; Rudolf, A. Hydrothermal liquefaction of biomass: a review of subcritical water technologies. *Energy* **2011**, *36*, 2328–2342.
25. Jena, U.; Das, K. C. Comparative evaluation of thermochemical liquefaction and pyrolysis for bio-oil production from microalgae. *Energy Fuels* **2011**, *25*, 5472–5482.
26. Zhang, Y.; Chen, W. T. Hydrothermal liquefaction of protein-containing feedstocks. In *Direct Thermochemical Liquefaction for Energy Applications*; Rosendahl, L., Ed.; Woodhead Publishing: Duxford, 2018; pp 127–168.
27. Hognon, C.; Delrue, F.; Texier, J.; Grateau, M.; Thiery, S.; Miller, H.; Roubaud, A. Comparison of pyrolysis and hydrothermal liquefaction of *Chlamydomonasreinhardtii*. Growth studies on the recovered hydrothermal aqueous phase. *Biomass Bioenergy* **2015**, *73*, 23–31.
28. Bach, Q. V.; Sillero, M. V.; Tran, K. Q.; Skjermo, J. Fast hydrothermal liquefaction of a Norwegian macro-alga: screening tests. *Algal Res.* **2014**, *6*, 271–276.
29. Chen, Y.; Mu, R.; Yang, M.; Fang, L.; Wu, Y.; Wu, K.; Liu, Y.; Gong, J. Catalytic hydrothermal liquefaction for bio-oil production over CNTs supported metal catalysts. *Chem. Eng. Sci.* **2017**, *161*, 299–307.
30. Remón, J.; Randall, J.; Budarin, V. L.; Clark, J. H. Production of bio-fuels and chemicals by microwave-assisted, catalytic, hydrothermal liquefaction (MAC-HTL) of a mixture of pine and spruce biomass. *Green Chem.* **2019**, *21*, 284–299.
31. Gollakota, A. R. K.; Kishore, N.; Gu, S. A review on hydrothermal liquefaction of biomass. *Renew. Sustain. Energy Rev.* **2018**, *81*, 1378–1392.
32. Ramirez, J. A.; Brown, R. J.; Rainey, T. J. A review of hydrothermal liquefaction bio-crude properties and prospects for upgrading to transportation fuels. *Energies* **2015**, *8*, 6765–6794.
33. Gollakota, A. R.; Reddy, M.; Subramanyam, M. D.; Kishore, N. A review on the upgradation techniques of pyrolysis oil. *Renew. Sustain. Energy Rev.* **2016**, *58*, 1543–1568.
34. Shi, Y.; Xing, E.; Wu, K.; Wang, J.; Yang, M.; Wu, Y. Recent progress on upgrading of bio-oil to hydrocarbons over metal/zeolite bifunctional catalysts. *Catal. Sci. Technol.* **2017**, *7*, 2385–2415.
35. Lyu, G.; Wu, S.; Zhang, H. Estimation and comparison of bio-oil components from different pyrolysis conditions. *Front. Energy Res.* **2015**, *3*, 28.
36. Kass, M. D.; Armstrong, B. L.; Kaul, B. C.; Connatser, R. M.; Lewis, S.; Keiser, J. R.; Jun, J.; Warrington, G.; Sulejmanovic, D. Stability, combustion, and compatibility of high-viscosity heavy fuel oil blends with a fast pyrolysis bio-oil. *Energy Fuels* **2020**, *34*, 8403–8413.
37. Nayak, S. K.; Nayak, B.; Mishra, P. C.; Noor, M. M.; Nanda, S. Effects of biodiesel blends and producer gas flow on overall performance of a turbocharged direct injection dual-fuel engine. *Energy Sour A: Recovery, Util. Environ. Effects* **2019**, DOI:10.1080/15567036.2019.1694101.
38. Lia, Y.; Zhang, C.; Liu, Y.; Tang, S.; Chen, G.; Zhang, R.; Tang, X. Coke formation on the surface of Ni/HZSM-5 and Ni-Cu/HZSM-5 catalysts during bio-oil hydrodeoxygenation. *Fuel* **2017**, *189*, 23–31.
39. Shemfe, M.; Gu, S.; Fidalgo, B. Techno-economic analysis of biofuel production via bio-oil zeolite upgrading: An evaluation of two catalyst regeneration systems. *Biomass Bioenergy* **2017**, *98*, 182–193.

40. Zhou, Q.; Zarei, A.; Girolamo, A. D.; Yan, Y.; Zhang, L. Catalytic performance of scrap tyre char for the upgrading of eucalyptus pyrolysis derived bio-oil via cracking and deoxygenation. *J. Anal. Appl. Pyrolysis* **2019**, *139*, 167–176.
41. Chen, G.; Yao, J.; Liu, J.; Yan, B.; Shan, R. Biomass to hydrogen-rich syngas via catalytic steam reforming of bio-oil. *Renew. Energy* **2016**, *91*, 315–322.
42. Singh, S.; Kumar, R.; Setiabudi, H. D.; Nanda, S.; Vo, D. V. N. Advanced synthesis strategies of mesoporous SBA-15 supported catalysts for catalytic reforming applications: A state-of-the-art review. *Appl. Catal. A: Gen.* **2018**, *559*, 57–74.
43. Sun, X.; Atiyeh, H. K.; Huhnke, R. L.; Tanner, R. S. Syngas fermentation process development for production of biofuels and chemicals: A review. *Bioresour. Technol. Rep.* **2019**, 7, 100279.
44. Nanda, S.; Rana, R.; Zheng, Y.; Kozinski, J. A.; Dalai, A. K. Insights on pathways for hydrogen generation from ethanol. *Sustain. Energy Fuels* **2017**, *1*, 1232–1245.
45. Kokayeff, P.; Zink, S.; Roxas, P. Hydrotreating in petroleum processing. In *Handbook of Petroleum Processing*; Treese, S. A., Pujadó, P. R., Jones, D. S. J., Eds.; Springer: Switzerland, 2014; pp 361–434.
46. Li, X.; Luo, X.; Jin, Y.; Li, J.; Zhang, H.; Zhang, A.; Xie, J. Heterogeneous sulfur-free hydrodeoxygenation catalysts for selectively upgrading the renewable bio-oils to second generation biofuels. *Renew. Sustain. Energy Rev.* **2018**, *82*, 3762–3797.
47. Zerva, C.; Karakoulia, S. A.; Kalogiannis, K. G.; Margellou, A.; Iliopoulou, E. F.; Lappas, A. A.; Papayannakos, N.; Triantafyllidis, K. S. Hydrodeoxygenation of phenol and biomass fast pyrolysis oil (bio-oil) over Ni/WO_3-ZrO_2 catalyst. *Catal. Today* **2020**, DOI:10.1016/j.cattod.2020.08.029.
48. Zhao, H.; Hu, X.; Hao, J.; Li, N.; Zhi, K.; He, R.; Wang, Y.; Zhou, H.; Liu, Q. An efficient bifunctional Ru-$NbOPO_4$ catalyst for the hydrodeoxygenation of aromatic ethers, phenols and real bio-oil. *Appl. Catal. A: Gen.* **2020**, *591*, 117378.
49. Remón, J.; Casales, M.; Gracia, J.; Callén, M. S.; Pinilla, J. L.; Suelves, I. Sustainable production of liquid biofuels and value-added platform chemicals by hydrodeoxygenation of lignocellulosic bio-oil over a carbon–neutral Mo_2C/CNF catalyst. *Chem. Eng. J.* **2021**, *405*, 126705.
50. Oh, S.; Lee, J. H.; Choi, I. H.; Choi, J. W. Enhancement of bio-oil hydrodeoxygenation activity over Ni-based bimetallic catalysts supported on SBA-15. *Renew. Energy* **2020a**, *149*, 1–10.
51. Oh, S.; Lee, J. H.; Choi, J. W. Hydrodeoxygenation of crude bio-oil with various metal catalysts in a continuous-flow reactor and evaluation of emulsion properties of upgraded bio-oil with petroleum fuel. *Renew. Energy* **2020b**, *160*, 1160–1167.
52. Wang, Y.; Lin, H.; Zheng, Y. Hydrotreatment of lignocellulosic biomass derived oil using a sulfidedNiMo/γ-Al_2O_3 catalyst. *Catal. Sci. Technol.* **2014**, *4*, 109–119.
53. Parapati, D. R.; Guda, V. K.; Penmetsa, V. K.; Tanneru, S. K.; Mitchell, B.; Steele, P. H. Comparison of reduced and sulfidedCoMo/γ-Al_2O_3 catalyst on hydroprocessing of pretreated bio-oil in a continuous packed-bed reactor. *Environ. Prog. Sustain. Energy* **2015**, *34*, 1174–1179.
54. Gutierrez, A.; Turpeinen, E. M.; Viljava, T. R.; Krause, O. Hydrodeoxygenation of model compounds on sulfidedCoMo/γ-Al_2O_3 and NiMo/γ-Al_2O_3 catalysts; Role of sulfur-containing groups in reaction networks. *Catal. Today* **2017**, *285*, 125–134.

55. Papa, V.; Cao, Y.; Spannenberg, A.; Junge, K.; Beller, M. Development of a practical non-noble metal catalyst for hydrogenation of N-heteroarenes. *Nature Catal.* **2020**, *3*, 135–142.
56. Wang, W.; Yunquan, Y.; Hean, L.; Huizuo, P.; Bing, H.; Wenying, L. Preparation of Ni(Co)–W–B amorphous catalysts for cyclopentanone hydro-deoxygenation. *Catal. Comm.* **2011a**, *12*, 1275–1279.
57. Wang, W.; Yunquan, Y.; Hean, L.; Tao, H.; Wenying, L. Amorphous Co–Mo–B catalyst with high activity for the hydrodeoxygenation of bio-oil. *Catal. Comm.* **2011b**, *12*, 436–4340.
58. Chen, Y.; Li, C.; Zhou, J.; Zhang, S.; Rao, D.; He, S.; Wei, M.; Evans, D. G.; Duan, X. Metal phosphides derived from hydrotalcite precursors toward the selective hydrogenation of phenylacetylene. *ACS Catal.* **2015**, *5*, 5756–5765.
59. Führer, M.; van Haasterecht, T.; Bitter, J. H. Molybdenum and tungsten carbides can shine too. *Catal. Sci. Technol.* **2020**, *10*, 6089–6097.
60. Usami, R.; Fujii, K.; Fushimi, C. Improvement of bio-oil and nitrogen recovery from microalgae using two-stage hydrothermal liquefaction with solid carbon and HCl acid catalysis. *ACS Omega* **2020**, *5*, 6684–6696.
61. Xiu, S.; Shahbazi, A. Bio-oil production and upgrading research: A review. *Renew. Sustain. Energy Rev.* **2012**, *16*, 4406–4414.
62. Santillan-Jimenez, E.; Pace, R.; Morgan, T.; Behnke, C.; Sajkowski, D. J.; Lappas, A.; Crocker, M. Co-processing of hydrothermal liquefaction algal bio-oil and petroleum feedstock to fuel-like hydrocarbons via fluid catalytic cracking. *Fuel Process. Technol.* **2019**, *188*, 164–171.
63. Gong, F.; Yang, Z.; Hong, C.; Huang, W.; Ning, S.; Zhang, Z.; Xu, Y.; Li, Q. Selective conversion of bio-oil to light olefins: Controlling catalytic cracking for maximum olefins. *Bioresour. Technol.* **2011**, *102*, 9247–9254.
64. Ma, W.; Liu, B.; Zhang, R.; Gu, T.; Ji, X.; Zhong, L.; Chen, G.; Ma, L.; Cheng, Z.; Li, X. Co-upgrading of raw bio-oil with kitchen waste oil through fluid catalytic cracking (FCC). *Appl. Energy* **2018**, *217*, 233–240.
65. Ibarra, Á.; Hita, I.; Azkoiti, M. J.; Arandes, J. M.; Bilbao, J. Catalytic cracking of raw bio-oil under FCC unit conditions over different zeolite-based catalysts. *J. Ind. Eng. Chem.* **2019**, *78*, 372–382.
66. Chen, Y. K.; Hsieh, C. H.; Wang, W. C. The production of renewable aviation fuel from waste cooking oil. Part II: Catalytic hydro-cracking/isomerization of hydro-processed alkanes into jet fuel range products. *Renew. Energy* **2020**, *157*, 731–740.
67. Hansen, S.; Mirkouei, A.; Diaz, L. A. A comprehensive state-of-technology review for upgrading bio-oil to renewable or blended hydrocarbon fuels. *Renew. Sustain. Energy Rev.* **2020**, *118*, 109548.
68. Xu, X.; Jiang, E.; Mingfeng, W.; Li, B. Rich hydrogen production from crude gas secondary catalytic cracking over Fe/γ-Al_2O_3. *Renew. Energy* **2012**, *39*, 126–131.
69. Schulzke, T.; Conrad, S.; Kaluza, S.; Van Loo, T. Upgrading of fast pyrolysis condensates via esterification with higher alcohols. *Biomass Bioenergy* **2017**, *103*, 11–20.
70. Ciddor, L.; Bennett, J. A.; Hunns, J. A.; Wilson, K.; Lee, A. F. Catalytic upgrading of bio-oils by esterification. *J. Chem. Technol. Biotechnol.* **2015**, *90*, 780–795.
71. Shagufta; Ahmad, I.; Dhar, R. Sulfonic acid-functionalized solid acid catalyst in esterification and transesterification reactions. *Catal. Surveys Asia* **2017**, *21*, 53–69.

72. Xu, Y.; Wang, T.; Ma, L.; Zhang, Q.; Liang, W. Upgrading of the liquid fuel from fast pyrolysis of biomass over MoNi/γ-Al_2O_3 catalysts. *Appl. Energy* **2010**, *87*, 2886–2891.
73. Xu, J.; Jiang, J.; Dai, W.; Zhang, T.; Xu, Y. Bio-oil upgrading by means of ozone oxidation and esterification to remove water and to improve fuel characteristics. *Energy Fuels* **2011**, *25*, 1798–1801.
74. Song, M.; Zhong, Z.; Jiajia, D. Different solid acid catalysts influence on properties and chemical composition change of upgrading bio-oil. *J. Anal. Appl. Pyrolysis* **2010**, *89*, 166–170.
75. Zhang, Q.; Chang, J.; Wang, T. J.; Xu, Y. Upgrading bio-oil over different solid catalysts. *Energy Fuels* **2006**, *20*, 2717–2720.
76. Lian, X.; Xue, Y.; Zhao, Z.; Xu, G.; Han, S.; Yu, H. Progress on upgrading methods of bio-oil: a review. *Int. J. Energy Res.* **2017**, *41*, 1798–1816.
77. Zhang, M.; Wu, H. Stability of emulsion fuels prepared from fast pyrolysis bio-oil and glycerol. *Fuel* **2017**, *206*, 230–238.
78. Murillo, G.; Ali, S. S.; Sun, J.; Yan, Y.; Bartocci, P.; El-Zawawy, N.; Azab, M.; He, Y.; Fantozzie, F. Ultrasonic emulsification assisted immobilized *Burkholderiacepacia* lipase catalyzed transesterification of soybean oil for biodiesel production in a novel reactor design. *Renew. Energy* **2019**, *135*, 1025–1034.
79. Taha, A.; Ahmed, E.; Ismaiel, A.; Ashokkumar, M.; Xu, X.; Pan, S.; Hua, H. Ultrasonic emulsification: An overview on the preparation of different emulsifiers-stabilized emulsions. *Trends Food Sci. Technol.* **2020**, *105*, 363–377.
80. Farooq, A.; Shafaghat, H.; Jae, J.; Jung, S. C.; Park, Y. K. Enhanced stability of bio-oil and diesel fuel emulsion using Span 80 and Tween 60 emulsifiers. *J. Environ. Manag.* **2019**, *231*, 694–700.
81. Du, M. T.; Huang, Y. C.; Shang, S.; Wang, P.; Wang, L.; Guo, H. J. Study on optimum HLB value and physicochemical properties of diesel and bio-oil emulsions. *J. Fuel Chem. Technol.* **2009**, *37*, 679e83.
82. Li, X. J.; Zhang, S. P.; Li, Q. Y.; Xu, Q. L.; Yan, Y. J. Emulsion of biomass pyrolysis oil in diesel oil. *Modern Chem. Ind.* **2010**, *9*, 7.
83. Isa, Y. M.; Ganda, E. T. Bio-oil as a potential source of petroleum range fuels. *Renew. Sustain. Energy Rev.* **2018**, *81*, 69–75.
84. Mei, Y.; Chai, M.; Shen, C.; Liu, B.; Liu, R. Effect of methanol addition on properties and aging reaction mechanism of bio-oil during storage. *Fuel* **2019**, *244*, 499–507.
85. Chen, X.; Ma, X.; Zeng, X.; Zheng, C.; Lu, X. Ethanol addition during aqueous phase recirculation for further improving bio-oil yield and quality. *Appl. Energy* **2020**, *262*, 114550.
86. Yang, Z.; Kumar, A.; Huhnke, R. L. Review of recent developments to improve storage and transportation stability of bio-oil. *Renew. Sustain. Energy Rev.* **2015**, *50*, 859–870.
87. Park, L. K. E.; Ren, S.; Yiacoumi, S.; Ye, X. P.; Borole, A. P.; Tsouris, C. Separation of switchgrass bio-oil by water/organic solvent addition and pH adjustment. *Energy Fuels* **2016**, *30*, 2164–2173.
88. Tang, Y.; Yu, W.; Mo, L.; Lou, H.; Zheng, X. One-step hydrogenation– esterification of aldehyde and acid to ester over bifunctional Pt catalysts: A model reaction as novel route for catalytic upgrading of fast pyrolysis bio-oil. *Energy Fuels* **2008**, *22*, 3484–3488.

89. Peng, J.; Chen, P.; Lou, H.; Zheng, X. Upgrading of bio-oil over aluminum silicate in supercritical ethanol. *Energy Fuels* **2008**, *22*, 3489–3492.

90. Xiong, W. M.; Zhu, M. Z.; Deng, L.; Fu, Y.; Guo, Q. X. Esterification of organic acid in bio-oil using acidic ionic liquid catalysts. *Energy Fuels* **2009**, *23*, 2278–2283.

91. Galadima, A.; Muraza, O. Hydrothermal liquefaction of algae and bio-oil upgrading into liquid fuels: Role of heterogeneous catalysts. *Renew. Sustain. Energy Rev.* **2018**, *81*, 1037–1048.

92. Meng, X.; Bian, J.; Li, J.; Ma, Z.; Long, Q.; Su, J. Porous aluminosilicates catalysts for low and medium matured shale oil in situ upgrading. *Energy Sci. Eng.* **2020**, *8*, 2859–2867.

93. Omar, S.; Yang, Y.; Wang, J. A review on catalytic & non-catalytic bio-oil upgrading in supercritical fluids. *Front. Chem. Sci. Eng.* **2021**, *15*, 4–17.

94. Reddy, S. N.; Nanda, S.; Dalai, A. K.; Kozinski, J. A. Supercritical water gasification of biomass for hydrogen production. *Int. J. Hydrogen Energy* **2014**, *39*, 6912–6926.

95. Cheng, S.; Wei, L.; Julson, J.; Muthukumarappan, K.; Kharel, P. R. Upgrading pyrolysis bio-oil to biofuel over bifunctional Co-Zn/HZSM-5 catalyst in supercritical methanol. *Energy Convers. Manag.* **2017**, *147*, 19–28.

96. Shafaghat, H.; Kim, J. M.; Lee, I. G.; Jae, J.; Jung, S. C.; Park, Y. K. Catalytic hydrodeoxygenation of crude bio-oil in supercritical methanol using supported nickel catalysts. *Renew. Energy* **2019**, *144*, 159–166.

97. Prajitno, H.; Insyani, R.; Park, J.; Ryu, C.; Kim, J. Non-catalytic upgrading of fast pyrolysis bio-oil in supercritical ethanol and combustion behavior of the upgraded oil. *Appl. Energy* **2016**, *172*, 12–22.

98. Lee, J. H.; Lee, I. G.; Park, J. Y.; Lee, K. Y. Efficient upgrading of pyrolysis bio-oil over Ni-based catalysts in supercritical ethanol. *Fuel* **2019**, *241*, 207–217.

99. Xu, X.; Zhang, C.; Zhai, Y.; Liu, Y.; Zhang, R.; Tang, X. Upgrading of bio-oil using supercritical 1-butanol over a Ru/C heterogeneous catalyst: role of the solvent. *Energy Fuels* **2014**, *28*, 4611–4621.

100. Kannapu, H. P. R.; Mullen, C. A.; Elkasabi, Y.; Boateng, A. A. Catalytic transfer hydrogenation for stabilization of bio-oil oxygenates: Reduction of *p*-cresol and furfural over bimetallic Ni–Cu catalysts using isopropanol. *Fuel Process. Technol.* **2015**, *137*, 220–228.

101. Omar, S.; Alsamaq, S.; Yang, Y.; Wang, J. Production of renewable fuels by blending bio-oil with alcohols and upgrading under supercritical conditions. *Front. Chem. Sci. Eng.* **2019**, *13*, 702–717.

102. Barreiro, D. L.; Gómez, B. R.; Ronsse, F.; Hornung, U.; Kruse, A.; Prins, W. Heterogeneous catalytic upgrading of biocrude oil produced by hydrothermal liquefaction of microalgae: State of the art and own experiments. *Fuel Process. Technol.* **2016**, *148*, 117–127.

103. Isa, K. M.; Snape, C. E.; Uguna, C.; Meredith, W.; Deng, H. Pyrolysis oil upgrading in high conversions using sub-and supercritical water above 400°C. *J. Anal. Appl. Pyrolysis* **2016**, *119*, 180–188.

104. Duan, P.; Xu, Y.; Wang, F.; Wang, B.; Yan, W. Catalytic upgrading of pretreated algal bio-oil over zeolite catalysts in supercritical water. *Biochem. Eng. J.* **2016**, *116*, 105–112.

105. Saha, N.; Saba, A.; McGaughy, K.; Reza, M. T. Effect of supercritical water temperature and Pd/C catalyst on upgrading fuel characteristics of gumweed-derived solvent-extracted biocrude. *Biomass Conv. Bioref.* **2020**, DOI: 10.1007/s13399-020-01050-z.

106. Cui, H. Y.; Wang, J. H.; Zhuo, S. P.; Li, Z. H.; Wang, L. H.; Yi, W. M. Upgrading bio-oil by esterification under supercritical CO_2 conditions. *J. Fuel Chem. Technol.* **2010**, *38*, 673–678.
107. Park, J. Y.; Jeon, W.; Lee, J. H.; Nam, N.; Lee, I. G. Effects of supercritical fluids in catalytic upgrading of biomass pyrolysis oil. *Chem. Eng. J.* **2019**, *377*, 120312.
108. Peng, J.; Chen, P.; Lou, H.; Zheng, X. Upgrading of bio-oil over aluminum silicate in supercritical ethanol. *Energy Fuels* **2008**, *22*, 3489–3492.
109. Zhang, J.; Luo, Z.; Dang, Q.; Wang, J.; Chen, W. Upgrading of bio-oil over bifunctional catalysts in supercritical monoalcohols. *Energy Fuels* **2012**, *26*, 2990–2995.
110. Xu, X.; Zhang, C.; Zhai, Y.; Liu, Y.; Zhang, R.; Tang, X. Upgrading of bio-oil using supercritical 1-butanol over a Ru/C heterogeneous catalyst: role of the solvent. *Energy Fuels* **2014**, *28*, 4611–4621.
111. Duan, P.; Savage, P. E. Upgrading of crude algal bio-oil in supercritical water. *Bioresour. Technol.* **2011**, *102*, 1899–1906.
112. Duan, P.; Zhang, C.; Wang, F.; Fu, J.; Lü, X.; Xu, Y.; Shi, X. Activated carbons for the hydrothermal upgrading of crude duckweed bio-oil. *Catal. Today* **2016**, *274*, 73–81.
113. Xu, Y.; Duan, P.; Wang, B. Catalytic upgrading of pretreated algal oil with a two-component catalyst mixture in supercritical water. *Algal Res.* **2015**, *9*, 186–193.
114. Naik, S.; Goud, V. V.; Rout, P. K.; Dalai, A. K. Supercritical CO_2 fractionation of bio-oil produced from wheat–hemlock biomass. *Bioresour. Technol.* **2010**, *101*, 7605–7613.
115. de Oliveira, M. S.; da Cruz, J. N.; Silva, S. G.; da Costa, W. A.; de Sousa, S. H. B.; Bezerra, F. W. F.; Teixeira, E.; da Silva, N. J. N.; de Aguiar Andrade, E. H.; de Jesus Chaves Neto, A. M.; de Carvalho Junior, R. N. Phytochemical profile, antioxidant activity, inhibition of acetylcholinesterase and interaction mechanism of the major components of the *Piper divaricatum* essential oil obtained by supercritical CO_2. *J. Supercrit. Fluids* **2019**, *145*, 74–84.
116. Leone, G. P.; Balducchi, R.; Mehariya, S.; Martino, M.; Larocca, V.; Di Sanzo, G.; Iovine, A.; Casella, P.; Marino, T.; Karatza, D.; Chianese, S.; Musmarra, D.; Molino, A. Selective Extraction of ω-3 fatty acids from *Nannochloropsis* sp. using supercritical CO_2 extraction. *Molecules* **2019**, *24*, 2406.
117. Rout, P. K.; Naik, M. K.; Naik, S. N.; Goud, V. V.; Das, L. M.; Dalai, A. K. Supercritical CO_2 fractionation of bio-oil produced from mixed biomass of wheat and wood sawdust. *Energy Fuels* **2009**, *23*, 6181–6188.
118. Cheng, T.; Han, Y.; Zhang, Y.; Xu, C. Molecular composition of oxygenated compounds in fast pyrolysis bio-oil and its supercritical fluid extracts. *Fuel* **2016**, *172*, 49–57.
119. Lam, C. H.; Das, S.; Erickson, N. C.; Hyzer, C. D.; Garedew, M.; Anderson, J. E.; Wallington, T. J.; Tamor, M. A.; Jackson, J. E.; Saffron, C. M. Towards sustainable hydrocarbon fuels with biomass fast pyrolysis oil and electrocatalytic upgrading. *Sustain. Energy Fuels* **2017**, *1*, 258–266.
120. Elangovan, S.; Larsen, D.; Bay, I.; Mitchell, E.; Hartvigsen, J.; Millett, B.; Elwell, J.; Santosa, D.; Elliott, D. C. Electrochemical upgrading of bio-oil. *ECS Trans.* **2017**, *78*, 3149.
121. Lister, T. E.; Diaz, L. A.; Lilga, M. A.; Padmaperuma, A. B.; Lin, Y. P.; Palakkal, V. M.; Arges, C. G. Low-temperature electrochemical upgrading of bio-oils using polymer electrolyte membranes. *Energy Fuels* **2018**, *32*, 5944–5950.

122. Pintauro, P. N.; Gil, M. P.; Warner, K.; List, G.; Neff, W. Electrochemical hydrogenation of soybean oil with hydrogen gas. *Ind. Eng. Chem. Res.* **2005**, *44*, 6188–6195.
123. Green, S. K.; Tompsett, G. A.; Kim, H. J.; Kim, W. B.; Huber, G. W. Electrocatalytic reduction of acetone in a proton-exchange-membrane reactor: a model reaction for the electrocatalytic reduction of biomass. *ChemSusChem* **2012**, *5*, 2410–2420.
124. Sáez, A.; García-García, V.; Solla-Gullón, J.; Aldaz, A.; Montiel, V. Electrocatalytic hydrogenation of acetophenone using a polymer electrolyte membrane electrochemical reactor. *Electrochim. Acta* **2013**, *91*, 69–74.
125. Li, Z.; Kelkar, S.; Raycraft, L.; Garedew, M.; Jackson, J. E.; Miller, D. J.; Saffron, C. M. A mild approach for bio-oil stabilization and upgrading: electrocatalytic hydrogenation using ruthenium supported on activated carbon cloth. *Green Chem.* **2014**, *16*, 844–852.
126. Diaz, L. A.; Lister, T. E.; Rae, C.; Wood, N. D. Anion exchange membrane electrolyzers as alternative for upgrading of biomass-derived molecules. *ACS Sustain. Chem. Eng.* **2018**, *6*, 8458–8467.
127. Jaeger, W. K.; Egelkraut, T. M. Biofuel economics in a setting of multiple objectives and unintended consequences. *Renew. Sustain. Energy Rev.* **2011**, *15*, 4320–4333.
128. Sharifzadeh, M.; Richard, C. J.; Liu, K.; Hellgardt, K.; Chadwick, D.; Shah, N. An integrated process for biomass pyrolysis oil upgrading: A synergistic approach. *Biomass Bioenergy* **2015**, *76*, 108–117.
129. Stefanidis, S. D.; Kalogiannis, K. G.; Lappas, A. A. Co-processing bio-oil in the refinery for drop-in biofuels via fluid catalytic cracking. *WIREs Energy Environ.* **2018**, *7*, e281.
130. Bhatt, A. H.; Zhang, Y.; Heath, G. Bio-oil co-processing can substantially contribute to renewable fuel production potential and meet air quality standards. *Appl. Energy* **2020**, *268*, 114937.
131. Badoga, S.; Alvarez-Majmutov, A.; Xing, T.; Gieleciak, R.; Chen, J. Co-processing of HTL biocrude with vacuum gas oil through hydrotreating and hydrocracking to produce low carbon fuels. *Energy Fuels* **2020**, *34*, 7160–7169.
132. Ruddy, D. A.; Schaidle, J. A.; Ferrell, J. R., III; Wang, J.; Moens, L.; Hensley, J. E. Recent advances in heterogeneous catalysts for bio-oil upgrading via "ex situ catalytic fast pyrolysis": catalyst development through the study of model compounds. *Green Chem.* **2014**, *16*, 454–490.
133. Lian, X.; Xue, Y.; Zhao, Z.; Xu, G.; Han, S.; Yu, H. Progress on upgrading methods of bio-oil: a review. *Int. J. Energy Res.* **2017**, *41*, 1798–1816.
134. Cheng, S.; Wei, L.; Zhao, X.; Julson, J. Application, deactivation, and regeneration of heterogeneous catalysts in bio-oil upgrading. *Catalysts* **2016**, *6*, 195.

Chapter 2

Valorization of Bio-Oils to Fuels and Chemicals

Ganapati D. Yadav* and Ujjal Mondal

Department of Chemical Engineering, Institute of Chemical Technology, Mumbai 400019, India

***Email: gd.yadav@ictmumbai.edu.in**

Because of the rapid depletion of fossil fuels, the concept of biorefineries has been promoted to meet the world's energy demands while achieving the goal of net-zero carbon emissions. One promising method for future fuel production is converting biomass into bio-oil and then upgrading it over suitable catalysts. Bio-oil synthesis can be accomplished by fast pyrolysis, which has been identified as one of the most feasible routes. Fast pyrolysis of biomass leads to the formation of a number of gas, liquid, and solid products, among which bio-oil can be further upgraded to improve the quality of the biofuel and bioproducts. However, bio-oil obtained by the fast pyrolysis process has a high acidity, high viscosity, high oxygen content, and low heating value, which make it incompatible for use as a transport fuel. To improve these properties, bio-oil needs to be upgraded over one or more catalysts, for which hydrotreating is an effective method. Steam reforming, hydrodeoxygenation (HDO), and catalytic cracking are the most important processes for hydrotreating using innovative catalysts and reactors. The catalytic steam reforming of bio-oil is well-known to produce biohydrogen and syngas, which are the main precursors for the synthesis of Fischer–Tropsch products. The HDO process is carried out at high pressure where hydrogen helps to remove oxygen from bio-oil to yield biofuels that are comparable to fuels obtained from crude oil. Catalytic cracking is an alternative route in which zeolites, such as HZSM-5, are used as catalysts for the deoxygenation reaction. This process is operated at atmospheric pressure and does not require hydrogen. However, extensive carbon deposition results in very short catalyst lifetimes. This chapter covers all of these aspects.

© 2021 American Chemical Society

Introduction

Energy and the environment are closely related in terms of the use of fossil fuels and emissions of carbon dioxide, as well as the emergence of the so-called biorefinery, which is ultimately intended to lead to net-zero carbon emissions from energy production across the world. The 2020 carbon dioxide concentration was ~410 ppm, corresponding to an increase in average global temperature of approximately 0.13 °C per decade; the Paris Agreement of 2015 envisages restricting CO_2 emissions to contain the temperature rise to less than 2.0 °C or preferably about 1.5 °C. This would mean that the CO_2 emissions of ~35 Gt in 2020 should be brought down to less than 10 Gt by 2050 (Figure 1) (*1*). In other words, the bioeconomy must be supplemented with green carbon-free energy sources, as well as green hydrogen from water splitting (2). There are technological challenges to achieving the goals of the Paris Agreement. Thus, the sequential upgrading of biomass into fuels and chemicals is becoming increasingly significant in the effort to alleviate global warming and to establish biomass as an alternative to fossil hydrocarbons. The main reasons for utilizing biomass, whether virgin or waste, for energy and materials are that it is renewable and carbon-neutral and that it releases less greenhouse gas upon burning than fossil sources. However, biomass needs to be processed using suitable catalytic or noncatalytic paths to crack and defunctionalize it into useful chemicals and products.

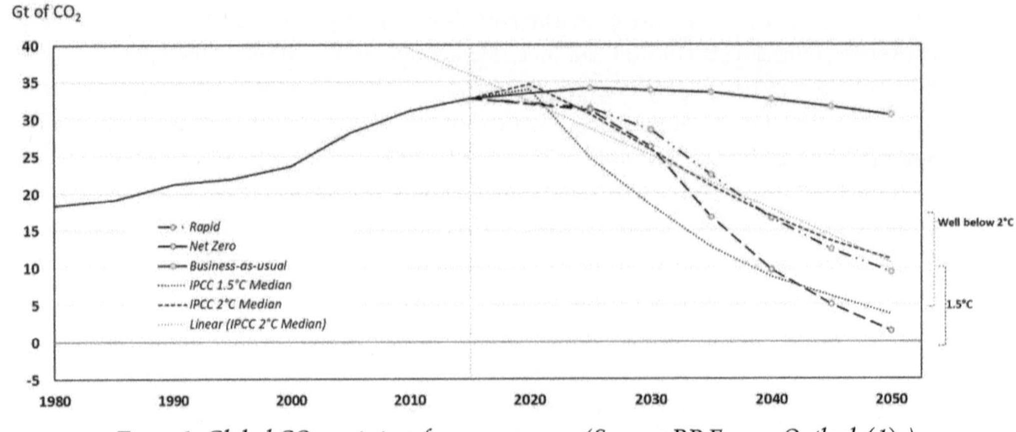

Figure 1. Global CO_2 emissions from energy use. (Source: BP Energy Outlook (1).)

The products of biomass pyrolysis in the absence of any oxygen consist of bio-oil; biochar; and gases such as carbon monoxide, carbon dioxide, methane, and hydrogen. Bio-oil or pyrolysis oil, which is typically produced by the pyrolysis of dried biomass at ~500 °C and subsequent cooling of the gaseous products to the liquid phase, is considered to be a substitute for fossil fuel. The high oxygen content of bio-oils is responsible for some negative properties such as their low heating value, high viscosity, and immiscibility with conventional fossil fuels, so bio-oil needs further upgrading to remove oxygen. The product distribution from pyrolysis is governed by the type of reactor, the thermal environment, and the ultimate temperature and consists mainly of biochar at low temperatures (<450 °C) for a very low rate of heating that takes several hours. In contrast, pyrolysis produces mostly gases at high temperatures (>800 °C) with rapid heating. Between 450 and 800 °C and at relatively high heating rates, bio-oil is the major product. Within a few seconds, fast pyrolysis produces ~60% bio-oil, ~20% char, and ~20% syngas. Fast pyrolysis is the most widely preferred method for obvious reasons. The aim of this chapter is to provide a perspective on bio-oil upgrading.

Upgrading of Biomass: Some Perspective

The world population, which stood at 7.8 billion in 2020, is projected to reach 9.8 billion in 2050 and 11.2 billion in 2100, according to the United Nations Department of Economic and Social Affairs, necessitating more energy and materials against the backdrop of a limited or finite source of fossil fuel. According to the U.S. Energy Information Administration (EIA), the total energy consumed by the United States alone at the end of 2016 was 88.34 quadrillion Btu, of which 71.19 quadrillion Btu was sourced from nonrenewable fossil hydrocarbon sources (3). The current consumption of crude oil is about 97.5 million barrels per day (MMBD), mainly going for energy, chemicals, and materials (2).

Liquid fossil fuels play a vital role in the transportation sector, and the total U.S. energy consumption was projected to increase from 97.1 quadrillion Btu in 2013 to 105.7 quadrillion Btu in 2040, of which the use of liquid fuels was projected to rise from 35.9 quadrillion Btu in 2013 to 37.1 quadrillion Btu in 2020 before falling back to 36.2 quadrillion Btu in 2040. Consequently, the transportation sector is the prime contributor of CO_2 to the environment, representing 26% of total CO_2 emissions and increasing at a rate of 2.5% annually since 2010 (*1*, *4*). However, this increase should reverse in the near future as a result of the use of alternative fuels and blending of biofuels into petrol (gasoline) and diesel. Biomass has good potential to play an active role as a renewable energy source to supply biofuels for transportation, electricity generation, and household heating (*5*, *6*). Comprehending the importance of bioenergy, several industrialized countries have rolled out initiatives such as the Multi-Year Plan of the U.S. Department of Energy and the Biomass Action Plan of the European Commission. Interestingly, one of the most ambitious claims, made in the *2016 Billion-Ton Report*, is that the United States has the potential to generate 1 billion dry tons of non-food-based biomass annually by 2030 (*7*). Moreover, Figure 2 shows that the United States and Europe are the leading producers and users of biofuel (*1*). The Asia-Pacific region also showed a significant annual growth rate of 20.1% from 2010 to 2016 (3). Indeed, the share of renewables in the total energy supply is projected to be ~60% by 2050 in the net-zero economy.

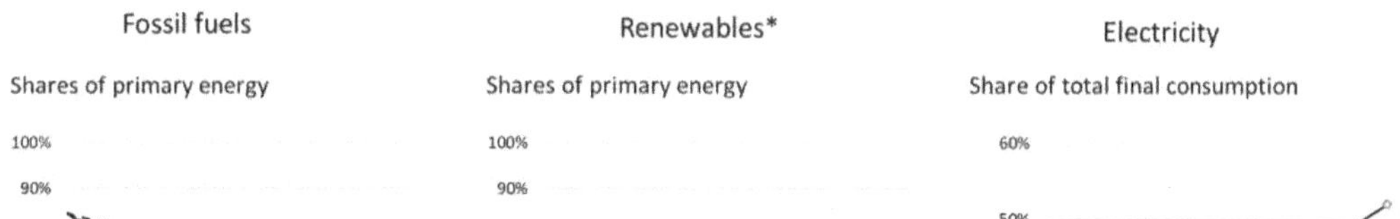

Figure 2. Changing the structure of global energy demand. (Source: BP Energy Outlook (1).)

Compared to fossil fuels, hemicellulosic and lignocellulosic biomass feedstocks provide 78% and 86% reductions of CO_2 emissions, respectively. The aim of this chapter is to provide a perspective on the portion of bio-oil that is upgraded for use as a fuel for turbines, boilers, truck and car engines, and modified diesel generators for electricity production. Generally, bio-oil is a mixture of char particles, oxygenated hydrocarbons, pyrolysis water, and water from biomass moisture. The relative amounts of these constituents generally vary with the biomass composition and the pyrolysis conditions. Although the aqueous and oil phases are easily separable after pyrolysis, the oil phase contains a high amount of dissolved moisture, reducing its heating value. Most of the chemical compounds in bio-oil are in their oxygenated forms, which is one reason for the low heating value of bio-oil, which is only 50% of the heating value of diesel fuel. The quality of bio-oil is upgraded by refining processes, such as evaporating the volatiles to remove all of the moisture. The addition of ethanol to the remaining oil has many benefits, such as enhancing stability and reducing viscosity. Raw bio-oil can be used directly in modern combustion engines, and it should be refined and upgraded before use as a diesel replacement. The upgrading process is carried out in dedicated downstream units or in the pyrolysis reactor. However, the upgrading of bio-oils to transport-grade fuels is done by a catalytic cracking or hydrotreating method. Generally, the catalytic cracking process is more economical than hydrotreating and can effectively remove oxygen from bio-oil.

Biofuels

Biomass feedstocks that are suitable for the synthesis of biofuel and the production of value-added chemicals include (1) sugary and starchy feedstock (e.g., sugar beet, sugar cane, sorghum, wheat, corn), (2) triglyceride feedstock (e.g., vegetable oils, microalgae oil, waste cooking oil, animal fat), and (3) lignocellulosic feedstock (e.g., wood, bagasse, rice straw, coconut leaves, mallee leaves, almond shell, cherry stones, napier grass, switchgrass). According to some reports, the world's total biomass reserves in land and aquatic sources comprise about 1.8 trillion tons and 4 billion tons, respectively. Among all of the naturally available biomass that can be converted into fuel and chemicals, 75% is carbohydrate in nature and abundantly found in the form of cellulose, hemicellulose, starch, and saccharose. An additional 20% is lignin, and the remaining 5% is a combination of oils, proteins, and other substances. Only 3–4% of the total biomass is used by humans for food and nonfood purposes (8).

From the point of view of harvesting energy, the available biomass in the world can provide 3,000 EJ of energy, which is more than 80 times the annual global energy requirement. However, only 14% of the total energy requirement is satisfied through biomass, which is roughly 56 million TJ/year [1230 million tonnes of oil equivalent (MTOE)/year]. A few northern and central European countries have prepared roadmaps for using the vast potential of biomass to replace all of their conventional fossil energy sources by setting up large-capacity plants to coproduce heat and energy from biomass. France, which has the most extensive agricultural land in Europe, has a unique plan to encourage the production of biodiesel and bioethanol through a comprehensive tax exemption policy based on the use of biomass. In contrast, the United Kingdom does not produce sufficient biofuels but has policies to encourage biogas recovery systems from landfills to produce energy and heat (9).

Biofuels are commonly divided into three generations based on their source and synthesis process. The important biomass sources are forestry and agricultural residues, waste from the wood processing industry (sawdust, shavings, etc.), animal residues from livestock farms, algae, sewage, and aquatic biomass (Figure 3). The biomass category also includes municipal solid wastes and other waste streams that originate from human activities and are not suitable for reuse even after subsequent processing.

(1) First-generation biofuels or conventional biofuels are derived from edible plant parts (beetroot, sugar cane, sorghum, corn grain, oilseeds, etc.).

(2) Second-generation biofuels are produced from nonedible biomass (including wood, food, organic waste, and biomass crops).
(3) Third-generation biofuels are generated from photosynthetic microorganisms such as microalgae.

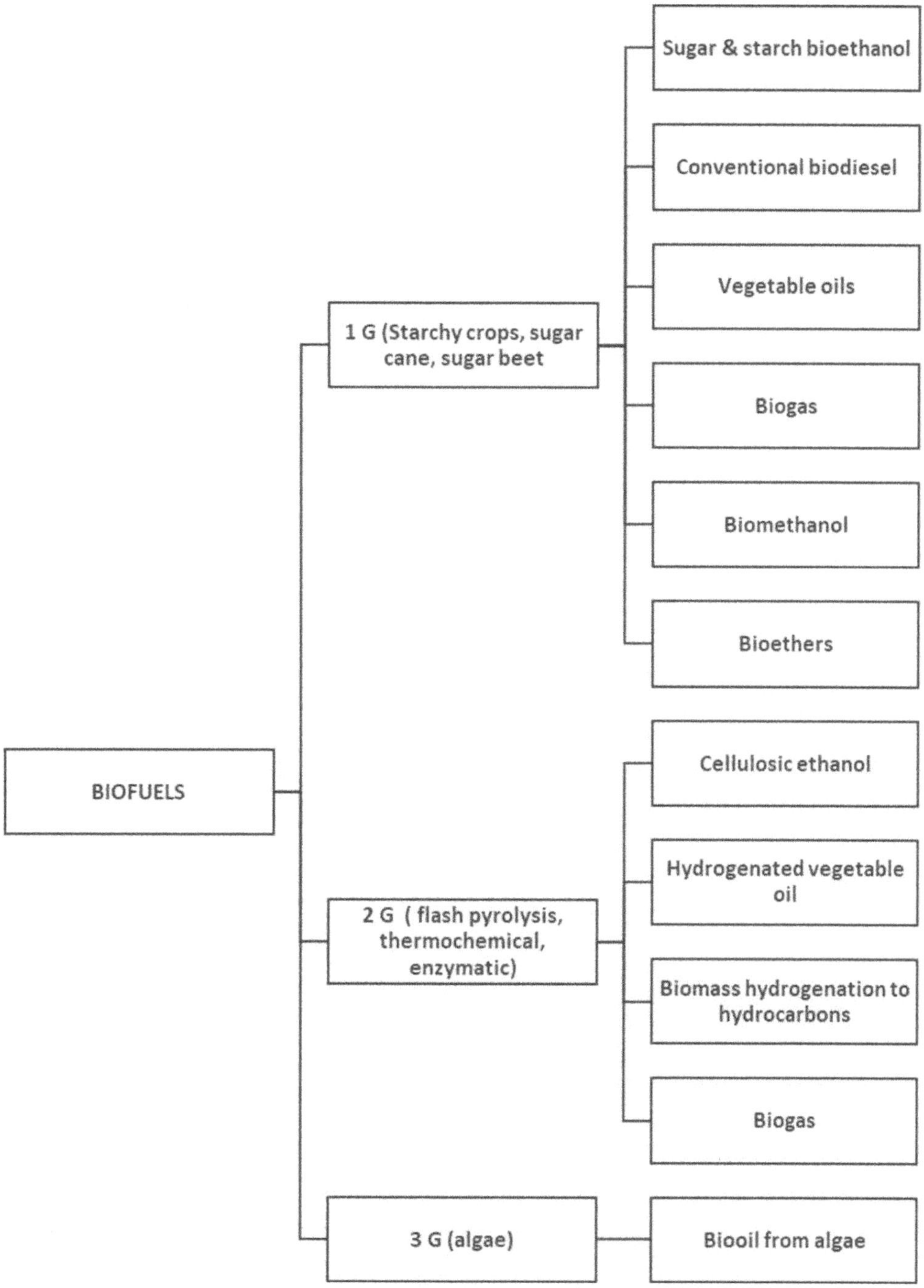

Figure 3. The three generations of biofuel and their sources.

Biofuel Production

First-Generation Biofuels

The most critical first-generation biofuels are biodiesel, bioethanol, and biogas; other biomass-derived products such as vegetable oils, bioethers, and biomethanol can also be included in this category. Starchy crops (mainly wheat and corn), sugar cane, and sugar beet are the main feedstocks for bioethanol. The United States is the largest producer of biofuels, generating ~1.2 MMBD of biofuels in 2020 according to the EIA, of which ~1 MMBD was bioethanol, compared to a worldwide production of 2.5 MMBD of biofuels (*10*).

Oil-based crops, such as sunflower, rapeseed, soybean, and palm oils, and waste edible oils are the most common feedstocks for biodiesel. According to a joint OECD-FAO Agricultural Outlook, global biodiesel production and consumption were expected to rise by more than 14% from 2016 to 2020, driven by changes in the biofuel policies in the United States, Brazil, Argentina, Indonesia, and the European Union (*11*). However, the coronavirus pandemic changed this trajectory. Among all oilseeds, rapeseed and sunflower seed are mainly processed to produce biodiesel (*12*).

Biogas is produced through the anaerobic digestion of mixtures of starch derived from corn, manure, and other organic wastes. Biogas derived from wastes and residues can be classified as a second-generation biofuel because it is not in competition with the food and feed industries. The use of biogas is increasing continuously, as many economic subsidies have been announced to promote electricity generation from biogas, particularly in European countries. Biogas is upgraded to biomethane for use in the transportation sector in Germany and Sweden. Sweden is the leading country in biogas production for the transportation sector, with several vehicles currently running on fuel containing 45% biomethane (*13, 14*).

Despite developments in the production of first-generation biofuels, such biofuels face a few social concerns related to the disempowerment of landowners and issues of food versus fuel. Even though first-generation biofuels are renewable in nature, the use of land for cultivating biofuel feedstocks is questioned. Problems mainly occur in maintaining the integrity of any land used for biofuel feedstock cultivation and in sustaining nearby critical ecosystem services. Poor management of the cultivation process can lead to irreversible damage to land, compromised reusability of the land, and detrimental downstream effects.

All of these limitations on first-generation biofuels can be overcome by the development of second-generation biofuel (*15*).

Second-Generation Biofuels

As mentioned previously, a greener alternative to first-generation biofuels is second-generation biofuels derived from wastes, residues, and non-food-based crops. The synthesis of second-generation biofuels by processing biomass in biorefineries ensures national energy security and keeps the price of biofuels competitive by producing high-value byproducts. Second-generation biofuels are generally produced from nonedible waste crops and foliage such as forestry, agricultural, and industrial residues, which are rich in lignocellulose material and do not affect society and the environment. The conversion processes, such as flash pyrolysis and thermochemical and enzymatic processes, and the generation of product gases and synthetic biofuels are more economical. Second-generation biofuels not only depend on above-ground biomass or edible biomass but also utilize the whole plant for biofuel production. Therefore, the land usage is much higher than for first-generation biofuels, resulting in environmentally superior performance (*16–18*). Several commercial and semicommercial operations can be found throughout the world.

Third-Generation Biofuels

Third-generation biofuels are synthesized from algal biomass, both microscopic and macroscopic, which generally contains a higher lipid fraction than most oil-containing oilseeds (*19*). Moreover, these biofuels do not compete with human or animal food sources, and the rapid growth of microalgae is a huge advantage for the biofixation of CO_2 (*20*). This advantage can be further improved by marine algae cultivation, whereby the availability of freshwater will remain unaltered for other activities. Table 1 provides a comparison of petrofuels and first-generation, second-generation, and third-generation biofuels (*12–20*).

Table 1. Sources of Biofuels (*12–20*)

	Fossil Fuel	*First-Generation*	*Second-Generation*	*Third-Generation*
Source(s)	Crude oil	(1) Starchy crops, sugar cane, and sugar beet (2) Oil-based crops (such as sunflower, rapeseed, soybean and palm oils, waste edible oils) (3) Starch derived from corn, manure, and other organic wastes	Forestry, agricultural, and industrial residues and wastes Woody crops	Algae
Advantages	Cheaper and better performance	Economical Environment-friendly Provide energy security	Environment-friendly Technologically advanced and cheaper than first-generation biofuels No threat to food security	Environment-friendly Can be produced all year Can be cultivated in marine water and on nonarable land
Products	Petrol Diesel Kerosine Jet fuel Liquefied petroleum gas Compressed natural gas	(1) Biodiesel (2) Bioethanol (3) Biogas	(1) Bioethanol (2) Lignocellulosic ethanol (3) Bio-oil, hydrotreating oil	(1) Bio-oil (2) Syngas (3) Bioethanol (4) Biomethane (5) Biohydrogen

Table 1. (Continued). Sources of Biofuels (*12–20*)

	Fossil Fuel	*First-Generation*	*Second-Generation*	*Third-Generation*
Problems	Greenhouse gas generation Environmental pollution Limited resource	Competition for land with food and feed crops Requires blending with conventional fuel	Requires technological advances for commercial application	Selection of highest lipid content Technological issues

Biorefining

Biorefining is defined as the sustainable processing of biomass into a variety of products and energy (*21–23*). Biomass can be profitably converted into different products using chemical, biochemical, and thermochemical processes (Figure 4).

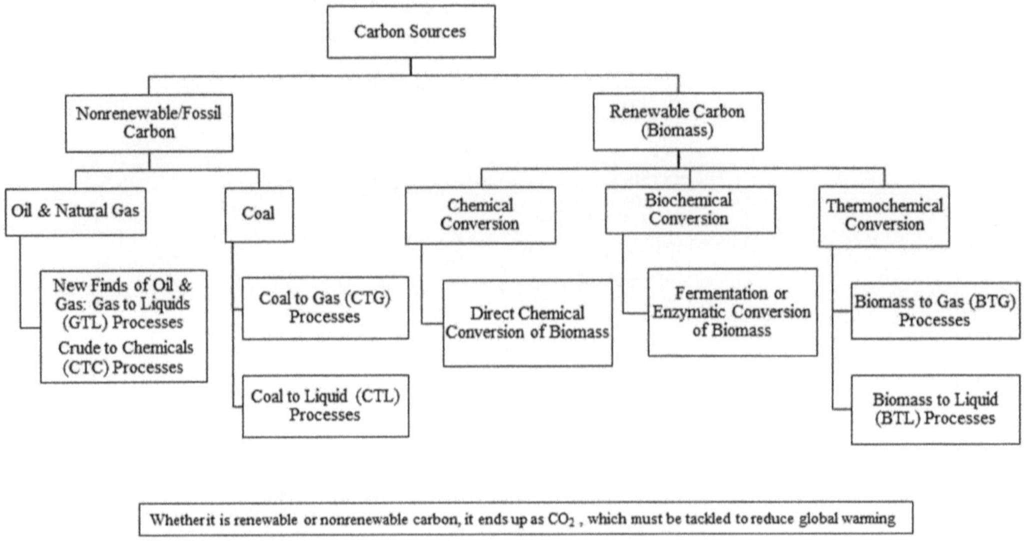

Figure 4. Comparison of different carbon sources. Reproduced with permission from reference (2). *Copyright 2020 Springer.*

Biorefineries are considered to be an encouraging substitute for oil refineries, as their main feedstocks are derived from starchy crops, sugary crops, perennial grasses, and lignocellulosic crops; lignocellulosic residues; oil crops; aquatic biomass; agricultural, forest, and industrial residues; and consumer-generated kitchen waste (*18, 24–26*). The sustainable conversion of biopolymers such as lignin, hemicellulose, starch, pectin, cellulose, chitin, and chitosan-based products into valuable products using (nano)catalysis has been critically analyzed in comparison to the fossil-fuel industry (*27*).

The composition and structure of the raw material for biorefineries is distinctly different from that used in oil refineries and is not homogeneous. Therefore, the elemental composition is a mixture of carbon, hydrogen, and oxygen, along with other minor components such as nitrogen, sulfur, and minute quantities of minerals (Table 2) (*23, 28, 29*). In comparison with crude oil, biomass feedstock has very little hydrogen, a high oxygen content, and a lower carbon fraction. This variation in the composition of biomass raw materials works as both an advantage and a disadvantage (*30–32*).

The desired biomass composition includes the following: (1) high volatile-matter fraction, hydrogen content, organic structural component, biomass activity and extractives, alkaline-earth content, and water-soluble nutrients; (2) low values of carbon, fixed carbon, ash, sulfur, nitrogen, and silicon contents; and (3) low initial ignition and combustion temperatures. Some of the significant disadvantages related to the composition and properties of biomass are (1) high moisture and oxygen contents, high water-soluble fraction, high halogen and alkaline contents; (2) low values of energy density (calorific value and bulk density); and (3) high variations in composition and properties.

An extensive range of products can be obtained from biorefineries because of the wide variety of raw biomaterials. However, to process this wide range of raw materials, advanced technologies are required (22). Biomass needs to be deoxygenated and depolymerized prior to being processed in a biorefinery. Deoxygenation is necessary because of the excessive presence of oxygen in biofuels, which compromises the heat content of biofuel and makes it more polar, hindering its blending with fossil fuels (*13*). Second-generation bioethanol is synthesized by converting nonedible plant biomass into sustainable liquid fuel. Lignin, which comprises up to 30% of plant biomass, is the main byproduct of this process. This huge quantity of lignin polymer is converted into fuel products by reductive depolymerization. This process involves a catalyst and hydrogen from the deoxygenation of the complex lignin polymer (33).

Table 2. Compositions of Various Lignocellulosic Feedstocks

Parameter[a]	*Units (dry)*	*Wheat Straw* (23)	*Switchgrass* (23)	*Softwood* (28)	*Corn Stover* (28)	*Fossil Fuel* (29)
Moisture	%	15	15	15	15	
Cellulose	%	32.6	35.4	44.5	38.1	
Glucan (C_6)	%	32.6	35.4	44.5	38.1	
LHV	MJ/kg	17.6	18.6	19.6	18.5	
Lignin	%	16.8	18.2	27.7	20.2	
Extractives	%	12.9	11.5	2.88	4.78	
Acids	%	2.24	2.15	2.67	4.84	
Ash	%	10.2	4.28	0.32	8.59	
Hemicellulose	%	22.6	26.5	21.9	25.3	
Mannan (C_6)	%	0.31	0.39	11.4	0.41	
Galactan (C_6)	%	0.75	0.96	2.56	0.74	
Xylan (C_5)	%	19.2	22.4	6.3	20.2	
Arabinan (C_5)	%	2.35	2.73	1.6	2.03	
C	%	43.9	46.9	50.3	46.7	83–87
O	%	38.7	42	42.1	38.4	0.1–1.5
S	%	0.16	0.7	0.01	0.1	0.5–;6
N	%	0.63	0.62	0.03	0.67	0.1–2

Table 2. (Continued). Compositions of Various Lignocellulosic Feedstocks

Parameter[a]		*Units (dry)*	*Wheat Straw (23)*	*Switchgrass (23)*	*Softwood (28)*	*Corn Stover (28)*	*Fossil Fuel (29)*
H		%	5.26	5.54	5.98	5.49	10–14
HHV		MJ/kg	17.51	18.64	18.064	17.65	–
Proximity analysis	Volatiles	%	71.30	81.36	70.55	75.17	–
	Ash	%	8.90	3.61	0.83	5.58	–
	FC	%	19.80	15.03	16.35	19.25	–
Ultimate analysis	C	%	43.20	47.45	45.66	43.65	–
	H	%	5	5.75	4.86	5.56	–
	O	%	39.40	42.37	34.94	43.41	–
	N	%	0.61	0.74	1.38	0.61	–
	S	%	0.11	0.08	0.06	0.01	–
	Cl	%	0.28	0.03		0.6	–
	Ash	%	11.40	3.50		6.26	–

[a] Abbreviations: FC, fixed carbon; HHV, higher heating value; LHV, lower heating value.

Because of the complex composition of biomass, the platform chemical to be selected at the beginning of the biorefining process will vary, as different processing strategies of the same chemical can generate other products. The future biorefinery will be based on a limited number of platforms from which all valuable biochemicals and fuels can be derived. The carbohydrate portion of the cellulose and hemicellulose fraction of lignocellulosic biomass will play the most crucial role for value-added products (Figure 5). Polysaccharides in biomass can be easily hydrolyzed to monosaccharides such as fructose, glucose, and xylose, which can serve as precursors for various chemical or fermentation processes to make a series of biomass platform molecules (BPMs) such as the benzene, toluene, and xylene (BTX) petroleum platform molecules of present oil refineries (Table 3) (*34–36*).

Another advantage of BPMs is their higher oxygen content compared to crude-oil–derived molecules. As a result, BPMs are further reduced by a greener pathway using hydrogenation over a suitable heterogeneous catalyst. In comparison with the fossil fuel-based chemical industry, where functional groups are externally added, these BPMs are mostly available with prefunctional groups or desired functionality (37). For example, levulinic acid ($C_5H_8O_3$) is synthesized by the acid hydrolysis of C_6 sugars and can be further converted into a large number of derivatives because of its highly reactive nature. Because it contains a single ketone carbonyl group and an acidic carboxyl group, levulinic acid can act both as a fatty acid and as a ketone. Cellulosic biomass can be acid hydrolyzed to C_6 sugars and then to 5-hydroxymethylfuran (5-HMF), which is an excellent platform molecule with huge potential for the synthesis of various commercially useful acids, aldehydes, alcohols, and amines, as well as the fuel 2,5-dimethylfuran and the renewable monomer furandicarboxylic acid (Figure 6) (35).

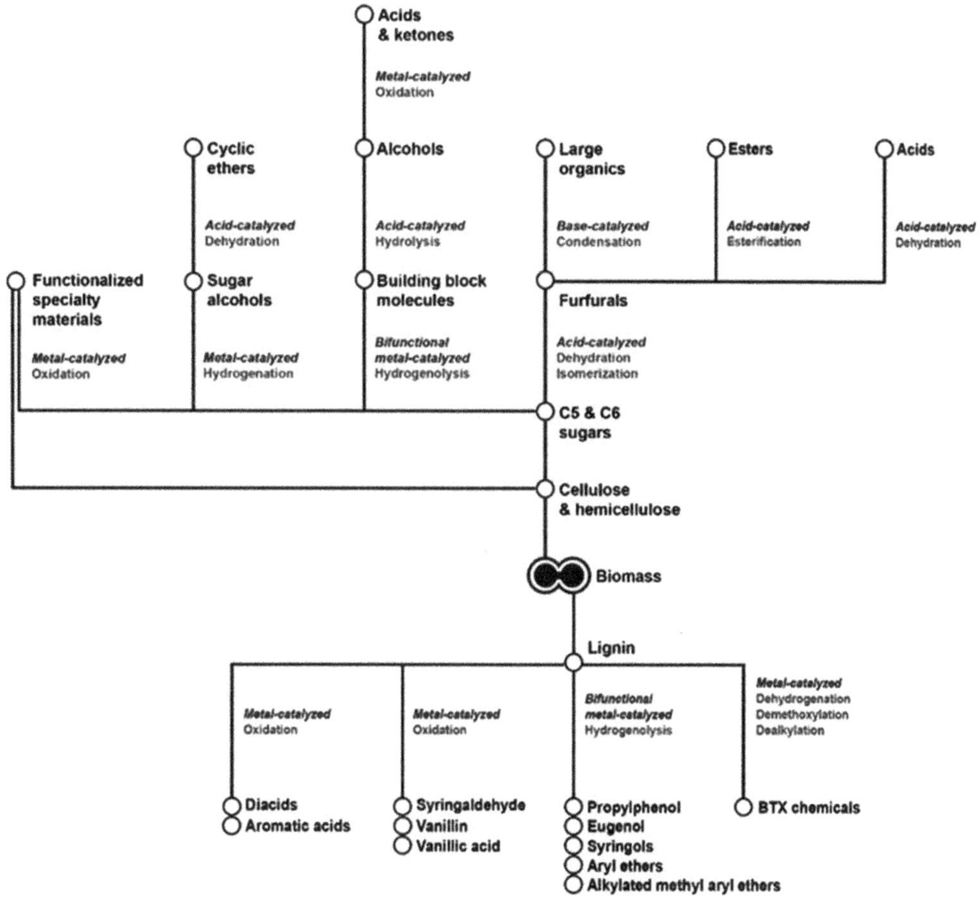

Figure 5. Variety of chemicals produced from biomass. Reproduced with permission from reference (2). Copyright 2020 Springer.

Table 3. Top 14 BPMs according to the U.S. Department of Energy

BPM	*Structure*	*BPM*	*Structure*
Fumaric acid	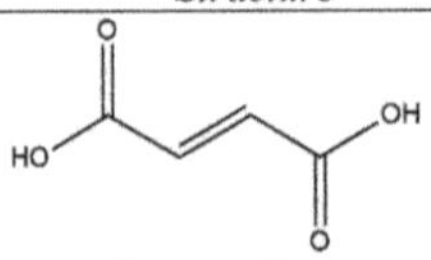	(*S*,*R*,*R*)-Xylitol	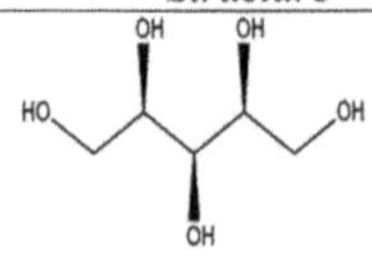
Glycerol	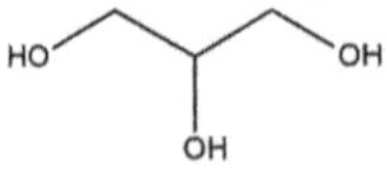	3-Hydroxypropionic acid	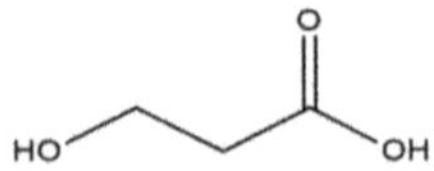
L-Glutamic acid	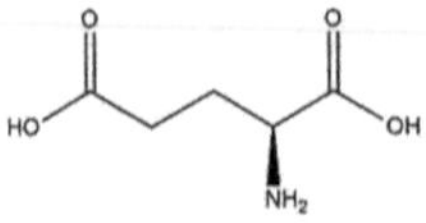	3-Hydroxybutyrolactone	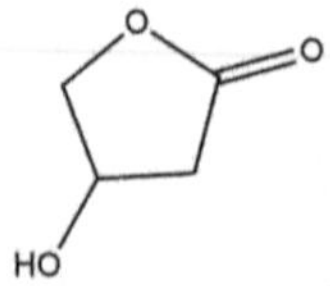
Itaconic acid	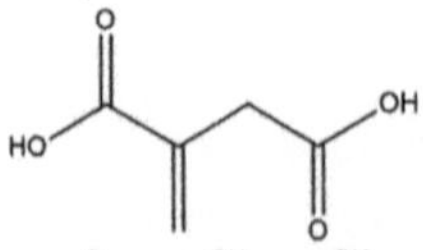	2,5-Furandicarboxylic acid	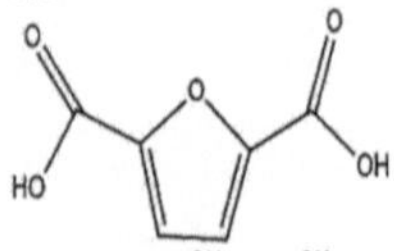
Glucaric acid	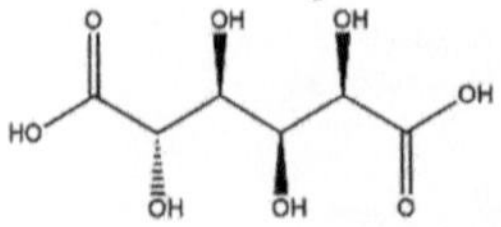	Sorbitol	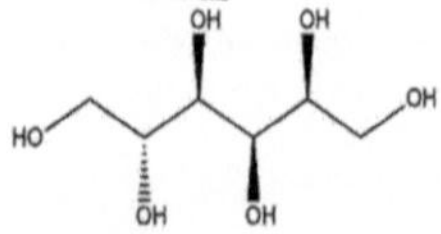
l-Malic acid	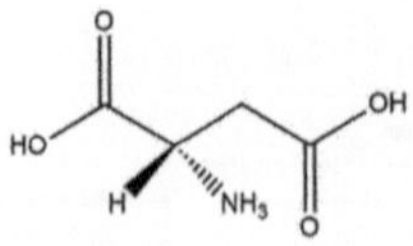	Succinic acid	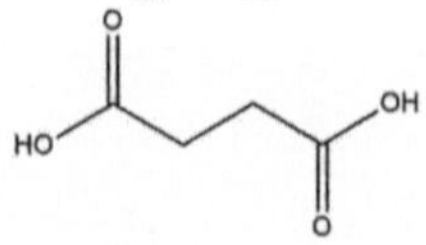
Levulinic acid	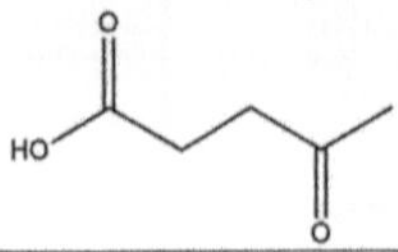	L-Aspartic acid	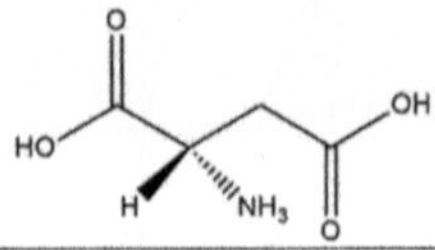

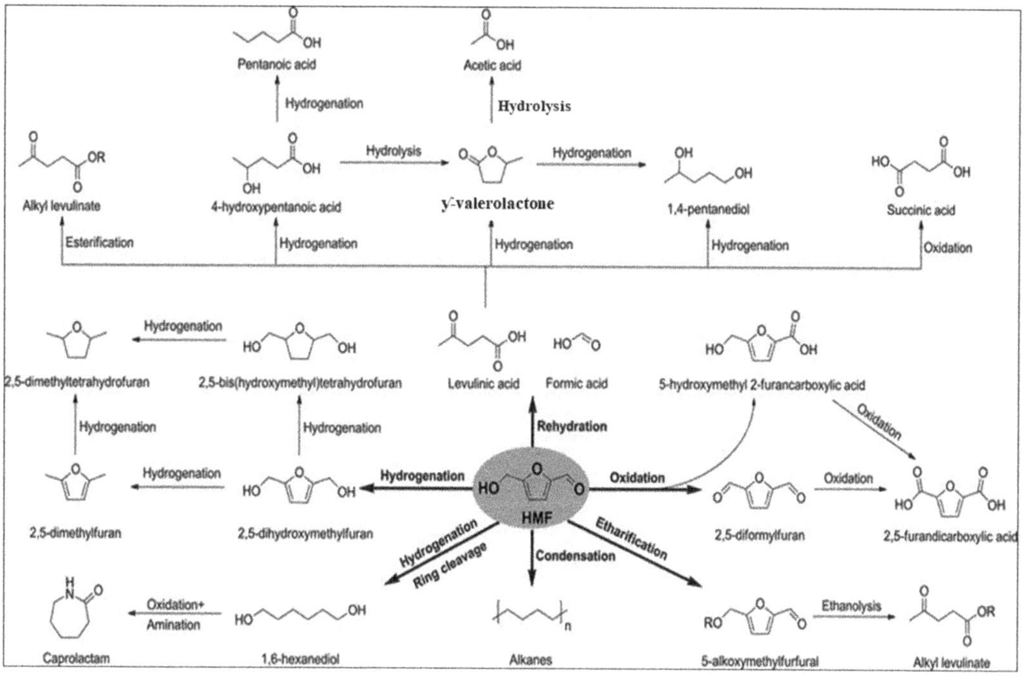

Figure 6. 5-HMF as a platform molecule for the production of different chemicals.

Biorefinery Products

Biorefinery products can be grouped into two large categories, namely, chemical products and energy products. The important energy products from biomass upgrading in biorefinery are as follows:

(1) gaseous biofuels (syngas, hydrogen, biogas, biomethane),
(2) liquid biofuels (Fischer–Tropsch fuels, bioethanol, biodiesel, and bio-oils), and
(3) solid biofuels (lignin, charcoal, pellets).

The important material and chemical products are as follows:

(1) organic acids (such as lactic, itaconic, and succinic acids and other sugar derivatives),
(2) chemicals (fine chemicals, bulk chemicals, and building blocks),
(3) biomaterials (cellulose, paper, pulp, wood panels),
(4) polymers and resins (phenol resins, furan resins, starch-based plastics),
(5) food and animal feeds, and
(6) fertilizers.

Biorefinery Processes

A variety of pathways are used for the production of energy compounds from biomass (34).

Direct Combustion of Biomass

Biomass combustion is a reaction in the presence of oxygen to produce thermal energy and flue gas. The thermal energy is used to generate steam for electricity generation. The main disadvantages of this process are the moisture content of the biomass and the percentage of ash produced as an end product.

Gasification

Through this thermal conversion process, biomass is gasified into syngas (CO_2, CO, H_2, CH_4) at high temperatures in the presence of a limited oxygen supply. Because any carbonaceous solid or liquid can be used in this process, it becomes very useful where waste biomass is plentiful. This is an exothermic process, and therefore, heat must be supplied at the start of the reaction and is regenerated throughout the process. The main disadvantage of this process is associated with reactor cleanup, as many undesirable compounds are produced. These tars and other unwanted compounds reduce the efficiency of the process.

Pyrolysis

Pyrolysis is a thermochemical process for converting biomass in the absence of oxygen into bio-oils, chars, and gases. When the biomass is heated at about 500 °C, it is converted into chars and condensable and noncondensable gases. Then, the condensable gases are condensed into a dark brown liquid, namely, bio-oil, by being passed through condenser trains. Typically, short residence times (usually seconds) are required for the synthesis of liquid bio-oils (fast pyrolysis), and when longer residence times are employed (slow pyrolysis), the main product of pyrolysis is solid biochar. Bio-oil generated by catalytic fast pyrolysis is composed of more aromatic compounds than that obtained by direct pyrolysis, which yields 20–30% aromatic compounds (38). The main problem with this process is that pyrolysis oil is highly acidic and corrosive, has a low heating value, and is unstable.

Liquefaction

Liquefaction is an alternative route for producing bio-oil. Specifically, the liquefaction of biomass is a catalytic thermal decomposition process in which large molecules of biomass are turned into shorter species that repolymerize into bio-oil. First, biomass is mixed with water and catalysts with basic properties (e.g., sodium carbonate). The liquefaction process is carried out at lower temperatures (524–725 K) but higher pressures (5–20 atm) for longer residence times. These operating conditions make this process expensive. However, the bio-oil produced contains less oxygen (12–14%) (39) than that obtained from pyrolysis and thus requires less further processing for use as a transportation fuel.

Catalytic Conversion

The bio-oil synthesized by biomass pyrolysis can be upgraded in sub- and supercritical ethanol using suitable catalysts. The supercritical upgrading process is the subject of intense research, as it is more efficient than the subcritical upgrading process. In supercritical ethanol, the acidic catalyst HZSM-5 can facilitate the esterification of the acids in crude bio-oil into esters.

Supercritical Gasification

Supercritical water gasification is an advanced step of conventional gasification that uses water as the main reaction medium to effectively decompose biomass into hydrogen-rich syngas. The yields and selectivities of the products of this process depend on different process parameters such as pressure, temperature, residence time, biomass particle size, feed concentration, reactor configuration, catalysts, and reaction pathways. These factors also play an essential role in determining the carbon conversion, gasification efficiency, and heating value of the gaseous products.

Biochemical Conversion

Another promising approach for the production of fuel and chemicals is the biochemical route, by which biobutanol and bioethanol can be synthesized through hydrolysis in the presence of enzymes. Bioethanol plays a significant role in countries such as India and United States as a gasoline substituent and also as a precursor in the chemical industry (for acetic anhydride, acetaldehyde, ethyl acetate, monoethylene glycol, etc.) (*40*). For example, 100 L of ethanol can be produced from 1 t of sugar cane. The effectiveness of enzymatic hydrolysis can be enhanced by varying the processing conditions, including the types of enzymes, solid loadings, conditioning and mixing methods, and pretreatment conditions. However, significant challenges need to be addressed because enzymes are costly, vigorous pretreatments of biomass are required to efficiently convert its fractions because of the slowness of biochemical reactions, and larger reactor sizes are needed. Research in association with leading commercial enzyme producers has recently been undertaken to apply recombinant DNA technology to fungi and bacteria to improve hemicellulose and cellulose enzymes with low-cost enzyme production methods (*41, 42*).

Fermentation

Fermentation is an anaerobic process mediated by microorganisms that break down the glucose in the biomass. It is a chain of chemical reactions that convert sugars into alcohols or acids (lactic acid and succinic acid) (*43*). Bacteria or yeast are added to the biomass feedstock, which feeds on the sugars to synthesize ethanol andCO_2. Later, ethanol is distilled and dehydrated to obtain pure alcohol to meet the criteria for an automobile-grade fuel. The remaining solid residue has various uses; for example, in the case of sugar cane, the bagasse can be used as a fuel for boilers or for further gasification.

Biofuel synthesized from feedstocks mainly composed of lignocellulosic biomass has a bright future as a sustainable alternative for growing energy demands because it has many advantages, such as abundant availability, low-cost cultivation, and noncompetition with food. Gasification of lignocellulosic biomass produces syngas, which predominantly consists of H_2 and CO. The conversion of biomass-derived syngas into biofuels by microbial catalysts (such as *Clostridium autoethanogenum*, *Clostridium ljungdahlii*, *Acetobacterium woodii*, *Peptostreptococcus productus*, and *Clostridium carboxidivorans*) has achieved considerable attention as a promising alternative for biofuel production. Gas–liquid mass-transfer limitations, microbial catalysts, syngas quality, and product recovery are the major issues associated with this process.

Bio-Oil

Bio-oil, a dark brown and viscous liquid, is the product of the fast pyrolysis of biomass. The thermal decomposition of biomass in the absence of oxygen forms char, aerosol/vapor, and noncondensable gases. The aerosol/vapor is condensed by rapid quenching to concentrate it in liquid form, which is generally referred to as biocrude or bio-oil. It is a mixture of more than 200 compounds, including sugars, acids, ketones, esters, aldehydes, phenols, furans, guaiacols, alcohols with H/C molar ratios greater than 1.5, and water (*43*). The composition of bio-oil is governed by various factors such as biomass quality, heating rates, residue time, and temperature. Bio-oil typically contains 14–30 wt % water, with the remaining 70–86 wt % consisting of oxygenated organic compounds (38–44 wt % oxygen in dry biomass) (*44, 45*).

Bio-oil production from renewable resources by the fast pyrolysis method can reach 0.3 billion barrels per day (roughly 4% of the total U.S. consumption) (*46*).However, the lack of refining technology is limiting the potential of the bio-oil industry. Bio-oil has a wide variety of uses, including as a fuel in boilers, gas turbines, furnaces, and diesel engines, and it can be upgraded for the synthesis of liquid fuels. The composition of bio-oil is distributed in several families based on their chemical composition and thermal behavior: C_2–C_4 compounds (mainly hydroxyacetaldehyde, acetol, and acetic acid), monophenols, monofurans, and monosugars (Figure 7) (*46*).

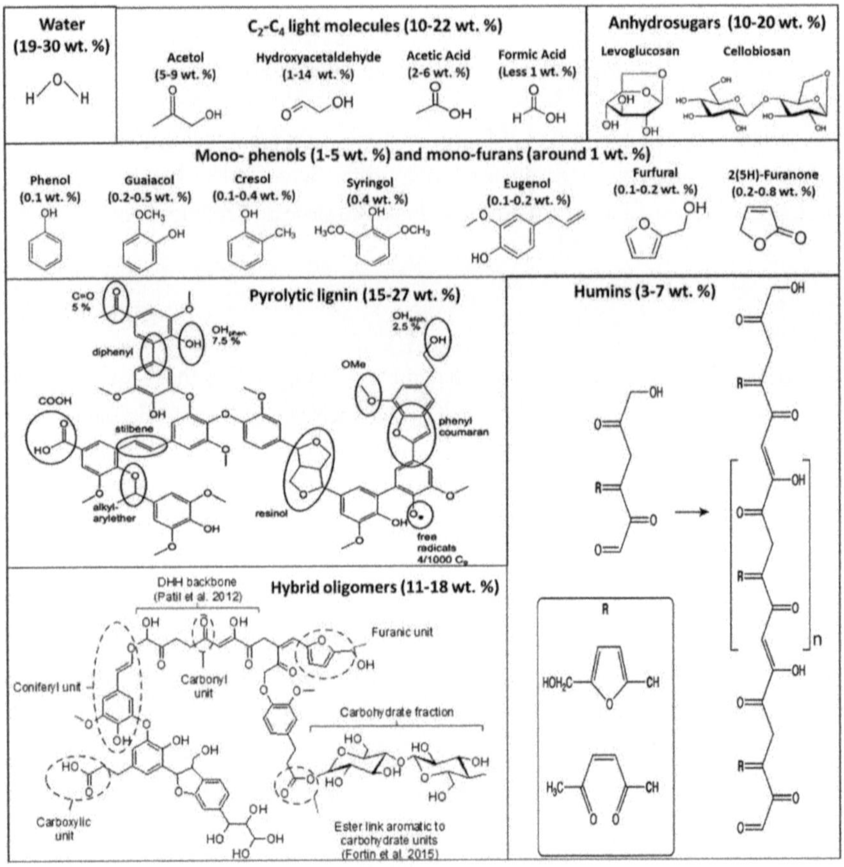

Figure 7. Representative constituents of each bio-oil fraction. Reproduced with permission from reference (46). Copyright 2019 American Chemical Society.

The use of bio-oil as obtained, without further treatment, encounters many technical problems. Specifically, bio-oil cannot be used as a fuel without further upgrading because of its acidity (pH 2–4) and low heating value of 16–19 MJ/kg compared to 44 MJ/kg for crude oil (Table 4). The olefins present in bio-oil tend to polymerize, and other compounds such as aldehydes, ketones, and acids tend to convert into ethers and acetals, thereby increasing the viscosity of bio-oil (*46, 47*).

Table 4. Selected Properties of Bio-oils Obtained from Different Biomass Feedstocks (*48–54*)

Bio-oil	*Water (wt %)*	*HHVa (MJ/kg)*	*pH*	*Viscosity (cSt, 40 °C)*	*Ref*
Nannochloropsis algae	7.63	33.4	6.61	–	(*49*)
Corn stover	15.2	185	2.87	17.51	(*50*)
Switch grass	21.8	19.6	2.97	25.6	(*51*)
Sludge	10.9	36.4	6.8	–	(*52*)
Pine	18.5	18.7	2.9	44	(*53*)
Rice husk	28	16.5	3.2	13.2	(*53*)
Poplar	29.7	19	3.1	22	(*54*)
Eucalyptus bark	32.86	12.23	2.87	16.28	(*55*)
Oak	16.1	18.9	2.8	115	(*54*)

[a] Abbreviation: HHV, higher heating value.

Bioproducts from Bio-Oil

Bio-oil can be fractioned into various intermediates and transformed into value-added chemicals by reactions with inorganic and organic compounds (*55–57*). Pyrolysis oil is composed of six major fractions (water, 15–30 wt %; light oxygenates, 8–26 wt %; monophenols, 2–7 wt %; water-insoluble oligomers derived from lignin, 15–25 wt %; and water-soluble molecules, 10–30 wt %). Thus, bio-oil can be considered equivalent to a petroleum crude oil, because the whole bio-oil, its fractions, and some of its components can be used for producing intermediate or final value-added products. Table 5 provides information about bioproducts that can be derived from bio-oil and its fractions as reported in the published literature (*46, 55*).

Table 5. Products Derived from Bio-Oils

Bio-Oil Source	*Bioproducts*	*Applications*
Whole bio-oil	Bio-oil phenols, carboxylic acids, and lime to produce calcium salt (BioLime)	Capturing acidic gases in refineries
	Slow-release nitrogen fertilizers (derived from carbonyl, carboxyl, hydroxyl, phenolic, and methoxyl groups with ammonia, urea, and other amino compounds)	High-value fertilizers containing humic-type matter (lignin), making them good soil conditioning materials
	Bio-oil from fast pyrolysis or wood vinegar from charcoal	Use as a pesticide (insecticide, fungicide, bactericide, or termiticide) and wood preservative
	Syngas	Variety of products, FischerTropsch synthesis
	Asphalt paving substitute	Bioblended asphalt binder

Table 5. (Continued). Products Derived from Bio-Oils

Bio-Oil Source	*Bioproducts*	*Applications*
Pyrolytic lignin and monophenols	Resins and adhesives	Substitute for phenol in phenol–formaldehyde resins, thermoresponsive materials with adhesive properties
	Vanilin, eugenol, and hydroquinone	Antioxidants, feedstock for other chemicals and materials
	Hydroquinone	Skin lightening agent
	Pyrolytic lignin	Carbon fiber
	Phenol alkylation with alcohols	Precursors of detergents, thermoplastic elastomers, antioxidants, fire retardants, fuel and lubricant additives, and components of phenolic resins
	mono- and oligophenols	Specialty polymers; fragrances, and intermediates in the synthesis of pharmaceuticals, bisphenol A, caprolactam, alkylphenols, adipic acid, thermal insulating material, and chemicals for leather tanning
Water-soluble fraction	Production of hydrogen through aqueous-phase reforming and steam reforming of the aqueous fraction or light oxygenates, such as acetic acid, ethanol, acetone, acetic acid, acetol, and propanal	Blue hydrogen, syngas, or more valuable compounds
	Aldol condensation and ketonization followed by deoxygenation	Mixture of hydrocarbons and aromatics useful for blending with petroleum-based liquid fuels and chemicals
	Ketone conversion to olefins	Variety of petrochemicals
	Glucoaldehyde, glyceraldehydes, pyruvaldehyde, dihydroxyacetone, acetone, and diacetyl	Browning agents to preserve fish and meat, food additives
	Anhydrosugars (levoglucosan, cellobiosan, and oligomers)	Production of polyethers
	Biogas production from torrefaction condensation (acetic acid, acetol, hydroxyacetaldehyde, and 2-methoxyphenol)	Fuel
	Products derived from pyrolytic humin byproducts of hydroxymethyl furfural, furfural, and furandicarboxylic acid	Semiductile thermoset matrix to be impregnated with cellulosic fibers for enhanced modulus and tensile strength

Table 5. (Continued). Products Derived from Bio-Oils

Bio-Oil Source	*Bioproducts*	*Applications*
Fermentable sugars	2-G[a] ethanol	Fuel additives, solvents, and chemicals
	Organic acids (e.g., citric acid, fumaric acid, lactic acid, propionic acid, succinic acid, itaconic acid, acetic acid, and butyric acid), amino acids, antibiotics, fatty acids, and others (e.g., glycerol, acetone, *n*-butanol, butanediol, isopropanol, and acetaldehyde)	Chemicals
	Hydrocarbons, biohydrogen, biomethane	Fuel
	Furfural, furans, glycols, methyl ethyl ketone, adipic acid, ethylene, propylene, 5-HMF	Feedstock for chemical industry
Other bio-oil fractions	Biopitch	Biocarbon electrode
	Surfactants from bio-oils	Higher-quality biosurfactants
	Benzene, toluene, xylene	Production of styrene; phenol; nylon; aniline; antiknock gasoline additives; *o*-, *m*-, and *p*-xylenes; PET[a] fibers; and resin films
Bio-oil mono-components	Acetic acid	Several products by ketonization, esterification, and steam reforming; ethanol by hydrogenation
	Acetol	Production of propylene glycol, acetone, acrolein, furan derivatives, and propionaldehyde by dehydration, hydrogenation, oxidation, and polymerization; used in the food, textile, and cosmetic industries; skin tanning agents; natural carbohydrates and unnatural polyhydroxylated molecules in medicine
	(Hydroxyacetaldehyde (glycolaldehyde)	Ethylene glycol (coolant and antifreeze, monomer in the production of polyester, solvent, and intermediate for chemical production), food browning agent
	Ethylene and propylene glycol	Polymer industry
	Levoglucosan	Production of pharmaceuticals, surfactants, biodegradable polymers, plasticizers, propellants, explosives, surfactants, resins, and plastics; chiral synthon for controlling stereoselective reactions during the synthesis of

Table 5. (Continued). Products Derived from Bio-Oils

Bio-Oil Source	*Bioproducts*	*Applications*
		antibiotics, antiparasitic agents, and other biologically active compounds

[a] Abbreviations: 2-G, second-generation; PET, poly(ethylene terephthalate).

Direct Bio-Oil–Derived Products

Different functional groups in the structure of bio-oil, such as carboxyl, carbonyl, hydroxyl, methoxyl, and phenol groups allow bio-oil to undergo chemical transformations into desirable compounds (*58*). BioLime is produced from the reaction between phenols, carboxylic acids, and lime. It is a calcium salt that can be used in the adsorption of acid gases such as SO_x (*59, 60*). Several slow-release nitrogen fertilizers (amides, imines, and Mannich bases) are produced from the reaction of bio-oil functional groups (carboxyl, carbonyl, hydroxyl, and methoxyl) with urea, ammonia, and other amino compounds. These fertilizers can act as slow-release fertilizers with 10% nitrogen (*22*). Surfactants are produced from bio-oil at various stages of the distillation of tar and the refining of raw acetic acids. Pesticides and wood preservatives can also be produced from bio-oils.

Depending on the source, bio-oils can be grouped into three categories. The first category comprises liquid products obtained from the destructive distillation of biomass, called wood tar, which is akin to coal tar containing polycyclic aromatic hydrocarbons and asphaltene. The second category consists of bio-oil derived from flash pyrolysis or liquidation, which mainly contains liquid fuel. However, compared to conventional crude oil, this type of bio-oil has several disadvantages, such as unsteady physicochemical properties, degradation with time and temperature, high viscosity, and high oxygen and water contents. The third category is tar produced from biomass gasification, which is an unavoidable byproduct that will alter the operation of systems once condensed and coked. The tar products contain acids such as pentanoic (valeric), 4-oxopentanoic, hexanoic (caproic), heptanoic, and hexadecanoic (palmitic) acids, as well alcohols such as methanol and ethanol. Moreover, because bio-oil is acidic in nature, acid catalysis during distillation results in the formation of esters, which are surfactants (*61*).

Bio-oils derived from tobacco leaf and dried coffee grounds can act as natural pesticides (*62, 63*). The presence of phenolic and terpenoid compounds in bio-oil provides fungicidal and insecticidal activities (*62*). Bio-oil can also be used as a substitute for fossil-fuel–based binders in asphalt-based emulsions (*62–64*).

Bioproducts Derived from Bio-Oil Fractions

The phenolic compounds in bio-oil can be used as partial substitutes for phenol in phenol–formaldehyde resins, such as resole and novolac resins (*65*). These phenol–formaldehyde resins are used in the manufacturing of various products such as billiard balls and laboratory equipment and as adhesives in the manufacturing of wood panels. Thermoresponsive materials with adhesive properties have also been developed from the phenolic fraction of bio-oils. Bio-oil is rich in antioxidants (eugenol, hydroquinone, vanillin) because of the presence of lignin-derived compounds. Vanillin, for example, plays an antioxidant role in protecting mitochondria from protein oxidation and lipid peroxidation (*66*). Lignin derived from bio-oil can be used to reinforce polymer matrixes, thereby improving their thermal conductivity and rigidity (*52*). Many other applications can be found for the monophenols and oligophenols from bio-oil, such as fragrances (*49*), specialty polymers (*48*), and pharmaceutical intermediates (*66*).

Products Derived from Water-Soluble Fractions

The steam reforming of bio-oil produces hydrogen, although this process requires a high energy input and is coupled with low purity hydrogen and its low atom economy makes this process economically unsuitable (*52*). The light oxygenated compounds present in the water-soluble portion of bio-oil can be modified by aldol condensation and ketonization reactions to form C–C bonds (*54*). Olefin synthesis by ketone conversion in the presence of Brønsted acid sites has also been reported (*53*). Food additives (e.g., browning agents) such as glyceraldehydes, glucoaldehyde, dihydroxyacetone, pyruvaldehyde, acetone, and diacetyl are also synthesized from bio-oil (*67, 68*).

Other bioproducts that can be synthesized from the fermentable sugars present in bio-oil by microbial fermentation include ethanol, amino acids, organic acids (e.g., fumaric acid, citric acid, lactic acid, succinic acid, propionic acid, itaconic acid, butyric acid, acetic acid), fatty acids, and antibodies, as well as acetone, glycol, *n*-butane, isopropanol, butanediol, and acetaldehyde (*69, 70*). Surfactants are produced from bio-oil at various stages of the distillation of tar and the refining of raw acetic acids (*71*).

Products Derived from Bio-Oil Monocomponents

Bio-oil–derived acetic acid can be converted into other value-added products by various methods, including esterification, ketonization, and steam reforming (*72, 73*). Even ethanol production from the hydrogenation of acetic acid has been extensively studied (*74, 75*). The highly active molecule acetol is also used as a key intermediate in the production of acetone, propylene glycol, acrolein, propionaldehyde, and furan derivatives (*76*). Hydroxyacetaldehyde, also known as glycolaldehyde, is a precursor of ethylene glycol, where the latter is produced by fermentation or hydrogenation (*77*). Ethylene glycol is used as a coolant and antifreeze agent. Ethylene and propylene glycols are also produced by the hydrogenations of acetol and hydroxyacetaldehyde (*74, 75*).

Bio-Oil Upgrading Process

Huge quantities of bio-oil are produced by the fast pyrolysis of lignocellulosic biomass, which is a promising source of biofuel. However, the chemical composition of the resulting bio-oil (C/H/O mass ratio of 0.54–0.58/0.055–0.07/0.35–0.40) makes it unsuitable for direct application (*78*). Bio-oil is upgraded using physical, chemical, and catalytic processes. Physical treatments of bio-oil include solvent addition, filtration, and emulsification, whereas the most common chemical treatments are reforming, deoxygenation, and esterification. To use biofuel as a drop-in transportation fuel, the upgrading process must be aimed at increasing its heating value, lowering its viscosity, reducing its corrosiveness, reducing its moisture content for blending with conventional fuels, and improving its stability.

Because bio-oil is immiscible with conventional fossil fuels, it can be emulsified using an appropriate surfactant, and the available infrastructure can be used for its transport and use. However, this approach encounters some drawbacks such as the high surfactant costs, high energy requirements for emulsification, low heating values, and equipment corrosion. Crude bio-oil can be upgraded to some degree to the desired properties by (partial) removal of the excessive oxygen content using decarboxylation, hydrodeoxygenation (HDO), catalytic cracking, or a combination of all three processes. The composition of bio-oil is the crucial factor determining its use, and therefore, upgrading strategies are determined according to the composition of the bio-oil (Figure 8).

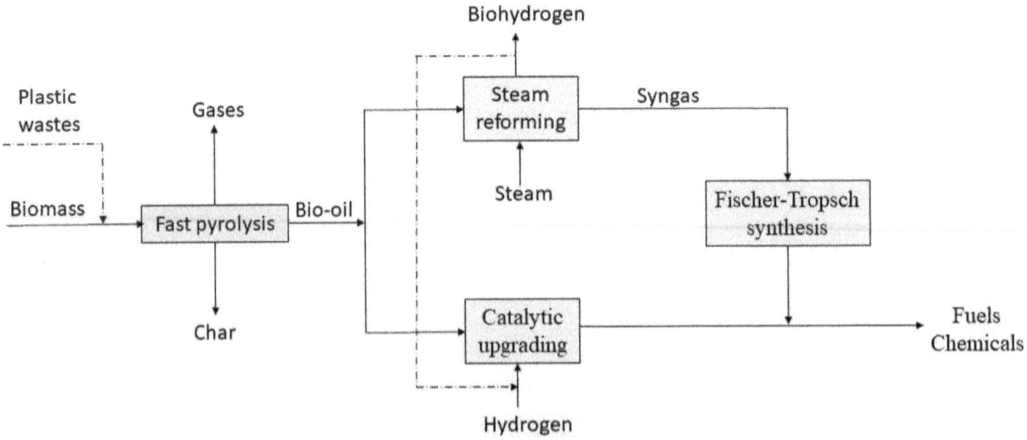

Figure 8. Pyrolysis route for upgrading biomass to chemicals.

Steam Reforming

The catalytic steam reforming of bio-oil is a well-known process for producing biohydrogen and syngas, which are the main precursors for the synthesis of Fischer–Tropsch products. Hydrogen is a clean energy carrier and can be used to run internal-combustion engines and fuel cells, releasing only water as a byproduct. Biohydrogen can also be used for the upgrading of bio-oil by HDO, or in ammonia synthesis, or in hydrotreating. There are a few disadvantages associated with the steam reforming of bio-oil, such as low hydrogen yield and rapid catalyst deactivation. Ni-based and noble-metal–based catalysts are widely used for the reactions of crude bio-oil and its aqueous fraction (*79*). Crude bio-oil, when mixed with water, forms two separate fractions: (1) a monomer-rich aqueous fraction (which contains 20% organic compounds) and (2) an organic fraction composed mainly of lignin and oligomers (*80*). The steam reforming reactions of crude bio-oil and its aqueous fractions have been tested with different catalyst systems (*81*). Steam reforming of the whole bio-oil generates a greater amount of hydrogen than the steam reforming of only the aqueous fraction.

The reaction mechanism and catalysts for the steam reforming of bio-oil are similar to those used in the conventional reforming of hydrocarbons and alcohols such as natural gas (methane), methanol, glycerol, ethanol, ethylene glycol, and *n*-butanol (*82–91*). Therefore, according to Le Chatelier's principle, this endothermic reaction is favored at high temperature and low pressure (*91*).

Because of the heterogeneous nature of bio-oils and the presence of thermally stable molecules, greater amounts of steam and higher temperatures are required to increase catalyst stability and reduce coke formation.

$$CH_{1.9}O_{0.7} + 1.3H_2O \leftrightarrow CO_2 + 2.25H_2 \quad (1)$$

Promoted nonprecious (Ni) and noble-metal (Pt, Pd, Rh) catalysts actively take part in bio-oil steam reforming to produce hydrogen (*92, 93*). The steam reforming of bio-oil occurs at 750–860 °C over Ni-based catalysts in a two-step process. The overall stoichiometry of the steam reforming reaction shows a maximum yield of (0.172 g of H_2)/(g of bio-oil) (11.2% based on wood) (*94–96*).One of the main constituents of bio-oils is propionic acid (HPAc), which has a C/H/O mass ratio of 0.49/0.08/0.48, similar to that of the whole bio-oil. Therefore, the thermochemical steam reforming of HPAc is considered as a good representative model for the steam reforming of bio-oil.

$$CH_3CH_2COOH + H_2O \leftrightarrow 3CO + 4H_2 \qquad H^o_{298} = +364 \text{ kJ/mol} \qquad (2)$$

Because of the occurrence of the reverse water–gas shift reaction, the H_2/CO_x ratio is changed during the reforming reaction.

$$CO_2 + H_2 \leftrightarrow CO + H_2O \qquad H^o_{298} = -41.2 \text{ kJ/mol} \qquad (3)$$

The methanation of CO is also observed.

$$CO + 3H_2 \leftrightarrow CH_4 + H_2O \qquad H^o_{298} = -206 \text{ kJ/mol} \qquad (4)$$

Cracking or thermal decomposition is among the other side reaction that generally occur at higher temperature.

$$CH_3CH_2COOH \leftrightarrow 2CO + 3H_2 + C \qquad H^o_{298} = +322 \text{ kJ/mol} \qquad (5)$$

Coke formation by the Boudouard reaction or carbon oxide reduction is also seen.

$$2CO \leftrightarrow CO_2 + C \qquad H^o_{298} = -172 \text{ kJ/mol} \qquad (6)$$

$$2CO + H_2 \leftrightarrow H_2O + C \qquad H^o_{298} = -132 \text{ kJ/mol} \qquad (7)$$

Generally, when bio-oil is considered as a carbon source for hydrocarbon synthesis through syngas, eq 4 is a desirable reaction path. If hydrogen yield is more important, then eq 8 is the most desirable path, which is the combination of eqs 2 and 3 and is similar to the reaction of crude bio-oil.

$$CH_3CH_2COOH + 4H_2O \leftrightarrow 3CO_2 + 7H_2 \qquad H^o_{298} = +240 \text{ kJ/mol} \qquad (8)$$

One of the advantages of bio-oil steam reforming is the separation of high-value products from low-value products, which can then be reformed. The effects of feed composition on equilibrium conversion and product distribution have been studied under operating conditions relevant for industrial operations, which are a temperature of 800 °C and a pressure of 1 MPa (92). The H_2 selectivity was highest (ca. 0.6) at a H_2O/HPAc molar ratio of 3, whereas the selectivity to CO_2 increased from 0.02 to 0.5 when the H_2O/HPAc ratio was changed to 8 from 3.

Reaction Mechanism

The reaction mechanism for the metal-catalyzed steam reforming of oxygenated organic molecules was proposed to be similar to the dual-site catalyst bifunctional mechanism of the steam reforming of other hydrocarbons (97). During the reaction, the organic molecules dissociatively adsorb on the metal catalyst sites (Ni), and the water molecules are adsorbed on the acidic support (Al) surface. H_2 is produced from the dehydrogenation of organic molecules adsorbed on the metallic sites and also from the reaction between the hydroxyl groups and the adsorbed organic fragments. These hydroxyl groups migrate from the acidic alumina support to the metal crystallite (Ni)/support (Al) interfaces. The latter reaction is responsible for the generation of CO, and side reactions also lead to the deposition of some carbon on the catalyst surface.

Catalyst deactivation is a major problem with the steam reforming of bio-oils, and bio-oils seem to have a greater deactivation problem than fossil-fuel–derived feedstocks (*98*). To avoid this problem, a number of strategies have been developed, such as modification of the catalyst and optimization of the reaction conditions. Among them, the most promising approach is the use of a dual catalytic system in which a guard catalyst, such as calcined dolomite, is applied before the active Ni catalyst to protect the latter from tar (*99*).

Deoxygenation Reactions

Deoxygenation treatment of bio-oil is required to improve its quality, wherein the H/C ratio is increased by removing oxygen, increasing the number of hydrogen atoms in bio-oil molecules, or a combination of the two. Because of the complex nature of crude bio-oil, catalytic upgrading involves a combination of complex reaction networks such as decarbonylation, cracking, decarboxylation, HDO, hydrocracking, polymerization, and hydrogenation (*78*, *100*, *101*). In most cases, coke formation is prevalent, which is the prime cause for catalyst deactivation. Total deoxygenation to hydrocarbons (by decarboxylation and HDO) is thermodynamically more suitable than partial deoxygenation to produce oxygenates with lower oxygen contents (*102*).

Hydrodeoxygenation (HDO)

The HDO of crude bio-oil, equivalent to the hydroprocessing of crude oil in the refining industry, can be represented by the simple equation

$$C_3H_8O_4 + 6H_2 \leftrightarrow 6CH_2 + 7H_2O \quad (9)$$

This is an exothermic reaction, with a heat of reaction of 2.4 MJ/kg (*35*). The optimum operating conditions involve high pressure (up to 8 MPa) and moderate temperature (up to 400 °C) with a large quantity of hydrogen (*74*), because of the different reactivities of different functional groups presents in bio-oil (*103*). This is a two-step process: heating of crude bio-oil at 250 °C, followed by the HDO reaction at higher temperature and pressure. The first step is necessary to avoid polymerization and coke formation in the bio-oil and to consume less hydrogen. The second step is a combination of HDO and complete deoxygenation. The HDO of bio-oil is followed by a phase separation step, which produces an aqueous phase and one or more oil phases. Complete hydrotreating leads to the formation of products such as naphtha, which can be further modified to form value-added chemicals and transport fuels. Oxygen removal involves several bond-breaking reactions, including the hydrogenolysis of C–O bonds and several hydrogenation reactions. The reactivities and strengths of C–O bonds are different for different compounds. For instance, the bond between oxygen and an aromatic carbon is 84 kJ/mol higher in energy than that between oxygen and an aliphatic carbon. Therefore, O removal is relatively easier for alcohols and aliphatic ethers than for aromatic ethers and phenols (*102*). On a catalyst, the deoxygenation activation energy is different for different model compounds, and therefore, the deoxygenations of different compounds occur at different temperatures. The rates of deoxygenation in terms of apparent activity are summarized as follows (*104*):

alcohol > ketone > alkylether > carboxylic acid ≈ *m*- and *p*-phenol ≈ naphthol > phenol > diarylether ≈ *o*-phenol ≈ alkylfuran > benzofuran > dibenzofuran

The main drawback of the HDO process is the requirement of large amounts of hydrogen, which is normally produced from methane reforming. This limitation can be easily overcome by the production of renewable hydrogen by bio-oil reforming or biomass.

Catalysts for HDO

The ideal HDO catalyst should target only the C–O bond without affecting the C–C bonds, and it should also consume less H_2 to produce unsaturated hydrocarbons as HDO products and should be highly stable under the reaction conditions (*105*). Several catalysts have been reported for HDO reactions, such as supported noble metals; sulfides; and transition-metal phosphides, nitrides, and carbides.

Sulfide and Oxide Catalysts

Generally, sulfided metal catalysts such as NiMoS and CoMoS are used in fossil-fuel refineries for the hydrodenitrogenation and hydrodesulfurization (HDS) reactions. These catalysts have exhibited similar activities for bio-oil HDO reactions, even though bio-oil does not contain high amounts of N or S atoms. These catalysts have achieved nearly complete deoxygenation of bio-oil to produce diesel- and gasoline-range hydrocarbons (*106*). Here, Ni or Co serves as a promoter, donating an electron to the molybdenum atoms and thereby weakening the bond between molybdenum and sulfur. Therefore, a sulfur vacancy site is generated, which is active for both the HDO and HDS reactions (*107–109*).

Romero et al. (*109*) proposed a mechanism for the HDO of 2-ethylphenol, a model compound for bio-oil, over a MoS_2-based catalyst (Figure 9). The first oxygen in the molecule is adsorbed onto the vacancy site of a MoS_2 slab edge, which activates the compound. S–H species, which are generated from the H_2 in the feed, will be present along the edge of the catalyst. Thereafter, proton donation takes place from sulfur to the attached molecule, leading to the formation of a carbocation. It undergoes direct C-O bond cleavage and forms a deoxygenated compound by removing O as H_2O.

Alumina (Al_2O_3) is a popular choice of support for these sulfide catalysts because it has a favorable combination of textural properties, such as high surface area, large pore volume, and unique pore-size distribution, and acid/base characteristics, which are related to its phase composition and surface chemical composition. However, in the highly aqueous environment of bio-oil, alumina becomes unstable and converts into boehmite. The Brønsted acid sites of the Al_2O_3 catalyst lead to coke formation (*110*). SiO_2 has been found to be a suitable catalyst because of its amphoteric nature and low affinity for coke formation. The stripping of sulfur from the catalyst surface deactivates this catalyst deactivated, so to reactivate it, various sulfur compounds, such as dimethyl disulfide ($C_2H_6S_2$) and hydrogen sulfide (H_2S), are passed through the reactor in gaseous form (*111*). However, these sulfur compounds also have some drawbacks because they are often corrosive and toxic and can contaminate the product. As a less toxic alternative, thiourea has been used in the HDO of phenol with and without a Ni-doped MoS_2 catalyst to obtain high selectivity toward the HDO of phenol compared to the hydrogenation of phenol (*112*).

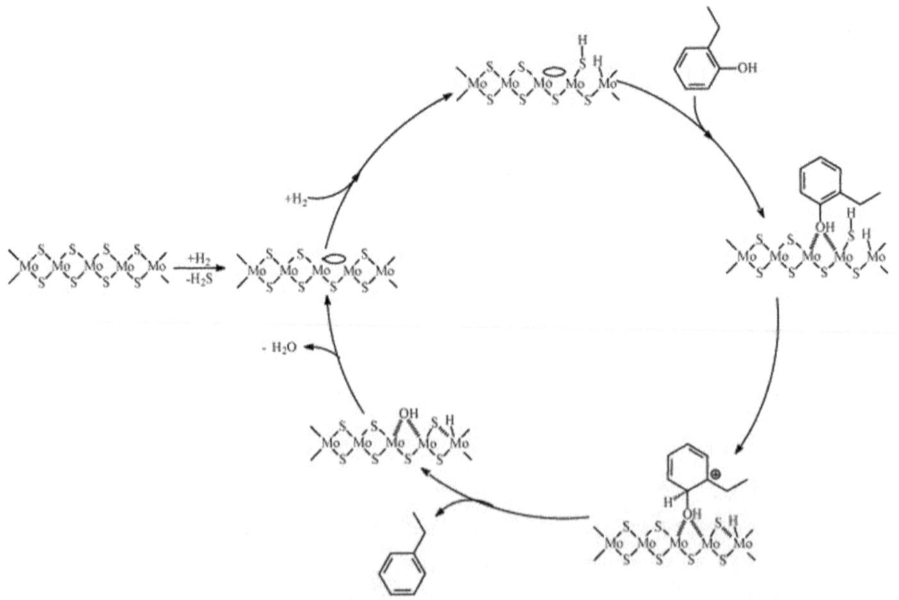

Figure 9. Proposed mechanism for the HDO of 2-ethylphenol on a schematic MoS_2-based catalyst. Reproduced with permission from reference (109). Copyright 2010 Elsevier.

Noble-Metal Catalysts

Supported transition metals (Ni and noble metals such as Pt, Ru, Pd, and Rh) are regarded as more active catalysts for HDO than sulfide catalysts (*109*). Ru/C catalyst was found to provide a 90 wt % deoxygenation level of bio-oil, which is superior to the performance of previous HDO catalysts such as sulfided Ni–Mo/Al_2O_3 (*113*) and Co–Mo/Al_2O_3 (*114*). Pt/Al_2O_3 catalyst was found to exhibit high activities for both the HDO and steam reforming of synthetic bio-oil at 350 °C under 10 bar of nitrogen pressure (*115*), resulting in the production of H_2 in situ. The oxygen content of the upgraded bio-oil decreased to 2.8 from 41.4 wt %. This process is more economical and more environment-friendly than the old HDO technology, which requires a considerable hydrogen supply. Carbon and metal oxides have been used as supports for these metals as well. Carbon is a much better support, because it is neutral in nature and provides a high surface area. Among the non-noble-metal catalysts, Ni has good activity and stability; however, it is less active than noble-metal catalysts such as Ru, Pt, and Rh. The addition of Cu to a Ni-based catalyst was found to reduce the temperature of HDO, thereby preventing the formation of coke on the catalyst surface and limiting the sintering (*116*). A study on bimetallic Ni–Cu catalysts demonstrated their activity for the syntheses of model compounds such as *o*-cresol (*117*), anisole, and crude bio-oil from lignocellulosic biomass (*118*). However, Ni–Cu catalysts are easily affected by sulfur compounds, which tend to chemisorb strongly on the metal surface, especially on nickel catalysts. As a result, reactant species are unable to become adsorbed on the active sites of the catalyst.

Transition-Metal Nitrides, Phosphides, and Carbides

The carbides, nitrides, and phosphides of transition metals, such as Co_2P, Ni_2P, Fe_2P, and WP, have been found to exhibit high activities in the HDO reactions of model compounds (*119*). Comparatively, carbide compounds are more active than nitrides and phosphides. Transition-metal carbides have received much attention because of their high reactivities in hydrogen- and oxygen-containing reactions (*120*). However, Mo_2C has a good HDS activity and is sulfur-tolerant (*121*). It has been speculated that this catalyst should be bifunctional. Specifically, it first activates the oxygen-containing compound using the valence of an oxide form of a transition metal associated with catalyst support. This should be complemented by hydrogen donation to the oxygen-containing compound, which takes place on the transition-metal sites. The transition-metal sites have the potential to activate H_2 (Figure 10) (*122–124*).

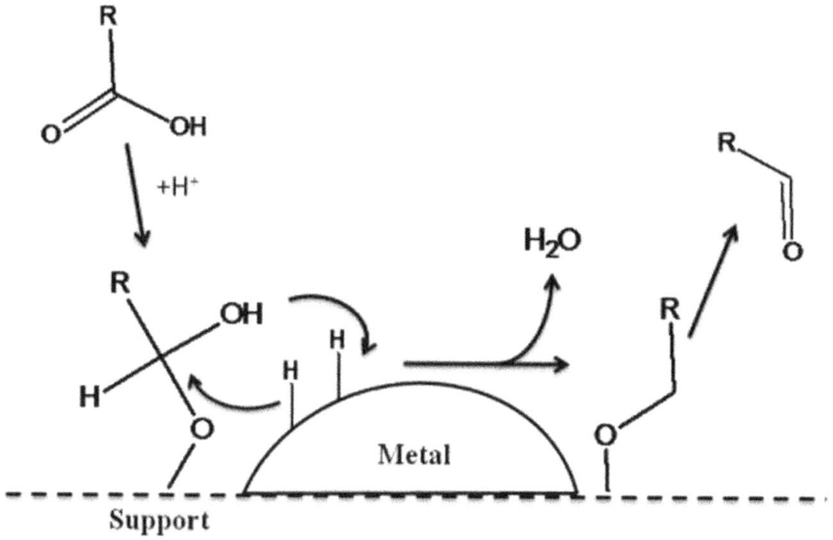

Figure 10. Mechanism for the HDO reaction as mediated by transition-metal catalysts. This reaction mechanism is based on information from reference (122), reference (123), and reference (124).

The activities of Rh, Pd, and Pt supported on ZrO_2 for the HDO reaction of 3 wt % guaiacols in hexadecane in a batch reactor at 100 °C and 80 bar were found to lie in the order (*125*)

$$Rh/ZrO_2 > Co\text{–}MoS_2/Al_2O_3 > Pd/ZrO_2 > Pt/ZrO_2$$

A comparison of TiO_2-supported molybdenum nitride, carbide, and phosphide catalysts for the HDO of phenol showed that 15% Mo_2C/TiO_2 had the highest activity (*126*).

Reaction Medium for HDO

HDO reactions are conducted in three-phase trickle-bed reactors, similarly to the HDS reactions in fossil fuel refining (*46*). In the HDO reaction, the choice of appropriate solvent is crucial, as it improves the HDO of bio-oil and reduces coke formation. The characteristics of good solvents are good hydrogen solubility, diffusivity, and decreased viscosity of the reaction mixture. Long-chain alkanes, isopropanol, and tetralin, for example, can function as H_2 donors to improve the HDO reaction yield, increase the H/C ratio in the product, and minimize the coke formation (*114*). Long-chain alkanes, isopropanol, and tetralin, among others, are used as solvents in HDO, which can improve the H/C ratio and yield of the product and inhibit coke formation. In recent times, supercritical fluids have been used because of their unique properties, such as their liquidlike densities, good dissolving power, and gaslike transport properties responsible for good heat and mass transfer. In the supercritical state, the solvent has excellent hydrogen solubility, and therefore, the feedstock and hydrogen are in a single phase. Different supercritical solvents such as ethanol (*127*), methanol (*128*), and 1-butanol (*129*) have been tested for bio-oil upgrading and found to be effective.

Catalytic Cracking

In the catalytic cracking of bio-oil, excess oxygen is removed as CO_2 by decarboxylation and decarbonylation and as H_2O by dehydration over acidic catalysts. Zeolites are popular catalysts for the catalytic cracking of bio-oil. Several aromatic-rich hydrocarbon products (H/C = 1–1.2) are formed by secondary reactions such as polymerizations, cracking, and aromatization (*130*).

$$C_3H_8O_4 \leftrightarrow 4.6CH_{1.2} + 1.4CO_2 + 1.2H_2O \quad (10)$$

These reactions are generally performed at temperatures of 350–500 °C and at atmospheric pressure. At higher temperatures, oxygen is removed in the form CO_2, and at lower temperatures, it is removed as H_2O (*131*). This route has the advantage of atmospheric-pressure operating conditions, whereas its drawbacks are low product yields and catalyst deactivation.

Catalysts for Cracking

The catalytic cracking of bio-oil is quite similar to the well-studied methanol-to-hydrocarbon synthesis reaction catalyzed by zeolites. A number of solid acid catalysts have been studied for the upgrading of bio-oil and model compounds, such as HZSM-5, HMOR, HY, silica–alumina, and silicates (*132*). Among the zeolite family, HZSM-5 exhibits the highest activity and stability. The main characteristics accounting for its high activity and product selectivity are shape selectivity and tuned acidity (moderate acidic strength). In comparison to silica–alumina catalysts (which are more selective toward aliphatic products), ZSM-5 promotes a high aromatic/aliphatic ratio in products in the bio-oil upgrading reaction (*133*).

Aluminosilicate minerals make up the three-dimensional microporous structure of zeolites. The addition of promoters to the zeolite structure can further increase its selectivity toward products by changing the acid site distribution. Ni and Ga addition in the framework greatly enhances the selectivity toward aromatic hydrocarbons. The metallic species promote the hydrogenation and dehydrogenation reactions, which changes the acidity strength of the zeolites (*134*).

Different types of zeolites (HZSM-5, H-mordenite, H-Y), silica–alumina, and silicalite were evaluated for bio-oil cracking in a fixed-bed reactor at reaction temperatures of 330–410 °C to determine their activities, which were found to lie in the following order (*135*, *136*):

HZSM-5 (5.4 Å) > H-mordenite (6.7 Å) > H-Y (7.4 Å) > silica-alumina (31.5 Å) > silicalite (5.4 Å)

In comparison to the HDO reaction, the catalyst cracking of bio-oil results in lower oil yield and deep deoxygenation. Further catalyst development must be done to address a few challenges associated with this reaction, namely, deep deoxygenation, deep cracking, and coke deposition on the catalyst surface. Among these difficulties, deep deoxygenation results in increased aromaticity of the product, and deep cracking results in overcracked gaseous products. The deep cracking notably increases above 400 °C (*137*). The HZSM-5–mediated bio-oil upgrading reaction generates 8–30 wt % coke (*138*). The deposited coke is a combination of two types of particles: (1) catalytic coke (deposited inside the zeolite crystals) and (2) thermal coke (pyrolytic lignin deposited on the macropores between the zeolite crystals). Higher concentrations of catalyst acid sites increase the formation of catalytic coke (*139*). Regeneration of the used catalyst can be achieved by heating at 450 °C for 2 h in the open air, although complete regeneration of all of the acid sites is not possible.

Moreover, dealumination during the reaction causes irreversible deactivation of the catalyst. Because of the loss of alumina, the tetrahedral alumina in the zeolite crystal frame becomes octahedrally coordinated alumina atoms. These can diffuse to the outer zeolite surface and lose their acidity, which decreases the number of available acidic sites. The typical cracking reaction atmosphere, specifically, the presence of steam at higher temperatures, promotes the dealumination process. However, this phenomenon can be avoided to some extent by adding promoters such as phosphorous (*140*). The use of zeolites for bio-oil cracking is still in the development stage, and there is a much wider spectrum of possibilities with mono- and multifunctional catalysts that need to be explored for this process (*141*).

Other Methods for Upgrading

Bio-oil contains large amounts of carboxylic acids (≥30 wt %), and hence, it is unstable and corrosive (*133*). Ketonization is a useful method for removing carboxylic acids without the use of hydrogen. Ketonic decarboxylation or ketonization is a catalytic reaction in which two carboxylic acid molecules are converted to form a larger ketone molecule, releasing a stoichiometric amount of water and CO_2.

$$RCOOOH + R'COOH \rightarrow RCOR' + CO_2 + H_2O \qquad (11)$$

Thereafter, these larger ketone molecules undergo an aldol condensation reaction to produce gasoline- and diesel-range molecules (*142*). For this ketonization reaction, a number of metal oxide catalysts have been employed, such as iron oxide supported on SiO_2, Al_2O_3, and TiO_2. The reaction conditions are moderate, as the reaction is carried out at atmospheric pressure and in the temperature range of 300–500 °C. Reducible oxides (ZrO_2, CeO_2, and TiO_2) are more active than pure acidic or basic oxides. The activities of these catalysts can be affected by the presence of CO_2, H_2O, and different organic compounds. As an example, furfural, which is a bio-oil component, demonstrated an inhibition effect on the ketonization of acetic acid in the presence of ceria/alumina catalysts (*142*). This ketonization reaction can act as a pretreatment for HDO, as the oxygen content is reduced without the use of hydrogen.

During bauxite refining for alumina production, red mud is produced as a waste material, but this red mud is a good source of Fe_2O_3 (12–60%), SiO_2 (2–22%), Al_2O_3 (8–48%), TiO_2 (5–20%), CaO (4–46%), and MgO (0.2–2%). All of these materials can be useful for the hydrogenation and ketonization reactions of carboxylic acid (*143*).

Conclusions

In the net-zero-carbon energy and materials scenario, biomass valorization will play an important role. Bio-oil produced by the fast pyrolysis of lignocellulosic biomass is a promising source of biofuel. The catalytic steam reforming of bio-oil can be accomplished to produce biohydrogen and syngas, which are the main feedstocks for the synthesis of Fischer–Tropsch products. Biohydrogen can also be used for the upgrading of bio-oil by HDO, or in ammonia synthesis, or in hydrotreating. Catalyst deactivation is a major issue with the bio-oil steam reforming process, and bio-oils seem to have high deactivation rates. This can be circumvented by a number of strategies, including modifying the catalyst, such as by using different supports and combinations of metals, and optimizing the reaction conditions. The most promising of these approaches is the use of dual catalytic systems. Improving the quality of bio-oil requires deoxygenation treatment, wherein the H/C ratio is increased by removing oxygen, increasing the number of hydrogen atoms in the bio-oil molecules, or a combination of the two. Catalytic upgrading involves a combination of complex reaction networks such as decarbonylation, cracking, decarboxylation, HDO, hydrocracking, polymerization, and hydrogenation. HDO and catalytic cracking have been widely studied to increase the quality of upgraded fuels and chemicals. However, more resources must be invested in the research and development of current research areas, as the existing transportation infrastructure and general consumer preferences all favor the continued use of liquid alkane fuels. These carbon-based biofuels serve as a short-term alternative to petroleum-derived fuels until more sustainable solutions based on solar energy, wind, geothermal, hydrogen, and hydrogen fuel cells achieve the capacity to fulfill worldwide energy needs.

There is a need to adopt a more sustainable biofuel strategy to utilize widely available biomass feedstocks to the largest extent possible, ultimately shifting to lignocellulosic biomass instead of only the starch and triglyceride fractions. We believe that a lignocellulosic biorefinery can be built with a combination of different technologies and biomass processing strategies for the flexible production of various fuels and chemicals. To achieve this goal, researchers should focus on (1) developing innovative routes to produce flexible chemical platform molecules such as γ-valerolactone and levulinic acid, (2) applying these platform chemicals for the synthesis of hydrocarbon fuels and specialty chemicals, (3) synergistically coupling thermochemical and hydrolytic methods into a more efficient complete biorefinery, (4) developing new and effective separation and purification strategies, (5) developing more efficient means of lignin utilization, and (6) performing a life-cycle assessment of the integrated processes to assess the environmental impacts associated with each of the stages of the biorefinery.

References

1. *bp Energy Outlook 2020* [Press release]; bp p.l.c.: London, Sept 14, 2020. https://www.bp.com/en/global/corporate/news-and-insights/press-releases/bp-energy-outlook-2020.html (accessed Dec 16, 2020).
2. Yadav, V. G.; Yadav, G. D.; Patankar, S. C. The production of fuels and chemicals in the new world: critical analysis of the choice between crude oil and biomass vis-à-vis sustainability and the environment. *Clean Technol. Environ. Policy* **2020**, 22, 1757–1774.
3. *U.S. energy facts explained - consumption and production - U.S. Energy Information Adminitration.* https://www.eia.gov/energyexplained/us-energy-facts/ (accessed Dec 09, 2020).
4. Ebadian, M.; van Dyk, S.; McMillan, J. D.; Saddler, J. Biofuels policies that have encouraged their production and use: An international perspective. *Energy Policy* **2020**, *147*, 111906.
5. Lauri, P.; Havlík, P.; Kindermann, G.; Forsell, N.; Böttcher, H.; Obersteiner, M. Woody biomass energy potential in 2050. *Energy Policy* **2014**, *66*, 19–31.
6. Lebaka, V. R. Potential bioresources as future sources of biofuels production: An overview. In *Biofuel Technologies: Recent Developments*; Gupta, V. K., Tuohy, M. G., Eds.; Springer: Berlin, 2013; Chapter 9, pp 223–258.
7. *2016 Billion-Ton Report: Advancing Domestic Resources for a Thriving Bioeconomy*; U.S. Department of Energy: Washington, DC, July 2016; Vol. 1. https://www.energy.gov/sites/prod/files/2016/12/f34/2016_billion_ton_report_12.2.16_0.pdf.

8. Corma, A.; Iborra, S.; Velty, A. Chemical routes for the transformation of biomass into chemicals. *Chem. Rev.* **2007**, *107*, 2411–2502 [Online]. https://doi.org/10.1021/cr050989d (accessed Dec 10, 2020).
9. Habert, G.; Bouzidi, Y.; Chen, C.; Jullien, A. Development of a depletion indicator for natural resources used in concrete. *Resour., Conserv. Recycl.* **2010**, *54*, 364–376.
10. *A Review of the Current State of Bioenergy Development in G8 +5 Countries*; FAO/GBEP: Rome, 2007. http://www.fao.org/3/a1348e/a1348e00.htm (accessed Dec 11, 2020).
11. *OECD/FAO (2016) OECD-FAO agricultural outlook 2016–2025*. Stat. OECD Publishing, Paris. http://www.fao.org/3/BO103e/BO103e.pdf (accessed Dec 11, 2020)
12. Jönsson, O.; Persson, M. Biogas as transportation fuel. In *Regenerative Kraftstoffe: Entwicklungstrends, Forschungs- und Entwicklungsansätze, Perspektiven*; ForschungsVerbund Erneuerbare Energien: Berlin, 2003; pp 99–111. https://www.fvee.de/fileadmin/publikationen/Workshopbaende/ws2003-2/ws2003-2_02_04.pdf (accessed Nov 23, 2020).
13. Lange, J. Lignocellulose conversion: an introduction to chemistry, process and economics. *Biofuels, Bioprod. Biorefining* **2007**, *1*, 39–48.
14. Marris, E. Drink the best and drive the rest. *Nature* **2006**, *444*, 670–672.
15. Cherubini, F.; Bird, N. D.; Cowie, A.; Jungmeier, G.; Schlamadinger, B.; Woess-Gallasch, S. Energy-and greenhouse gas-based LCA of biofuel and bioenergy systems: Key issues, ranges and recommendations. *Resour., Conserv. Recycl.* **2009**, *53*, 434–447.
16. Searcy, E.; Flynn, P. C. Processing of straw/corn stover: comparison of life cycle emissions. *Int. J. Green Energy* **2008**, *5*, 423–437.
17. Kamm, B.; Gruber, P. R.; Kamm, M. Biorefineries–industrial processes and products. *Ullmann's Encyclopedia of Industrial Chemistry*, 6th ed.; Wiley-VCH: Weinheim, Germany, 2000; pp 1–38.
18. Fleming, J. S.; Habibi, S.; MacLean, H. L. Investigating the sustainability of lignocellulose-derived fuels for light-duty vehicles. *Transp. Res., Part D: Transp. Environ.* **2006**, *11*, 146–159.
19. Chisti, Y. Biodiesel from microalgae. *Biotechnol. Adv.* **2007**, *25*, 294–306.
20. Brennan, L.; Owende, P. Biofuels from microalgae—A review of technologies for production, processing, and extractions of biofuels and co-products. *Renewable Sustainable Energy Rev.* **2010**, *14*, 557–577.
21. Cherubini, F.; Mandl, M.; Philips, C.; Wellisch, M.; Jørgensen, H.; Skiadas, I. V.; Boniface, L.; Dohy, M.; Pouet, J.-C.; Wile, T. *IEA Bioenergy Task 42 on Biorefineries: Co-production of fuels, chemicals, power and materials from biomass: IEA Bioenergy Task 42 – Countries Report*; IEA Bioenergy, International Energy Agency: Paris, 2007.
22. Dale, B. E.; Kim, S. Biomass refining global impact–The biobased economy of the 21st century. In *Biorefineries – Industrial Processes and Products: Status Quo and Future Directions*; Kamm, B., Gruber, P. R., Kamm, M., Eds.; Wiley-VCH: Weinheim, Germany, 2006; Chapter 2, pp 41–66.
23. Cherubini, F. The biorefinery concept: using biomass instead of oil for producing energy and chemicals. *Energy Convers. Manage.* **2010**, *51*, 1412–1421.
24. Pandey, A., Bhaskar, T., Stöcker, M., Sukumaran, R., Eds. *Recent Advances in Thermochemical Conversion of Biomass*; Elsevier: Amsterdam, 2015.

25. Lago, C., Caldés, N., Lechón, Y. , Eds. *The Role of Bioenergy in the Emerging Bioeconomy: Resources, Technologies, Sustainability and Policy*; Elsevier: Amsterdam, 2018.
26. Bhaskar, T., Pandey, A., Rene, E. R., Tsang, D. C. W., Eds. *Waste Biorefinery: Integrating Biorefineries for Waste Valorisation*; Elsevier, Amsterdam, 2020.
27. Varma, R. S. Biomass-Derived Renewable Carbonaceous Materials for Sustainable Chemical and Environmental Applications. *ACS Sustainable Chem Eng* **2019**, *7*, 6458–6470 [Online]. https://pubs.acs.org/doi/10.1021/acssuschemeng.8b06550 (accessed Nov 26, 2020).
28. Hamelinck, C. N.; Van Hooijdonk, G.; Faaij, A. P. C. Ethanol from lignocellulosic biomass: techno-economic performance in short-, middle-and long-term. *Biomass Bioenergy* **2005**, *28*, 384–410.
29. Speight, J. G. *The Chemistry and Technology of Petroleum*, 5th ed.; CRC Press: Boca Raton, FL, 2014.
30. Vassilev, S. V.; Vassileva, C. G.; Vassilev, V. S. Advantages and disadvantages of composition and properties of biomass in comparison with coal: An overview. *Fuel* **2015**, *158*, 330–350.
31. Vassilev, S. V.; Vassileva, C. G.; Vassilev, V. S. An overview of the behaviour of biomass during combustion: Part I. Phase-mineral transformations of organic and inorganic matter. *Fuel* **2013**, *112*, 391–449.
32. Vassilev, S. V.; Vassileva, C. G.; Vassilev, V. S. An overview of the behaviour of biomass during combustion: Part II. Ash fusion and ash formation mechanisms of biomass types. *Fuel* **2014**, *117*, 152–183.
33. Zakzeski, J.; Bruijnincx, P. C. A.; Jongerius, A. L.; Weckhuysen, B. M. The catalytic valorization of lignin for the production of renewable chemicals. *Chem. Rev.* **2010**, *110*, 3552–3599.
34. Clark, J. H., Deswarte, F., Eds. *Introduction to Chemicals from Biomass*, 2nd ed.; Wiley Series in Renewable Resources; John Wiley and Sons: New York, 2015.
35. Hayes, D. J.; Fitzpatrick, S.; Hayes, M. H. B.; Ross, J. R. H. The biofine process – Production of levulinic acid, furfural, and formic acid from lignocellulosic feedstocks. In *Biorefineries – Industrial Processes and Products: Status Quo and Future Directions*; Kamm, B., Gruber, P. R., Kamm, M., Eds.; Wiley-VCH: Weinheim, Germany, 2006; Chapter 7, pp 139–164.
36. Bozell, J. J.; Moens, L.; Elliott, D. C.; Wang, Y.; Neuenscwander, G. G.; Fitzpatrick, S. W.; Bilski, R. J.; Jarnefeld, J. L. Production of levulinic acid and use as a platform chemical for derived products. *Resour., Conserv. Recycl.* **2000**, *28*, 227–239.
37. Kamm, B.; Schönicke, P.; Hille, Ch. Green biorefinery – Industrial implementation. *Food Chem., Part B* **2016**, *197*, 1341–1345[Online]. https://europepmc.org/article/med/26675876 (accessed Nov 28, 2020).
38. Bhaumik, P.; Dhepe, P. L. Solid Acid Catalyzed Synthesis of Furans from Carbohydrates. *Catal. Rev.* **2016**, *58* (1), 36–112.
39. Carlson, T. R.; Tompsett, G. A.; Conner, W. C.; Huber, G. W. Aromatic Production from Catalytic Fast Pyrolysis of Biomass-Derived Feedstocks. *Top. Catal.* **2009**, 52Article 241.
40. Elliott, D. C. Historical Developments in Hydroprocessing Bio-oils. *Energy Fuels* **2007**, *21*, 1792–1815.
41. Demirbas, A. Biorefineries: current activities and future developments. *Energy Convers. Manage.* **2009**, *50*, 2782–2801.

42. Byrd, A. J.; Pant, K. K.; Gupta, R. Hydrogen production from glucose using Ru/Al_2O_3 catalyst in supercritical water. *Ind. Eng. Chem. Res.* **2007**, *46*, 3574–3579.
43. Aho, A.; Kumar, N.; Eranen, K.; Salmi, T.; Hupa, M.; Murzin, D. Yu. Catalytic pyrolysis of woody biomass in a fluidized bed reactor: influence of the zeolite structure. *Fuel* **2008**, *87*, 2493–2501.
44. Isahak, W. N. R. W.; Hisham, M. W. M.; Yaro, M. A.; Hin, T. Y. A review on bio-oil production from biomass by using pyrolysis method. *Renewable Sustainable Energy Rev.* **2012**, *16*, 5910–5923.
45. Stankovikj, F.; McDonald, A. G.; Helms, G. L.; Garcia-Perez, M. Quantification of bio-oil functional groups and evidences of the presence of pyrolytic humins. *Energy Fuels* **2016**, *30*, 6505–6524.
46. Pinheiro Pires, A. P.; Arauzo, J.; Fonts, I.; Domine, M. E.; Arroyo, A. F.; Garcia-Perez, M. E.; Montoya, J.; Chejne, F.; Pfromm, P.; Garcia-Perez, M. Challenges and Opportunities for Bio-oil Refining: A Review. *Energy Fuels* **2019**, 33, 4683–4720.
47. Mortensen, P. M.; Grunwaldt, J.-D.; Jensen, P. A.; Knudsen, K. G.; Jensen, A. D. A review of catalytic upgrading of bio-oil to engine fuels. *Appl. Catal. A: Gen.* **2011**, *407*, 1–19.
48. Oasma, A.; Czernik, S. Fuel oil quality of biomass pyrolysis oils – State of the art for the end users. *Energy Fuels* **1999**, *13*, 914–921 [Online]. https://doi.org/10.1021/ef980272b (accessed Dec 17, 2020).
49. Salema, A. A. Extraction of phenolic compounds from crude pyrolysis oil. *Postdoc J.* **2014**, *2*, 47–50. http://www.postdocjournal.com/archives/691/extraction-of-phenolic-compounds-from-crude-pyrolysis-oil.htm (accessed Dec 15, 2020).
50. de Wild, P.; Reith, H.; Heeres, E. Biomass pyrolysis for chemicals. *Biofuels* **2011**, *2*, 185–208.
51. Shabaker, J. W.; Huber, G. W.; Dumesic, J. A. Aqueous-phase reforming of oxygenated hydrocarbons over Sn-modified Ni catalysts. *J. Catal.* **2004**, *222*, 180–191.
52. Murwanashyaka, J. N.; Pakdel, H.; Roy, C. Seperation of syringol from birch wood-derived vacuum pyrolysis oil. *Sep. Purif. Technol.* **2001**, *24*, 155–165.
53. Qu, W.; Xue, Y.; Gao, Y.; Rover, M.; Bai, X. Repolymerization of pyrolytic lignin for producing carbon fiber with improved properties. *Biomass Bioenergy* **2016**, *95*, 19–26.
54. Tago, T.; Konno, H.; Sakamoto, M.; Nakasaka, Y.; Masuda, T. Selective synthesis for light olefins from acetone over ZSM-5 zeolites with nano- and macro-crystal sizes. *Appl. Catal. A: Gen.* **2011**, *403*, 183–191.
55. Jacobson, K.; Maheria, K. C.; Dalai, A. K. Bio-oil valorization: A review. *Renewable Sustainable Energy Rev.* **2013**, *23*, 91–106.
56. Lyu, G.; Wu, S.; Zhang, H. Estimation and comparison of bio-oil components from different pyrolysis conditions. *Front. Energy Res.* **2015**, *3* [Online]. https://doi.org/10.3389/fenrg.2015.00028 (accessed Nov 27, 2020).
57. Zhang, X.-S.; Yang, G.-X.; Jiang, H.; Liu, W.-J.; Ding, H.-S. Mass production of chemicals from biomass-derived oil by directly atmospheric distillation coupled with co-pyrolysis. *Sci. Rep.* **2013**, *3*, Article 1120. [Online]. https://pubmed.ncbi.nlm.nih.gov/23350028/ (accessed Nov 14, 2020).
58. Pham, T. N.; Shi, D.; Resasco, D. E. Evaluating strategies for catalytic upgrading of pyrolysis oil in liquid phase. *Appl. Catal. B: Environ.* **2014**, *145*, 10–23.

59. Czernik, S.; Bridgwater, A. V. Overview of applications of biomass fast pyrolysis oil. *Energy Fuels* **2004**, *18*, 590–598.

60. Oehr, K. H.; Scott, D. S.; Czernik, S. (Energy Mines & Resources Canada). *Method of producing calcium salts from biomass*. U.S. Patent 5,264,623 A, 1993.

61. Milne, T. A.; Evans, R. J.; Abatzoglou, N. *Biomass Gasifier "Tars": Their Nature, Formation, and Conversion*; Report NREL/TP-570-25357; National Renewable Energy Laboratory: Golden, CO, 1998. https://www.nrel.gov/docs/fy99osti/25357.pdf (accessed Nov 13, 2020).

62. Bedmutha, R.; Booker, C. J.; Ferrante, L.; Briens, C.; Berruti, F.; Yeung, K. K.-C.; Scott, I.; Conn, K. Insecticidal and bactericidal characteristics of the bio-oil from the fast pyrolysis of coffee grounds. *J. Anal. Appl. Pyrolysis* **2011**, *90*, 224–231.

63. Booker, C. J.; Bedmutha, R.; Vogel, T.; Gloor, A.; Xu, R.; Ferrante, L.; Yeung, K. K.-C.; Scott, I. M.; Conn, K. L.; Berruti, F.; Briens, C. Experimental investigations into the insecticidal, fungicidal, and bactericidal properties of pyrolysis bio-oil from tobacco leaves using a fluidized bed pilot plant. *Ind. Eng. Chem. Res.* **2010**, *49*, 10074–10079.

64. Radlein, D. S. A. G.; Piskorz, J. K.; Majerski, P. A. (AST Research Inc.). *Method of producing slow-release nitrogenous organic fertilizer from biomass*. U.S. Patent 5,676,727, 1997.

65. Effendi, A.; Gerhauser, H.; Bridgwater, A. V. Production of renewable phenolic resins by thermochemical conversion of biomass: a review. *Renewable Sustainable Energy Rev.* **2008**, *12*, 2092–2116.

66. Stankovikj, F.; McDonald, A. G.; Helms, G. L.; Olarte, M. V.; Garcia-Perez, M. Characterization of the water-soluble fraction of woody biomass pyrolysis oils. *Energy Fuels* **2017**, *31*, 1650–1664.

67. Shabaker, J. W.; Huber, G. W.; Dumesic, J. A. Aqueous-phase reforming of oxygenated hydrocarbons over Sn-modified Ni catalysts. *J. Catal.* **2004**, *222*, 180–191.

68. Mohan, D.; Pittman, C. U., Jr.; Steele, P. H. Pyrolysis of wood/biomass for bio-oil: A critical review. *Energy Fuels* **2006**, *20*, 848–889.

69. Dale, B. E.; Artzen, C. E. *Biobased Industrial Products: Research and Commercialization Priorities*; National Academies Press: Washington, D.C., 1999.

70. Lian, J.; Chen, S.; Zhou, S.; Wang, Z.; O'Fallon, J.; Li, C.-Z.; Garcia-Perez, M. Separation, hydrolysis and fermentation of pyrolytic sugars to produce ethanol and lipids. *Bioresour. Technol.* **2010**, *101*, 9688–9699.

71. Emrich, W. *Handbook of Charcoal Making: The Traditional and Industrial Methods*; Springer: Dordrecht, The Netherlands, 1985 .

72. Gliński, M.; Zalewski, G.; Burno, E.; Jerzak, A. Catalytic ketonization over metal oxide catalysts. XIII. Comparative measurements of activity of oxides of 32 chemical elements in ketonization of propanoic acid. *Appl. Catal. A: Gen.* **2014**, *470*, 278–284.

73. Pulido, A.; Oliver-Tomas, B.; Renz, M.; Boronat, M.; Corma, A. Ketonic decarboxylation reaction mechanism: a combined experimental and DFT study. *ChemSusChem* **2013**, *6*, 141–151.

74. Ito, Y.; Kawamoto, H.; Saka, S. Efficient and selective hydrogenation of aqueous acetic acid on Ru–Sn/TiO_2 for bioethanol production from lignocellulosics. *Fuel* **2016**, *178*, 118–123.

75. Hedlund, J. K.; Cronauer, D. C.; Jacobs, G.; Kropf, A. J.; Libera, J. A.; Elam, J. W.; Marshall, C. L.; Pendyala, V. R. R.; Davis, B. H. Titania Supported Ru Nanoclusters as Catalysts for Hydrodeoxygenation of Pyrolysis Oils. *Catal. Lett.* **2016**, *146*, 525–539.
76. Dasari, M. A.; Kiatsimkul, P.-P.; Sutterlin, W. R.; Suppes, G. J. Low-pressure hydrogenolysis of glycerol to propylene glycol. *Appl. Catal. A: Gen.* **2005**, *281*, 225–231.
77. Puckette, T. A.; Devon, T. J. *Process for the preparation of glycolaldehyde*, U.S. Patent 7,420,093 B2, 2008.
78. Bridgwater, A. V. Review of fast pyrolysis of biomass and product upgrading. *Biomass Bioenergy* **2012**, *38*, 68–94.
79. Liu, S.; Chen, M.; Chu, L.; Yang, Z.; Zhu, C.; Wang, J.; Chen, M. Catalytic steam reforming of bio-oil aqueous fraction for hydrogen production over Ni–Mo supported on modified sepiolite catalysts. *Int. J. Hydrogen Energy* **2013**, *38*, 3948–3955.
80. Yan, C.-F.; Cheng, F.-F.; Hu, R.-R. Hydrogen production from catalytic steam reforming of bio-oil aqueous fraction over Ni/CeO_2–ZrO_2 catalysts. *Int. J. Hydrogen Energy* **2010**, *35*, 11693–11699.
81. Vagia, E. C.; Lemonidou, A. A. Hydrogen production via steam reforming of bio-oil components over calcium aluminate supported nickel and noble metal catalysts. *Appl. Catal. A: Gen.* **2008**, *351*, 111–121.
82. Ranjekar, A. M.; Yadav, G. D. Steam Reforming of Methanol for Hydrogen Production: A Critical Analysis of Catalysis, Processes and Scope. *Ind. Eng. Chem. Res.* **2021**, *60*, 89–113.
83. García, L. Hydrogen production by steam reforming of natural gas and other nonrenewable feedstocks. In *Compendium of Hydrogen Energy*; Subramani, V., Basile, A., Veziroğlu, T. N., Eds.; Woodhead Publishing Series in Energy; Woodhead Publishing: Cambridge, U.K., 2015; Chapter 4, pp 83–107. https://doi.org/10.1016/B978-1-78242-361-4.00004-2 (accessed Dec 20, 2020).
84. Davidson, S. D.; Zhang, H.; Sun, J.; Wang, Y. Supported metal catalysts for alcohol/sugar alcohol steam reforming. *Dalton Trans.* **2014**, *43*, 11782–11802 [Online]. https://doi.org/10.1039/C4DT00521J (accessed Dec 20, 2020).
85. Epron, F.; Bion, N.; Duprez, D.; Batiot-Dupeyrat, C. Steam Reforming of Alcohols from Biomass Conversion for H_2 Production. In *Perovskites and Related Mixed Oxides*; Granger, P., Parvulescu, V. I., Prellier, W., Eds.; Wiley-VCH: Weinheim, Germany, 2016; Chapter 24, pp 539–558. https://doi.org/10.1002/9783527686605.ch24 (accessed Dec 20, 2020).
86. Frusteri, F.; Bonura, G. Hydrogen production by reforming of bio-alcohols. In *Compendium of Hydrogen Energy*; Subramani, V., Basile, A., Veziroğlu, T. N., Eds.; Woodhead Publishing Series in Energy; Woodhead Publishing: Cambridge, U.K., 2015; Chapter 5, pp 109–136. https://doi.org/10.1016/B978-1-78242-361-4.00005-4 (accessed Dec 21, 2020).
87. Shejale, A. D.; Yadav, G. D. Cu promoted Ni–Co/hydrotalcite catalyst for improved hydrogen production in comparison with several modified Ni-based catalysts via steam reforming of ethanol. *Int. J. Hydrogen Energy* **2017**, *42*, 11321–11332 [Online]. https://doi.org/10.1016/j.ijhydene.2017.03.052 (accessed Dec 21, 2020).
88. Shejale, A. D.; Yadav, G. D. Ni–Cu and Ni–Co supported on La–Mg based metal oxides prepared by co-precipitation and impregnation for superior hydrogen production via steam reforming of glycerol. *Ind. Eng. Chem. Res.* **2018**, 57, 4785–4797.

89. Li, Y.; Zhang, L.; Zhang, Z.; Lium, Q.; Zhang, S.; Liu, Q.; Hu, G.; Wang, Y.; Hu, X. Steam reforming of the alcohols with varied structures: Impacts of acidic sites of Ni catalysts on coking. *Appl. Catal. A: Gen.* **2019**, *584*, 117162.
90. Shejale, A. D.; Yadav, G. D. Noble metal promoted Ni–Cu/La_2O_3–MgO catalyst for renewable and enhanced hydrogen production via steam reforming of bio-based *n*-butanol: Effect of promotion with Pt, Ru and Pd on catalytic activity and selectivity. *Clean Technol. Environ. Policy* **2019**, *21*, 1323–1339. https://doi.org/10.1007/s10098-019-01708-x (accessed Dec 25, 2020).
91. Trane, R.; Dahl, S.; Skjøth-Rasmussen, M. S.; Jensen, A. D. Catalytic steam reforming of bio-oil. *Int. J. Hydrogen Energy* **2012**, *37*, 6447–6472.
92. Lin, Y.-C.; Huber, G. W. The critical role of heterogeneous catalysis in lignocellulosic biomass conversion. *Energy Environ. Sci.* **2009**, *2*, 68–80.
93. Garcia, L.; French, R.; Czernik, S.; Chornet, E. Catalytic steam reforming of bio-oils for the production of hydrogen: effects of catalyst composition. *Appl. Catal. A: Gen.* **2000**, *201*, 225–239.
94. Zinoviev, S.; Müller-Langer, F.; Das, P.; Bertero, N.; Fornasiero, P.; Kaltschmitt, M.; Centi, G.; Miertus, S. Next-generation biofuels: survey of emerging technologies and sustainability issues. *ChemSusChem* **2010**, 3, 1106–1133.
95. Nath, K.; Das, D. Hydrogen from biomass. *Curr. Sci.* **2003**, *85*, 265–271.
96. Sahebdelfar, S. Steam reforming of propionic acid: Thermodynamic analysis of a model compound for hydrogen production from Bio-oil. *Int. J. Hydrogen Energy* **2017**, *42*, 16386–16395.
97. Czernik, S.; Evans, R.; French, R. Hydrogen from biomass-production by steam reforming of biomass pyrolysis oil. *Catal. Today* **2007**, *129*, 265–268.
98. Melligan, F. J. *Pyrolysis of biomass and biorefinery residual materials for production of advanced biofuels*. Doctoral dissertation, University of Limerick, Limerick, Ireland, 2012. https://ulir.ul.ie/handle/10344/2841 (accessed Dec 25, 2020).
99. Lónyi, F.; Valyon, J.; Someus, E.; Hancsók, J. Steam reforming of bio-oil from pyrolysis of MBM over particulate and monolith supported Ni/γ-Al_2O_3 catalysts. *Fuel* **2013**, *112*, 23–30.
100. Mortensen, P. M.; Grunwaldt, J.-D.; Jensen, P. A.; Knudsen, K. G.; Jensen, A. D. A review of catalytic upgrading of bio-oil to engine fuels. *Appl. Catal. A: Gen.* **2011**, *407*, 1–19.
101. Butler, E.; Devlin, G.; Meier, D.; McDonnell, K. A review of recent laboratory research and commercial developments in fast pyrolysis and upgrading. *Renewable Sustainable Energy Rev.* **2011**, *15*, 4171–4186.
102. Sahebdelfar, S.; Ravanchi, M. T. Deoxygenation of propionic acid: Thermodynamic equilibrium analysis of upgrading a bio-oil model compound. *Renewable Energy* **2017**, *114*, 1113–1122.
103. Elliott, D. C.; Hart, T. R.; Neuenschwander, G. G.; Rotness, L. J.; Zacher, A. H. Catalytic hydroprocessing of biomass fast pyrolysis bio-oil to produce hydrocarbon products. *Environ. Prog. Sustainable Energy* **2009**, *28*, 441–449.
104. Furimsky, E. Catalytic hydrodeoxygenation. *Appl. Catal. A: Gen.* **2000**, *199*, 147–190.

105. Ren, H.; Yu, W.; Salciccioli, M.; Chen, Y.; Huang, Y.; Xiong, K.; Vlachos, D. G.; Chen, J. G. Selective hydrodeoxygenation of biomass-derived oxygenates to unsaturated hydrocarbons using molybdenum carbide catalysts. *ChemSusChem* **2013**, *6*, 798–801.
106. Topsøe, H.; Clausen, B. S.; Massoth, F. E. Hydrotreating Catalysis. In *Catalysis: Science and Technology*; Anderson, J. R., Boudart, M. , Eds.; 1996; Vol. 11, pp 1–269.
107. Massoth, F. E.; Politzer, P.; Concha, M. C.; Murray, J. S.; Jakowski, J.; Simons, J. Catalytic hydrodeoxygenation of methyl-substituted phenols: Correlations of kinetic parameters with molecular properties. *J. Phys. Chem. B* **2006**, *110*, 14283–14291.
108. Ryymin, E.-M.; Honkela, M. L.; Viljava, T.-R.; Krause, A. O. I. Competitive reactions and mechanisms in the simultaneous HDO of phenol and methyl heptanoate over sulphided NiMo/γ-Al_2O_3. *Appl. Catal. A: Gen.* **2010**, *389*, 114–121.
109. Romero, Y.; Richard, F.; Brunet, S. Hydrodeoxygenation of 2-ethylphenol as a model compound of bio-crude over sulfided Mo-based catalysts: Promoting effect and reaction mechanism. *Appl. Catal. B: Environ.* **2010**, *98*, 213–223.
110. Popov, A.; Kondratieva, E.; Goupil, J. M.; Mariey, L.; Bazin, P.; Gilson, J.-P.; Travert, A.; Maugé, F. Bio-oils hydrodeoxygenation: Adsorption of phenolic molecules on oxidic catalyst supports. *J. Phys. Chem. C* **2010**, *114*, 15661–15670.
111. Iino, A.; Cho, A.; Takagaki, A.; Kikuchi, R.; Oyama, S. T. Kinetic studies of hydrodeoxygenation of 2-methyltetrahydrofuran on a Ni_2P/SiO_2 catalyst at medium pressure. *J. Catal.* **2014**, *311*, 17–27.
112. Itthibenchapong, V.; Ratanatawanate, C.; Oura, M.; Faungnawakij, K. A facile and low-cost synthesis of MoS_2 for hydrodeoxygenation of phenol. *Catal. Commun.* **2015**, *68*, 31–35.
113. Niquille-Röthlisberger, A.; Prins, R. Hydrodesulfurization of 4,6-dimethyldibenzothiophene and dibenzothiophene over alumina-supported Pt, Pd, and Pt-Pd catalysts. *J. Catal.* **2006**, *242*, 207–216.
114. Guo, C. *Novel Carbon-Supported Transition Metal Phosphide Catalysts for Hydrodeoxygenation of Fast Pyrolysis Oil*. M.S. Thesis, University of Western Ontario, London, Ontario, Canada, 2015. https://ir.lib.uwo.ca/etd/3086/ (accessed Dec 25, 2020).
115. Fisk, C. A.; Morgan, T.; Ji, Y.; Crocker, M.; Crofcheck, C.; Lewis, S. A. Bio-oil upgrading over platinum catalysts using in situ generated hydrogen. *Appl. Catal. A: Gen.* **2009**, *358*, 150–156.
116. Guo, Q.; Wu, M.; Wang, K.; Zhang, L.; Xu, X. Catalytic hydrodeoxygenation of algae bio-oil over bimetallic Ni–Cu/ZrO_2 catalysts. *Ind. Eng. Chem. Res.* **2015**, *54*, 890–899.
117. Dickinson, J. G.; Savage, P. E. Development of NiCu catalysts for aqueous-phase hydrodeoxygenation. *ACS Catal.* **2014**, *4*, 2605–2615.
118. Ardiyanti, A. R.; Khromova, S. A.; Venderbosch, R. H.; Yakovlev, V. A.; Melián-Cabrera, I. V.; Heeres, H. J. Catalytic hydrotreatment of fast pyrolysis oil using bimetallic Ni–Cu catalysts on various supports. *Appl. Catal. A: Gen.* **2012**, *449*, 121–130.
119. Moon, J.-S.; Kim, E.-G.; Lee, Y.-K. Active sites of Ni_2P/SiO_2 catalyst for hydrodeoxygenation of guaiacol: a joint XAFS and DFT study. *J. Catal.* **2014**, *311*, 144–152.
120. Mortensen, P. M.; de Carvalho, H. W. P.; Grunwaldt, J.-D.; Jensen, P. A.; Jensen, A. D. Activity and stability of Mo_2C/ZrO_2 as catalyst for hydrodeoxygenation of mixtures of phenol and 1-octanol. *J. Catal.* **2015**, *328*, 208–215.

121. Furimsky, E. Metal carbides and nitrides as potential catalysts for hydroprocessing. *Appl. Catal. A: Gen.* **2003**, *240*, 1–28.
122. Mendes, M. J.; Santos, O. A. A.; Jordao, E.; Silva, A. M. Hydrogenation of oleic acid over ruthenium catalysts. *Appl. Catal. A: Gen.* **2001**, *217*, 253–262.
123. Stakheev, A. Y.; Kustov, L. M. Effects of the support on the morphology and electronic properties of supported metal clusters: modern concepts and progress in 1990s. *Appl. Catal. A: Gen.* **1999**, *188*, 3–35.
124. Vannice, M. A.; Sen, B. Metal-support effects on the intramolecular selectivity of crotonaldehyde hydrogenation over platinum. *J. Catal.* **1989**, *115*, 65–78.
125. Gutierrez, A.; Kaila, R. K.; Honkela, M. L.; Slioor, R.; Krause, A. O. I. Hydrodeoxygenation of guaiacol on noble metal catalysts. *Catal. Today* **2009**, *147*, 239–246.
126. Boullosa-Eiras, S.; Lødeng, R.; Bergem, H.; Stöcker, M.; Hannevold, L.; Blekkan, E. A. Catalytic hydrodeoxygenation (HDO) of phenol over supported molybdenum carbide, nitride, phosphide and oxide catalysts. *Catal. Today* **2014**, *223*, 44–53.
127. Dang, Q.; Luo, Z.; Zhang, J.; Wang, J.; Chen, W.; Yang, Y. Experimental study on bio-oil upgrading over Pt/SO_4^{2-}/ZrO_2/SBA-15 catalyst in supercritical ethanol. *Fuel* **2013**, *103*, 683–692.
128. Pan, W.; Li, C.; Zhang, Q.; Liu, Z.; Peng, J.; Chen, P.; Lou, H.; Zheng, X. Upgrading of low-boiling fraction of bio-oil in supercritical methanol and reaction network. *Bioresour. Technol.* **2011**, *102*, 4884–4889.
129. Zhang, X.; Xu, C.; Zhai, Y.; Liu, Y.; Zhang, R.; Tang, X. Upgrading of bio-oil using supercritical 1-butanol over a Ru/C heterogeneous catalyst: role of the solvent. *Energy Fuels* **2014**, *28*, 4611–4621.
130. Huber, G. W.; Iborra, S.; Corma, A. Synthesis of transportation fuels from biomass: chemistry, catalysts, and engineering. *Chem. Rev.* **2006**, *106*, 4044–4098.
131. Williams, P. T.; Horne, P. A. Analysis of aromatic hydrocarbons in pyrolytic oil derived from biomass. *J. Anal. Appl. Pyrolysis* **1995**, *31*, 15–37.
132. Graça, I.; Lopes, J. M.; Cerqueira, H. S.; Ribeiro, M. F. Bio-oils upgrading for second generation biofuels. *Ind. Eng. Chem. Res.* **2013**, *52*, 275–287.
133. Serrano-Ruiz, J. C.; Dumesic, J. A. Catalytic production of liquid hydrocarbon transportation fuels. In *Catalysis for Alternative Energy Generation*; Guczi, L., Erdôhelyi, A., Eds.; Springer: Dordrecht, The Netherlands, 2012; Chapter 2, pp 29–56.
134. Park, H. J.; Heo, H. S.; Jeon, J.-K.; Kim, J.; Ryoo, R.; Jeong, K.-E.; Park, Y.-K. Highly valuable chemicals production from catalytic upgrading of radiata pine sawdust-derived pyrolytic vapors over mesoporous MFI zeolites. *Appl. Catal. B: Environ.* **2010**, 95, 365–373.
135. Adjaye, J. D.; Bakhshi, N. N. Upgrading of a wood-derived oil over various catalysts. *Biomass Bioenergy* **1994**, 7, 201–211.
136. Katikaneni, S. P. R.; Adjaye, J. D.; Idem, R. O.; Bakhshi, N. N. Performance studies of various cracking catalysts in the conversion of canola oil to fuels and chemicals in a fluidized-bed reactor. *J. Am. Oil Chem. Soc.* **1998**, 75, 381–391.
137. Wei, Y.; Lei, H.; Zhu, L.; Zhang, X.; Liu, Y.; Yadavalli, G.; Zhu, X.; Qian, M.; Yan, D. Hydrocarbon produced from upgrading rich phenolic compound bio-oil with low catalyst coking. *Fuel* **2016**, *178*, 77–84.

138. Fan, Y.; Cai, Y.; Li, X.; Yin, H.; Chen, L.; Liu, S. Regeneration of the HZSM-5 zeolite deactivated in the upgrading of bio-oil via non-thermal plasma injection (NTPI) technology. *J. Anal. Appl. Pyrolysis* **2015**, *111*, 209–215.
139. Gayubo, A. G.; Valle, B.; Aguayo, A. T.; Olazar, M.; Bilbao, J. Pyrolytic lignin removal for the valorization of biomass pyrolysis crude bio-oil by catalytic transformation. *J. Chem. Technol. Biotechnol.* **2010**, *85*, 132–144.
140. Blasco, T.; Corma, A.; Martínez-Triguero, J. Hydrothermal stabilization of ZSM-5 catalytic-cracking additives by phosphorus addition. *J. Catal.* **2006**, *237*, 267–277.
141. Serrano-Ruiz, J. C.; West, R. M.; Dumesic, J. A. Catalytic conversion of renewable biomass resources to fuels and chemicals. *Annu. Rev. Chem. Biomol. Eng.* **2010**, *1*, 79–100.
142. Deng, L.; Fu, Y.; Guo, Q. X. Upgraded acidic components of bio-oil through catalytic ketonic condensation. *Energy Fuels* **2009**, *23*, 564–568.
143. Kastner, J. R.; Hilten, R.; Weber, J.; McFarlane, A. R.; Hargreaves, J. S. J.; Batra, V. S. Continuous catalytic upgrading of fast pyrolysis oil using iron oxides in red mud. *RSC Adv.* **2015**, *5*, 29375–29385.

Chapter 3

Subcritical and Supercritical Water Treatments for Bio-Oil Production and Upgrading

Priyanka Yadav,[1] Sivamohan N. Reddy,[*,1] Sonil Nanda,[2] and Ajay K. Dalai[2]

[1]Department of Chemical Engineering, Indian Institute of Technology Roorkee, Roorkee, Uttarakhand, India 247667

[2]Department of Chemical and Biological Engineering, University of Saskatchewan, Saskatoon, Saskatchewan, Canada S7N 5A9

[*]Email: n.reddy@ch.iitr.ac.in

Because of growing energy demands and environmental concerns, different clean technologies are being developed for harnessing energy from waste biomass, which can be commercially feasible. Liquid biofuels obtained from thermochemical processes cannot be directly used in combustion engines because of certain undesirable properties such as high water content, viscosity, acidity, and high-oxygenated compounds, thus requiring upgrading. This chapter focuses on the comparison of some notable bio-oil upgrading technologies such as emulsification, esterification, hydrotreatment, hydrocracking, solvent addition, steam reforming, supercritical fluids, and zeolite cracking as well as subcritical and supercritical water treatment. The upgrading of oils in the presence of different homogenous and heterogeneous catalysts are elucidated. The effects of operating parameters such as temperature, pressure, reaction time, feedstock concentration, biomass/water ratio, catalyst loading, biomass composition, and steam/carbon ratio on bio-oil upgrading are discussed.

Introduction

Bio-oil is a brownish organic liquid with a characteristic smoky odor derived from various biomass thermochemical conversion techniques including pyrolysis, hydrothermal liquefaction, and gasification. These thermochemical conversion technologies also result in a solid product (biochar) and gaseous products (producer gas and syngas) as shown in Figure 1. Biomass-derived bio-oil is a promising substitute for liquid fossil fuels because it is a cleaner fuel with less pollution and negligible greenhouse gas emissions (*1*). Bio-oil is a composite mixture of numerous organic components consisting of acids, alcohols, esters, ketones, and phenolic compounds along with high aqueous content (e.g., water, acids, and alcohols) (*1*–*3*). Biochar is a carbon-rich porous coproduct of biomass pyrolysis, carbonization, gasification, and liquefaction that can be used for a wide variety of

© 2021 American Chemical Society

applications as a solid fuel, biocomposite, adsorbent, catalyst support, precursor for activated carbon and graphene, gas storage agent, and soil amendment agent (4). On the other hand, the producer gas obtained from the pyrolysis of biomass contains a mixture of H_2, CO, CO_2, CH_4, and C_{2+} gases, which can be conditioned and used for combined heat and power. Moreover, synthesis gas or syngas generated from biomass gasification contains predominantly H_2 and CO, which can be used for several gas-to-liquid processes such as Fischer-Tropsch catalysis to synthesize liquid hydrocarbon fuels and green diesel (5).

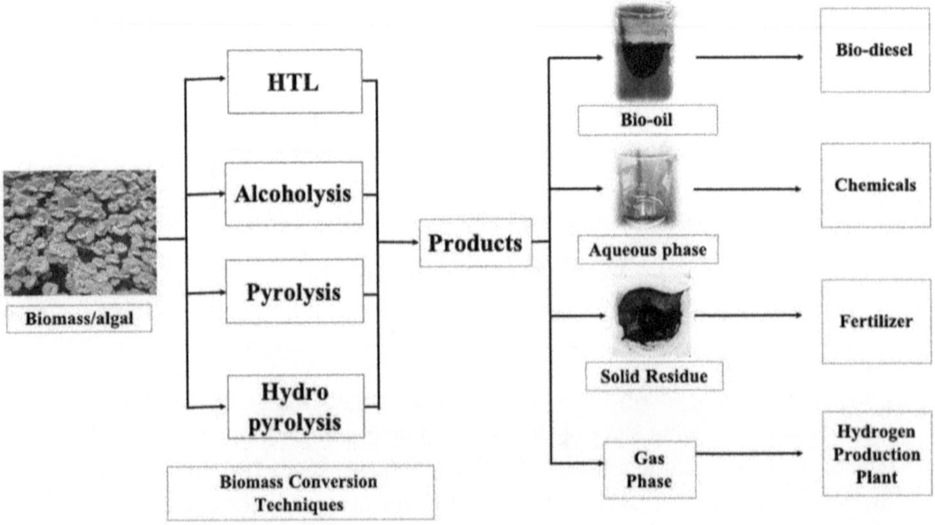

Figure 1. Overview of thermochemical biomass conversion technologies and product applications.

Some of the undesirable properties of crude bio-oil are high water content, inferior heating value, high content of oxygenated compounds, high viscosity, low pH, corrosiveness, high nitrogen content, and thermochemical instability due to polymerization (*1, 3, 6–8*). The presence of highly oxygenated compounds in bio-oil increases the water content, which usually ranges from 20 to 30 wt % and results in lower heating value, lower combustion rate, and increased ignition delay of bio-oil (*8, 9*).

Bio-oil is a highly nonhomogeneous mixture of various organic compounds, which leads to partial separation and filtration problems. The presence of higher alkali metals in bio-oils also results in solid deposition, thus compromising equipment efficiency. During the transportation of bio-oil, high-pressure drops need to be maintained, which results in higher maintenance of equipment (*3*). These properties of bio-oils have limited their application as a drop-in transportation fuel. Therefore, crude bio-oil must be upgraded to improve its properties while increasing its commercial and industrial usability.

This chapter evaluates some notable bio-oil upgrading technologies such as emulsification, esterification, hydrotreatment, hydrocracking, solvent addition, steam reforming, supercritical fluids, and zeolite cracking as well as subcritical and supercritical water treatment. It should be noted that the yield, quality, and composition of upgraded bio-oil depends on several process conditions of the upgrading technologies. With this objective, this chapter reviews the impacts of upgrading process parameters such as temperature, pressure, reaction time, feedstock concentration, biomass/water ratio, catalyst loading, biomass composition, and steam/carbon ratio on the fuel quality of bio-oils.

Promising Bio-Oil Upgrading Techniques

Bio-oil upgrading techniques can be broadly classified into physical and chemical treatment methods. Some notable oil upgrading methods include emulsification, esterification or alcoholysis, hydrotreating, hydrocracking or catalytic cracking, solvent addition, steam reforming, zeolite cracking, and supercritical fluids as shown in Figure 2 (*1*). Table 1 summarizes some key aspects of bio-oil upgrading technologies.

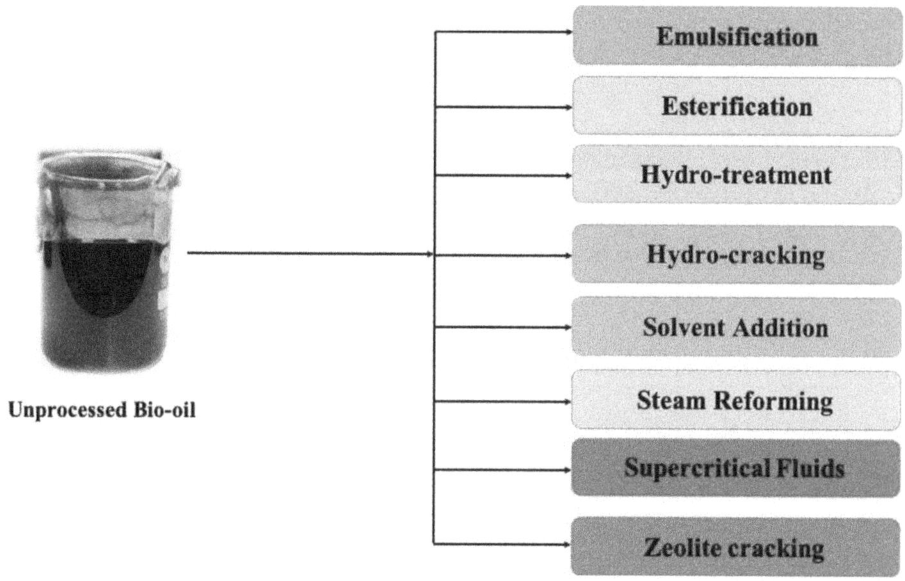

Figure 2. Bio-oil upgrading techniques.

Emulsification

Bio-oils can be emulsified by diesel or biodiesel with or without surfactants. However with the usage of surfactants it will increase the cost (*10*). Bio-oil emulsions have the desired ignition properties. Emulsification leads to limited conversion rates and miscibility issues resulting in the utilization of a fraction of upgraded bio-oil (*11*).

Esterification or Alcoholysis

In esterification, the reaction between alcohol and fatty acids results in the formation of alkyl ester or biodiesel in the presence of an acidic catalyst (*12*). However, both heterogeneous and homogeneous catalysts can be used for esterification. Heterogeneous catalysts are preferable because they are relatively easier to separate from the product mixture. Esterification of crude oil can be performed for the upgrading of bio-oil with zeolites.

Table 1. Comparison of Bio-Oil Upgrading Techniques. Reference: Data from reference (*1*). Copyright 2016 Elsevier.

Upgrading Techniques	*Operating Conditions*	*Reaction Pathways*	*Advantages*	*Disadvantages*
Emulsification	In milder operating conditions, a surfactant is added	Crude bio-oil emulsification with diesel/biodiesel by addition of surfactant	Relatively easy process	Energy consumption is high and no reaction mechanism for the extraction of undesirable fractions from bio-oil
Esterification	In milder operating conditions, alcohol is added	Esterification followed by acetalization	Relatively easy and economical if methanol used	Inefficient in removing the nitrogenous compounds
Hydrocracking and Catalytic Cracking	High temperature and high hydrogen pressure are required along with catalysts	Simultaneous hydrogenation and catalytic cracking	Converts heavier fractions into lighter fractions	A high-pressure reactor is required, and there are chances of catalyst deactivation and coking
Hydrotreating	High temperature and high hydrogen pressure are required	Hydrodeoxygenation and hydrogenation reactions without cracking	Effectively removes the impurities in the form of NH_3, H_2O, and H_2S while enhancing the heating value	A high-pressure reactor is required
Solvent Addition	In milder operating conditions, a polar solvent is added	Lowers the viscosity and enhances the heating value	Relatively easy process	No reaction mechanism for the extraction of undesirable fractions from bio-oil
Steam Reforming	High operating temperature and requires catalyst	Water-gas shift reaction and catalytic reforming	Hydrogen is produced as an energy source	A high-pressure reactor is required
Supercritical Fluids	High temperature and high pressure are required along with organic solvents	Advances the reaction mechanism by liquidlike density and gaslike diffusivity	Efficiently enhances the heating value and lowers viscosity	A high-pressure reactor and expensive solvents are required
Zeolite Cracking	High temperature and atmospheric pressure are required along with zeolite catalysts	Cracking of heavier compounds by zeolites	Converts the heavier fractions into lighter fractions	Carbon deposition results in catalyst deactivation and a low hydrogen/carbon ratio

Hydrotreatment

Hydrotreating requires moderate temperatures (i.e., 300–450 °C) and high hydrogen pressure process (15 MPa) in the presence of a catalyst for the reduction of oxygen, sulfur, and nitrogen heteroatoms from the crude oil and bio-oil feedstocks as water, hydrogen sulfide, and ammonia, respectively. The catalytic hydrotreating technologies for the removal of oxygen, sulfur, and nitrogen heteroatoms are called hydrodeoxygenation, hydrodesulfurization, and hydrodenitrogenation, respectively (*13*). Upon the removal of water molecules from the bio-oil, its heating value increases, whereas the removal of sulfur and nitrogenous compounds eliminates the changes of SO_x and NO_x emissions upon combustion.

Hydrocracking, Catalytic Cracking, and Zeolite Cracking

Hydrotreating of bio-oil is accompanied by catalytic cracking for the conversion of heavier molecules to lighter molecules in the presence of moderate temperature and high hydrogen pressures (*14*). Catalytic cracking does not usually involve hydrogen, unlike catalytic hydrotreating, which requires hydrogen. The heavy crude oil and larger molecules of fatty acids can be converted to gasoline and other lighter hydrocarbons using catalytic cracking (*15*). Through catalytic cracking and hydrotreating, oxygen is evicted as water, CO_2, and CO (*16*). Zeolites are microporous aluminosilicates containing acidic sites having a high surface area with significant adsorption capacity. Zeolite cracking is performed at a relatively slightly higher temperatures (350–500 °C) and atmospheric pressure resulting in the cracking of heavy components to lighter components (*17*). Bio-oil undergoes deoxygenation reaction followed by zeolite cracking, leading to the formation of coke, oil-soluble organics, water-soluble organics, and water.

Solvent Addition

Solvent addition results in chemical reactions between solvents and bio-oil restricting the aging and polymerization of bio-oil during storage. This leads to a reduction in viscosity and an increment in heating values. Ethyl acetate, acetone, methanol, and ethanol are some of the solvents used for bio-oil upgrading (*1*). Solvents also play an important role in hydrothermal liquefaction because they decompose the biomass, which results in active hydrogen. This active hydrogen formed by the decomposition of intermediates enhances the bio-oil yield. Ethanol also decomposes the biomass and acts as a hydrogen donor to promote bio-oil formation (*18*). The classification of solvents is done according to their polarity, such as polar protic, dipolar aprotic, and nonpolar. Polar protic solvents, such as alcohol and water, correspond to molecules with the bonding of hydrogen atoms attached to an electronegative oxygen atom. Dipolar aprotic solvents, such as 1,4-dioxane and acetone, correspond to compounds bonded with a dipole attached to C–O and C–N groups without any –OH bonding. Nonpolar solvents, such as benzene and diethyl ether, correspond to a mixture with a lower dielectric constant that is immiscible with water. A rise in biomass-to-solvent ratio reduces the possibility of a solvent dissolving and stabilizing in a mixture, resulting in reduced bio-oil yield (*18*).

Supercritical Fluids

Supercritical fluids can be alternatives to organic solvents for bio-oil upgrading because they have a high rate of heat transfer, mass transfer, viscosity, liquidlike density, and gaslike diffusivity (*19*). The critical temperature and critical pressure determine the homogeneity of the fluid phase because beyond the critical points, the liquid and vapor phases coexist (*20*). The application of supercritical fluids in bio-oil upgrading is less energy-intensive and cost-effective and has a faster reaction time (*21*).

Steam Reforming

Reforming of unprocessed bio-oil in fixed- and fluidized-bed reactors under steam results in syngas production, which can be further converted to a wide range of fuels through gas-to-liquid technologies (e.g., Fischer-Tropsch catalysis (22)). The operating parameters influencing the steam reforming of bio-oil are temperature, catalyst/feed ratio, and steam/carbon ratio. Steam reforming leads to the simultaneous production of upgraded bio-oil and gaseous hydrogen, which adds value to the process. One of the major bottlenecks of this process is catalyst deactivation by coke formation, which can increase the maintenance cost (*23*).

Application of Subcritical and Supercritical Water in Bio-Oil Production and Upgrading

Subcritical water is the fluid phase of water occurring at temperatures and pressure below water's critical points (i.e., 374 °C and 22.1 MPa (*24*)). Supercritical water occurs at temperatures and pressures above water's critical points (*25*). Subcritical water is mostly applied for hydrothermal liquefaction, whereas supercritical water is used for hydrothermal gasification (*18*). Both subcritical and supercritical water are considered green solvents due to their promising thermophysical properties. Supercritical fluids are organic solvents such as acetones, butanol, 1,4-dioxane, ethanol, methanol, *n*-hexanol, and 2-propanol that are used for upgrading bio-oil. Supercritical fluids result in constraining the production of coke or tar.

Xiu and Shahbazi (*26*) reported that in the presence of subcritical and supercritical water, milder operating conditions are required to advance the upgrading reactions, which increases bio-oil yield and decreases oxygen content and viscosity. Toor et al. (*27*) conducted hydrothermal liquefaction experiments on algae (*Nannochloropsis salina* and *Spirulina platensis*) under subcritical and supercritical water conditions of 220–375 °C and 2–25.5 MPa pressure. The authors suggested that subcritical conditions are more effective than are supercritical conditions for bio-oil production from wet algae through hydrothermal liquefaction. *S. platensis* and *N. salina* resulted in the highest bio-crude oil yields of 38% and 46% at optimal reaction conditions of 310 °C under 11.5 MPa and 350 °C under 17.5 MPa, respectively.

Zhang et al. (*28*) investigated hydrothermal liquefaction of green algae (*Chlorella pyrenoidosa*) catalyzed by Raney Ni and HZSM-5 at subcritical and supercritical conditions of ethanol. They reported that the catalysts did not have a significant effect on the bio-crude yield. In contrast, ethanol addition resulted in 50–70 wt % of bio-crude yield with 36.2 MJ/kg of heating value at 300 °C. Chan et al. (*29*) examined the hydrothermal liquefaction of oil palm biomass using water as a solvent under subcritical and supercritical conditions. The highest bio-oil yields of 38.5%, 37.4%, and 34.3% were obtained from palm kernel shell, empty fruit bunch, and palm mesocarp fiber, respectively, at optimal conditions of 390 °C and 25 MPa.

Xu et al. (*30*) demonstrated the role of supercritical solvent 1-butanol for upgrading bio-oil over a Ru/C catalyst at 250–300 °C and 8.8–11.5 MPa pressure. In the presence of supercritical solvents, the properties of bio-oil improved, including an increase in heating value (from 30.5 MJ/kg to 32 MJ/kg), a decrease in oxygen content (from 21.4 to 14.5%), and a reduction in carbon deposition (from 9.9 to 0.2%). The authors also concluded that acetalization, etherification, esterification, hydrodeoxygenation, and hydrogenation were the primary reactions involved in the supercritical upgrading of bio-oils.

Yin et al. (*31*) examined the catalytic upgrading obtained by fast pyrolysis of oils in the presence of porous metal oxide doped with copper in supercritical mono-alcohols (i.e., ethanol and methanol). They concluded that the obtained product mixture consisted of aliphatic mono-alcohols (51%), diols (22%), ethers (9%), esters (15%), and furanics (3%) at 275 °C and 5 MPa hydrogen pressure in supercritical methanol and ethanol.

Chang (*32*) compared fast pyrolysis along with hydrothermal liquefaction methods for the bio-oils derived by palm empty fruit bunches. The author suggested the optimum conditions for higher bio-oil yield from fast pyrolysis were 300–600 °C, 0.1 MPa pressure, and a biomass particle size of <1 mm. In contrast, optimal conditions for hydrothermal liquefaction were 270–400 °C, 2–29.1 MPa pressure, a biomass particle size <1 mm, and solvents such as water, ethylene glycol, methanol, and a water–ethanol mixture with a biomass/solvent ratio of 1:2 to 1:17.

Patil et al. (*33*) extracted the bio-oil from microalgae (*N. salina*) in the presence of supercritical CO_2 and an azeotropic mixture of acetone and ethanol. The maximum total algal lipid yield of 31.4% was obtained at optimum conditions of 80 °C, 34 MPa pressure, and hexane and acetone in a 1:1 ratio under CO_2 flow rate. The total fatty acid extraction was approximately 70–75% contributed by eicosapentaenoic acid, oleic acid, palmitic acid, and palmitoleic acid compounds.

Mechanism of Biomass Hydrolysis in Subcritical and Supercritical Water

Lignocellulosic biomass comprises hemicellulose, cellulose, lignin, proteins, lipids, and ash. Under subcritical conditions, as the temperature increases from 50 to 100 °C, hemicellulose and amorphous cellulose units decompose to their monomeric units along with the formation of intermediates at 100–200 °C (*19*). A further increase in temperature between 200 and 300 °C results in the formation of bio-oil, water-soluble fraction, gases, and bio-residue (tar and char) during hydrothermal liquefaction. The long-chain compounds of lignin and cellulose in biomass are depolymerized to monomers, followed by hydrothermal decomposition via dehydration, decarboxylation, and deamination to produce bio-oil. The repolymerization of volatiles vapors and bio-oil results in the formation of char.

Cellulose and hemicellulose degrade at lower temperatures to mostly produce liquid products. As the temperature increases to >350 °C, the dominant reaction mechanisms of repolymerization along with cyclization of lignin starts to take place, which results in char formation. Hence, biomasses with higher cellulosic and hemicellulosic fraction improve bio-oil yield, whereas those with higher lignin fraction increase the formation of char and tar (*18*). Figure 3 illustrates the mechanisms of hydrothermal degradation of biomass components.

Lipids

At low temperatures, lipids undergo hydrolysis reaction to form glycerol and fatty acids (*34*). A rise in temperature results in glycerol decomposition to aldehydes and alcohols. The fatty acids degrade to alkanes and alkenes, which further undergo halogenation reactions to form bio-oil. Some parts of the fatty acids result in the formation of amides and esters.

Proteins

Hydrolysis of protein results in the formation of monomeric units of amino acids. With the increase in temperature, amino acids undergo deamination reaction to form ammonia, acetic acid, and CO_2. As the temperature increases, the condensation of ammonia takes place from cyclic amides. Some part of amino acids undergoes the cyclization reaction to form ring diamide intermediates. The instability of ring diamide results in the dehydration reaction to form nitrogen and sulfur heterocyclic compounds such as indoles and quinolines.

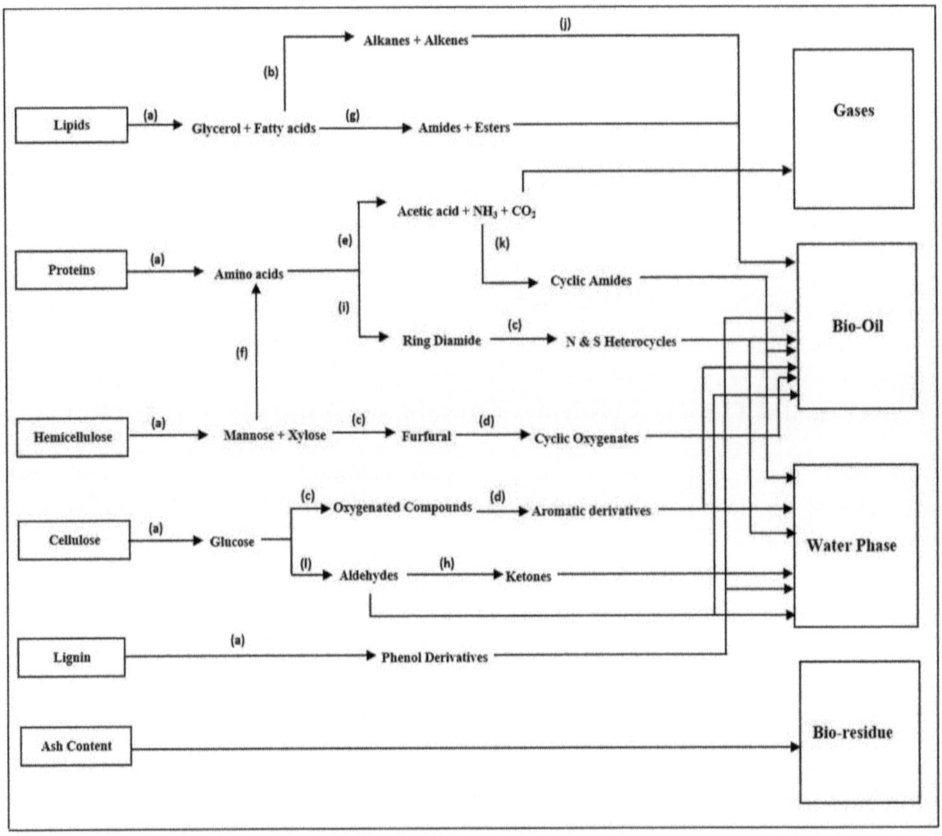

Figure 3. General reaction mechanism of biomass/algal to products. References: Data from reference (36). Copyright 2014 Elsevier; reference (37). Copyright 2017 Royal Society of Chemistry; reference (38). Copyright 2017 Royal Society of Chemistry.

Hemicellulose

Hemicellulose is a heteropolymer mainly composed of pentose sugars (i.e., xylose and arabinose), hexose sugars (i.e., glucose, rhamnose, galactose, and mannose), and sugar acids (i.e., glucuronic acid and galacturonic acid) (35). As the temperature increases, hemicellulose is depolymerized into sugar monomers via hydrolysis. Mannose is further converted to amino acids by the Maillard reaction, whereas dehydration of xylose results in furfural formation. With temperature, furfural is repolymerized to form cyclic oxygenates and coke.

Cellulose

Hydrolysis of cellulose results in glucose formation. Decomposition of glucose proceeds by two pathways that depend on the reaction temperature. At low temperatures, dehydration of glucose results in intermediates of C_5-C_6 compounds, which are further dehydrogenated from oxygenated compounds. At high temperatures, it follows the retro-aldol reaction. In the retro-aldol reaction of glucose, C_1-C_4 intermediates are formed, and they further undergo condensation reaction to form aldehydes and ketones. Oxygenated aromatic derivatives repolymerize to form aromatic compounds, whereas some aldehydes isomerize to form ketones.

Impact of Operating Parameters on Bio-Oil Yield and Upgrading during Hydrothermal Liquefaction of Bio-Oil or Biomass

Duan et al. (*39*) investigated the effects of four variables—temperature (270–380 °C), reaction time (10–120 min), catalyst, and reactor loading—on product yields by hydrothermal liquefaction of aquatic biomass (duckweed). During the hydrothermal liquefaction of duckweed, an increase in the temperature resulted in an increase in bio-oil quantity, with the highest yield of 17.5% at 350 °C. A further rise in temperature led to a decrease in bio-oil yield. The gas and water-soluble phases increased as the temperature rose, whereas the solid char phase showed a decreasing trend. Moreover, as the reaction time increased from 5 to 120 min, a maximum bio-oil yield of 18.2 wt % was obtained in 30 min. Increasing the reaction time beyond the optimum point can result in a constant yield of bio-oil. The solid residue and water-soluble phase decreased with the increase in reaction time, whereas the gas phase increased. Reactor loading was also varied from 0.5 to 5.5 g to study the effects on product yields.

Yang et al. (*40*) conducted experiments on hydrothermal liquefaction of undried macroalgae (*Enteromorpha prolifera*) to study the impacts of reaction conditions, namely temperature, reaction time, biomass/water ratio, and acid catalysts, on product yields. They concluded that a maximum bio-oil yield of 28.4 wt % was obtained at 290 °C in 20 min with a biomass/water ratio of 1:3. The researchers found that as the temperature increased from 230 to 290 °C, the bio-oil yield also improved within the range of 10.9 to 28.4%. However, a further rise in temperature to 310 °C resulted in a reduction in bio-oil yield up to 18.6 wt %. In terms of char production, there was a decrease in the yield from 21.9 wt % at 230 °C to 8.8 wt % at 270 °C. The authors studied the effects of residence time at 290 °C and biomass/water ratio of 1:3. The authors reported that at 20 min, the bio-oil yield reached its maximum value of 28.4 wt %, which decreased to 15.5 wt % as the residence time increased to 100 min.

Gong et al. (*41*) investigated the esterification reaction of oleic acid in the presence of 12-tungstenphosphoric acid for bio-oil upgrading. Near-complete conversion of oleic acid was obtained with negligible catalyst deposition as well as better thermal stability and acidity. The authors examined the effects of variable temperatures (40–100 °C) by performing the experiments at a 10:1 molar ratio with 7 wt % of the catalyst. It was found that the oleic acid conversion rate increases with the increase in temperature within the range of 40–60 °C with 100% conversion being achieved at 80 °C. The influence of temperature was reported by conducting the esterification reaction at 160–220 °C for 1 h. As the reaction time increased, the oleic acid conversion rate increased, with 100% conversion obtained at 5 h. A further increase in reaction time resulted in almost constant conversion. Complete conversion of oleic acid was obtained by utilizing the active catalyst at 80 °C with 7 wt % of catalyst/oleic acid ratio with a reaction time of 5 h at 10:1 alcohol/acid mole ratio.

Milina et al. (42) examined the impacts of temperature, residence time, and size of zeolite micropores on the upgrading of bio-oil by catalytic esterification of alcohol and acids. Maximum upgraded bio-oil yields were obtained over large size BEA 220 (beta) and FAU 15 (faujasite) catalysts. With the application of large size zeolites at 180 °C to moderate-size zeolites at 200 °C, a decrease in acidity and oxygen content with significant ester content can be obtained in the upgraded bio-oil. In the presence of FAU 15 (large pore zeolite), the acetic acid conversion increased to more than 60 wt % at 220 °C. The cresol acetate conversion also showed a trend similar to the acetic acid conversion curve except at higher temperatures at which the coking reaction was more dominant. In the case of MFI 40 (medium pore zeolite), acetic acid conversion varied from 20 wt % to more than 40 wt % at 160 °C and 220 °C, respectively. Cresol acetate conversion also increased to more than 20 wt % at 220 °C. To evaluate the impact of reaction time (0–1.5 h), esterification reactions were conducted at an optimal temperature of 180 °C utilizing large-pore-size zeolite FAU 15. The acetic acid conversion was maximum initially and decreased to 50 wt % after nearly 1 h. A further increase in the reaction time resulted in a constant conversion rate. The cresol acetate conversion increased to 50 wt % after 1 h and remained constant thereafter. The study also suggested that lower conversion rates were obtained for zeolites compared with homogeneous catalysts because of the catalyst deactivation, blockage of active sites by coke, and strong adsorption of water formation.

Yang et al. (43) conducted the hydrodeoxygenation reaction of unprocessed bio-oil derived by hydrothermal liquefaction of corn stalk over a dispersed bimetallic ammonium nickel molybdate catalyst at 280–370 °C, 4 MPa pressure, and 60 min of residence time under supercritical ethanol. The authors studied the effects of temperature on the upgrading of bio-oil and residue by varying the temperature from 280 to 370 °C. The authors concluded that processed bio-oil resulted in a decreased yield, from 26.7 to 16.3 wt %, with a temperature rise. The water content decreased from 25.6 to 1.6 wt %, whereas the viscosity decreased from 37.3 to 8.8 mm^2/s. Meanwhile, there was an increment in the heating value from 32.1 to 38.3 MJ/kg.

In a study by Remón et al. (44), increasing the catalyst/bio-oil ratio up to 0.25 g/g improved the gas composition while reducing the bio-oil and char yield via reforming and cracking reactions. The authors investigated the impact of reaction parameters, namely temperature and pressure, on liquid biofuels along with gaseous yields by the upgrading of bio-oil produced by fast pyrolysis of pinewood at subcritical and supercritical conditions. A rise in catalyst/bio-oil ratio within a temperature range of 310–380 °C and 20 MPa resulted in an enhancement of gas yield along with the reduction in the liquid yield in contrast to the temperature range of 380–450°C. Hence, temperature showed a significant kinetic effect on the gas yields.

Jo et al. (45) examined the effects of operating parameters of temperature, concentration, and reaction time on bio-oil quantity during the upgrading of the low boiling fractions of bio-oil by fast pyrolysis at the supercritical condition. The authors determined the upgrading of low boiling point bio-oil fraction obtained by fast pyrolysis using supercritical methanol. A maximum bio-oil yield of 78.4 wt % was obtained at 400 °C in 30 min with a decrease in the total acid number of 4 mg-KOH/g-oil and a heating value of 29.9 MJ/kg. The energy efficiency and energy recovery were obtained within the range of 4–131% and 79–109%, respectively. The authors reported that an increase in temperature led to methanol consumption from 16.2 to 22.1 wt %, respectively. As the reaction time increased from 0 min to 120 min, the yield of heavy bio-oil decreased from 18.9 to 13 wt %. A similar trend for the decrease in light bio-oil yield was also obtained from 75.6 wt % (at 0 min) to 60.8 wt % (at 120 min). However, reaction time did not have a significant effect on gaseous or char yields, and methanol consumption was slightly increased from 21.9 to 26.4 wt %. The authors also concluded that, in contrast to supercritical ethanol and supercritical isopropanol, supercritical methanol was more efficient in bio-oil upgrading, with higher yields and fuel properties.

Baloch et al. (*18*) examined the effects of temperature (240–280 °C) and solvents by hydrothermal liquefaction of sugarcane bagasse at subcritical and supercritical conditions. The authors investigated the subcritical and supercritical extraction of sugarcane bagasse in the presence of solvents such as water and ethanol on bio-oil quantity and quality at variable temperatures of 240–280 °C. The bio-oil and solid residue yield obtained in the presence of water was in the range of 46–51.8% and 36.3–28.2%, respectively. When ethanol was used as a solvent, bio-oil and char yields obtained were in the range of 54.6–62.3 wt % and 39.3–24.7 wt %, respectively. In both cases, a rise in the temperature promoted the yield of bio-oil because of the higher fragments of biomass that undergo repolymerization followed by a hydrolysis reaction to form bio-oil. The authors reported that a rise in temperature from 240 to 280 °C using water as a solvent resulted in an increase in bio-oil yield from 46 to 51.8 wt % and a decrease in char yield from 36.3 to 28.2 wt %. In contrast, with ethanol solvent, the yield of bio-oil increased from 54.7 to 62.3% and char yield decreased from 39.3 to 24.7%. However, this process also enhanced bio-oil properties including greater carbon content, higher heating value, lower oxygen content, and better combustion performance. Heterogeneous catalysts such as $FeSO_4$, $Ca(OH)_2$, and Rb_2CO_3 enhanced bio-oil yield as 63%, 65%, and 30%, respectively (*18*).

Effects of Homogeneous Catalysts

Catalysts reduce the processing temperature of the upgrading process along with a decrease in oxygen content by deoxygenation, decarboxylation, oligomerization, and cracking reactions to produce high-quality upgraded bio-oil and chemicals (*7*). Catalysts play an essential role in enhancing the yields of bio-oils and gases while reducing the char formation through hydrothermal liquefaction. Catalysts can be classified into homogeneous and heterogeneous catalysts. Some examples of homogeneous catalysts are Na_2CO_3, NaOH, K_2CO_3, KOH, LiOH, RbOH, Rb_2CO_3, and CsOH. Examples of heterogeneous catalysts are Ni-based ($Ni/SiO_2/Al_2O_3$), Fe-based (FeS and $FeSO_4$), Pt-supported on carbon (Pt/C), Pd-supported on carbon (Pd/C), Ru-supported on carbon (Ru/C), and $CoMo/Al_2O_3$.

Biller and Ross (*46*) conducted hydrothermal liquefaction of microalgae (*Chlorella vulgaris*, *Nannochloropsis occulata*, and *Porphyridium cruentum*), cyanobacteria (*Spirulina*), and biochemical model compounds, namely albumin, asparagine, glucose, glutamine, soya protein, starch, and sunflower oil, at 350 °C in 1M Na_2CO_3, 1M HCOOH, and water separately. The bio-crude consisted of a maximum fraction of 55–80 wt % lipids followed by 11–18 wt % protein and 6–15 wt % carbohydrates. The bio-oil resulted in 4–25 wt % higher yield with respect to the lipid content of microalgae.

Jena et al. (*47*) compared the effects of $Ca_3(PO_4)_2$, Na_2CO_3, and NiO catalysts on bio-oil yield by thermochemical liquefaction of microalgae (*S. platensis*). The highest yield of bio-crude oil (i.e., 51.6%) was obtained using Na_2CO_3, which was 29% higher than that of noncatalytic reaction conditions. Barreiro et al. (*48*) reported different hydrothermal conversion processes for microalgae at variable operating conditions. They reported a maximum bio-oil yield of 64% with negligible nitrogen content (0.7 wt %) obtained from the hydrothermal liquefaction of *Botryococcus braunii* with 5% Na_2CO_3 catalyst.

Peng et al. (*49*) demonstrated the effect of alkaline catalysts (e.g., 0–20 wt % KOH, NaOH, K_2CO_3, and Na_2CO_3) on pyrolysis of lignin and reported bio-oil yield of 30% with more than

80% production of phenols in bio-oil. They also proposed that the alkaline additive supported decarboxylation and decarbonylation reactions. Moreover, the extraction of unsaturated alkyl compounds and alkaline hydroxide addition resulted in enhanced deoxygenation of methoxy compounds.

Auta et al. (*50*) investigated the effect of homogeneous catalysts, namely $Ca(OH)_2$, K_2CO_3, and MgO, on pyrolysis product yields of empty fruit bunch. They reported the highest bio-oil yield of 42.6 wt % in the presence of 10 wt % $Ca(OH)_2$ at 600 °C with the 200 mL/min flow rate of N_2. The pretreatment of bio-oil by oxidation accompanied by esterification for upgrading was evaluated in the presence of a heterogeneous catalyst (i.e., Ru/γ-Al_2O_3 (*51*)). The authors reported increased production of carboxylic acids, which resulted in improved esterification reaction with a reduction in alcohol requirement by 20%. Comparison of different homogeneous and heterogeneous catalysts are summarized in Table 2.

Effects of Heterogeneous Catalysts

Gong et al. (*52*) studied bio-oil conversion catalyzed by La/HZSM-5 in a fixed-bed reactor resulting in lighter olefinic compounds. The addition of La into HZSM-5 effectively enhanced the total acid number with a significant percentage of medium acid sites of 35% in the catalyst. Park et al. (*53*) investigated the pyrolysis of *Miscanthus* at 400–500 °C using different catalysts such as Al-MCM-41, Al-MCM-48, HZSM-5, and Meso-MFI. HZSM-5 and Meso-MFI were found to be more efficient catalysts for the production of aromatics because of high acidity. Grilc et al. (*54*) performed hydrodeoxygenation of solvolyzed lignocellulosic biomass at 300 °C and 8 MPa catalyzed by unsupported Mo (molybdenum carbide, dioxide, and disulfide) and WS_2 (tungsten disulfide). Molybdenum showed the highest selectivity and activity with its sulfide form. Lu and Heyden (*55*) examined the density functional theory calculations and microkinetic modeling for the formation of aromatics by hydrodeoxygenation reaction of guaiacol catalyzed by Ru (0001) model surface. In agreement with theoretical and experimental data, phenol was a major product along with the dehydrogenation of methoxy compounds in guaiacol.

Xu et al. (*56*) studied the effects of catalyst mixture of Ru/C + Mo_2C, Ru/C + Pt/γ-Al_2O_3, and Ru/C + Pt/C on bio-oil upgrading in supercritical water at 400 °C for a reaction time of 4 h. The bio-oil yield was in the range of 63.2–77.2% using Ru/C and Mo_2C catalysts. The catalysts also reduced the oxygen (0.1 wt %), nitrogen (1.8 wt %), and sulfur (0.065 wt %) contents in the bio-oil.

Duan et al. (*57*) investigated the upgrading of bio-crude obtained by hydrothermal liquefaction of algae (*C. pyrenoidosa*) using Pt/γ-Al_2O_3 catalyst in supercritical water at 400 °C for 1 h. It was found that catalyst loading did not have much effect on bio-oil yield, whereas a decrease in the density of water and formic-acid loading produced high-quality bio-oil. Duan et al. (*58*) analyzed different thermochemical conversion routes such as hydrothermal liquefaction, alcoholysis, pyrolysis, and hydropyrolysis followed by Ru/C- and Mo_2C-catalyzed bio-oil upgrading. It was found that hydrothermal liquefaction of *Chlorella* at 400 °C under 6 MPa pressure in 4 h with 10 wt % of Ru/C and Mo_2C resulted in the highest bio-oil yield of 51.5%.

Duan et al. (*59*) examined the upgrading of bio-oil by duckweed by a two-stage hydrothermal treatment using 10 wt % of activated carbon from apricot pit, bamboo stem, coal, coconut shell, peach pit, and pine wood at supercritical water conditions of 400 °C, 6 MPa hydrogen pressure for a residence time of 1 h. The bamboo-stem-activated carbon resulted in 76.3 wt % of upgraded bio-oil, with the highest boiling fraction of 90.1% and a heating value of 44.1 MJ/kg. The authors determined the effects of nine zeolite catalysts on the upgrading of bio-oil in supercritical water at 400 °C for 240 min under 6 MPa pressure (*60*). The zeolite catalysts were of different proportions as Hβ and HY with varying composition of Na_2O of 5% and 0.8%, HZSM-5 with variations of SiO_2:Al_2O_3 ratio (i.e., 25:1, 50:1, and 170:1), MCM-41 with varying composition of Si of 50% and 100%, and SAPO-11. The maximum bio-oil yield of 54.5 wt % resulted from MCM-41 (100% Si).

Table 2. Comparison of Different Catalysts for Bio-Oil Yield

Catalyst	*Biomass*	*Bio-Oil Yield*	*Reference*
Na_2CO_3, HCOOH, and Water	Microalgae	25%	Biller et al. (*46*)
Na_2CO_3	Microalgae	51.6%	Jena et al. (*47*)
Na_2CO_3	*B. braunii*	64%	Barreiro et al. (*48*)
KOH, NaOH, K_2CO_3, and Na_2CO_3	Lignin	30%	Peng et al. (*49*)
$Ca(OH)_2$	Empty fruit bunch	42.6%	Auta et al. (*50*)
Ru/C + Mo_2C		77.2%	Xu et al. (*56*)
Ru/C and Mo_2C	Algae	51.5%	Duan et al. (*58*)
Activated Carbon	Duckweed	76.3%	Duan et al. (*59*)
Zeolite Catalysts (MCM-41 with 100% Si)	*C. pyrenoidosa*	54.5%	Duan et al. (*60*)
Ni/nano-Al_2O_3	Hydroxyacetone	98%	Wang et al. (*61*)
Ni/C	Algae	61%	Shakya et al. (*62*)
Iron Oxide	Water hyacinth	38.1%	Yadav et al. (*63*)

Wang et al. (*61*) analyzed the effects of temperature (500–800 °C) via catalytic reforming of acetic acid, phenol, and hydroxyacetone with Ni/nano-Al_2O_3 catalyst. A rise in the temperature increased hydrogen yields and feedstock conversion rates. Because the steam reforming of oxygenated compounds is endothermic at higher temperatures, the larger molecules dissociated into smaller molecules, which react with OH^- and H^+ ions to generate hydrogen. Among the bio-oil model compounds, phenol had the lowest hydrogen yield and conversion rate. The conversion rate and hydrogen yield of acetic acid were 82.9% and 80.4%, respectively, at lower temperatures of 500 °C. As the temperature rose to 800 °C, both the conversion rates and hydrogen yield increased to 98.5% and 87.4%, respectively. In the case of catalytic reforming of phenol, a similar trend was observed for conversion rates and hydrogen yield. Hydroxyacetone conversion increased from 72.8% to 98% as the temperature escalated from 500 °C to 800 °C. To evaluate the effects of the steam/carbon ratio (1–9.2) on the conversion rate and hydrogen yield, the catalytic reforming of acetic acid was performed at 500 °C, weight hour space velocity of 1 h^{-1}, and 1 g of 15% Ni/nano-Al_2O_3 catalyst loading. At the steam/carbon ratios of 1, 2.5, and 9.2, hydrogen yields of 38.3%, 52.3%, and 80.4%, respectively, were obtained. This was mainly because of the higher extent of the water–gas shift reaction at a higher steam/carbon ratio.

Shakya et al. (*62*) investigated the effects of different heterogeneous catalysts, such as Ni-supported on carbon, Ni-supported on ZSM-5, Ru-supported on carbon, Pt-supported on carbon, and ZSM-5 on bio-oil upgrading obtained by hydrothermal liquefaction of algal biomass (*Nannochloropsis* sp.). The maximum processed bio-oil yield of 61% was obtained from Ni/C as a catalyst at 350 °C. However, high-quality upgraded bio-oils were obtained from Ru/C and Pt/C catalysts.

Yadav and Reddy (*63*) investigated hydrothermal liquefaction of aquatic biomass (water hyacinth) at 280 °C along with the formation of nanometal carbon hybrids. They reported that impregnation of iron oxide particles on water hyacinth acted as a nanocatalyst, thus enhancing the yield of bio-oil to 38.1% along with the in situ reduction to iron nanometal carbon hybrids. The resulting bio-oil samples were quantified with 6.6% light oil and 31.5% heavy oil.

Performance of Upgraded Bio-Oil

The performance of upgraded bio-oil with or without blending has been investigated by different authors. The blending can be done with bio-diesel or petroleum diesel. The performance parameters analyzed included heating and calorific values, ignition delay, brake-specific fuel consumption, brake-specific energy consumption, and CO and NO_x emissions. Yuan et al. (*64*) investigated the effects of bio-oil compositions on engine performance and emissions for applications in a direct injection diesel engine. Bio-oil was emulsified with biodiesel/diesel for its application in the direct injection diesel engine. Synthetic bio-oil was prepared by blending acetaldehyde (12 wt %), acetic acid (14 wt %), acetol (8 wt %), furfural (4 wt %), guaiacol (17 wt %), glucose (8 wt %), glyoxal (4 wt %), methanol (5 wt %), and vanillin (8 wt %) with deionized water (20 wt %). It was classified into one of three categories—hydroxy compounds, aldehydes, or sugars—depending on the composition. The blends of the mixture were prepared by adding diesel with 5%, 10%, and 15% proportions to the crude bio-oil on a volumetric basis, and they were emulsified and designated as SD-05, SD-10, and SD-15, respectively. It was found that emulsion fuels increased the brake-specific fuel consumption along with the brake-specific energy consumption and hydroxyl compounds solution in diesel emulsion. This resulted in negligible CO and smoke emissions with clean combustion properties because of the presence of higher hydroxide radical concentration and the least viscosity as compared with diesel. In contrast, the synthetic bio-oil in diesel emulsion and glucose solution in diesel emulsion resulted in a higher NO_x emission. Based on the engine performance and emissions, the authors concluded that bio-oils with a higher concentration of hydroxy radicals were more suited for a diesel engine as compared with bio-oil with a high sugar concentration, as the latter results in higher NO_x emissions.

Krutof and Hawboldt (*8*) suggested applications for upgraded bio-oil derived by blending pyrolysis oil, petroleum fraction, and other biomass-based fuels. This study presented the impact of mixing pyrolysis-derived oil with vegetable oil, biodiesel, emulsion with biodiesel, a blend of petroleum diesel fraction with co-solvents, emulsion with petroleum diesel fraction, a blend of fish oil with petroleum diesel, and a blend of fish oil with No. 2 and No. 6 fuel oils. The calorific value of pyrolysis oil was lower because of higher water content and oxygen content but was enhanced by the mixing of bio-oil with fish oil. The ignition properties of pyrolysis oil also improved by blending with fish oil. A decrease in the water fraction of the mixture with an increment in fish oil fraction was also reported. Moreover, mixing with fish oil resulted in conditioning the pH of the pyrolysis oil, thus reducing the chances of corrosion. Promising results were obtained by adding 10 vol % of methanol as a surfactant to increase the miscibility of pyrolysis oil fraction with fish oil in a 50:50 blend. The researchers suggested applications of bio-oil for furnaces and boilers, diesel engines, turbines, and Stirling engines.

Huang et al. (65) examined the engine performance and fuel properties of bio-oil and biodiesel blended fuel. The authors upgraded the bio-oil by extracting the fraction of bio-oil by solvent extraction and blending with biodiesel. It was designated as EPBB (i.e., extraction phase liquid of bio-oil to biodiesel blends). The fuel properties of EPBB and biodiesel resulted in higher density and lower surface tension, viscosity, and heating value (38.9 MJ/kg) as compared with biodiesel (39.8 MJ/kg). The performance of EPBB and biodiesel on diesel engines was studied. It was concluded that the ignition delay and premixed combustion duration of EPBB were higher, whereas the duration of total combustion was lesser when compared with that of biodiesel.

Conclusions

Biomass-derived bio-oil cannot be used directly because of undesirable fuel properties such as higher contents of water, oxygen, nitrogen, and sulfur, high viscosity, low acidity, low heating value, acidic pH, and thermal instability. These detrimental properties restrict the direct use of bio-oil in various applications, especially for transportation, storage, and use as a drop-in transportation fuel. The heteroatoms from the bio-oil should be removed and the fuel properties enhanced through necessary upgrading technologies. There are different techniques for upgrading bio-oil, such as solvent addition, emulsification, esterification, supercritical fluids, hydrotreating, hydrocracking, zeolite cracking, and steam reforming, that offer many advantages. Of all the processes, hydrotreatment has gained more attention because of superiority in terms of conversion and the quality of the upgraded bio-oil. Operating conditions such as residence time, temperature, catalyst loading, feed/solvent ratio, and several others affect the quality and yield of bio-oil. Concisely, different reactions are controlled at favorable conditions to enhance the quality of bio-oil. Based on experimental studies, an optimum temperature of around 200–350 °C with lower residence time yields high-quality bio-oil. Further, it was found that Ru-based catalysts show maximum activity along with improvement in the quality of bio-oil. It was found that blending of upgraded bio-oil improved the engine performance in terms of combustion and reduction in emissions.

References

1. Saber, M.; Nakhshiniev, B.; Yoshikawa, K. A review of production and upgrading of algal bio-oil. *Renew. Sust. Energy Rev.* **2016**, *58*, 918–930.
2. Nanda, S.; Mohammad, J.; Reddy, S. N.; Kozinski, J. A.; Dalai, A. K. Pathways of lignocellulosic biomass conversion to renewable fuels. *Biomass Conv. Bioref.* **2014a**, *4*, 157–191.
3. Gollakota, A. R. K.; Reddy, M.; Subramanyam, M. D.; Kishore, N. A review on the upgradation techniques of pyrolysis oil. *Renew. Sust. Energy Rev.* **2016**, *58*, 1543–1568.
4. Nanda, S.; Dalai, A. K.; Berruti, F.; Kozinski, J. A. Biochar as an exceptional bioresource for energy, agronomy, carbon sequestration, activated carbon and specialty materials. *Waste Biomass Valor.* **2016**, *7*, 201–235.
5. Okolie, J. A.; Nanda, S.; Dalai, A. K.; Berruti, F.; Kozinski, J. A. A review on subcritical and supercritical water gasification of biogenic, polymeric and petroleum wastes to hydrogen-rich synthesis gas. *Renew. Sust. Energy Rev.* **2020**, *119*, 109546.

6. Grac`a, I.; Lopes, J. M.; Cerqueira, H. S.; Ribeiro, M. F. Bio-oils upgrading for second generation biofuels. *Ind. Eng. Chem. Res.* **2013**, *52*, 275–287.
7. Kabir, G.; Hameed, B. H. Recent progress on catalytic pyrolysis of lignocellulosic biomass to high-grade bio-oil and bio-chemicals. *Renew. Sust. Energy Rev.* **2017**, *70*, 945–967.
8. Krutof, A.; Hawboldt, K. Blends of pyrolysis oil, petroleum, and other bio-based fuels: A review. *Renew. Sust. Energy Rev.* **2016**, *59*, 406–419.
9. Nanda, S.; Mohanty, P.; Kozinski, J. A.; Dalai, A. K. Physico-chemical properties of bio-oils from pyrolysis of lignocellulosic biomass with high and slow heating rate. *Energy Environ. Res.* **2014b**, *4*, 21–32.
10. Liu, K.; Zhao, W.; Guo, T.; Lei, Q.; Guan, Y.; Wang, D.; Cui, M.; Fu, S.; Zhao, J.; Zong, Z.; Wei, X. Emulsification and performance measurement of bio-oil with diesel. *Waste Biomass Valor.* **2019**, DOI: 10.1007/s12649-019-00917-1.
11. Leng, L.; Li, H.; Yuan, X.; Zhou, W.; Huang, H. Bio-oil upgrading by emulsification/microemulsification: A review. *Energy* **2018**, *161*, 214–232.
12. Li, L.; Yan, B.; Li, H.; Yu, S.; Liu, S.; Yu, H.; Ge, X. SO_4^{2-}/ZrO_2 as catalyst for upgrading of pyrolysis oil by esterification. *Fuel* **2018**, *226*, 190–194.
13. Han, Y.; Gholizadeh, M.; Tran, C. C.; Kaliaguin, S.; Li, C. Z.; Olarte, M.; Garcia-Perez, M. Hydrotreatment of pyrolysis bio-oil: A review. *Fuel Process. Technol.* **2019**, *195*, 106140.
14. Tanneru, S. K.; Steele, P. H. Direct hydrocracking of oxidized bio-oil to hydrocarbons. *Fuel* **2015**, *154*, 268–274.
15. Wang, H.; Lin, H.; Feng, P.; Han, X.; Zheng, Y. Integration of catalytic cracking and hydrotreating technology for triglyceride deoxygenation. *Catal. Today* **2017**, *291*, 172–179.
16. Corma, A.; Huber, G. W.; Sauvanaud, L.; O'onnor, P. Processing biomass-derived oxygenates in the oil refinery:Catalytic cracking (FCC) reaction pathways and role of catalyst. *J. Catal.* **2007**, *247*, 307–327.
17. Ibarra, Á.; Hita, I.; Azkoiti, M. J.; Arandes, J. M.; Bilbao, J. Catalytic cracking of raw bio-oil under FCC unit conditions over different zeolite-based catalysts. *J. Ind. Eng. Chem.* **2019**, *78*, 372–382.
18. Baloch, H. A.; Nizamuddin, S.; Siddiqui, M. T. H.; Mubarak, N. M.; Dumbre, D. K.; Srinivasan, M. P.; Griffin, G. J. Sub-supercritical liquefaction of sugarcane bagasse for production of bio-oil and char: Effect of two solvents. *J. Environ. Chem. Eng.* **2018**, *6*, 6589–6601.
19. Reddy, S. N.; Nanda, S.; Dalai, A. K.; Kozinski, J. A. Supercritical water gasification of biomass for hydrogen production. *Int. J. Hydrogen Energy* **2014**, *39*, 6912–6926.
20. Okolie, J. A.; Rana, R.; Nanda, S.; Dalai, A. K.; Kozinski, J. A. Supercritical water gasification of biomass: A state-of-the-art review of process parameters, reaction mechanisms and catalysis. *Sustain. Energy Fuels* **2019**, *3*, 578–598.
21. Omar, S.; Yang, Y.; Wang, J. A review on catalytic & non-catalytic bio-oil upgrading in supercritical fluids. *Front. Chem. Sci. Eng.* **2020**, 1–14, DOI: 10.1007/s11705-020-1933-x.
22. Singh, S.; Kumar, R.; Setiabudi, H. D.; Nanda, S.; Vo, D. V. N. Advanced synthesis strategies of mesoporous SBA-15 supported catalysts for catalytic reforming applications: A state-of-the-art review. *Appl. Catal. A: Gen.* **2018**, *559*, 57–74.

23. Zhang, J.; Chen, W. T.; Zhang, P.; Luo, Z.; Zhang, Y. Hydrothermal liquefaction of *Chlorella pyrenoidosa* in sub- and supercritical ethanol with heterogeneous catalysts. *Bioresour. Technol.* **2013**, *133*, 389–397.
24. Sun, J.; Xu, L.; Dong, G. H.; Nanda, S.; Li, H.; Fang, Z.; Kozinski, J. A.; Dalai, A. K. Subcritical water gasification of lignocellulosic wastes for hydrogen production with Co modified Ni/Al_2O_3 catalyst. *J. Supercrit Fluids* **2020**, *162*, 104863.
25. Okolie, J. A.; Nanda, S.; Dalai, A. K.; Kozinski, J. A. Hydrothermal gasification of soybean straw and flax straw for hydrogen-rich syngas production: Experimental and thermodynamic modeling. *Energy Convers. Manag.* **2020**, *208*, 112545.
26. Xiu, S.; Shahbazi, A. Bio-oil production and upgrading research: A review. *Renew. Sust. Energy Rev.* **2012**, *16*, 4406–4414.
27. Toor, S. S.; Reddy, H.; Deng, S.; Hoffmann, J.; Spangsmark, D.; Madsen, L. B.; Holm-Nielsen, J. B.; Rosendahl, L. A. Hydrothermal liquefaction of *Spirulina* and *Nannochloropsis salina* under subcritical and supercritical water conditions. *Bioresour. Technol.* **2013**, *131*, 413–419.
28. Zhang, L.; Liu, R.; Yin, R.; Mei, Y. Upgrading of bio-oil from biomass fast pyrolysis in China: A review. *Renew. Sust. Energy Rev.* **2013**, *24*, 66–72.
29. Chan, Y. H.; Yusup, S.; Quitain, A. T.; Uemura, Y.; Sasaki, M. Bio-oil production from oil palm biomass via subcritical and supercritical hydrothermal liquefaction. *J. Supercrit. Fluids* **2014**, *95*, 407–412.
30. Xu, X.; Zhang, C.; Zhai, Y.; Liu, Y.; Zhang, R.; Tang, X. Upgrading of bio-oil using supercritical 1-butanol over a Ru/C heterogeneous catalyst: Role of the solvent. *Energy Fuels* **2014**, *28*, 4611–4621.
31. Yin, W.; Venderbosch, R. H.; Bottari, G.; Krawzcyk, K. K.; Barta, K.; Heeres, H. J. Catalytic upgrading of sugar fractions from pyrolysis oils in supercritical mono-alcohols over Cu doped porous metal oxide. *Appl. Catal. B: Environ.* **2015**, *166–167*, 56–65.
32. Chang, S. H. Bio-oil derived from palm empty fruit bunches: Fast pyrolysis, liquefaction and future prospects. *Biomass Bioenergy* **2018**, *119*, 263–276.
33. Patil, P. D.; Dandamudi, K. P. R.; Wang, J.; Deng, Q.; Deng, S. Extraction of bio-oils from algae with supercritical carbon dioxide and co-solvents. *J. Supercrit. Fluids* **2018**, *135*, 60–68.
34. Nanda, S.; Rana, R.; Hunter, H. N.; Fang, Z.; Dalai, A. K.; Kozinski, J. A. Hydrothermal catalytic processing of waste cooking oil for hydrogen-rich syngas production. *Chem. Eng. Sci.* **2019**, *195*, 935–945.
35. Okolie, J. A.; Nanda, S.; Dalai, A. K.; Kozinski, J. A. Chemistry and specialty industrial applications of lignocellulosic biomass. *Waste Biomass Valor.* **2020**, 1–25, DOI: 10.1007/s12649-020-01123-0.
36. Chen, W. T.; Zhang, Y.; Zhang, J.; Yu, G.; Schideman, L. C.; Zhang, P.; Minarick, M. Hydrothermal liquefaction of mixed-culture algal biomass from wastewater treatment system into bio-crude oil. *Bioresour. Technol.* **2014**, *152*, 130–139.
37. Wu, X.; Liang, J.; Wu, Y.; Hu, H.; Huang, S.; Wu, K.; Zhang, B.; Lin, Q.; Zhang, Q.; Wu, K.; Pu, W.; Yang, M.; Wu, Y.; Wu, X.; Liang, J.; Wu, Y.; Hu, H.; Huang, S.; Wu, K. *RSC Adv.* **2017**, *7*, 8944–8951.

38. Zhang, B.; Lin, Q.; Zhang, Q.; Wu, K.; Pu, W.; Yang, M.; Wu, Y. Catalytic hydrothermal liquefaction of *Euglena* sp. microalgae over zeolite catalysts for the production of bio-oil. *RSC Adv.* **2017**, 7, 8944–8951.
39. Duan, P.; Chang, Z.; Xu, Y.; Bai, X.; Wang, F.; Zhang, L. Hydrothermal processing of duckweed: Effect of reaction conditions on product distribution and composition. *Bioresour. Technol.* **2013**, *135*, 710–719.
40. Yang, W.; Li, X.; Liu, S.; Feng, L. Direct hydrothermal liquefaction of undried macroalgae *Enteromorpha prolifera* using acid catalysts. *Energy Convers. Manag.* **2014**, *87*, 938–945.
41. Gong, S. W.; Lu, J.; Wang, H. H.; Liu, L. J.; Zhang, Q. Biodiesel production via esterification of oleic acid catalyzed by picolinic acid modified 12-tungstophosphoric acid. *Appl. Energy* **2014**, *134*, 283–289.
42. Milina, M.; Mitchell, S.; Pérez-Ramírez, J. Prospectives for bio-oil upgrading via esterification over zeolite catalysts. *Catal. Today* **2014**, *235*, 176–183.
43. Yang, T.; Jie, Y.; Li, B.; Kai, X.; Yan, Z.; Li, R. Catalytic hydrodeoxygenation of crude bio-oil over an unsupported bimetallic dispersed catalyst in supercritical ethanol. *Fuel Process. Technol.* **2016**, *148*, 19–27.
44. Remón, J.; Arcelus-Arrillaga, P.; García, L.; Arauzo, J. Production of gaseous and liquid bio-fuels from the upgrading of lignocellulosic bio-oil in sub- and supercritical water: Effect of operating conditions on the process. *Energy Convers. Manag.* **2016**, *119*, 14–36.
45. Jo, H.; Prajitno, H.; Zeb, H.; Kim, J. Upgrading low-boiling-fraction fast pyrolysis bio-oil using supercritical alcohol: Understanding alcohol participation, chemical composition, and energy efficiency. *Energy Convers. Manag.* **2017**, *148*, 197–209.
46. Biller, P.; Ross, A. B. Potential yields and properties of oil from the hydrothermal liquefaction of microalgae with different biochemical content. *Bioresour. Technol.* **2011**, *102*, 215–225.
47. Jena, U.; Das, K. C.; Kastner, J. R. Comparison of the effects of Na_2CO_3, $Ca_3(PO_4)_2$, and NiO catalysts on the thermochemical liquefaction of microalga *Spirulina platensis*. *Appl. Energy* **2012**, *98*, 368–375.
48. Barreiro, D. L.; Prins, W.; Ronsse, F.; Brilman, W. Hydrothermal liquefaction (HTL) of microalgae for biofuel production: State of the art review and future prospects. *Biomass Bioenergy* **2013**, *53*, 113–127.
49. Peng, C.; Zhang, G.; Yue, J.; Xu, G. Pyrolysis of lignin for phenols with alkaline additive. *Fuel Process. Technol.* **2014**, *124*, 212–221.
50. Auta, M.; Ern, L. M.; Hameed, B. H. Fixed-bed catalytic and non-catalytic empty fruit bunch biomass pyrolysis. *J. Anal. Appl. Pyrolysis* **2014**, *107*, 67–72.
51. Tanneru, S. K.; Parapati, D. R.; Steele, P. H. Pretreatment of bio-oil followed by upgrading via esterification to boiler fuel. *Energy* **2014**, 73, 214–220.
52. Gong, F.; Yang, Z.; Hong, C.; Huang, W.; Ning, S.; Zhang, Z.; Xu, Y.; Li, Q. Selective conversion of bio-oil to light olefins: Controlling catalytic cracking for maximum olefins. *Bioresour. Technol.* **2011**, *102*, 9247–9254.
53. Park, H. J.; Park, K. H.; Jeon, J. K.; Kim, J.; Ryoo, R.; Jeong, K. E.; Park, S. H.; Park, Y. K. Production of phenolics and aromatics by pyrolysis of miscanthus. *Fuel* **2012**, 97, 379–384.

54. Grilc, M.; Veryasov, G.; Likozar, B.; Jesih, A.; Levec, J. Hydrodeoxygenation of solvolysed lignocellulosic biomass by unsupported MoS_2, MoO_2, Mo_2C and WS_2 catalysts. *Appl. Catal. B: Environ.* **2015**, *163*, 467–477.
55. Lu, J.; Heyden, A. Theoretical investigation of the reaction mechanism of the hydrodeoxygenation of guaiacol over a Ru(0 0 0 1) model surface. *J. Catal.* **2015**, *321*, 39–50.
56. Xu, Y.; Duan, P.; Wang, B. Catalytic upgrading of pretreated algal oil with a two-component catalyst mixture in supercritical water. *Algal Res.* **2015**, *9*, 186–193.
57. Duan, P.; Bai, X.; Xu, Y.; Zhang, A.; Wang, F.; Zhang, L.; Miao, J. Catalytic upgrading of crude algal oil using platinum/gamma alumina in supercritical water. *Fuel* **2013**, *109*, 225–233.
58. Duan, P.; Wang, B.; Xu, Y. Catalytic hydrothermal upgrading of crude bio-oils produced from different thermo-chemical conversion routes of microalgae. *Bioresour. Technol.* **2015**, *186*, 58–66.
59. Duan, P.; Zhang, C.; Wang, F.; Fu, J.; Lü, X.; Xu, Y.; Shi, X. Activated carbons for the hydrothermal upgrading of crude duckweed bio-oil. *Catal. Today* **2016**, *274*, 73–81.
60. Duan, P.; Xu, Y.; Wang, F.; Wang, B.; Yan, W. Catalytic upgrading of pretreated algal bio-oil over zeolite catalysts in supercritical water. *Biochem. Eng. J.* **2016**, *116*, 105–112.
61. Wang, S.; Cai, Q.; Zhang, F.; Li, X.; Zhang, L.; Luo, Z. Hydrogen production via catalytic reforming of the bio-oil model compounds: Acetic acid, phenol and hydroxyacetone. *Int. J. Hydrogen Energy* **2014**, *39*, 18675–18687.
62. Shakya, R.; Adhikari, S.; Mahadevan, R.; Hassan, E. B.; Dempster, T. A. Catalytic upgrading of bio-oil produced from hydrothermal liquefaction of *Nannochloropsis* sp. *Bioresour. Technol.* **2018**, *252*, 28–36.
63. Yadav, P.; Reddy, S. N. Hydrothermal liquefaction of Fe-impregnated water hyacinth for generation of liquid bio-fuels and nano Fe carbon hybrids. *Bioresour. Technol.* **2020**, *313*, 123691.
64. Yuan, X.; Ding, X.; Leng, L.; Li, H.; Shao, J.; Qian, Y.; Huang, H.; Chen, X.; Zeng, G. Applications of bio-oil-based emulsions in a DI diesel engine: The effects of bio-oil compositions on engine performance and emissions. *Energy* **2018**, *154*, 110–118.
65. Huang, Y.; Li, Y.; Han, X.; Zhang, J.; Luo, K.; Yang, S.; Wang, J. Investigation on fuel properties and engine performance of the extraction phase liquid of bio-oil/biodiesel blends. *Renew. Energy* **2020**, *147*, 1990–2002.

Chapter 4

Microalgae Oil Upgrading over Zeolite-Based Catalysts

Kalpana C. Maheria,* Aayushi Lodhi, and Gunjan Sharma

Department of Chemistry, Sardar Vallabhbhai National Institute of Technology, Ichchhanath, Surat 395 007, Gujarat, India

***Email: kcm@chem.svnit.ac.in. Telephone: +919426386339.**

The world population has depended heavily on fossil fuels and their derivatives. Hence, in the current scenario of rapid economic growth there is a growing trend in demand for energy due to the intensive use of fuels mainly originated from coal, oil, and natural gas. Such intensive use of fossil fuels and their derivatives has led to emissions of greenhouse gases such as methane, carbon dioxide, and nitrous oxide, causing climate change and thereby a global threat to the environment. In this context, the last few decades have seen a growing interest in the utilization of alternative renewable energy sources to mitigate environmental pollution. Researchers have sought to produce renewable alternative and sustainable fuels to meet this increasing energy demand. Biomass is a promising candidate in this regard, as it is a carbon-neutral, abundant, and cheap renewable resource. Bioenergy produced from biomass is an essential substitute for fossil energy and has boosted widespread attention globally. In this context, algae biomass is a promising alternative among several tree- and crop-based feedstock for bioenergy production because of its low cost of cultivation, sustainable nature, and fast growth rate, as well as the presence of high lipid content, carbohydrates, and high photosynthetic efficiency. Most importantly it does not compete with any kind of arable land, as it is suitable for growth in nonarable lands and in fresh, salted, or brackish water. Microalgae are useful in the mitigation of elevated CO_2 levels and treatment of wastewater. Moreover, the upgradation of algal fuel and bioproducts technology from pilot scale to commercial level is possible by overcoming the associated challenges and limitations. In this chapter, a brief introduction is followed by discussions of its significance and sources of microalgae. The significance of algae oil and methods of microalgae oil upgrading are discussed. A thermochemical conversion route (pyrolysis) is briefly discussed, and the importance of heterogeneous catalysts and the effects of catalysts on pyrolysis products are summarized. The emphasis is put mainly on zeolite- and hierarchical zeolite-based catalysts for algae oil upgrading due their extraordinary physicochemical properties such as shape selectivity, high surface area, tunable acidity, bimodal porosity, and thermal stability.

© 2021 American Chemical Society

Introduction

Fossil fuels contribute around 80% of the global energy supply. Moreover, due to rapid industrialization and increasing population, petroleum fuel consumption has increased in recent decades, leading to fast fossil fuel depletion. Intensive use of such nonrenewable energy resources has led to serious issues such as climate change and global warming due to extensive greenhouse gas emissions. Transportation along with rapid industrialization contribute significantly to CO_2 emissions and global warming. The transportation sector contributes significantly to greenhouse gas (GHG) emissions, accounting for 28% of total global primary energy consumption (mostly fossil fuels) and 71% of total crude oil consumption (*1*). Therefore, the demand for sustainable alternative fuel has increased extremely. At present, the world is experiencing a major paradigm shift from fossil fuel biomass utilization for energy production, and biomass usage for energy production has emerged as a potential alternative to address several issues associated with fossil fuel utilization for energy generation (2). The major advantages that are associated with biomass are its nontoxic, biodegradable, and renewable nature, along with its low cost. Biomass is recognized as a carbon-neutral source; the combustion of fuel originated from biomass generates a lesser amount of carbon monoxide, sulfur dioxide, and particulate matter than from coal. The valuable biofuels and chemicals obtained from biomass are associated with a low carbon footprint. Biomass-derived biofuel/biodiesel also competes favorably with fossil–fuel-derived petroleum fuel (3). The subsequent text discusses the significance and advantages of algae-based feedstock, looking at the current aim of this chapter, which is centered on microalgae upgrading via catalytic pyrolysis involving zeolite-based catalysts.

The exploitation of edible oils from either plants or animals can threaten hunger and cause a rise in the price of food. Therefore, algae species emerged as a significant option to reduce injurious particulate emissions and to solve energy problems. Algae can be a potential renewable feedstock for biodiesel based on the belief that oil originally stemmed from algae itself. Algae biomass is a very promising option for the production of several renewable biofuels. These biofuels have the potential to replace nonrenewable petroleum fuels. Carbohydrates, proteins, and natural oils are major components of algae biomass (*4*). Algae has been divided into two subcategories, microalgae and macroalgae, on the basis of their cellular organization. Microalgae have the advantage of consuming less space and can be easily grown even on nonagricultural land. Further, due to the presence of approximately 20,000 to 800,000 species, microalgae possess more diversity than any other crops or terrestrial plants (*5*, *6*). Thorough understanding of the behavior of algae species under varied conditions and judicious selection of specific strains among 40,000 different strains of algae to get the required end product is significant for the success of algal biotechnology. Table 1 summarizes the lipid, protein, and carbohydrate content of various microalgae strains.

Oil and lipids are major constituents of microalgae and they produce more oil (*14*). Besides oil and lipids, microalgae contain fatty acids as membrane components, storage products, and metabolites, and they act as a bioenergy source (*15*). When algal oil is mixed with either acid or base and alcohol, it generates fatty acid methyl esters (i.e., so-called biodiesel) via transesterification reaction (*16*). The kind of feedstock utilized for bio-oil production directs the composition of bio-oil. Microalgae are basically the eukaryotic or prokaryotic photosynthetic microorganisms that can live in harsh conditions and can grow rapidly due to their unicellular or simple multicellular structure, whereas macroalgae are eukaryotic macrophytes that exist both in the freshwater and marine ecosystem and are also photosynthetically inclined (*17*). The distinction of microalgae and macroalgae into prokaryotes and eukaryotes is based mainly on organelles present in it (*18*). Some examples of prokaryotic and eukaryotic microorganisms are cyanobacteria (*Cyanophyceae*) and green algae (*Chlorophyta*), respectively (*19*) (Figure 1).

Table 1. Lipid, Protein, and Carbohydrate Content of Various Microalgae Strains

Microalgae strains	*[a]Lipid content (%)*	*[a]Protein content (%)*	*[a]Carbohydrate content (%)*	*References*
Spirogyra	11–21	6–20	33–64	(*7*)
Chlorella sp.	14–22	51–58	12–17	(*8*)
Nannochloropsis sp.	12–53	40.0	19.20	(*8*)
Dunaliella sp.	6–8	49–57	4–32	(*7*)
Ankistrodesmus falcatus	24–31	47–70	5	(*9*)
Spirulina maxima	6–7	60–71	13–16	(*7*)
Pavlova	9–14	24–29	6–9	(*11*)
Scendesmus	35.66	7.15	50.40	(*12*)
Anabaena sp.	4–7	48	25–30	(*11*)
Euglena sp.	14–20	39–61	14–18	(*9, 11*)
Pavlovalutheri	9–14	24–29	6–9	(*13, 11*)
Porphyridium sp.	28–39	28–39	50–57	(*11*)
Tetraselmis	16–45	52	15	(*7*)

[a] % in dry weight biomass

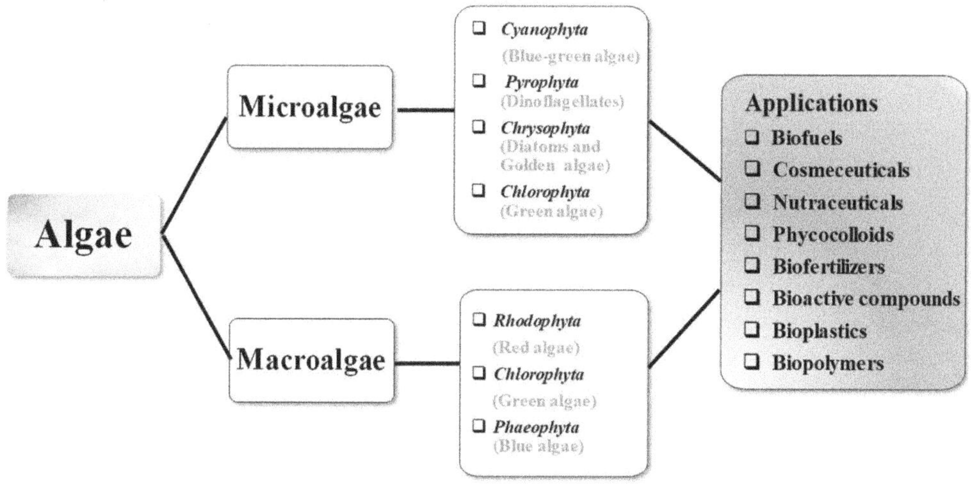

Figure 1. Classification and applications of algae.

Microalgae have proven to be the best class of algae for the production of biodiesel. Utilizing microalgae as a source of biodiesel has various advantages over other biomass sources:

- Microalgae are cost beneficial, as the cost associated with the harvesting and transportation is much lower compared to those of other biomass sources such as trees and crops (*20*).
- Microalgae do not compete with any kind of arable land, as it is suitable to grow in nonarable lands or in fresh, salted, or brackish water (*21*).
- Microalgae normally double their biomass in 24 hr, but exponential growth rates in a period as short as 3.5 hr can result in a two-fold increase in their growth (*22*).
- Effective CO_2 remediation, rapid growth, and a high content of lipids and carbohydrates without lignin are some of the key advantages of microalgae. Further, coproducts of microalgae such as carbohydrates, biopolymer, and proteins can be used as the source of feed or fertilizer (*23*).
- Microalgae are highly photosynthetic efficient, as they utilize only solar energy for their growth and do not require any fertilizer, pesticide, or herbicide (*24*).

Also, microalgae significantly contribute toward a variety of industrial chemical scaffolds, intermediates, and value-added chemical products for items such as cosmetics, biofertilizers, food, nutraceuticals, and bioplastics (*17*) (Figure 1). A characteristic of marine algae is the abundance of sulfated polysaccharides in their cell walls. Several commercial products are derived from algae such as iodine, algin, and mannitol. The lignin-related fraction of algae was suggested as a significant ingredient for making specialty plastics, adhesives, pharmaceuticals, or pesticides (*25*). Macroalgae are a prime source for phycocolloids like alginates and agars. These polymers are located either within the cells or in cell walls, and comprise major industrial products made from algae (*26, 27*). The significance of algae biomass can be realized from its wide applications in diversified areas as depicted in Figure 1.

Due to global energy needs, to furnish adequate renewable feedstock for the growing production of biodiesel there is a need to pay attention to the utilization of algal biomass as a bioenergy source (*28*). Owing to a rich source of oil, easy adaptability to growth conditions, feasibility for growing in either marine water or fresh water, and suitability for growing in nonarable land, algae biomass can become the best option as a renewable feedstock for biodiesel (*29*). Moreover, biodiesel obtained from algae biomass offers several benefits such as no sulfur content, nontoxic nature, high biodegradability, and the potential for harvesting on a daily basis (Figure 2).

Moreover, the presence of a higher level of polysaturates in algal biodiesel enables its use in cold climate conditions. As estimated, 0.53 billion m^3 of biodiesel would be required to replace the current U.S. transportation consumption of all petroleum fuels (*30*). Algae oil feedstock seems to have a very high potential to meet this biodiesel demand compared to waste oil and any seed–oil-based feedstock. As far as physicochemical properties of algae bio-oil and algae biodiesel are concerned, the viscosity and density of algal biodiesel production from algae biomass are low compared to algae oil obtained from algae biomass due to conversion of free fatty acids present in oil into fatty acid methyl esters through transesterification (*31*). It is noteworthy to mention that the oil content of some microalgae exceeds 80% of dry wt. of algae biomass and is responsible for the higher density and viscosity of algae oil (*32*). The comparison of properties of algae oil and algae biodiesel is given in Table 2.

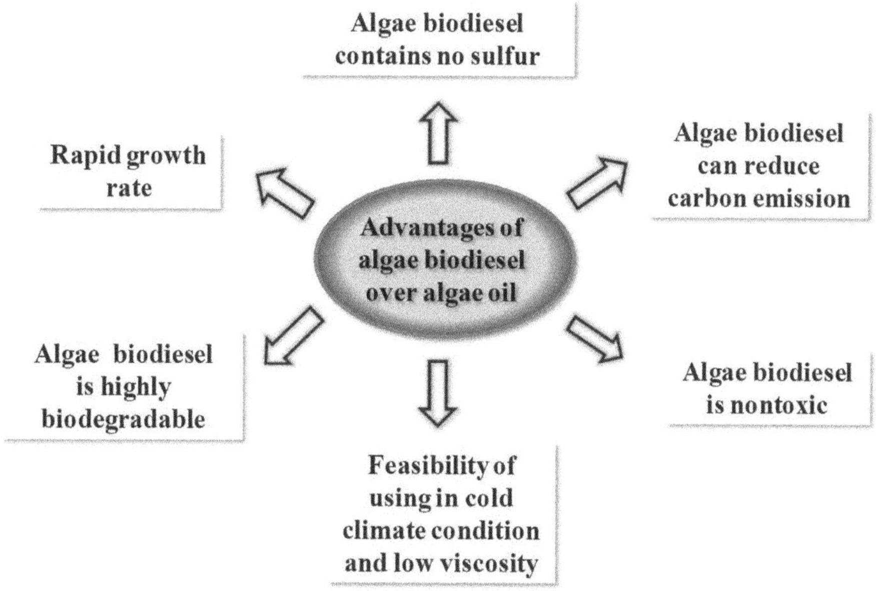

Figure 2. Advantages of algae biodiesel over algae oil.

Table 2. Comparison of Physicochemical Properties of Algal Oil and Algal Biodiesel (*31, 33, 34*)

Sr. no.	*Properties*	*Unit*	*Microalgal oil*	*Algal biodiesel*
1	Density at 15°C	Kg m^{-3}	**919**	864
2	Flash point	°C	65–115	115
3	Heating value	MJ Kg^{-1}	41	41
4	Cetane number	–	51	–
5	Kinematics viscosity at 40°C	mm^2s^{-1}	**33.06**	5.2
6	Saponification value	mg KOH g^{-1}	**152**	–
7	Peroxide value	Meg kg^{-1}	1.00 ± 0.007	–
8	Oxidation stability, 110°C	Hours	**8.83**	–
9	Iodine value	g I_2.100 g^{-1}	**119.1**	–
10	Acid value	mg KOH g^{-1}	**0.13**	–
11	CFPP	°C	18	–
12	Solidifying point	°C	–12	–

For large-scale commercial success of algal technology and better sustainability, feasibility, and economic viability of algal biotechnology, it is essential to develop successful culture technologies. The success of culture technology is associated with which kind of cultivation system and culture medium (sea water, marine water, or fresh water) is used. The types of microalgae cultivation are discussed in the next section.

Cultivation of Microalgae

Microalgae show the remarkable ability to grow on land and water, as well as being able to survive even in the harshest environmental conditions. Microalgae can be grown with the least amount of nutrients and water compared to land biomass sources (*35*). For instance, 333 liters of water is required to produce 1 kg of algal biomass whereas the same amount of biomass from other land plants requires 2,204 liters of water (*36, 37*). Another merit of microalgae is that they can utilize the excess nutrients of industrial wastewater and can grow on it. The growth characteristics and constitution of microalgae depend on the conditions followed during cultivation. However, the oil production is less in this case. The lipid and other content of produced oil largely depend on the type of cultivation used. The chemical composition of microalgae species obtained from different cultivation determines their end-use applications as each species has different properties and morphology (*38*). Microalgae have four major types of cultivation conditions:

1. ***Phototrophic cultivation*:** In this type of cultivation, microalgae utilize light energy sources as the major source of energy to convert it into chemical energy. This cultivation condition is most commonly used for microalgae growth (*21*).
2. ***Heterotrophic cultivation*:** Some microalgae can use the light energy source, but they also utilize organic carbon under dark conditions. This type of cultivation is known as heterotrophic cultivation, where the microalgae use organic carbon as both the energy and carbon source (*38*).
3. ***Mixotrophic cultivation*:** In this type of cultivation, the microalgae utilize both inorganic carbon and organic compounds as a carbon source. Hence, this type of microalgae can live heterotrophically, phototrophically, or both (*17*).
4. ***Photoheterotrophic cultivation*:** In this type of cultivation, microalgae require light to utilize organic compounds as a carbon source (*38*).

Thus, mixotrophic and phototrophic cultivation can be distinguished only on the basis of the kind of energy source they utilize. In photoheterotrophic cultivation, light is required as an energy source whereas, in the case of mixotrophic cultivation, microalgae utilize organic compounds to fulfill the purpose (*39*).

Haematococcus pluvialis, *Chlorella vulgaris*, and *Arthrospira (Spirulina) platensis* are examples of strains that can be grown under different conditions such as photoautotrophic, heterotrophic, and mixotrophic. Other examples of strains that can grow under photoautotrophic, heterotrophic, or photoheterotrophic conditions are *Scendesmus acutus* and *Selenastrum capricornutum* (*39*).

It is reported that heterotrophic growth could furnish better oil production than other cultivation conditions (*39*). However, heterotrophic growth has issues with large-scale production. The use of phototrophic cultivation is more frequent due to easy scale-up and could capture CO_2 from flue gases of industries and convert it into oil (*39*). In the growth of microalgae, not only do carbon source, vitamins, and other nutrients play a vital role, but also growth depends on the equilibrium between oxygen, carbon dioxide, pH, light intensity, temperature, and other operational parameters (*40*). Microalgae biomass is either disintegrated or converted into a variety of biofuels as discussed earlier in the chapter, via different thermochemical and biochemical conversion routes (Figure 3). These routes include hydrochemical liquefaction, pyrolysis, anaerobic digestion, alcoholic fermentation, acid/alkali/enzymatic hydrolysis, acid/alkali/enzyme-catalyzed transesterification, and photobiological (*41*).

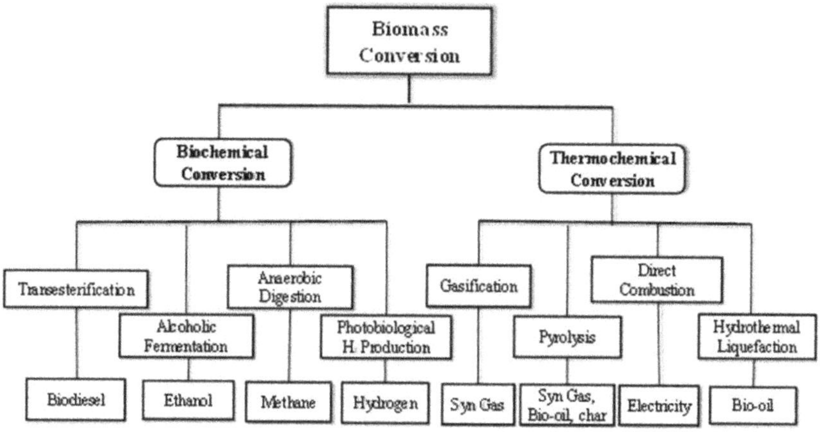

Figure 3. Biomass conversion routes.

In the thermochemical process, biomass is heated and disintegrated into solid, liquid, and gas bio-fuels in the presence of either air or oxygen (*42*). Thermochemical routes are favored over biochemical ones due to the high cost associated with the production, low conversion efficiency, and lengthy reaction steps that are involved in biochemical conversion processes (*43*). Among all thermochemical routes established for microalgae conversion to biofuels, pyrolysis is favored as it quickly produces bio-oil with high fuel-to-feed ratio in a simpler manner (*43*). Catalytic pyrolysis is extensively studied in conventional biomass conversion (*44*). However, not much attention has been paid so far to the microalgae biomass pyrolysis thermochemical route, although there has been a rapid increase of publications on the synthesis of many value-added chemicals through pyrolysis using microalgae as sustainable feedstock. In general, the use of catalyst facilitates deoxygenation of pyrolytic vapors to afford more aromatic hydrocarbons like benzene, toluene, xylene, and phenol (*45, 46*). In this chapter, the emphasis has been put on catalytic upgrading of algae bio-oil via pyrolysis using zeolite-based catalysts.

Harvesting Microalgae

In the past decade, a wide variety of microalgae has drawn much attention of researchers due to the microalgae's commercial applications. Harvesting biomass of microalgae requisite is a productive solid–liquid separation technique since microalgae are cultivated in liquid medium as a suspension at a concentration of around 0.5 g L^{-1} or 0.05 % (*47, 48*). A broad range of solid–liquid segregating methods are available. Coagulation, flocculation, centrifugation, and filtration were found to be effective among diverse harvesting techniques (*47*). To enhance harvesting efficiency, these methods can be used separately or in combination.

Microalgae are part of a heterogeneous group of microorganisms composed of eukaryotic members from different lines of evolution. As a consequence, the scale, shape, cell surface properties, and cell rigidity vary widely (*49*). Furthermore, microalgae excrete organic matter; the volume of excreted matter and its characteristics vary between species in the culture medium (*49, 50*). Some harvesting techniques are explained in the following text. It is necessary to note that harvesting should not lead to contamination (e.g., due to the addition of chemical flocculants) or harm to biomass. The quantity of contamination or harm that is reasonable depends on the final use of the biomass, so the harvesting choice would depend on its specific use.

1. ***Centrifugation*:** Centrifugation is the most commonly used technique for harvesting microalgae. For suspension with a low algal dry weight (solid) content (0.01%–20%) a disc bowl centrifuge is used, whereas for higher algal dry weight solid content (10%–50%), a decanter centrifuge is required for suspension (*51*, *52*). The centrifugation process is time-consuming, and it is expensive to process a large volume of culture using centrifugation (*52*).
2. ***Flocculation*:** A number of individual microalgal cells are accumulated during flocculation into larger particles called flocs. These flocs can be isolated from the culture broth more quickly than the individual cells; this is the case for separation based on gravity and filtration (*53*, *54*).
3. ***Filtration*:** Micro- and ultrafiltration membrane-type filtrations are the most frequently used techniques for microalgae filtration. A filtration unit is primarily governed by the fluxes that can be achieved and the quantity of biomass of the algae that can be retained. Fluxes are defined as permeate volume per unit time per unit membrane surface ($Lh^{-1}m^{-2}$) and are regulated by algae morphology, cell integrity, algae cell density, speed of cross-flow, pore size, transmembrane strain, form of membrane, and medium composition (*54*, *55*). Generally, microfiltration allows higher fluxes in the short term, whereas ultrafiltration has been observed to achieve better in the longer term because of higher fouling resistance (*56*).
4. ***Flotation*:** This is a harvesting technique that is based on gravity separation (*57*). In this process, tiny air bubbles communicate with microalgal cells in flotation and take these cells to the surface of the water where they produce a scum that can be skimmed off. Flotation is based on the attraction between microalgal cells and air bubbles (*58*). Due to the presence of negative charge on both microalgal cells and an air bubble, they do not interact with each other, and flotation usually does not work well for microalgae harvesting unless additives are used (*59*).

When biomass is processed under high temperature in the absence of oxygen, products are produced in three phases: the vapor phase, the liquid phase, and the solid phase. The liquid phase is a complex mixture called bio-oils. They are liquid or gaseous fuels made from either agricultural biomass materials, municipal wastes, or forestry by-products. As mentioned earlier, microalgae are a rich source of oil. Algal oil is one of the potential biofuels produced from algae biomass. The per-unit area yield of oil from algae is estimated to be between 18,927 and 75,708 L/acre/year; this is 7 to 31 times greater than the next best crop, palm oil, at 2,404 L/acre/year. For utilization of biomass in remote locations, it is more economical to convert the biomass into bio-oil and then transport the bio-oil. Bio-oil can be used in vehicle engines, either totally or partially in a blend (*30*). Next we cover synthesis pathways of algae oil production from algae biomass.

Algae Oil Synthesis Pathways

The techniques that are used to obtain oil from microalgae are classified into two methods, chemical and mechanical (Figure 4). Soxhlet extraction, supercritical fluid extraction, and rapid solvent extraction are the chemical methods of lipid extraction; mechanical methods include oil expeller, microwave-assisted extraction, and ultrasound-assisted extraction (UAE) (*60*).

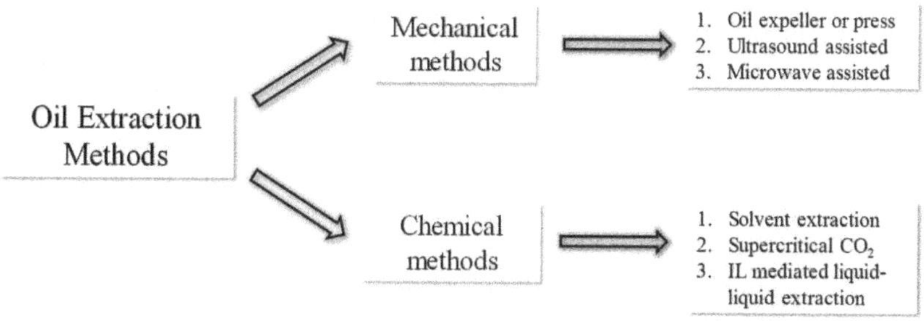

Figure 4. Methods for the oil extraction from microalgae.

The commonly used methods for the extraction of oil from biomass are briefly described in the subsequent text.

Cell Disruption Methods (Mechanical)

The first approach is the mechanical cell disruption of algal cells to extract oil without contaminating other chemicals. Mechanical pressing or French pressing is similar to the extraction of oil from seeds or nuts by mechanical pressing. It requires high-pressure pressurization of algal biomass wherein the cell walls are broken to release the oil. This process of cell disruption depends on several process variables such as residence time, concentration of cells, volume of the chamber, volume of beads, and rotation number (*60*, *61*).

Solvent Extraction (Chemical)

A modern method used to extract oils from the algal biomass is solvent extraction. Solvents should be chosen while keeping in mind the effectiveness, selectivity for different lipid classes, and solvents' ability to avoid potential lipid degradation. Lipids and other algae cell organelles are linked with each other through either H-bonding or covalent bonding or held together by weak van der Waal's interactions. These linkages need to be broken for complete extraction (*62*). Bligh and Dyer, Folch, and Soxhlet extraction are the traditional solvent extraction techniques. When the algal cells interact with the organic solvents, these solvents diffuse through the cell wall and interact with the selective class of lipids depending on their dielectric constant to form a solvent-lipid complex (*60*). Due to the concentration gradient, this complex diffuses into the bulk solvent and continues until this phase reaches equilibrium (*27*).

Supercritical CO_2 Extraction (Chemical)

As an extraction solvent, the supercritical fluid is considered to be ideal since the solvent strength of supercritical fluids depends on their densities, which can be changed by varying the temperature and pressure of the extraction process. The basic concept is that when the critical pressure and critical temperature are crossed by the fluids, they possess the property of both gases and liquids. This specific state of the fluid is known as a supercritical state of the fluid, which shows the solvent characteristic and mass transfer characteristic of liquid and gas, respectively (*63*). Due to lower critical temperature (31.1°C) and lower critical pressure (72.9 atm) of carbon dioxide, supercritical CO_2 is preferred for the extraction process (*64*). The supercritical extraction process is used worldwide in various food, pharmaceutical, and nutraceutical industries because of lower toxicity, high selectivity, high purity, and chemical inertness of the extracted compound (*61*).

Ultrasound-Assisted Extraction

UAE is a mechanical oil extraction method that eliminates the problems of traditional mechanical destruction. The presence of a thick cell wall inhibits the release of intracellular lipids in the case of microalgae (*65*). UAE methods involve generation of sound waves that disperse in media and causes cell structure disruptions. UAE makes use of acoustic waves (in KHz range) which generates cavitation bubbles while travelling through the solvent. Due to the shockwaves generated by bursting of these cavitation bubbles the cell structure disruptions occurs. This leads to enhanced mass transfer of organic compounds (from plants) across the cell membranes into solution. The busting of cavitation bubbles causes pressure and temperature changes. As a result, the shear strength across the cell is generated, which contributes to cell structure disturbances (*66*). For high lipid recovery, the use of polar and nonpolar solvents in UAE has been recorded, in which the researcher has found improved amplitude of time and the mass transfer rate. UAE is more commonly used than mechanical extraction, as the UAE method is employed at low temperatures with reduced thermal denaturation of biological molecules (*67*). UAE approach has key advantages: product purity, short running time, and low cost.

Microwave-Assisted Extraction

Microwave-assisted extraction methods have been used for processing crop, food, fodder, and soil of lipids and pesticides. This method is useful in extraction because it is reasonably safe, easy, and economical to use on microalgal crops. The frictional heat gradually destroys the algae cell wall for the release of intracellular lipid. The microwaves combined with ultrasonic therapy are found to improve the overall *C. vulgaris* lipid yield (*68*). High-temperature extraction for a longer period led to better efficiencies in processing oil than Soxhlet extraction. The approach of the microwave-assisted method is currently being employed as a cost-effective tool for extracting wet lipids from microalgae in view of benefits of fast reaction times and high-quality lipids. However, maintenance costs are considered a limiting factor in commercial scaling up (*69*).

Ionic Liquid Mediated Liquid–Liquid Extractions

Ionic liquid (IL)-based liquid–liquid extraction (LLE) approaches have been studied as a follow-up step in the purification and separation of the target biocompounds. Since most biomass extracts are water rich, water-immiscible hydrophobic ILs designated as IL- liquid liquid extraction (IL-LLE) are initially applied (*70*). The IL-based aqueous biphasic system is another approach, which involves the addition of a salting agent to the water-miscible ILs. In certain cases, IL-based LLE may facilitate the combination of extraction, cleaning, and concentration within a single stage, although it has not been thoroughly investigated (*71*).

Pyrolysis of Algae Oil

Pyrolysis is the conversion of algae biomass into algae oil, biochar, and pyro-gas (72). It occurs in the absence of air and involves high temperatures (300–700°C). Product yield depends on the temperature of the pyrolysis, heating rate, and time of residence. There are three categories of pyrolysis based on these operating parameters: slow, fast, and flash. Saber and coworkers have mentioned the reaction parameters and expected product percentage yields for pyrolysis processes, depicted in Figure 5 (*73*). Through pyrolysis, bio-oil, biochar, and gaseous products are produced from biomass (*74*). Slow pyrolysis leads to a larger amount of char, whereas fast pyrolysis yields more liquid phase compared to slow pyrolysis. As far as bio-oil is concerned, flash pyrolysis offers the best results in terms of high percentage yields of bio-oil, up to 80 wt.% by dry biomass feedstock (*75*).

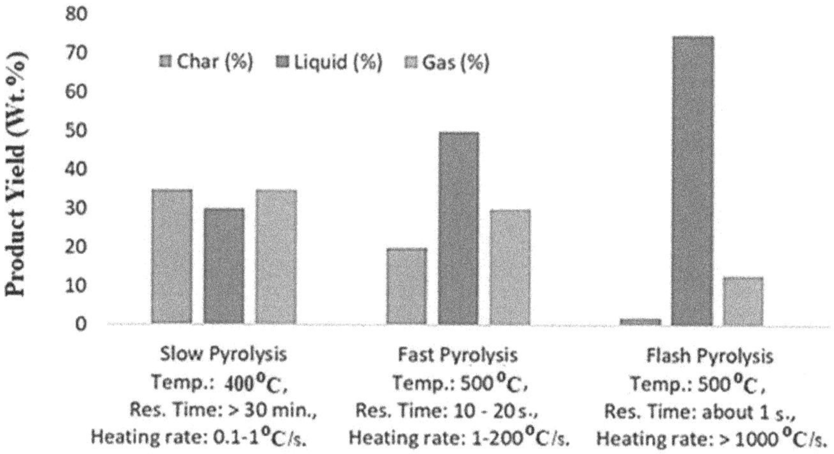

Figure 5. Reaction parameters and expected product % yields for pyrolysis processes. Data from reference (73). Copyright 2016 Elsevier Ltd.

In general, the bio-oil obtained through pyrolysis possesses higher oxygen content and thus lower heating value in comparison with liquefaction oils (*76, 77*). Pyrolysis needs dry feedstock, which requires more energy to dry the biomass. However, pyrolysis conversion technologies are commercially viable in view of its low pressure operating conditions of 1–5 atm compared to the liquefaction process (*78*).

Many typical processes of slow and quick pyrolysis use fixed and fluidized bed reactors, which are heated by hot surfaces, sands, and so on. Microwave pyrolysis has recently been used as an effective method for biomass conversion. In microwave technology, the electromagnetic waves move through the material and cause its molecules to oscillate, which produces heat (*79*). Microwave heating can be applied to large biomass particles as it is uniform. It is a mature technology and is easy to scale up. Bioenergy such as pyro-gas produced with microwave technology possesses high heating value since the carrier gas used for fluidized bed pyrolysis is not diluted (*80*). The products are cleaner because no agitation or fluidization is involved in microwave heating (*79*).

Several simultaneous chemical reactions are involved in biomass pyrolysis such as depolymerization, repolymerization, dehydration, fragmentation, rearrangement reactions, and others (*81, 82*). Crude bio-oil is one of the end products of these reactions that find use as a liquid transportation fuel. However, such crude bio-oil carries around 300 individual organic compounds and several large complex oxygenates that cause adverse effects to the properties of bio-oil such as high viscosity, instability, high corrosively, and high heating value (*82*).

These negative properties make such bio-oil unfavorable as transportation liquid fuel/advanced biofuels; hence, bio-oil must be upgraded before it can be used to fuel engines. Upgrading crude bio-oil is nothing but the conversion of bio-oil into hydrocarbons or other intermediates or the direct use of crude bio-oil in a petroleum refinery to produce green gasoline, diesel, and several significant industrial chemicals (*81*). Figure 6 shows the general processes that are involved in the upgrading of bio-oil through pyrolysis. Oxygenated compounds are degraded along with the simultaneous formation of water, CO, and CO_2 during the upgrading process, leading to alteration of chemical structures and reduction of molecular weight, thereby decreasing the viscosity of crude bio-oil (*83*).

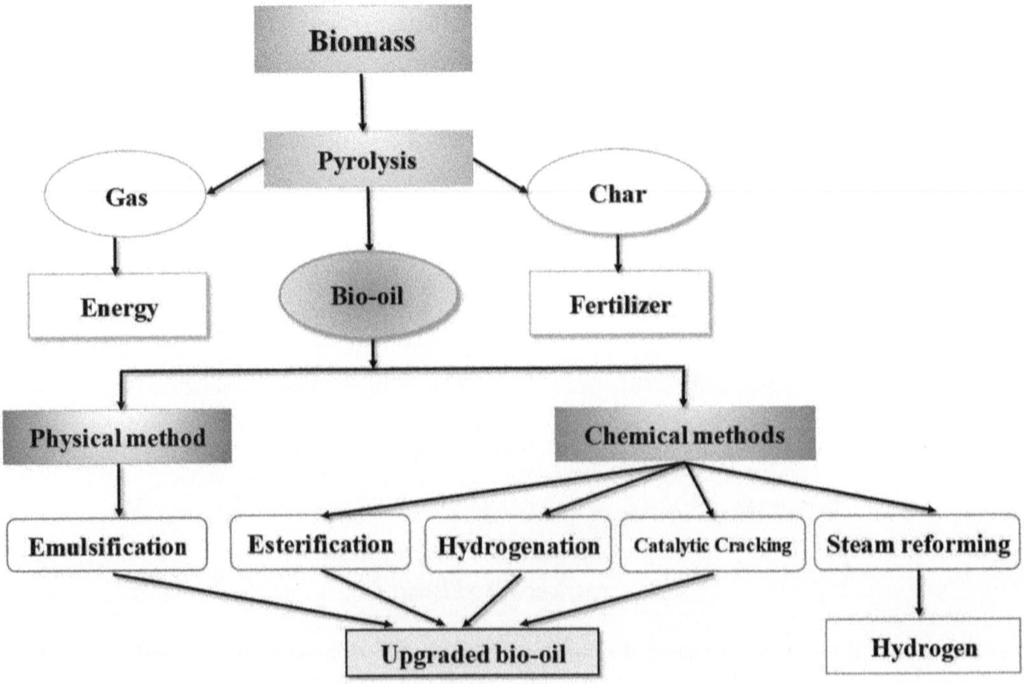

Figure 6. Upgrading of bio-oil via pyrolysis.

These upgrading procedures may involve either single or multiple organic simultaneous reactions such as hydrocracking, catalytic cracking, hydrodeoxygenation (HDO), decarbonylation, decarboxylation, hydrogenation, and so on (Figure 7) (*84*). As a consequence, several key intermediates and products are simultaneously produced during this bio-oil upgrading.

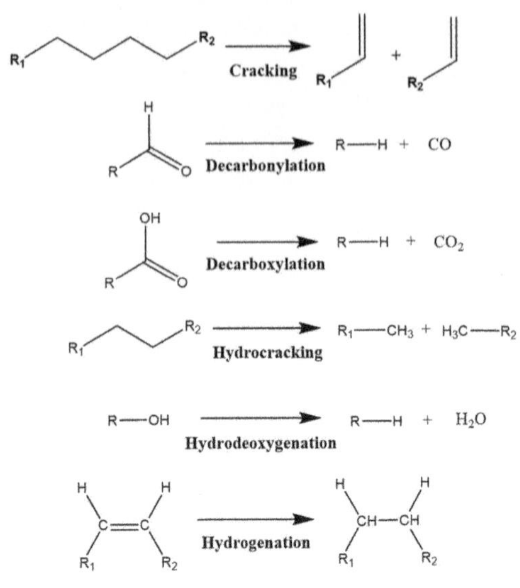

Figure 7. Reactions involved in pyrolysis.

Catalyst cracking and HDO are promising key processes for crude bio-oil valorization (*81*). These processes are capable of significantly reducing oxygenates and produce appreciable yields of hydrocarbons (*85*). Catalytic cracking seems economical as it involves operation under atmospheric pressure without the utilization of hydrogen (*86*). During bio-oil upgrading, several reactions occur between the products of individual components in the presence of alkali and alkaline earth metals (*87*). These metals are acting as catalysts and cause alteration of product distribution significantly and create complex reaction networks. This will lead to difficulty in product separation and purification. Such reactions and impurities in bio-oil deactivate the employed catalyst (*87*). Catalyst deactivation is a major issue in bio-oil upgrading. The efficiency of bio-oil upgrading and the processes is significantly driven by catalyst activity, selectivity, and energy efficiency. The pyrolytic pathway can be modified with the help of a suitable catalyst to target selected products with improved product quality.

As far as pyrolysis of biomass is concerned, more attention is paid to catalytic pyrolysis of conventional biomass rather than microalgae. However, for the last three years, there has been a rapid rise in publications addressing catalytic pyrolysis of microalgae. Lee and coworkers (*88*) have reviewed recent achievements in the field of catalytic pyrolysis of microalgae and reported on current challenges and potential fields. Several solid acid catalysts such as zeolites, metal oxides, supported metal oxides [TiO_2 supported metal (Ce, Ni, and Co) oxide catalysts, Mg-Ce/Al_2O_3], and a combination of zeolites and metal oxides have been employed for catalytic pyrolysis of microalgae (*89, 90*). The type of metal/support used, the kind of metal/support interaction involved, and the extent of metal dispersion affects the type of products and properties [stability, higher heating value (HHV), bio-oil yield, oxygenate, viscosity, amount of aromatic and aliphatic compound, etc.] of pyrolysis products.

Among several solid acid catalysts, zeolites are well established as heterogeneous catalysts in several industrial processes owing to their high surface area, high hydrothermal stability, tuneable acidity, and well-defined uniform microporosity (*87*). Researchers have reviewed catalytic pyrolysis of a variety of microalgae over different zeolites and nonzeolite catalysts (*88*). The literature revealed that the product distribution in pyrolysis is significantly affected by pore size, shape selectivity, number, and strength of active acid centers (*91, 92*). In this context, the next section includes a detailed discussion on the use and future scope of zeolite- and mesozeolite-based catalysts in the area of catalytic pyrolysis of microalgae in view of their significance and potential in this area.

Zeolites

In 1765, a Swedish mineralogist, Axel Fredrik Cronstedt, discovered the mineral stilbite and described zeolite as "boiling stones," which originates from the Greek words *zeo* (to boil) and *lithos* (stone) (*93*). Zeolites are inorganic microporous anhydrous crystalline aluminosilicate materials having very well-defined regular pore structures of molecular dimensions (*94*). The framework structure of zeolites reveals regular arrays of channel systems and cavities. Silicon (Si) and aluminum (Al), the so-called T atoms, contribute significantly to the rigid framework of zeolites. The general formula of a zeolite may be written as $M_{x/n}$ [(Al_x Si_y O_2 (x+y))] p H_2O, where M = Na^+, K^+, or Li^+; or M = Ca^{2+}, Mg^{2+}, Ba^{2+}, or Sr^{2+}; *n* is cation charge; y/x = 1–6; and p/x = 1–4 (*95*). An isomorphous substitution of Si^{4+} by Al^{3+} in tetrahedrons leads to the development of a negative charge on the zeolite framework. This causes charge imbalance, which is compensated by monovalent or divalent cations in the structure (*96*). The structures of zeolites, therefore, consist of three-dimensional frameworks of linked $[SiO_4]^{4-}$ and $[AlO_4]^{5-}$ tetrahedral (*97*). In this tetrahedral Si^{4+} or Al^{3+} ions are located at the center of the tetrahedron, and four O^{2-} ions at the vertices (*98*). The schematic of the zeolite framework is depicted in Figure 8. These materials are distinguished in terms of the framework "corner atoms" and have open cavities as channels and cages in their design (*99, 100*).

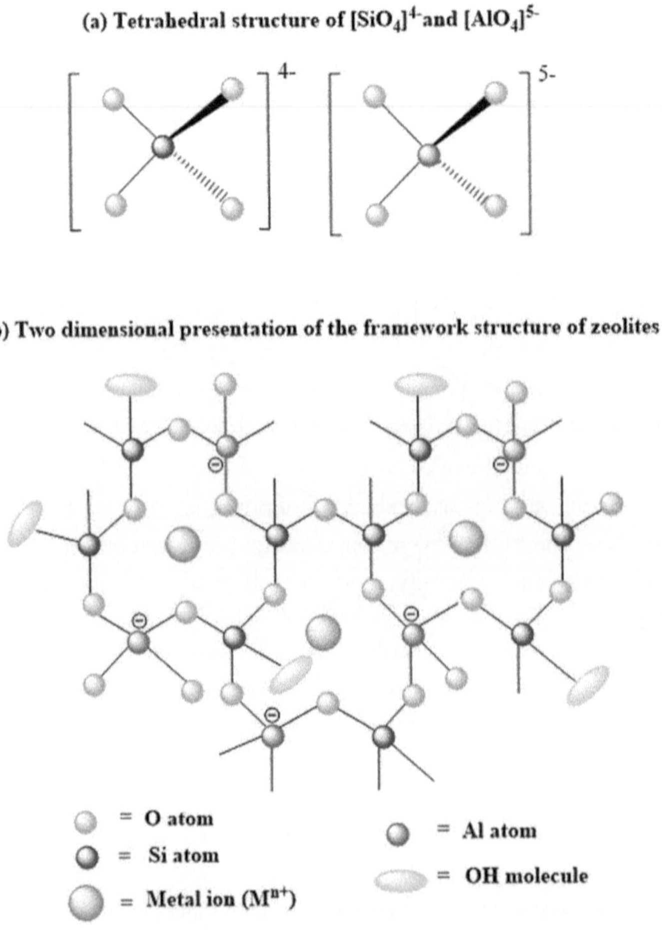

Figure 8. Building blocks and framework of zeolites.

Zeolite possesses an intricate and delicate porous structure (*99*). The catalytic activity of zeolites is caused by the presence of both Lewis and Brønsted acidic sites in their 3D framework. When the Na form of any zeolite is converted into H form or proton form, the zeolite is said to be acidic (*101*), and when this zeolite exhibits the property to act as a proton donor, this zeolite is referred to as Brønsted acidic (*102*). Acidity depends on the number of Al^{3+} present in the framework; zeolite with a low silica/alumina ratio (SAR) is considered to be highly acidic. When zeolite is heated at 823 K, Si-O-Al linkages break, giving rise to a generation of Lewis acid sites and water (*101*).

Different constituent atoms and their subsequent arrangements lead to a plethora of structures (e.g., zeolites, metal organic frameworks, silico alumino phosphates [SAPOs], aluminum phosphate molecular sieves [AlPOs], a few other silicates, etc.) that are formally attributed with a nomenclature by the Structure Commission of the International Zeolite Association (IZA) (*103*). Due to the crystalline nature of zeolites, we can introduce one or more active sites in either framework or extra framework positions. The presence of compensating cations in its structure adds to the properties of ion exchangeability and catalysis (*104*). They are successfully employed as ion exchangers with an intention to soften the water, nutrient release agents in horticulture, and odor control agents in pet litter, and for separation of radioactive isotopes such as ^{137}Cs and ^{90}Sr from polluted water (*105*). Also, they are extensively used as heterogeneous catalysts in various organic transformations involved in oil refinery and pharmaceutical industries for petrochemicals and fine chemical synthesis, respectively (*106*). The cavities in these porous materials can effectively accommodate guest species and embed small molecules, and hence can act as reactors (*107*). They contribute greatly in fast replacing corrosive liquid mineral acids and unstable and less performing nonzeolite catalysts in several refining and platforming processes (*108*). The petrochemical and refining industry depends heavily on zeolites as catalysts (*109*, *110*) for several key chemical processes such as cracking, cyclization, isomerization, dehydrogenation, transalkylation, and others (*109*, *111*, *112*). Various processes such as fluidized crystal cracking (FCC) and hydrocracking are commercialized, making use of zeolites. The significant contribution of zeolites in industrial processes is realized by the fact that 74 out of 180 heterogeneous catalysts employed in 127 industrial processes are zeolite-based catalysts (*113*, *114*). Moreover, the heterogeneous nature of the catalyst adds the benefits of easy separation and reusability. For the most part, the world's gasoline is produced by the FCC process of petroleum using zeolites due to their key properties such as high intrinsic acidity, shape selectivity, high hydrothermal stability, and specific and uniform pore structure of molecular size dimensions (*115*). The use of zeolites as solid acid catalysts for organic reactions began in the early 1960s (*116*). Until now, zeolite-based catalysts were investigated for applications in the petrochemical industries. In 1968, Venuto and Landis provided the first comprehensive overview of zeolite catalysts for organic transformations within and outside the area of fuel synthesis (*117*).

Claire Deville attempted to synthesize zeolites in the laboratory in 1862 (*93*). The most significant milestone in zeolite synthesis was achieved by Milton and Breck in the 1950s at Union Carbide, which was based on reactive gel crystallization (*93*). This was the breakthrough, and even today this synthesis is recognized as the standard synthesis method of zeolites. The history of series of developments in zeolite preparation and landmark industrial processes that were developed over the years are depicted in Figure 9.

Barrer et al. (*118*) have described the growth and crystallization mechanism of zeolite as a bottom-up phenomenon wherein small preformed framework secondary building units (SBUs) condense into the three-dimensional microporous structures. The Structural Commission of the IZA has approved around 248 different structure types of zeolites (*119*). In zeolite structures, $(SiO_4)^{4+}$ and $(AlO_4)^{5+}$ tetrahedral are the primary building units. These units are combined via oxygen linkages with adjacent tetrahedra and generate simple geometric forms called SBUs. There are currently such 23 different SBUs known to exist and all zeolite structures can be visualized by a varied combination of these SBUs (*98*). These SBUs exist either as single rings, double rings, polyhedra, or even more complex units. These SBUs combine together in a different way to make unique cage structures and channel systems (*98*). Extra framework cations and water molecules occupy open cavities in the form of channels and cages. Based on effective pore diameter, zeolites are classified as small-, medium-, large-, and extra-large-pore zeolites. Small-pore zeolites like chabazite and linde–Y involve eight-membered rings (8-MR) and generally contain pores of diameter $d < 5$ Å. Medium-pore zeolites like ZSM-5, ZSM-11, and SAPO-11 involve 10-MR and generally contain pores of diameter $d < 5.0–6.5$ Å. Large-pore zeolites like Beta (BEA), SAPO-5, AlPO-5, and ZSM-22 involve 12-MR and generally contain pores of diameter $d < 6.5–13.0$ Å. Extra-large-pore zeolites like cloverite, ITQ-15, and ITQ-21 involve 14-MR, 18-MR, and 24-MR, and generally contain pores of diameter d = 13.0–20 Å (*120*). As per literature, zeolite channels can permit an access for reactants and products to diffuse with pore diameters ranging from 1.7 to 12.1 Å (*121*, *122*).

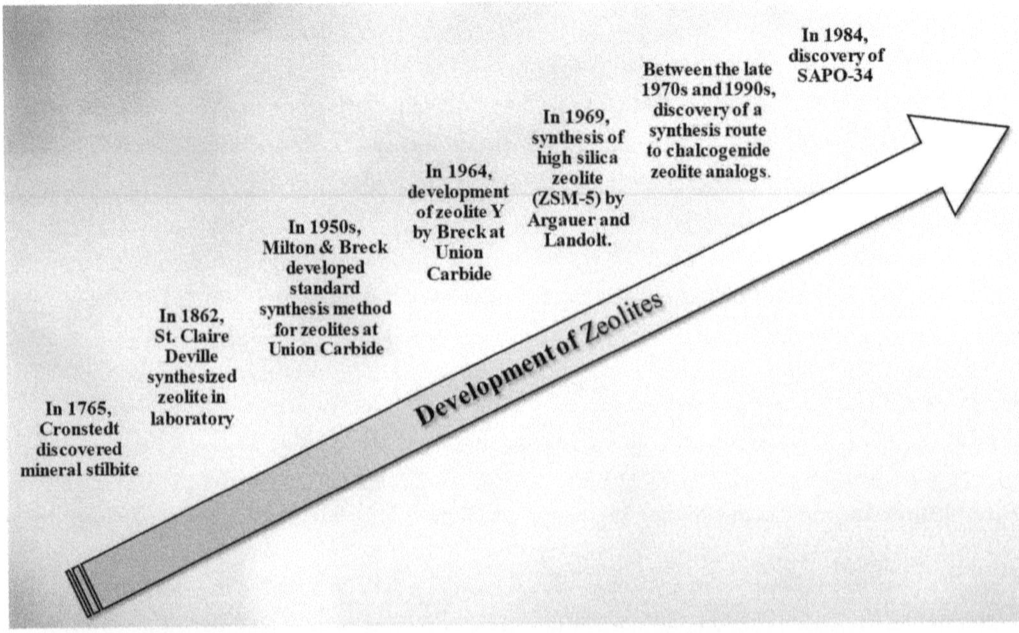

Figure 9. Zeolite synthesis and industrial processes: development over the years.

The zeolite synthesis begins with the first step of crystallization from inhomogeneous gel originated from the alumina and silica sources, combined with water in the presence of NaOH mineralizer (i.e., high pH concentration). In the next step alumina and silica are polycondensed using the sol-gel process. The physicochemical characteristics and properties of the final product depend on several parameters such as SAR in the starting medium, temperature, pH, nucleating agents, water content, ageing period, stirring, and presence of various inorganic and organic cations. The method often involves the use of structure-directing compounds commonly termed templates, which are mostly organics of amines, quaternary ammonium cations, cyclic ethers, and coordination compounds (or the organometallics) (*123*). In the last few decades researchers have shown a growing interest for developing new applications of zeolites for several reasons. With advances in zeolite synthetic routes, new knowledge on zeolite chemistry and emerging new zeolite structures with extraordinary physicochemical properties has created new possibilities for techno-innovation. However, there are several shortcomings pertaining to microporous zeolites such as issues with complex pore and diffusion limitations. In order to overcome such drawbacks, research on the development of zeolites with large and hierarchical pores (mesozeolites) is growing and has grabbed significant attention. More details pertaining to mesozeolites are provided next.

Mesozeolites, Hierarchical Zeolites, Micro-Meso Composite Materials

Porous materials are conveniently classified into three broad classes. These materials can have porosity in the size ranges of 2 nm and lower, 2–50 nm, and 50 nm and higher; respectively known as microporous, mesoporous, and macroporous materials (*124*). Zeolites are microporous materials and have several benefits attributed to their name as mentioned earlier in the text. All existing zeolites are based on very few framework types, mainly mobile five, mordentite (MOR), faujasite, and BEA.

These framework types belong to medium- and large-pore zeolites and possess free dimensions of the pore opening in the range 0.5–0.8 nm (*124*). However, the accessibility to the pores is tough and is limited only to small molecules due to small pore size. Trends in the development of solid acid catalysts with less hindered and more active sites have led to the development of multiple porosites within zeolites, often termed mesozeolites, hierarchical zeolites, or micro-mesoporous composites.

The mesozeolites are promising materials due to their enhanced accessibility generated from the hierarchical pore networks and high acidity. Hierarchical zeolites possess bimodal porosity (*125*). They have their intrinsic microporosity integrated with additional (secondary) porosity, in the mesopore range 2–50 nm (*126*). This integrated porosity leads to enhanced catalytic performance and pore accessibility. Mesoporous zeolites are reported to have better anticoking potential. In hierarchical zeolites, mesopores allow the diffusion of large molecules and direct the conversion of pyrolysis vapors into targeted chemicals and fuels by making use of active sites of micropores (*127, 128*).

Hierarchization enhances external surface area, which leads to slow rate of deactivation of active sites because coking is prone to occur at the external crystal surface (*129*). Channels of molecular-sized dimensions are interconnected in hierarchical porosity and connected to the external surface via mesopores, removing mass transfer limitations and facilitating access of reactants to the active sites (*130*). In mesozeolites, products spend less time at active sites of micropores, which causes fewer secondary reactions to occur and hence leads to less coke product formation and less deactivation of catalysts (*131*). Thus, it can be said that mesoporosity plays a vital role in the improvement of a catalyst lifetime. In the case of mesozeolites, wall thickness increases due to the introduction of microporous zeolite particles onto mesopore walls; this is a responsible factor for higher catalytic stability and an enhanced catalyst lifetime of hierarchical zeolites (*132*). As far as biomass pyrolysis is concerned, the mesoporosity of zeolites helps to improve the selectivity of deoxygenation, convert pyrolysis vapors to more C_8-C_9 mono aromatics, and give more percentage yield of aromatic hydrocarbons (*133*). Thus, it can be said that the use of zeolites with bimodal porosity largely improves the quality of bio-oil (*134*).

Demetallation of zeolites helps in suppressing coke formation during bio-oil pyrolysis. It facilitates high accessibility of reactants to active sites (*128, 135*). The molecules, having a larger size than zeolite pores such as syringyl-aromatics, levoglucosan, and their oxygenated derivatives, suffer diffusion limitations during catalytic pyrolysis (*136*). Such scaffolds are responsible for coke formation on the surface of zeolite catalysts. Desilication and dealumination are two top-down approaches that involve the removal of silicon and aluminum atoms, respectively. These procedures involve the treatment of premade microporous zeolites with varying alkali (for desilication) and acid (for dealumination) concentrations (*137*). Dealumination severely damages the zeolite framework as it involves severe acid treatments. Moreover, it involves the generation of isolated pores and reduction in the number of interconnecting channels from the external surface to the pores, thus suffering from the drawback of poor accessibility (*94*). In the case of desilication postsynthetic strategy, a large number of mesopores are readily created in the initial stage, which can be significantly slowed down by employing bulky quaternary ammonium hydroxides (TPAOH, TBAOH, etc.) (*138, 139*). The use of such desilication protectants led to the generation of mesopores, which tend to be more closed. The presence of such interconnected mesopores generated during desilication is preferable over closed or isolated mesopores for a better catalyst lifetime. The use of desilication protectant like piperidine led to the preparation of controllable mesopores with preserved maximum micropores and acid sites (*139, 140*). Figure 10 depicts the controlled desilication route for the synthesis of mesoporous zeolite BEA from the filtrate of alkali-treated microporous zeolite BEA (as Si/Al source) nanocrystal precursors. Cationic surfactant cetyltrimethylammonium bromide is employed as a mesotemplate. The mesopores formation efficiency via desilication strongly depends on the Si/Al ratio of originally taken microporous zeolite. The mesopore size is optimum in the Si/Al range 25–50 (*141*). In the case of Si/Al < 15, silicon leaching is prevented due to the presence of framework aluminum, which leads to the formation of limited mesopores (*142*). There is excessive and uncontrolled Si extraction in the case of desilication of microporous zeolite with Si/Al ratio > 200, which results in generation of macropores and loss of intrinsic micropores (*142*). Desilication promotes carbon content in bio-oil and reduces oxygenating compounds due to the presence of bimodal porosity and enhanced acid sites (*134*).

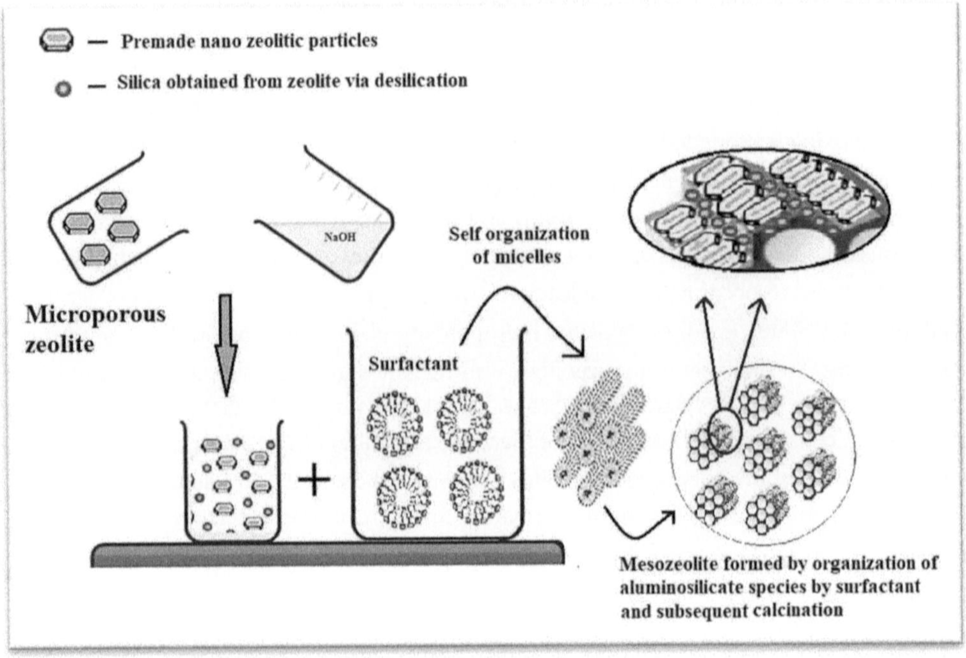

Figure 10. Postsynthetic controlled desilication for the preparation of mesozeolite H-BEA.

The micro-meso composite possesses improved diffusion of reactants provided by mesopores, therefore enhanced accessibility of the acidic sites possess shorter diffusion path length, and therefore the reduced retention time of products in the microporous, which leads to a lower probability of secondary reactions, improving the selectivity of primary products. Another advantage of these zeolites is the improved resistance of the catalyst to deactivation of the sites due to enhanced transport of coke precursors out of the catalyst, and henceforth an enhancement in usability (*143*). Advantages of mesozeolites over mesoporous materials (MCM-41, SBA-15, etc.) include higher hydrothermal/thermal stability due to reinforcement of mesoporous walls by zeolite building blocks, thereby facilitating repeated use of the catalyst and high zeolite acidity, which opens new horizons for the applications of mesoporous zeolites in the catalytic reactions demanding strong acidity (*130, 144*).

Good interconnectivity is one of the salient features of hierarchical zeolite, which facilitates better desorption of products generated on microporous active sites, and thus eliminates carbon deposition and deactivation of active sites (*144, 145*). When bulkier entities diffuse into hierarchical pore structure, they initially start approaching large pores and start disintegrating upon interaction with mild acid sites. Thereafter, intermediates continue to diffuse into small pores and are cracked due to strong acid sites. Extensive literature is now available on the use of mesozeolites and their better performance as promising catalysts in several organic transformations involving bulkier entities. Zhou et al. have observed high selectivity for targeted p-methylation reaction by making use of hierarchical faujasite zeolite catalyst without the occurrence of methyl to hydrocarbon and isomerization (*146*). Yan and coworkers have reported a successful pilot scale-up synthesis of hierarchical zeolites following a synthetic strategy of using sequential alkali dissolution and mesopore reassembly (*130*). They have found high stability and reproducibility of synthesized hierarchical zeolites and observed similar performance in FCC. Ru/Meso-H-ZSM-5 showed better activity and excellent alkane selectivity than microporous H-ZSM-5 due to abundant exposed acid sites (*147*). Lee and coworkers have studied guaiacol conversion over several Pt loaded zeolites such as microporous H-ZSM-5, H-BEA, meso-H-BEA, and mesoporous Al-MCM 48 (*148*). In this study, authors have observed better performance of Pt/H-BEA and Pt/Meso –H-BEA, which may be due to large pores and strong acid sites. Further, the authors suggested that mesopores of zeolites provide more opportunities for making small particles and their uniform distribution. Maheria and coworkers have prepared several micro-meso composite materials (meso H-ZSM-5 and meso H-BEA) using different surface-active agents as mesotemplates and rice husk, by a postsynthetic controlled desilication route (*149*, *150*). They successfully demonstrated their applications as heterogeneous catalysts in the esterification of biomass-derived levulinic acid esterification (to produce value-added fuel additive, n-butyl levulinate) and in multicomponent organic reactions such as dihydropyrimidinones, amidoalkyl naphthols, and such for the synthesis of bioactive drug-like molecules (*151*, *152*).

Zeolite-Based Catalysts Mediated Algae Biomass/Bio-Oil Upgrading

Catalysts play an important role in the pyrolysis of algae biomass. Catalysts affect the selectivity of the products due to change in pyrolytic pathways. As far as catalytic pyrolysis of algae biomass is concerned, the bio-oil yield is a crucial parameter, along with nitrogen and oxygen content, which assess the catalyst performance. So far several metal oxides and supported metal catalysts have also been reported for catalytic pyrolysis of algae. Zeolites are referred to as promising solid acid catalysts for algae oil upgrading due to their tuneable acidity, peculiar pore characteristics, ability to eliminate oxygen without hydrogen, and ability to resist carbon deposition. The degradation of macromolecules of algae to smaller molecular-sized compounds is facilitated with the aid of acid sites on zeolites. In the subsequent steps, the resultant small molecular-sized compounds are subjected to deoxygenation/aromatization, forming hydrocarbons.

Among several zeolites, ZSM-5 zeolite was reported to be widely used due to its better performance in deoxygenation, decarboxylation, and decarbonylation (*153*). Some of the illustrations of zeolite catalyzed microalgae pyrolysis have been summarized in Table 1. Several patents and papers have been published on catalytic microalgae pyrolysis. A patent related to direct synthesis of biogasoline from fast pyrolysis of lignocellulosic biomass using zeolite ZSM-5 is available (*154*). Phosphorus-treated ZSM-5 is reported to give higher bio-oil yield and lower coke formation. Several other zeolites such as mordenite, zeolite BEA, and zeolite HY are also used for bio-oil production (*155*). A study of Ga-ZSM-5 catalyst revealed the promotion of both decarboxylation and olefin aromatization by the presence of Ga (*156*). In another study, Ni/ZSM-5 catalyst was found effective in promoting HDO reactions (*153*). Upgrading of bio-oil is also studied using H-MOR, HY, HZSM-5, SAPO-11, and SAPO-5 catalysts. The authors have found the order of catalytic activity as HZSM-5 > H-MOR > HY > silica–alumina > silicalite-1. They suggested that zeolite cracking is highly dependent on the availability of acid sites (*157–159*). Another study is carried out in fixed-bed reactors using high silica zeolites [HY (100 SAR), ZSM-5 (1500 SAR), and BEA (100 SAR)] for selective conversion of bio-oils (*153*). In this study, zeolite BEA showed higher selectivity to hydrocarbon products, while ZSM-5 was found more selective to aromatics. Zeolite BEA was also found effective in upgrading oxygenated products to give olefins and aromatic hydrocarbons. Also, the authors concluded that the aromatics' yield per measured active site is inversely proportional to SAR, indicating greater separation of acid sites that render the catalyst more efficient at converting the pyrolysis vapors. Zeolite beta with low SAR was found to be deactivated at a slower rate. An assessment of catalytic activity of different Co-Mo loaded HZSM-5 catalysts on the product (biofuel, biochar, and gas) yields and product distribution is carried out and HHV, water content, density, viscosity, and pH of catalytic biofuel were found to be improved as compared to noncatalytic biofuel (*153*).

In hierarchical zeolites, mesopores usually facilitate mass transport to the crystalline nanodomains, the reaction site, where the reaction takes place and micropores assure easy elution of the products from one side. Several mesoporous zeolites are already contributing toward the formulation of industrial catalysts like zeolite beta (BEA) nanocrystals (active phase of PBE-1 catalyst) for the cumene and ethylbenzene processes and mesoporous ultrastable Y as an active part of commercial FCC catalyst (*153*). In one of the reports, authors found hierarchical zeolite HY as an efficient catalyst in bio-oil cracking. The authors have obtained mesozeolite HY by desilication postsynthetic route using NaOH (*160*). This catalyst was found more active than its parent analog in bio-oil cracking, and more selective toward olefins than aromatics. Studies related to HDO and hydrocracking conversion of methyl palmitate into jet biofuel were undertaken in batch mode by Cheng et al. over Ni/desilicated meso-Y catalyst (*74*). The Ni-supported meso-Y catalyst showed the highest jet fuel range alkane selectivity (44.5%). Lanzafame et al. performed isomerization and hydrocracking of n-hexadecane and concluded that mesoporous H-BEA zeolite reduced carbon deposits and improved stability of the catalyst (*161*). The conversion rate was found to be increased from 63.6% to 83%. Tarach et al. compared several mesoporous ZSM-5 catalysts desilicated by using only NaOH. Meso-ZSM-5 zeolites desilicated by NaOH and both NaOH and tetrabutylammonium hydroxide treatment (*162*). The authors found less surface area of meso-ZSM-5 as conventional zeolite; however, the pore diameter distribution was found to be wider. Tarach et al. evaluated hydrocracking activity by taking n-decane as the model compound, which was smaller in scale than bio-oils. Further, it is reported that the dealumination is not a preferred approach for microalgae biodiesel, HDO, and hydrocracking as this acid leaching approach significantly reduces both Bronsted and Lewis acidity (*163, 164*). Thus, modifying parent microporous zeolite by a simple and controllable method in order to adjust its mesoporosity for microalgae jet biofuel is of great research interest.

Coking is a severe problem for catalytic pyrolysis, particularly with zeolite catalysts. The rate of coke formation on zeolites is strongly associated with the pore shape and size, which is available in proximity to its active sites characteristics and pyrolysis conditions (*176*). Coke formation can be reduced by controlling the operation conditions carefully to limit the formation of coke-indicator molecules such as polycyclic aromatic hydrocarbons (PAHs) and olefins (*179*). The use of a zeolite catalyst with low-density acid sites would help to reduce coking. Coked zeolites can be regenerated by the oxidation of the coke in air or oxygen. On the other hand, water has detrimental effects on the catalytic sites of zeolites at high temperatures. Therefore, new methods to regenerate zeolites under mild conditions are needed (*136*).

Table 3. Catalytic Pyrolysis of Algae over Different Zeolite-Based Catalysts

Feedstock	*Catalyst*	*Condition*	*Catalytic performance*	*Ref.*
Spirulina	Fe/HMS-ZSM5	500°C, catalyst/algae = 0.5	Highest bio-oil yield: 37.8%, hydrocarbon yield: 33.04%	(*165*)
Tetraselmis suecica	CBV zeolite	400°C, catalyst/algae = 0.2	Zeolite (Si/Al = 30) showed good activity in deoxygenation	(*166*)
Cyanobacteria	MgAl4-LDO /ZSM-5	550°C, catalyst/algae = 0.75	Maximum liquid yield: 41.1%, O/C: 0.09, HHV: 37.164 $MJ.kg^{-1}$	(*167*)
Spirulina	HZSM-5 and H-BEA with different Si/Al ratio	450°C, 30 s catalyst/algae = 0.75	Maximum of aromatic (6.54%) hydrocarbons were obtained over HZSM-5 (Si/Al = 23), but lower acidity catalysts increased the production of aliphatic hydrocarbons	(*168*)
C. vulgaris	HZSM-5 (Si/Al = 23)	800°C; catalyst/feed ratio = 20	Carbon yield of aromatics = 24% (total selectivity toward aromatics = 75%)	(*13*)
Nannochloropsis sp. + *scum*	HZSM-5 (Si/Al = 30)	800°C (microwave heating); feed/catalyst ratio = 0.5	Yield of bio-oil = 25.2 wt.% (the maximum proportion of aromatic species = 73.2 area %)	(*4*)
Isochrysis sp.	Li-LSX-zeolite	500°C; feed/catalyst ratio = 1	Yield of bio-oil = 29 wt.% (23.1% aliphatics and 11.8% aromatics)	(*169*)
C. vulgaris	5 wt.% Ni/zeolite-Y (Si/Al = 30)	400°C; feed/catalyst ratio = 0.5	Yield of bio-oil = 10% (selectivity toward hydrocarbons = 36 area %)	(*170*)
Arthrospira plantensis	10% Fe/HMS-ZSM5	500°C; acid washing pretreatment	Yield of bio-oil = 30 wt.% (selectivity toward aromatics and hydrocarbons = ~50 area %)	(*165*)
Nannochloropsis salina and *polypropylene*	HZSM-5 (Si/Al = 25)	800°C; microalgae/polypropylene ratio = 1	Yield of bio-oil = 45 wt.% (yield of monoaromatic hydrocarbons = 0.1223 g/g of feed)	(*171*)
Nannochloropsis	HZSM-5 (Si/Al = 80)	500°C (microwave heating); catalyst ratio = 0.5	Yield of bio-oil = 59 wt.%	(*172*)
A. platensis	Zeolite-β hydrogen (Si/Al = 360)	600°C; feed/catalyst ratio = 0.02	Relative composition of monoaromatics = 25 area %	(*173*)
C. vulgaris	HZSM-5 (Si/Al = 50)	500°C; feed/ catalyst ratio = 0.11	Yield of aromatics = 25 wt.%	(*174*)
C. pyrenoidosa	ZSM5 (Si/Al = 45)	650°C; feed/catalyst ratio = 0.05	Carbon yield of aromatics-rich liquid = 19.4%	(*175*)

Table 3. (Continued). Catalytic Pyrolysis of Algae over Different Zeolite-Based Catalysts

Feedstock	*Catalyst*	*Condition*	*Catalytic performance*	*Ref.*
C. vulgaris	HZSM-5 (Si/Al = 30)	550°C; feed/catalyst ratio = 0.2	Yield of BTX = 10 wt.% of feed	(176)
T. suecica	CBV 720 (Si/Al = 30)	400°C; feed/catalyst ratio = 5	Relative composition of hydrocarbons = 9.8 area %	(166)
Desmodesmus communis	HZSM-5 (Si/Al = 45)	600°C; feed/catalyst ratio = 0.05	Relative composition of BTEX = 61.8 area %	(177)
C. vulgaris	5 wt.% Cu/HZSM-5 (Si/Al = 30)	250°C; feed/catalyst ratio = 0.2	Yield of aromatics = 21.2 wt.% of feed	(178)

Despite many advantages, there are certain issues with hierarchical zeolites such as cost of templates used, hydrothermal stability, the large-scale production of these advanced materials, and so on (*114*). Moreover, understanding nucleation and crystallization mechanisms, locating active sites, synthesis of hierarchical zeolite with predicted properties, modeling, understanding reaction mechanism occurring in micro- and mesopores, and characterization of micro- and mesopores are some challenges associated with these hierarchical materials (*81*).

Conclusion

To reduce dependency on fossil–fuel-based nonrenewable feedstock and to conceptualize biorefinery, considerable research interest is growing for use of biomass-based renewable feedstock for advanced biofuels and biomass-based value-added chemical production. The algae biomass emerged as a significant option to reduce injurious particulate emissions and solve energy problems. Algae oil is one of the important products used to produce algae biofuels. The algae bio-oil obtained from pyrolysis possesses high viscosity, and high oxygen and nitrogen content. Such properties negatively affect the quality of biofuel obtained from algae oil, and the produced biofuel is considered unsuitable as transportation fuel.

There are several algae oil upgrading pathways; among them, catalytic pyrolysis offers higher yield and better quality of algae bio-oil, which are two main advantages of catalytic pyrolysis. Among several heterogeneous catalysts, zeolites are well-established industrial catalysts due to their robust properties, including peculiar shape selective property, unique micropore structure of molecular dimensions, tuneable acidity, good thermal/hydrothermal stability, and so on. They are successfully and widely used in petrochemical and fine chemical industries in order to make fuels from oils, and in numerous chemical and drug intermediates' synthesis. Also, they play a pivotal role in the elimination of air pollutants.

Zeolites can act as bifunctional catalysts, and this feature plays an important role in making a plethora of chemical compounds from pyrogas and even from carbon dioxide and hydrogen. Zeolite cracking is one of the key processes for biomass/bio-oil valorization. The macromolecules of algae degraded in small molecular-sized compounds by making use of acid sites of zeolites. Small-sized compounds then converted to reduced compounds like hydrocarbon via aromatization/deoxygenation. However, coking, catalyst deactivation, and poor quality oil are major issues with microporous zeolite cracking. Lignin disintegration via catalytic pyrolysis using a catalyst with a large pore opening affords more liquid and limited coke. The hierarchical zeolites can be advanced catalytic materials and have immense future potential to become promising candidates for catalytic pyrolysis of algae biomass to produce chemicals and advance biofuels. Research efforts are needed for designing catalysts for chemical selectivity, developing more environmentally benign processes, and realizing a stable framework with a larger pore size. The use of hierarchical zeolite with bimodal porosity is a better option compared to the monomodal ones, which combine advantages of both microporous and mesoporous materials like high acidity, thermal stability, expanded pore size, and high surface area. However, there is an issue of hydrothermal stability with hierarchical zeolites, which may be addressed by using silylation/functionalization of mesopores after performing desilication.

Very limited literature is available on algae-oil upgrading using hierarchical zeolites, and there is a need for sincere research efforts to explore more about applicability of hierarchical zeolites in the research area of algae oil upgrading for advanced biofuels and value-added bio-based chemical production.

References

1. Leite, G. B.; Abdelaziz, A. E. M.; Hallenbeck, P. C. Algal Biofuels: Challenges and Opportunities. *Bioresour.Technol.* **2013**, *145*, 134–141. https://doi.org/10.1016/j.biortech.2013.02.007.
2. Parry, M.; Arnell, N.; McMichael, T.; Nicholls, R.; Martens, P.; Kovats, S.; Livermore, M.; Rosenzweig, C.; Iglesias, A.; Fischer, G. Millions at Risk. *Glob. Environ. Chang.* **2001**, *11* (3), 1–3.
3. Jahirul, M. I.; Rasul, M. G.; Chowdhury, A. A.; Ashwath, N. Biofuels Production through Biomass Pyrolysis—A Technological Review. *Energies* **2012**, *5* (12), 4952–5001. https://doi.org/10.3390/en5124952.
4. Xie, Q.; Addy, M.; Liu, S.; Zhang, B.; Cheng, Y.; Wan, Y.; Li, Y.; Liu, Y.; Lin, X.; Chen, P.; Ruan, R. Fast Microwave-Assisted Catalytic Co-Pyrolysis of Microalgae and Scum for Bio-Oil Production. *Fuel* **2015**, *160*, 577–582. https://doi.org/10.1016/j.fuel.2015.08.020.
5. Hu, Q.; Sommerfeld, M.; Jarvis, E.; Ghirardi, M.; Posewitz, M.; Seibert, M.; Darzins, A. Microalgal Triacylglycerols as Feedstocks for Biofuel Production: Perspectives and Advances. *Plant J.* **2008**, *54* (4), 621–639. https://doi.org/10.1111/j.1365-313X.2008.03492.x.
6. Oncel, S. S. Microalgae for a Macroenergy World. *Renew. Sustain. Energy Rev.* **2013**, *26*, 241–264. https://doi.org/10.1016/j.rser.2013.05.059.
7. Milledge, J. J.; Nielsen, B. V; Maneein, S.; Harvey, P. J. A Brief Review of Anaerobic Digestion of Algae for Bioenergy. *Energies* **2019**. https://doi.org/10.3390/en12061166.
8. Juneja, A.; Ceballos, R. M.; Murthy, G. S. Effects of Environmental Factors and Nutrient Availability on the Biochemical Composition of Algae for Biofuels Production: A Review. *Energies* **2013**, *6* (9), 4607–4638. https://doi.org/10.3390/en6094607.
9. Collyer, D. M.; Fogg, G. E. Studies on Fat Accumulation by Algae. *J. Exp. Bot.* **1955**, *6* (2), 256–275. https://doi.org/10.1093/jxb/6.2.256.
10. Brown, M. R. The Amino-Acid and Sugar Composition of 16 Species of Microalgae Used in Mariculture. *J. Exp. Mar. Bio. Ecol.* **1991**, *145* (1), 79–99. https://doi.org/10.1016/0022-0981(91)90007-J.
11. Becker, E. W. *Microalgae: Biotechnology and Microbiology*; Cambridge; New York: Cambridge University Press, 1994.
12. Shakya, R.; Adhikari, S.; Mahadevan, R.; Shanmugam, S. R.; Nam, H.; Hassan, E. B.; Dempster, T. A. Influence of Biochemical Composition during Hydrothermal Liquefaction of Algae on Product Yields and Fuel Properties. *Bioresour. Technol.* **2017**, *243*, 1112–1120. https://doi.org/10.1016/j.biortech.2017.07.046.
13. Wang, K.; Brown, R. C. Catalytic Pyrolysis of Microalgae for Production of Aromatics and Ammonia. *Green Chem.* **2013**, *15* (3), 675–681. https://doi.org/10.1039/C3GC00031A.
14. Chen, Z.; Wang, L.; Qiu, S.; Ge, S. Determination of Microalgal Lipid Content and Fatty Acid for Biofuel Production. *Biomed Res. Int.* **2018**, *2018*, 1503126. https://doi.org/10.1155/2018/1503126.
15. Dolganyuk, V.; Andreeva, A.; Budenkova, E.; Sukhikh, S.; Babich, O.; Ivanova, S.; Prosekov, A.; Ulrikh, E. Study of Morphological Features and Determination of the Fatty Acid Composition of the Microalgae Lipid Complex. *Biomolecules* **2020**, *10* (11), 1–15. https://doi.org/10.3390/biom10111571.

16. Huber, G. W.; Iborra, S.; Corma, A. *Synthesis of Transportation Fuels from Biomass: Chemistry, Catalysts, and Engineering*. **2006**, *2*, 4044–4098.
17. Mata, T. M.; Martins, A. A.; Caetano, N. S. Microalgae for Biodiesel Production and Other Applications: A Review. *Renew. Sustain. Energy Rev.* **2010**, *14* (1), 217–232. https://doi.org/10.1016/j.rser.2009.07.020.
18. Bhalamurugan, G. L.; Valerie, O.; Mark, L. Valuable Bioproducts Obtained from Microalgal Biomass and Their Commercial Applications: A Review. *Environ. Eng. Res.* **2018**, *23* (3), 229–241. https://doi.org/10.4491/eer.2017.220.
19. Li, Y.; Horsman, M.; Wang, B.; Wu, N.; Lan, C. Q. Effects of Nitrogen Sources on Cell Growth and Lipid Accumulation of Green Alga Neochloris Oleoabundans. *Appl. Microbiol. Biotechnol.* **2008**, *81* (4), 629–636. https://doi.org/10.1007/s00253-008-1681-1.
20. Li, Y.; Horsman, M.; Wu, N.; Lan, C. Q.; Dubois-Calero, N. Biofuels from Microalgae. *Biotechnol. Prog.* **2008**, *24* (4), 815–820. https://doi.org/10.1021/bp070371k.
21. Huang, G.; Chen, F.; Wei, D.; Zhang, X.; Chen, G. Biodiesel Production by Microalgal Biotechnology. *Appl. Energy* **2010**, *87* (1), 38–46. https://doi.org/10.1016/j.apenergy.2009.06.016.
22. Chisti, Y. Biodiesel from Microalgae. *Biotechnol. Adv.* **2007**, *25* (3), 294–306. https://doi.org/10.1016/j.biotechadv.2007.02.001.
23. Rodolfi, L.; Chini Zittelli, G.; Bassi, N.; Padovani, G.; Biondi, N.; Bonini, G.; Tredici, M. R. Microalgae for Oil: Strain Selection, Induction of Lipid Synthesis and Outdoor Mass Cultivation in a Low-Cost Photobioreactor. *Biotechnol. Bioeng.* **2009**, *102* (1), 100–112. https://doi.org/10.1002/bit.22033.
24. Vonshak, A. Recent Advances in Microalgal Biotechnology. *Biotechnol. Adv.* **1990**, *8*, 709–727.
25. Chynoweth, D. P.; Turick, C. E.; Owens, J. M.; Jerger, D. E.; Peck, M. W. Biochemical Methane Potential of Biomass and Waste Feedstocks. *Biomass and Bioenergy* **1993**, *5* (1), 95–111. https://doi.org/10.1016/0961-9534(93)90010-2.
26. Radmer, R. J. Algal Diversity and Commercial Algal Products. *Bioscience* **1996**, *46* (4), 263–270. https://doi.org/10.2307/1312833.
27. Pulz, O.; Gross, W. Valuable Products from Biotechnology of Microalgae. *Appl. Microbiol. Biotechnol.* **2004**, *65* (6), 635–648. https://doi.org/10.1007/s00253-004-1647-x.
28. Medipally, S. R.; Yusoff, F. M.; Banerjee, S.; Shariff, M. Microalgae as Sustainable Renewable Energy Feedstock for Biofuel Production. *Biomed Res. Int.* **2015**, *2015*, 519513. https://doi.org/10.1155/2015/519513.
29. Khan, M. I.; Shin, J. H.; Kim, J. D. The Promising Future of Microalgae: Current Status, Challenges, and Optimization of a Sustainable and Renewable Industry for Biofuels, Feed, and Other Products. *Microb. Cell Fact.* **2018**, *17* (1), 1–21. https://doi.org/10.1186/s12934-018-0879-x.
30. Demirbas, A.; Fatih Demirbas, M. Importance of Algae Oil as a Source of Biodiesel. *Energy Convers. Manag.* **2011**, *52* (1), 163–170. https://doi.org/10.1016/j.enconman.2010.06.055.
31. Kumar, M.; Sharma, M. Potential Assessment of Microalgal Oils for Biodiesel Production: A Review. *J. Mater. Environ. Sci.* **2014**, *5*, 757–766.
32. Patel, P.; Patel, S.; Krishnamurthy, R. Microalgae: Future Biofuel. *Indian J. Geo-Marine Sci.* **2016**, *45* (7), 823–829.

33. Kumar, P.; Suseela, M. R.; Toppo, K. Physico-Chemical Characterization of Algal Oil: A Potential Biofuel. *Asian J. Exp. Biol. Sci.* **2011**, *2* (3), 493–497.
34. Schlagermann, P.; Göttlicher, G.; Dillschneider, R.; Rosello-Sastre, R.; Posten, C. Composition of Algal Oil and Its Potential as Biofuel. *J. Combust.* **2012**, *2012*. https://doi.org/10.1155/2012/285185.
35. Benedetti, M.; Vecchi, V.; Barera, S.; Dall'Osto, L. Biomass from Microalgae: The Potential of Domestication towards Sustainable Biofactories. *Microb. Cell Fact.* **2018**, *17* (1), 173. https://doi.org/10.1186/s12934-018-1019-3.
36. Hannon, M.; Gimpel, J.; Tran, M.; Rasala, B.; Mayfield, S. Biofuels from Algae: Challenges and Potential. *Biofuels* **2010**, *1* (5), 763–784. https://doi.org/10.4155/bfs.10.44.
37. Ercin, A. E.; Aldaya, M. M.; Hoekstra, A. Y. The Water Footprint of Soy Milk and Soy Burger and Equivalent Animal Products. *Ecol. Indic.* **2012**, *18*, 392–402. https://doi.org/10.1016/j.ecolind.2011.12.009.
38. Chojnacka, K., Marquez-Rocha, F. J. *Kinetic and Stoichiometric Relationships of the Energy and Carbon Metabolism in the Culture of Microalgae. Biotechnology*; Asian Network for Scientific Information, 2004.
39. Chen, C.-Y.; Yeh, K.-L.; Aisyah, R.; Lee, D.-J.; Chang, J.-S. Cultivation, Photobioreactor Design and Harvesting of Microalgae for Biodiesel Production: A Critical Review. *Bioresour. Technol.* **2011**, *102* (1), 71–81. https://doi.org/10.1016/j.biortech.2010.06.159.
40. Zhao, B.; Su, Y. Process Effect of Microalgal-Carbon Dioxide Fixation and Biomass Production: A Review. *Renew. Sustain. Energy Rev.* **2014**, *31*, 121–132. https://doi.org/10.1016/j.rser.2013.11.054.
41. Lee, S. Y.; Sankaran, R.; Chew, K. W.; Tan, C. H.; Krishnamoorthy, R.; Chu, D.-T.; Show, P.-L. Waste to Bioenergy: A Review on the Recent Conversion Technologies. *BMC Energy* **2019**, *1* (1), 4. https://doi.org/10.1186/s42500-019-0004-7.
42. Canabarro, N.; Soares, J. F.; Anchieta, C. G.; Kelling, C. S.; Mazutti, M. A. Thermochemical Processes for Biofuels Production from Biomass. *Sustain. Chem. Process.* **2013**, *1* (1), 22. https://doi.org/10.1186/2043-7129-1-22.
43. Morgan, T. J.; Youkhana, A.; Turn, S. Q.; Ogoshi, R.; Garcia-Pérez, M. Review of Biomass Resources and Conversion Technologies for Alternative Jet Fuel Production in Hawai'i and Tropical Regions. *Energy & Fuels* **2019**, *33* (4), 2699–2762. https://doi.org/10.1021/acs.energyfuels.8b03001.
44. Iliopoulou, E. F.; Triantafyllidis, K. S.; Lappas, A. A. Overview of Catalytic Upgrading of Biomass Pyrolysis Vapors toward the Production of Fuels and High-Value Chemicals. *WIREs Energy Environ.* **2019**, *8* (1), e322. https://doi.org/10.1002/wene.322.
45. Cheng, Y.-T.; Huber, G. W. Chemistry of Furan Conversion into Aromatics and Olefins over HZSM-5: A Model Biomass Conversion Reaction. *ACS Catal.* **2011**, *1* (6), 611–628. https://doi.org/10.1021/cs200103j.
46. Jae, J.; Tompsett, G. A.; Foster, A. J.; Hammond, K. D.; Auerbach, S. M.; Lobo, R. F.; Huber, G. W. Investigation into the Shape Selectivity of Zeolite Catalysts for Biomass Conversion. *J. Catal.* **2011**, *279* (2), 257–268. https://doi.org/10.1016/j.jcat.2011.01.019.
47. Uduman, N.; Qi, Y.; Danquah, M. K.; Forde, G. M.; Hoadley, A. Dewatering of Microalgal Cultures: A Major Bottleneck to Algae-Based Fuels. *J. Renew. Sustain. Energy* **2010**, *2* (1), 12701. https://doi.org/10.1063/1.3294480.

48. Moheimani, N. R.; Parlevliet, D.; McHenry, M. P.; Bahri, P. A.; de Boer, K. *Past, Present and Future of Microalgae Cultivation Developments*, 2015. https://doi.org/10.1007/978-3-319-16640-7_1.
49. Eldridge, R. J.; Hill, D. R. A.; Gladman, B. R. A Comparative Study of the Coagulation Behaviour of Marine Microalgae. *J. Appl. Phycol.* **2012**, *24* (6), 1667–1679. https://doi.org/10.1007/s10811-012-9830-4.
50. Henderson, R. K.; Parsons, S. A.; Jefferson, B. Surfactants as Bubble Surface Modifiers in the Flotation of Algae: Dissolved Air Flotation That Utilizes a Chemically Modified Bubble Surface. *Environ. Sci. Technol.* **2008**, *42* (13), 4883–4888. https://doi.org/10.1021/es702649h.
51. Dassey, A. J.; Theegala, C. S. Harvesting Economics and Strategies Using Centrifugation for Cost Effective Separation of Microalgae Cells for Biodiesel Applications. *Bioresour. Technol.* **2013**, *128*, 241–245. https://doi.org/10.1016/j.biortech.2012.10.061.
52. Milledge, J. J.; Heaven, S. A Review of the Harvesting of Micro-Algae for Biofuel Production. *Rev. Environ. Sci. Bio/Technology* **2013**, *12* (2), 165–178. https://doi.org/10.1007/s11157-012-9301-z.
53. Phasey, J.; Vandamme, D.; Fallowfield, H. Harvesting of Algae in Municipal Wastewater Treatment by Calcium Phosphate Precipitation Mediated by Photosynthesis, Sodium Hydroxide and Lime. *Algal Res.* **2017**, *27*, 115–120. https://doi.org/10.1016/j.algal.2017.06.015.
54. Al Hattab, M.; Ghaly, A.; Hammouda, A. Microalgae Harvesting Methods for Industrial Production of Biodiesel: Critical Review and Comparative Analysis Fundamentals of Renewable Energy and Applications. *J. Fundam. Renewable Energy Appl.* **2015**, 5 (2), 1–26. https://doi.org/10.4172/20904541.1000154.
55. Mo, W.; Soh, L.; Werber, J. R.; Elimelech, M.; Zimmerman, J. B. Application of Membrane Dewatering for Algal Biofuel. *Algal Res.* **2015**, *11* (C), 1–12. https://doi.org/10.1016/j.algal.2015.05.018.
56. Akhondi, E.; Zamani, F.; Tng, K. H.; Leslie, G.; Krantz, W. B.; Fane, A. G.; Chew, J. W. The Performance and Fouling Control of Submerged Hollow Fiber (HF) Systems: A Review. *Appl. Sci.* **2017**, 7 (8), 1–39. https://doi.org/10.3390/app7080765.
57. Chen, Y. M.; Liu, J. C.; Ju, Y.-H. Flotation Removal of Algae from Water. *Colloids Surfaces B Biointerfaces* **1998**, *12* (1), 49–55. https://doi.org/10.1016/S0927-7765(98)00059-9.
58. Sandbank, E.; Shelef, G. Harvesting of Algae from High-Rate Ponds by Flocculation-Flotation. *Water Sci. Technol.* **1987**, *19* (12), 257–263. https://doi.org/10.2166/wst.1987.0154.
59. Garg, S.; Li, Y.; Wang, L.; Schenk, P. M. Flotation of Marine Microalgae: Effect of Algal Hydrophobicity. *Bioresour. Technol.* **2012**, *121*, 471–474. https://doi.org/10.1016/j.biortech.2012.06.111.
60. Raut, N.; Al-Balushi, T.; Panwar, S.; Vaidya, R. S.; Shinde, G. B. Microalgal Biofuel. *Biofuels - Status Perspect.* **2015**. https://doi.org/10.5772/59821.
61. Mercer, P.; Armenta, R. E. Developments in Oil Extraction from Microalgae. *Eur. J. Lipid Sci. Technol.* **2011**, *113* (5), 539–547. https://doi.org/10.1002/ejlt.201000455.
62. Pyle, D. J.; Garcia, R. A.; Wen, Z. Producing Docosahexaenoic Acid (DHA)-Rich Algae from Biodiesel-Derived Crude Glycerol: Effects of Impurities on DHA Production and Algal Biomass Composition. *J. Agric. Food Chem.* **2008**, *56* (11), 3933–3939. https://doi.org/10.1021/jf800602s.

63. Satyanarayana, K. G.; Mariano, A. B.; Vargas, J. V. C. A Review on Microalgae, a Versatile Source for Sustainable Energy and Materials. *Int. J. Energy Res.* **2011**, 35 (4), 291–311. https://doi.org/10.1002/er.1695.
64. Hampannavar, U.; Shivayogimath, C. Anaerobic Treatment of Sugar Industry Wastewater by Upflow Anaerobic Sludge Blanket Reactor at Ambient Temperature. *Int. J. Environ. Sci.* **2010**, *1* (4), 631–639.
65. Halim, R.; Gladman, B.; Danquah, M. K.; Webley, P. A. Oil Extraction from Microalgae for Biodiesel Production. *Bioresour. Technol.* **2011**, *102* (1), 178–185. https://doi.org/10.1016/j.biortech.2010.06.136.
66. Adam, F.; Abert-Vian, M.; Peltier, G.; Chemat, F. "Solvent-Free" Ultrasound-Assisted Extraction of Lipids from Fresh Microalgae Cells: A Green, Clean and Scalable Process. *Bioresour. Technol.* **2012**, *114*, 457–465. https://doi.org/10.1016/j.biortech.2012.02.096.
67. Metherel, A. H.; Taha, A. Y.; Izadi, H.; Stark, K. D. The Application of Ultrasound Energy to Increase Lipid Extraction Throughput of Solid Matrix Samples (Flaxseed). *Prostaglandins. Leukot. Essent. Fatty Acids* **2009**, *81* (5–6), 417–423. https://doi.org/10.1016/j.plefa.2009.07.003.
68. Šoštarič, M.; Klinar, D.; Bricelj, M.; Golob, J.; Berovič, M.; Likozar, B. Growth, Lipid Extraction and Thermal Degradation of the Microalga *Chlorella Vulgaris*. *N. Biotechnol.* **2012**, *29* (3), 325–331. https://doi.org/10.1016/j.nbt.2011.12.002.
69. Balasubramanian, S.; Allen, J. D.; Kanitkar, A.; Boldor, D. Oil Extraction from Scenedesmus Obliquus Using a Continuous Microwave System—Design, Optimization, and Quality Characterization. *Bioresour. Technol.* **2011**, *102* (3), 3396–3403. https://doi.org/10.1016/j.biortech.2010.09.119.
70. Absalan, G.; Akhond, M.; Sheikhian, L. Extraction and High Performance Liquid Chromatographic Determination of 3-Indole Butyric Acid in Pea Plants by Using Imidazolium-Based Ionic Liquids as Extractant. *Talanta* **2008**, *77* (1), 407–411. https://doi.org/10.1016/j.talanta.2008.06.049.
71. Ventura, S. P. M.; E Silva, F. A.; Quental, M. V.; Mondal, D.; Freire, M. G.; Coutinho, J. A. P. Ionic-Liquid-Mediated Extraction and Separation Processes for Bioactive Compounds: Past, Present, and Future Trends. *Chem. Rev.* **2017**, *117* (10), 6984–7052. https://doi.org/10.1021/acs.chemrev.6b00550.
72. Yang, C.; Li, R.; Cui, C.; Liu, S.; Qiu, Q.; Ding, Y.; Wu, Y.; Zhang, B. Catalytic Hydroprocessing of Microalgae-Derived Biofuels: A Review. *Green Chem.* **2016**, *18* (13), 3684–3699. https://doi.org/10.1039/C6GC01239F.
73. Saber, M.; Nakhshiniev, B.; Yoshikawa, K. A Review of Production and Upgrading of Algal Bio-Oil. *Renew. Sustain. Energy Rev.* **2016**, *58*, 918–930. https://doi.org/10.1016/j.rser.2015.12.342.
74. Cheng, S.; Wei, L.; Zhao, X.; Kadis, E.; Julson, J. Conversion of Prairie Cordgrass to Hydrocarbon Biofuel over Co-Mo/HZSM-5 Using a Two-Stage Reactor System. *Energy Technol.* **2016**, *4* (6), 706–713. https://doi.org/10.1002/ente.201500452.
75. Czernik, S.; Bridgwater, T. Overview of Applications of Biomass Fast Pyrolysis Oil. *Energy* **2004**, *18*, 590. https://doi.org/10.1021/ef034067u.
76. Duan, P.; Savage, P. E. Upgrading of Crude Algal Bio-Oil in Supercritical Water. *Bioresour. Technol.* **2011**, *102* (2), 1899–1906. https://doi.org/10.1016/j.biortech.2010.08.013.

77. Peterson, A. A.; Vogel, F.; Lachance, R. P.; Fröling, M.; Antal Michael J., J.; Tester, J. W. Thermochemical Biofuel Production in Hydrothermal Media: A Review of Sub- and Supercritical Water Technologies. *Energy Environ. Sci.* **2008**, *1* (1), 32–65. https://doi.org/10.1039/B810100K.
78. Biswas, S.; Katiyar, R.; Gurjar, B. R.; Pruthi, V. Biofuels and Their Production through Different Catalytic Routes. *Chem. Biochem. Eng. Q.* **2017**, *31* (1), 47–62. https://doi.org/10.15255/CABEQ.2016.897.
79. Zhang, R.; Li, L.; Tong, D.; Hu, C. Microwave-Enhanced Pyrolysis of Natural Algae from Water Blooms. *Bioresour. Technol.* **2016**, *212*, 311–317. https://doi.org/10.1016/j.biortech.2016.04.053.
80. Mohd, N. A. *Conventional and Microwave Pyrolysis of Empty Fruit Bunch and Rice Husk Pellets.* 2017, February, 1–163.
81. Cheng, S. *Development of Heterogeneous Catalysts for Upgrading Biomass Pyrolysis Bio-Oils into Advanced Biofuels.* Ph.D Dissertation, South Dakota State University, 2017.
82. Mohan, D.; Pittman, C. U.; Steele, P. H. Pyrolysis of Wood/Biomass for Bio-Oil: A Critical Review. *Energy Fuels* **2006**, *20* (3), 848–889. https://doi.org/10.1021/ef0502397.
83. Masnadi, M. S.; Grace, J. R.; Bi, X. T.; Lim, C. J.; Ellis, N. From Fossil Fuels towards Renewables: Inhibitory and Catalytic Effects on Carbon Thermochemical Conversion during Co-Gasification of Biomass with Fossil Fuels. *Appl. Energy* **2015**, *140* (C), 196–209.
84. Mohammed, I. Y.; Kazi, F. K.; Yusup, S.; Alaba, P. A.; Sani, Y. M.; Abakr, Y. A. Catalytic Intermediate Pyrolysis of Napier Grass in a Fixed Bed Reactor with ZSM-5, HZSM-5 and Zinc-Exchanged Zeolite-A as the Catalyst. *Energies* **2016**. https://doi.org/10.3390/en9040246.
85. Campanella, A.; Harold, M. P. Fast Pyrolysis of Microalgae in a Falling Solids Reactor: Effects of Process Variables and Zeolite Catalysts. *Biomass Bioenergy* **2012**, *46*, 218–232. https://doi.org/10.1016/j.biombioe.2012.08.023.
86. Speight, J. G. Catalytic Cracking. In *Heavy and Extra-Heavy Oil Upgrading Technologies;* Speight, J. G., Ed.; Gulf Professional Publishing: Boston, 2013; pp 39–67. https://doi.org/10.1016/B978-0-12-404570-5.00003-X.
87. Serrano, D. P.; Melero, J. A.; Morales, G.; Iglesias, J.; Pizarro, P. Progress in the Design of Zeolite Catalysts for Biomass Conversion into Biofuels and Bio-Based Chemicals. *Catal. Rev. - Sci. Eng.* **2018**, *60* (1), 1–70. https://doi.org/10.1080/01614940.2017.1389109.
88. Lee, J.; Kwon, E. E.; Park, Y. K. Recent Advances in the Catalytic Pyrolysis of Microalgae. *Catal. Today* **2020**, *355* (December 2018), 263–271. https://doi.org/10.1016/j.cattod.2019.03.010.
89. Aysu, T.; Ola, O.; Maroto-Valer, M. M.; Sanna, A. Effects of Titania Based Catalysts on In-Situ Pyrolysis of Pavlova Microalgae. *Fuel Process. Technol.* **2017**, *166*, 291–298. https://doi.org/10.1016/j.fuproc.2017.05.001.
90. Aysu, T.; Maroto-Valer, M. M.; Sanna, A. Ceria Promoted Deoxygenation and Denitrogenation of Thalassiosira Weissflogii and Its Model Compounds by Catalytic In-Situ Pyrolysis. *Bioresour. Technol.* **2016**, *208*, 140–148. https://doi.org/10.1016/j.biortech.2016.02.050.
91. Nishu; Liu, R.; Rahman, M. M.; Sarker, M.; Chai, M.; Li, C.; Cai, J. A Review on the Catalytic Pyrolysis of Biomass for the Bio-Oil Production with ZSM-5: Focus on Structure. *Fuel Process.*

Technol. **2020**, *199* (November 2019), 106301. https://doi.org/10.1016/j.fuproc.2019.106301.
92. Hu, C.; Zhang, H.; Xiao, R. Catalytic Fast Pyrolysis of Biomass over Core-Shell HZSM-5@silicalite-1 in a Bench-Scale Two-Stage Fluidized-Bed/Fixed-Bed Reactor. *J. Anal. Appl. Pyrolysis* **2018**, *136*, 27–34. https://doi.org/10.1016/j.jaap.2018.11.005.
93. Zimmermann, N. E. R.; Haranczyk, M. History and Utility of Zeolite Framework-Type Discovery from a Data-Science Perspective. *Cryst. Growth Des.* **2016**, *16* (6), 3043–3048. https://doi.org/10.1021/acs.cgd.6b00272.
94. Serrano, D. P.; Escola, J. M.; Pizarro, P. Synthesis Strategies in the Search for Hierarchical Zeolites. *Chem. Soc. Rev.* **2013**, *42* (9), 4004–4035. https://doi.org/10.1039/c2cs35330j.
95. Ovhal, S.; Butler, I. S.; Xu, S. The Potential of Zeolites to Block the Uptake of Radioactive Strontium-90 in Organisms. *Contemporary Chemistry* **2018**, *1* (1)
96. Corma, A.; Navarro, M. T.; Rey, F.; Rius, J.; Valencia, S. Pure Polymorph C of Zeolite Beta Synthesized by Using Framework Isomorphous Substitution as a Structure-Directing Mechanism. *Angew. Chemie Int. Ed.* **2001**, *40* (12), 2277–2280. https://doi.org/10.1002/1521-3773(20010618)40:12<2277::AID-ANIE2277>3.0.CO;2-O.
97. Bedioui, F. Zeolite-Encapsulated and Clay-Intercalated Metal Porphyrin, Phthalocyanine and Schiff-Base Complexes as Models for Biomimetic Oxidation Catalysts: An Overview. *Coord. Chem. Rev.* **1995**, *144*, 39–68. https://doi.org/10.1016/0010-8545(94)08000-H.
98. Moshoeshoe, M.; Silas Nadiye-Tabbiruka, M.; Obuseng, V. A Review of the Chemistry, Structure, Properties and Applications of Zeolites. *Am. J. Mater. Sci.* **2017**, *2017* (5), 196–221. https://doi.org/10.5923/j.materials.20170705.12.
99. Coombs, D. S.; Alberti, A.; Armbruster, T.; Artioli, G.; Colella, C.; Galli, E.; Grice, J. D.; Liebau, F.; Mandarino, J. A.; Minato, H.; Nickel, E. H.; Passaglia, E.; Peacor, D. R.; Quartieri, S.; Rinaldi, R.; Ross, M.; Sheppard, R. A.; Tillmanns, E.; Vezzalini, G. Recommended Nomenclature for Zeolite Minerals: Report of the Subcommittee on Zeolites of the International Mineralogical Association, Commission on New Minerals and Mineral Names. *Can. Mineral.* **1997**, *35* (6), 1571–1606.
100. Hashimoto, S. Zeolite Photochemistry: Impact of Zeolites on Photochemistry and Feedback from Photochemistry to Zeolite Science. *J. Photochem. Photobiol. C Photochem. Rev.* **2003**, *4* (1), 19–49. https://doi.org/10.1016/S1389-5567(03)00003-0.
101. Kraushaar, B. *Characterization and Modification of Zeolites and Related Materials*; Technische Universiteit Eindhove, 1989. https://doi.org/10.6100/IR302523.
102. Bronsted, J. N. The Acid-Basic Function of Molecules and Its Dependency on the Electric Charge Type. *J. Phys. Chem.* **1926**, *30* (6), 777–790. https://doi.org/10.1021/j150264a007.
103. Baerlocher, C., McCusker, L. B.; Olson, D. H. *Atlas of Zeolite Framework Types*; Elsevier, 2007.
104. Akimkhan, A. M. Structural and Ion-Exchange Properties of Natural Zeolite. In *Ion Exchange Technologies*; Kilislioğlu, A., Ed.; IntechOpen: Rijeka, 2012. https://doi.org/10.5772/51682.
105. Iwai, Y.; Yamanishi, T. Influence of Framework Silica-to-Alumina Ratio on the Tritiated Water Adsorption and Desorption Characteristics of NaX and NaY Zeolites. *J. Nucl. Sci. Technol.* **2008**, *45* (6), 532–540. https://doi.org/10.1080/18811248.2008.9711877.
106. Védrine, J. C. Heterogeneous Catalysis on Metal Oxides. *Catalysts* **2017**. https://doi.org/10.3390/catal7110341.

107. Slater, A. G.; Cooper, A. I. Function-Led Design of New Porous Materials. *Science* **2015**, *348* (6238), aaa8075. https://doi.org/10.1126/science.aaa8075.
108. Fernandes, D. R.; Rocha, A. S.; Mai, E. F.; Mota, C. J. A.; Teixeira, V. Levulinic Acid Esterification with Ethanol to Ethyl Levulinate Production over Solid Acid Catalysts. *Applied Catal. A: Gen.* **2012**, *425–426*, 199–204. https://doi.org/10.1016/j.apcata.2012.03.020.
109. Yilmaz, B.; Müller, U. Catalytic Applications of Zeolites in Chemical Industry. *Top. Catal.* **2009**, *52* (6–7), 888–895. https://doi.org/10.1007/s11244-009-9226-0.
110. Choudary, N. V.; Newalkar, B. L. Use of Zeolites in Petroleum Refining and Petrochemical Processes: Recent Advances. *J. Porous Mater.* **2011**, *18* (6), 685–692. https://doi.org/10.1007/s10934-010-9427-8.
111. Choudary, N. V; Newalkar, B. L. Use of Zeolites in Petroleum Refining and Petrochemical Processes: Recent Advances. *J. Porous Mater.* **2011**, *18* (6), 685–692. https://doi.org/10.1007/s10934-010-9427-8.
112. Millini, R.; Bellussi, G. Zeolite Science and Perspectives. In *Zeolites in Catalysis: Properties and Applications*; The Royal Society of Chemistry, London, 2017; Chapter 1, pp 1–36. https://doi.org/10.1039/9781788010610
113. Kubička, D.; Kikhtyanin, O. Opportunities for Zeolites in Biomass Upgrading–Lessons from the Refining and Petrochemical Industry. *Catal. Today* **2015**, *243*, 10–22. https://doi.org/10.1016/j.cattod.2014.07.043.
114. Wang, L.; Xiao, F. S. Nanoporous Catalysts for Biomass Conversion. *Green Chem.* **2015**, *17* (1), 24–39. https://doi.org/10.1039/c4gc01622j.
115. De Klerk, A. Zeolites as Catalysts for Fuels Refining after Indirect Liquefaction Processes. *Molecules* **2018**, 23 (1), 1–17. https://doi.org/10.3390/molecules23010115.
116. Pavelić, K.; Hadžija, M.; Bedrica, L.; Pavelić, J.; Đikić, I.; Katić, M.; Kralj, M.; Bosnar, M. H.; Kapitanović, S.; Poljak-Blaži, M.; Križanac, Š.; Stojković, R.; Jurin, M.; Subotić, B.; Čolić, M. Natural Zeolite Clinoptilolite: New Adjuvant in Anticancer Therapy. *J. Mol. Med.* **2001**, *78* (12), 708–720. https://doi.org/10.1007/s001090000176.
117. Tanabe, K.; Misono, M.; Ono, Y.; Hattori, H. *New Solid Acids and Bases*; Elsevier: Amsterdam, 1989. ISBN 0-444-98800-9
118. Barrer, R. M. *Hydrothermal Chemistry of Zeolites*; Academy Press: London, New York, 1982.
119. Čejka, J.; Millini, R.; Opanasenko, M.; Serrano, D. P.; Roth, W. J. Advances and Challenges in Zeolite Synthesis and Catalysis. *Catal. Today* **2020**, *345*, 2–13. https://doi.org/10.1016/j.cattod.2019.10.021.
120. Boal, B. W.; Schmidt, J. E.; Deimund, M. A.; Deem, M. W.; Henling, L. M.; Brand, S. K.; Zones, S. I.; Davis, M. E. Facile Synthesis and Catalysis of Pure-Silica and Heteroatom LTA. *Chem. Mater.* **2015**, 27 (22), 7774–7779. https://doi.org/10.1021/acs.chemmater.5b03579.
121. Gendy, T. S. Simulation of Liquid and Vapour Phase Xylene Isomerization over Deactivating H-Y-Zeolite. *J. Chem. Technol. Biotechnol.* **1998**, 73 (2), 109–118. https://doi.org/10.1002/(SICI)1097-4660(1998100)73:2<109::AID-JCTB876>3.0.CO;2-I.
122. Zheng, S.; Jentys, A.; Lercher, J. A. Xylene Isomerization with Surface-Modified HZSM-5 Zeolite Catalysts: An in Situ IR Study. *J. Catal.* **2006**, *241* (2), 304–311. https://doi.org/10.1016/j.jcat.2006.04.026.

123. Pérez-Page, M.; Yu, E.; Li, J.; Rahman, M.; Dryden, D. M.; Vidu, R.; Stroeve, P. Template-Based Syntheses for Shape Controlled Nanostructures. *Adv. Colloid Interface Sci.* **2016**, *234*, 51–79. https://doi.org/10.1016/j.cis.2016.04.001.
124. Traa, Y.; Weitkamp, J. Characterization of the Pore Width of Zeolites and Related Materials by Means of Molecular Probes. *Handbook of Porous Solids.* **2002**, October 10, 1015–1057. https://doi.org/10.1002/9783527618286.ch18f.
125. Verboekend, D.; Milina, M.; Mitchell, S.; Pérez-Ramírez, J. Hierarchical Zeolites by Desilication: Occurrence and Catalytic Impact of Recrystallization and Restructuring. *Cryst. Growth Des.* **2013**, *13*, 5025–5035. https://doi.org/10.1021/cg4010483.
126. Holm, M.; Taarning, E.; Egeblad, K.; Christensen, C. Catalysis with Hierarchical Zeolites. *Catal. Today* **2011**, *168*, 3–16. https://doi.org/10.1016/j.cattod.2011.01.007.
127. Nishu; Liu, R.; Rahman, M. M.; Sarker, M.; Chai, M.; Li, C.; Cai, J. A Review on the Catalytic Pyrolysis of Biomass for the Bio-Oil Production with ZSM-5: Focus on Structure. *Fuel Process. Technol.* **2020**, *199*, 106301.
128. Li, J.; Li, X.; Zhou, G.; Wang, W.; Wang, C.; Komarneni, S.; Wang, Y. Catalytic Fast Pyrolysis of Biomass with Mesoporous ZSM-5 Zeolites Prepared by Desilication with NaOH Solutions. *Appl. Catal. A Gen.* **2014**, *470*, 115–122. https://doi.org/10.1016/j.apcata.2013.10.040.
129. Zhao, T.; Li, F.; Yu, H.; Ding, S.; Li, Z.; Huang, X.; Li, X.; Wei, X.; Wang, Z.; Lin, H. Synthesis of Mesoporous ZSM-5 Zeolites and Catalytic Cracking of Ethanol and Oleic Acid into Light Olefins. *Appl. Catal. A Gen.* **2019**, *575*, 101–110. https://doi.org/10.1016/j.apcata.2019.02.011.
130. Verboekend, D.; Pérez-Ramírez, J. Design of Hierarchical Zeolite Catalysts by Desilication. *Catal. Sci. Technol.* **2011**, *1* (6), 879–890. https://doi.org/10.1039/c1cy00150g.
131. Ben Tayeb, K.; Hamieh, S.; Canaff, C.; Nguyen, H.; Vezin, H.; Pinard, L. The Radical Internal Coke Structure as a Fingerprint of the Zeolite Framework. *Microporous Mesoporous Mater.* **2019**, *289*, 109617. https://doi.org/10.1016/j.micromeso.2019.109617.
132. Fu, X.; Sheng, X.; Zhou, Y.; Fu, Z.; Zhao, S.; Bu, X.; Zhang, C. Design of Micro–Mesoporous Zeolite Catalysts for Alkylation. *RSC Adv.* **2016**, *6* (56), 50630–50639. https://doi.org/10.1039/C6RA08099E.
133. Kelkar, S.; Saffron, C. M.; Li, Z.; Kim, S.-S.; Pinnavaia, T. J.; Miller, D. J.; Kriegel, R. Aromatics from Biomass Pyrolysis Vapour Using a Bifunctional Mesoporous Catalyst. *Green Chem.* **2014**, *16* (2), 803–812. https://doi.org/10.1039/C3GC41350K.
134. Ding, K.; Zhong, Z.; Wang, J.; Zhang, B.; Addy, M.; Ruan, R. Effects of Alkali-Treated Hierarchical HZSM-5 Zeolites on the Production of Aromatic Hydrocarbons from Catalytic Fast Pyrolysis of Waste Cardboard. *J. Anal. Appl. Pyrolysis* **2017**, *125*, 153–161. https://doi.org/10.1016/j.jaap.2017.04.006.
135. Stefanidis, S.; Kalogiannis, K.; Iliopoulou, E. F.; Lappas, A. A.; Triguero, J. M.; Navarro, M. T.; Chica, A.; Rey, F. Mesopore-Modified Mordenites as Catalysts for Catalytic Pyrolysis of Biomass and Cracking of Vacuum Gasoil Processes. *Green Chem.* **2013**, *15* (6), 1647–1658. https://doi.org/10.1039/C3GC40161H.
136. Jia, L. Y.; Raad, M.; Hamieh, S.; Toufaily, J.; Hamieh, T.; Bettahar, M. M.; Mauviel, G.; Tarrighi, M.; Pinard, L.; Dufour, A. Catalytic Fast Pyrolysis of Biomass: Superior Selectivity of Hierarchical Zeolites to Aromatics. *Green Chem.* **2017**, *19* (22), 5442–5459. https://doi.org/10.1039/C7GC02309J.

137. Na, K.; Somorjai, G. A. Hierarchically Nanoporous Zeolites and Their Heterogeneous Catalysis: Current Status and Future Perspectives. *Catal. Letters* **2015**, *145* (1), 193–213. https://doi.org/10.1007/s10562-014-1411-5.
138. Verboekend, D.; Mitchell, S.; Pérez-Ramírez, J. Hierarchical Zeolites Overcome All Obstacles: Next Stop Industrial Implementation. *Chimia (Aarau).* **2013**, *67* (5), 327–332. https://doi.org/10.2533/chimia.2013.327.
139. Sadowska, K.; Wach, A.; Olejniczak, Z.; Kuśtrowski, P.; Datka, J. Hierarchic Zeolites: Zeolite ZSM-5 Desilicated with NaOH and NaOH/Tetrabutylamine Hydroxide. *Microporous Mesoporous Mater.* **2013**, *167*, 82–88. https://doi.org/10.1016/j.micromeso.2012.03.045.
140. Pérez-Ramírez, J.; Verboekend, D.; Bonilla, A.; Abelló, S. Zeolite Catalysts with Tunable Hierarchy Factor by Pore-Growth Moderators. *Adv. Funct. Mater.* **2009**, *19* (24), 3972–3979. https://doi.org/10.1002/adfm.200901394.
141. Groen, J. C.; Moulijn, J. A.; Pérez-Ramírez, J. Desilication: On the Controlled Generation of Mesoporosity in MFI Zeolites. *J. Mater. Chem.* **2006**, *16* (22), 2121–2131. https://doi.org/10.1039/B517510K.
142. Groen, J.; Jansen, J.; Moulijn, J. A.; Pérez-Ramírez, J. Optimal Aluminum-Assisted Mesoporosity Development in MFI Zeolites by Desilication. *ChemInform* **2004**, *35*. https://doi.org/10.1021/jp047194f.
143. Serrano, D.; Aguado, J.; Rodríguez, J. M.; Peral, A. Catalytic Cracking of Polyethylene over Nanocrystalline HZSM-5: Catalyst Deactivation and Regeneration Study. *J. Anal. Appl. Pyrolysis - J ANAL APPL PYROL* **2007**, *79*, 456–464. https://doi.org/10.1016/j.jaap.2006.11.013.
144. Peng, P.; Gao, X.-H.; Yan, Z.-F.; Mintova, S. Diffusion and Catalyst Efficiency in Hierarchical Zeolite Catalysts. *Natl. Sci. Rev.* **2020**, *7* (11), 1726–1742. https://doi.org/10.1093/nsr/nwaa184.
145. Chuang, Y. Y.; Ou, M. F.; Ho, G.; Du, K. J.; Liu, S. Bin; Tsai, T. C. Selective Mono-Alkylbenzene Disproportionation over Silylated MFI Zeolite. *Catal. Today* **2020**, July, *1–7*. https://doi.org/10.1016/j.cattod.2020.08.020.
146. Zhou, J.; Wang, Y.; Zou, W.; Wang, C.; Li, L.; Liu, Z.; Zheng, A.; Kong, D.; Yang, W.; Xie, Z. Mass Transfer Advantage of Hierarchical Zeolites Promotes Methanol Converting into Para-Methyl Group in Toluene Methylation. *Ind. Eng. Chem. Res.* **2017**, *56* (33), 9310–9321. https://doi.org/10.1021/acs.iecr.7b02187.
147. Peng, P.; Wang, Y.; Zhang, Z.; Qiao, K.; Liu, X.; Yan, Z.; Subhan, F.; Komarneni, S. ZSM-5-Based Mesostructures by Combined Alkali Dissolution and Re-Assembly: Process Controlling and Scale-Up. *Chem. Eng. J.* **2016**, *302*, 323–333. https://doi.org/10.1016/j.cej.2016.05.027.
148. Lee, E. H.; Park, R.; Kim, H.; Park, S. H.; Jung, S.-C.; Jeon, J.-K.; Kim, S. C.; Park, Y.-K. Hydrodeoxygenation of Guaiacol over Pt Loaded Zeolitic Materials. *J. Ind. Eng. Chem.* **2016**, *37*, 18–21. https://doi.org/10.1016/j.jiec.2016.03.019.
149. Morawala, D. H.; Dalai, A. K.; Maheria, K. C. Synthesis of N-Butyl Levulinate Using Mesoporous Zeolite H-BEA Catalysts with Different Catalytic Characteristics. *Catal. Letters* **2019**, *0123456789*. https://doi.org/10.1007/s10562-019-03005-0.

150. Morawala, D.; Dalai, A.; Maheria, K. Rice Husk Mediated Synthesis of Meso-ZSM-5 and Its Application in the Synthesis of n-Butyl Levulinate. *J. Porous Mater.* **2019**, *26* (3), 677–686. https://doi.org/10.1007/s10934-018-0664-6.
151. Morawala, D. H.; Lathiya, D. R.; Dalai, A. K.; Maheria, K. C. TTAB Mediated Synthesis of Meso-H-BEA and Its Application in the Production of n-Butyl Levulinate. *Catal. Today* **2019**. https://doi.org/10.1016/j.cattod.2019.10.009.
152. Kalpana C. Maheria, S. R. M. Synthesis, Characterization of Mesoporous Zeolite BEA and Its Application in Multicomponent Reactions. *J. Catal. Catal.* **2014**, *1*.
153. Widayatno, W. B.; Guan, G.; Rizkiana, J.; Du, X.; Hao, X.; Zhang, Z.; Abudula, A. Selective Catalytic Conversion of Bio-Oil over High-Silica Zeolites. *Bioresour. Technol.* **2015**, *179*, 518–523. https://doi.org/10.1016/j.biortech.2014.12.046.
154. Huber, G. W.; Cheng, Y.-T.; Carlson, T.; Vispute, T.; Jae, J.; Tompsett, G. *Catalytic Pyrolysis of Solid Biomass and Related Biofuels, Aromatic, and Olefin Compounds*. U.S. Patent US20090227823A1, 2009.
155. Thilakaratne, R.; Brown, T.; Li, Y.; Hu, G.; Brown, R. Mild Catalytic Pyrolysis of Biomass for Production of Transportation Fuels: A Techno-Economic Analysis. *Green Chem.* **2014**, *16* (2), 627–636. https://doi.org/10.1039/C3GC41314D.
156. Uslamin, E. A.; Kosinov, N. A.; Pidko, E. A.; Hensen, E. J. M. Catalytic Conversion of Furanic Compounds over Ga-Modified ZSM-5 Zeolites as a Route to Biomass-Derived Aromatics. *Green Chem.* **2018**, *20* (16), 3818–3827. https://doi.org/10.1039/C8GC01528G.
157. Mortensen, P. M.; Grunwaldt, J. D.; Jensen, P. A.; Knudsen, K. G.; Jensen, A. D. A Review of Catalytic Upgrading of Bio-Oil to Engine Fuels. *Appl. Catal. A Gen.* **2011**, *407* (1–2), 1–19. https://doi.org/10.1016/j.apcata.2011.08.046.
158. Adjaye, J. D.; Bakhshi, N. N. Upgrading of a Wood-Derived Oil over Various Catalysts. *Biomass Bioenergy* **1994**, *7* (1), 201–211. https://doi.org/10.1016/0961-9534(94)00060-7.
159. Adjaye, J. D.; Bakhshi, N. N. Production of Hydrocarbons by Catalytic Upgrading of a Fast Pyrolysis Bio-Oil. Part II: Comparative Catalyst Performance and Reaction Pathways. *Fuel Process. Technol.* **1995**, *45* (3), 185–202. https://doi.org/10.1016/0378-3820(95)00040-E.
160. Bertero, M.; García, J. R.; Falco, M.; Sedran, U. Hydrocarbons from Bio-Oils: Performance of the Matrix in FCC Catalysts in the Immediate Catalytic Upgrading of Different Raw Bio-Oils. *Waste Biomass Valorization* **2017**, *8* (3), 933–948. https://doi.org/10.1007/s12649-016-9624-z.
161. Lanzafame, P.; Perathoner, S.; Centi, G.; Heracleous, E.; Iliopoulou, E. F.; Triantafyllidis, K. S.; Lappas, A. A. Effect of the Structure and Mesoporosity in Ni/Zeolite Catalysts for n-Hexadecane Hydroisomerisation and Hydrocracking. *ChemCatChem* **2017**, *9* (9), 1632–1640. https://doi.org/10.1002/cctc.201601670.
162. Tarach, K. A.; Góra-Marek, K.; Martinez-Triguero, J.; Melián-Cabrera, I. Acidity and Accessibility Studies of Desilicated ZSM-5 Zeolites in Terms of Their Effectiveness as Catalysts in Acid-Catalyzed Cracking Processes. *Catal. Sci. Technol.* **2017**, *7* (4), 858–873. https://doi.org/10.1039/C6CY02609E.
163. Matias, P.; Lopes, J. M.; Ayrault, P.; Laforge, S.; Magnoux, P.; Guisnet, M.; Ribeiro, F. R. Effect of Dealumination by Acid Treatment of a HMCM-22 Zeolite on the Acidity and Activity of the Pore Systems. *Appl. Catal. A Gen.* **2009**, *365* (2), 207–213. https://doi.org/10.1016/j.apcata.2009.06.014.

164. Gac, W.; Greluk, M.; Słowik, G.; Millot, Y.; Valentin, L.; Dzwigaj, S. Effects of Dealumination on the Performance of Ni-Containing BEA Catalysts in Bioethanol Steam Reforming. *Appl. Catal. B Environ.* **2018**, *237*, 94–109. https://doi.org/10.1016/j.apcatb.2018.05.040.
165. Jafarian, S.; Tavasoli, A. A Comparative Study on the Quality of Bioproducts Derived from Catalytic Pyrolysis of Green Microalgae Spirulina (Arthrospira) Plantensis over Transition Metals Supported on HMS-ZSM5 Composite. *Int. J. Hydrogen Energy* **2018**, *43* (43), 19902–19917. https://doi.org/10.1016/j.ijhydene.2018.08.171.
166. Zainan, N. H.; Srivatsa, S. C.; Bhattacharya, S. Catalytic Pyrolysis of Microalgae Tetraselmis Suecica and Characterization Study Using in Situ Synchrotron-Based Infrared Microscopy. *Fuel* **2015**, *161*, 345–354. https://doi.org/10.1016/j.fuel.2015.08.030.
167. Gao, L.; Sun, J.; Xu, W.; Xiao, G. Catalytic Pyrolysis of Natural Algae over Mg-Al Layered Double Oxides/ZSM-5 (MgAl-LDO/ZSM-5) for Producing Bio-Oil with Low Nitrogen Content. *Bioresour. Technol.* **2017**, *225*, 293–298. https://doi.org/10.1016/j.biortech.2016.11.077.
168. Chagas, B. M. E.; Dorado, C.; Serapiglia, M. J.; Mullen, C. A.; Boateng, A. A.; Melo, M. A. F.; Ataíde, C. H. Catalytic Pyrolysis-GC/MS of Spirulina: Evaluation of a Highly Proteinaceous Biomass Source for Production of Fuels and Chemicals. *Fuel* **2016**, *179*, 124–134. https://doi.org/10.1016/j.fuel.2016.03.076.
169. Rahman, N. A. A.; Fermoso, J.; Sanna, A. Effect of Li-LSX-Zeolite on the in-Situ Catalytic Deoxygenation and Denitrogenation of Isochrysis Sp. Microalgae Pyrolysis Vapours. *Fuel Process. Technol.* **2018**, *173*, 253–261. https://doi.org/10.1016/j.fuproc.2018.01.020.
170. Zainan, N. H.; Srivatsa, S. C.; Li, F.; Bhattacharya, S. Quality of Bio-Oil from Catalytic Pyrolysis of Microalgae Chlorella Vulgaris. *Fuel* **2018**, *223* (March 2017), 12–19. https://doi.org/10.1016/j.fuel.2018.02.166.
171. Qi, P.; Chang, G.; Wang, H.; Zhang, X.; Guo, Q. Production of Aromatic Hydrocarbons by Catalytic Co-Pyrolysis of Microalgae and Polypropylene Using HZSM-5. *J. Anal. Appl. Pyrolysis* **2018**, *136*, 178–185. https://doi.org/10.1016/j.jaap.2018.10.007.
172. Borges, F. C.; Xie, Q.; Min, M.; Muniz, L. A. R.; Farenzena, M.; Trierweiler, J. O.; Chen, P.; Ruan, R. Fast Microwave-Assisted Pyrolysis of Microalgae Using Microwave Absorbent and HZSM-5 Catalyst. *Bioresour. Technol.* **2014**, *166*, 518–526. https://doi.org/10.1016/j.biortech.2014.05.100.
173. Anand, V.; Sunjeev, V.; Vinu, R. Catalytic Fast Pyrolysis of Arthrospira Platensis (Spirulina) Algae Using Zeolites. *J. Anal. Appl. Pyrolysis* **2016**, *118*, 298–307. https://doi.org/10.1016/j.jaap.2016.02.013.
174. Thangalazhy-Gopakumar, S.; Adhikari, S.; Chattanathan, S. A.; Gupta, R. B. Catalytic Pyrolysis of Green Algae for Hydrocarbon Production Using H+ZSM-5 Catalyst. *Bioresour. Technol.* **2012**, *118*, 150–157. https://doi.org/10.1016/j.biortech.2012.05.080.
175. Dong, X.; Chen, Z.; Xue, S.; Zhang, J.; Zhou, J.; Liu, Y.; Xu, Y.; Liu, Z. Catalytic Pyrolysis of Microalga Chlorella Pyrenoidosa for Production of Ethylene, Propylene and Butene. *RSC Adv.* **2013**, *3* (48), 25780–25787. https://doi.org/10.1039/C3RA43850C.
176. Du, Z.; Hu, B.; Ma, X.; Cheng, Y.; Liu, Y.; Lin, X.; Wan, Y.; Lei, H.; Chen, P.; Ruan, R. Catalytic Pyrolysis of Microalgae and Their Three Major Components: Carbohydrates, Proteins, and Lipids. *Bioresour. Technol.* **2013**, *130*, 777–782. https://doi.org/10.1016/j.biortech.2012.12.115.

177. Conti, R.; Pezzolesi, L.; Pistocchi, R.; Torri, C.; Massoli, P.; Fabbri, D. Photobioreactor Cultivation and Catalytic Pyrolysis of the Microalga Desmodesmus Communis (Chlorophyceae) for Hydrocarbons Production by HZSM-5 Zeolite Cracking. *Bioresour. Technol.* **2016**, *222*, 148–155. https://doi.org/10.1016/j.biortech.2016.10.002.
178. Du, Z.; Ma, X.; Li, Y.; Chen, P.; Liu, Y.; Lin, X.; Lei, H.; Ruan, R. Production of Aromatic Hydrocarbons by Catalytic Pyrolysis of Microalgae with Zeolites: Catalyst Screening in a Pyroprobe. *Bioresour. Technol.* **2013**, *139*, 397–401. https://doi.org/10.1016/j.biortech.2013.04.053.
179. Sirous Rezaei, P.; Shafaghat, H.; Daud, W. M. A. W. Suppression of Coke Formation and Enhancement of Aromatic Hydrocarbon Production in Catalytic Fast Pyrolysis of Cellulose over Different Zeolites: Effects of Pore Structure and Acidity. *RSC Adv.* **2015**, *5* (80), 65408–65414. https://doi.org/10.1039/C5RA11332F.

Chapter 5

Glucose Decomposition with Internal Hydrogen Production and Deoxygenation via Nanocasted Ceria–Zirconia Catalysts

Kyle A. Rogers[1] and Ying Zheng[*,1,2]

[1]Department of Chemical Engineering, University of New Brunswick, 15 Dineen Drive, Fredericton, NB, Canada E3B 5A3

[2]Department of Chemical and Biochemical Engineering, Western University, 1150 Richmond Street, London, ON, Canada N6A 3K7

[*]Email: ying.zheng@uwo.ca

Catalysts supported on ceria–zirconia nanocasted from MCM-41 (CZ-41) were used to perform internal hydrogen production and deoxygenation during the decomposition of glucose in a horizontally stirred reactor. The two catalysts tested were NiFe/CZ-41 and trimetallic NiCoFe/CZ-41. The addition of Co was found to make the overall catalyst more reducible, as CoFe interacts strongly with the support. Although the catalysts provided comparable deoxygenation results, the trimetallic catalyst yielded the most liquid product. Meanwhile, CZ-41 was found to significantly affect glucose decomposition by promoting dehydration. For the batch reactor, the optimal reaction temperature was determined to be 350 °C.

Introduction

The thermolysis of biomass material can lead to the production of useful gases, such as H_2, H_2O, CO, and CH_4 (*1*, *2*). The bio-oil produced from biomass decomposition requires upgrading via hydrodeoxygenation processes, which requires hydrogen. Therefore, it is advantageous to utilize the gases produced during the initial thermolysis stage to perform some preliminary deoxygenation, thereby reducing the hydrogen requirements of downstream processes.

Previous work demonstrated that in situ deoxygenation during the decomposition of glucose (as a model for cellulose) is possible using monometallic and bimetallic catalysts supported on SiO_2 (*3*, *4*). In addition, the catalysts demonstrated the potential of increasing the overall solid conversion to useful liquid products. However, such work was conducted using atmospheres containing externally sourced H_2. Herein, attention is given to catalysts that promote the production of additional H_2 internally and use it for in situ deoxygenation as the sole source of H_2. As in previous studies, there is a high demand for the catalysts to promote selective deoxygenation, that is, to selectively use H_2 to remove oxygen rather than to saturate carbon double bonds.

© 2021 American Chemical Society

Potential methods to consider for internal H_2 production include the water–gas shift reaction (WGSR), steam reforming, and tar cracking. Florin and Harris (5) and Widyawati et al. (*6*) both demonstrated the possibility of using the WGSR to produce H_2 during biomass pyrolysis using CaO to help capture the CO_2, thermodynamically pushing the WGSR to produce more H_2. The same research group studied the use of SiO_2-supported Ni and Co catalysts for H_2 production (*7–10*) during cellulose pyrolysis via tar cracking and steam reforming. Reportedly, this is enabled by a buildup of steam between 300 and 425 °C, resulting in elevated H_2 production via reforming (and the WGSR) as temperatures approach 400 °C (*8*, *10*). However, excessive promotion of the cracking and reforming reactions may lead to a reduction in liquid product yield, as H_2 may be produced at the cost of lighter components. Thus, a balance must be met in terms of producing internal H_2 and maintaining liquid product yields. This can be done by using the WGSR option at reduced temperatures to avoid significant reforming.

Ideally, the catalyst that is used for promoting the WGSR should also be active toward in situ deoxygenation. Ceria and ceria–zirconia catalysts, which use ZrO_2 to provide stability to CeO_2, have been shown to be active toward both deoxygenation (*11–13*) and the WGSR (*14–17*). Watanabe et al. (*15*), for example, used a bimetallic NiFe catalyst supported on CeO_2–ZrO_2 (CZ) nanocasted from KIT-6 and reported that the oxygen storage and transport capabilities of the CZ assisted in the bifunctional roles of Ni and Fe. CZ has been shown to be effective in deoxygenating guaiacol because of its Lewis acid sites (oxygen vacancies). However, Schimming et al. (*11*) suggested that Lewis acidity may also have an influence on oligomerization. Vasconcelos et al. (*18*) reported that CZ was useful for the gas-phase dehydration of glycerol within the temperature range of 150–350 °C because of the acid–base cooperativity of moderately basic sites and Lewis acid sites on $Ce_{0.8}Zr_{0.2}O_2$. During glucose decomposition, dehydration reactions may be important for the initial deoxygenation of intermediates (*19*, *20*).

In previous studies, Ni-based catalysts were determined to be valuable for their ability to promote dehydration in the early stages of glucose decomposition and to reduce organic solid residue by promoting mild tar cracking (3, 4). Therefore, the catalyst's effects on biomass decomposition must be seriously considered. Chloride salts acting as Lewis acids have been shown to significantly affect biomass decomposition (*21*, *22*). Carvalho et al. (*21*), for example, demonstrated that the addition of $ZnCl_2$ to biomass reduced the temperatures at which high decomposition rates were observed. As a Lewis acid, $ZnCl_2$ promoted early C–C and C–O bond breakage, which resulted in heavy dehydration and char formation.

Although scarce, some work has been conducted using heterogeneous catalysts to improve biomass pyrolysis. A NiCu/γ-Al_2O_3 catalyst was shown to have an effect on the pyrolysis of coffee grounds by Kan et al. (23). Introducing the catalyst resulted in an increase in CO and CO_2 production and a change in decomposition rates within the range of 360–513 °C. Ding et al. (*24*) reported that CeO_2 was capable of removing oxygen from the vapors produced during the pyrolysis of corn stover and attributed this effect to the catalyst's ketonization ability. In their study, the original biomass material was separated from the catalyst by quartz wool, and therefore, there were no immediate interactions between the biomass and the catalyst. Jeon et al. (25) similarly performed catalytic pyrolysis using mesoporous catalysts separated from the biomass using quartz and reported that the acidic sites of AlSBA-15 helped promote dehydration and cracking of the pyrolysis vapors.

In the present study, ceria–zirconia catalysts nanocasted from MCM-41 (denoted as CZ-41) with a Ce/Zr molar ratio of 67:33 were used in a horizontally stirred batch reactor for the thermal decomposition of glucose, internal production of H_2 (via the WGSR), and in situ deoxygenation. Catalysts with the compositions 2%Ni4%Fe/CZ-41 and 2%Ni0.5%Co4%Fe/CZ-41 were selected for tests of their ability to perform internal deoxygenation and hydrogen production during the decomposition of glucose. No H_2 was added to the system, so the catalysts had to utilize gases formed from the decomposition reactions to produce internal H_2. This study also focused on the effects of the catalysts on glucose decomposition. As in previous works (*3, 4*), furanic products such as 2-methylfuran, 2,5-dimethylfuran (DMF), and furfural were used as indicators for in situ deoxygenation activity because of their ease of identification and the difficulty with which they undergo deoxygenation. The selected catalysts and reaction conditions were also used for cellulose decomposition to glean the effects of the catalyst on real biomass decomposition.

Experimental Details

Catalyst Preparation

Ceria–zirconia nanocasted from MCM-41 based on a modified templating protocol (*26, 27*), denoted as CZ-41, was used because it was shown to have superior physical properties (203 m^2/g Brunauer–Emmett–Teller surface area). The catalysts 2%Ni4%Fe/CZ-41 and 2%Ni0.5%Co4%Fe/CZ-41 were prepared by sequential addition using the incipient wetness impregnation method. For these catalysts, the precursors for Fe or CoFe (iron nitrate and cobalt nitrate) were added to the support first and then calcined (600 °C for 6 h), after which Ni (in the form of nickel nitrate) was added and the samples were calcined again. Prior to activity tests, impregnated catalysts were reduced at 550 °C for 2 h in flowing hydrogen.

Catalyst Characterization

X-ray diffraction (XRD) analyses were performed on prereduced catalysts using a Bruker AXS D8 XRD apparatus within the 2θ range of 20–80° (0.02°/s) at the University of New Brunswick Geochemical and Spectrographic Facility. Temperature-programmed reduction (TPR) analyses of the catalysts were performed using an Autosorb 1-C instrument paired with an RGA200 mass spectrometer. For each analysis, 150 mg of catalyst was first vacuum dried in situ at 300 °C for 20 min under helium flowing at 50 nmL/min in a quartz U-cell (where nmL/min represents units of milliliters per minute at normal temperature and pressure conditions of 20 °C and 1 atm absolute). TPR analysis was then performed by flowing a mixture of He and H_2 through the U-cell (50 nmL/min He with 1% H_2) and heating from room temperature to 900 °C at 10 °C/min.

Catalytic Activity Tests

Activity tests for glucose decomposition with in situ deoxygenation using internally produced H_2 were conducted using a horizontally stirred (200 rpm) 55 mL 316L stainless steel reactor with 800 mg of glucose. Catalyst was added at a 1:5 catalyst-to-glucose mass ratio, and a solvent, 1-methylnaphthalene (1MN), was added at a 10:1 solvent-to-glucose mass ratio. The reactor was flushed with He to remove oxygen and then filled with He to a pressure of 310 kPaA (kilopascals, absolute). The reactor was then heated to the desired temperature (300, 325, or 350 °C) and held for the desired reaction time (0.5, 1, of 1.5 h). Cellulose decomposition was tested in a similar manner using the same reactor, feed ratios, and conditions. The reactor was fitted with internal baffles that allowed solids to the carried up the wall, similar to a ball mill, allowing for appropriate contact with both the liquid and gas phases. The validity of the reactor design was verified in previous work (*3, 4*), where the WGSR was shown to occur inside the reactor loaded with 1MN, water, and CO in the presence of catalyst.

Testing of the product gas was performed using an RGA200 mass spectrometer once the reactor had cooled to 22.5 °C. An HP-PLOT/U column held at 30 °C was used to ensure separation of gases such as CO and CO_2. Solids were separated from the liquid product using a centrifuge, washed with acetone, dried overnight at 80 °C, and then weighed. The H/C molar ratio of the solid residue was determined using a Perkin Elmer 2400 CHNS/O Series II elemental analyzer at the Limerick Pulp and Paper Center at the University of New Brunswick. Liquid products were subjected to qualification and quantification analyses by gas chromatography–mass spectrometry (GC–MS; Shimadzu GCMS QP5000) and GC with flame ionization detection (Varian GC450), respectively.

Supplementary analyses of the effects of the catalysts on glucose decomposition were conducted by thermogravimetric analysis (TGA) performed using a TA Instruments TGA Q500 instrument. For these analyses, glucose and catalysts were loaded onto platinum TGA pans at a 1:5 catalyst-to-glucose mass ratio. The TGA sample chamber was heated from room temperature to 950 °C at a rate of 5 °C/min under flowing He (50 nmL/min). The use of alumina pans with CZ-41 catalysts is not recommended, as the inherent small particle sizes of CZ-41 (the CZ-41 structure resembles an agglomeration of nanoparticles) can permanently damage the alumina pans.

At the end of the activity tests, solid residue was found to be bound to the walls of the reactor. As a result, the solid residue was extremely difficult to remove, especially when no catalysts were added. Rigorous cleaning had to be performed to successfully remove all solid content from the reactor. This involved loading the reactor with water, sodium bicarbonate, and hydrogen peroxide and then heating the reactor to 95–100 °C for 2 h. Upon completion of this heating process, the solid content could be easily removed. To account for the solids that were not collected previously with the liquid sample, the reactor itself was dried at 80 °C (after being rinsed with acetone) and weighed.

Results

Catalyst Characterization

XRD analyses of CZ-41, 2%Ni4%Fe/CZ-41, and 2%Ni0.5%Co4%Fe/CZ-41 (see Figure 1) showed no significant changes to the spectrum of CZ-41 upon addition of Ni, Fe, and Co. No peaks were detected representing the presence of either metal or alloy, suggesting that the metals were highly dispersed on the surface.

TPR analysis of the catalysts showed significant differences among the three tested materials. As can be seen in Figure 2, CZ-41 is reducible, generating oxygen vacancies as low as ~380 °C and reaching significant hydrogen consumption by ~630 °C. Two peaks were observed for CZ-41 reduction. The first reduction peak (~630 °C) was likely due to surface reduction, and the second was attributed to bulk reduction (~850 °C) (*11*). The addition of Ni, Fe, and Co decreased the overall catalyst reduction temperature and appeared to increase the overall hydrogen consumption. Compared to the support alone, the NiFe catalyst started to consume hydrogen earlier, and the overall peaks became broader as reduction of nickel oxide started at lower temperatures than that of the CZ-41 itself.

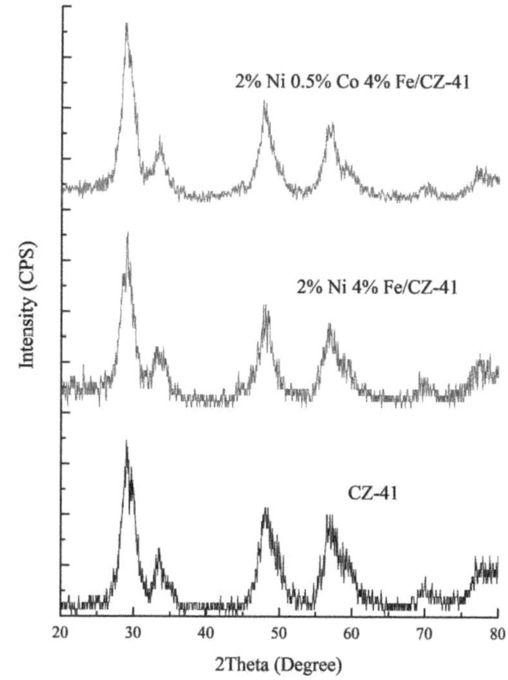

Figure 1. XRD spectra of CZ-41–supported catalysts after reduction.

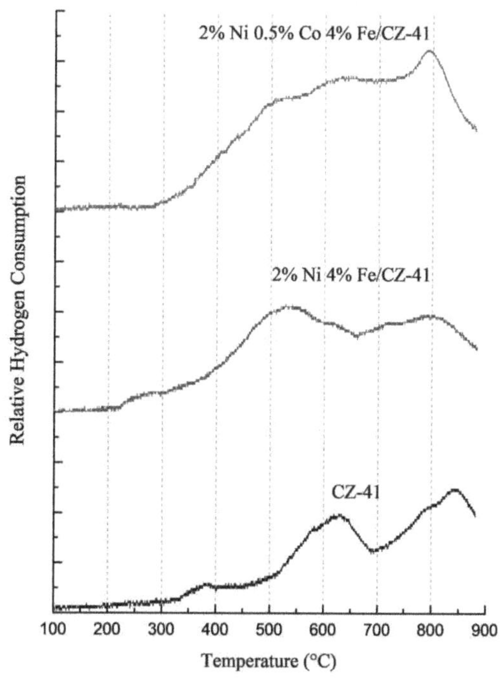

Figure 2. TPR hydrogen consumption profiles for CZ-41–supported catalysts. Conditions: catalyst, 150 mg; gas flow, 50 nmL/min at 1% H_2 in He; temperature ramp, 10 °C/min.

A significant difference can be observed in the TPR profile of the NiCoFe catalyst, which had a large and very broad hydrogen consumption profile with a small peak on top at ~890 °C. This demonstrates that the addition of 0.5 wt % Co to the Fe significantly altered the reducibility of the catalyst, as intended. With such a large effect on the overall reducibility, it is suspected that the active metal phases, Ni and CoFe, have strong interactions with the support. These results are similar to those of Tepamatr et al. (*28*), who reported that the addition of Re to a Cu catalyst supported on Gd-doped ceria helped the Cu reduce the ceria and produce oxygen vacancies. For the NiCoFe catalyst presented herein, the proposed reduction mechanism involves the following steps: Ni in general facilitates the activation of hydrogen by dissociation and donates the hydrogen to the CoFe phase, where it is used to remove oxygen as water. The ceria–zirconia support, in addition to undergoing hydrogen dissociation and oxygen removal itself, readily transfers oxygen to the CoFe phase, where it is again removed as water with the help of Ni, allowing for a rapid, deep reduction of the entire catalyst.

The significant increase in catalyst reducibility highlights more facile oxygen transfer (and subsequent removal) between the cobalt-modified iron phase and the CZ-41 support. More facile oxygen transfer between active metals and ceria–zirconia makes better use of the support's oxygen storage capabilities, improving its activity and preventing its deactivation (*11*, *12*). The ability to transfer oxygen to the support for storage may allow the iron phase to be readily available for adsorbing oxygenated compounds, such as furfural, permitting deoxygenation.

Catalyst Activity Tests

NiFe/CZ-41 versus NiCoFe/CZ-41

For the process of selecting between 2%Ni4%Fe/CZ-41 and 2%Ni0.5%Co4%Fe/CZ-41, activity tests were performed with and without addition of the catalysts for 1 h at 350 °C. These catalysts demonstrated their ability to reduce organic solid material. As can be seen in Figure 3, the NiCoFe catalyst was superior in terms of both solid conversion and liquid product yield. The NiFe catalyst had a liquid yield similar to that of the base case without catalyst because of higher gas yields. The NiCoFe, which had gas yields (see Table 1) similar to those of the NiFe catalyst, achieved a higher liquid yield because of the higher overall conversion of organic solids. In addition to reducing the solid content, the catalysts increased the H/C molar ratio of the organic solids (see Figure 3), which altogether explains why solid material was easier to remove from the reactor when catalysts were involved. This could demonstrate the addition of internally produced hydrogen to tar leading to a reduction in char formation.

The improved reduction in solid residue with the NiCoFe catalyst is likely due to improved catalyst stability rather than improved tar cracking activity. Tar cracking activity has largely been attributed to monometallic Ni phases (*7–9*, *29*). In studies of the effects of both monometallic (*3*) and bimetallic (*4*) catalysts, neither Co nor CoFe catalysts have been found to display significant activity toward tar cracking. Therefore, the improved reduction in solid residue is due to the strong interaction between the NiCoFe and the CZ-41 support. As can be seen in the TPR results, the NiCoFe catalyst is more reducible than the NiFe catalyst because of an improved oxygen transfer mechanism between the CZ support and the CoFe phases. The ability to transfer and store oxygen on the support makes CoFe less prone to deactivation by oxidation.

As can be seen in Table 1, the catalysts did not have a huge effect on the overall output of H_2. H_2 could have been produced and readily consumed. Along with a very slight increase in H_2, a slight decrease in CO and an increase in CO_2 were observed upon catalyst addition, suggesting WGSR activity. Ultimately, it was difficult to confirm that significant WGSR activity occurred because H_2 could be rapidly consumed.

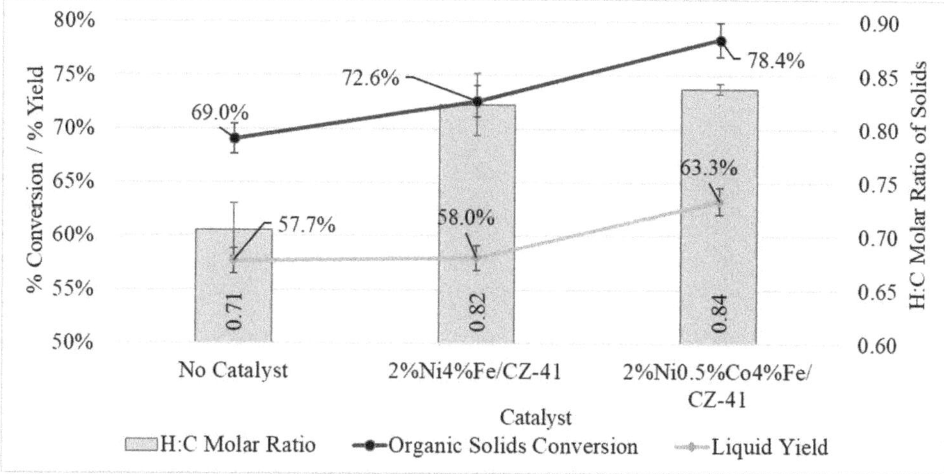

Figure 3. Organic solid conversion, liquid yield, and H/C molar ratio (solid residue) results from tests using CZ-41–supported catalysts. Reaction temperature, 350 °C; reaction time, 1 h.

Table 1. Product Gas Composition Results from Tests Using CZ-41–Supported Catalysts[a]

	Product Gas Partial Pressure (kPaA) at 22.5°C			
Catalyst	H_2	CO_2	CO	CH_4
None	6	78	37	5
2%Ni4%Fe/CZ-41	8	112	30	6
2%Ni0.5%Co4%Fe/CZ-41	9	116	30	6

[a] Reaction temperature, 350 °C; reaction time, 1 h.

Addition of the catalysts significantly reduced the relative selectivity toward the furanic intermediates furfural and 5-methylfurfural, instead favoring the net production of the desired 2-methylfuran (2MF) product (see Figure 4). This demonstrated that both catalysts were capable of producing hydrogen and performing in situ deoxygenation. The addition of the catalysts also netted small portions of furfuryl alcohol (FA), a product of furfural hydrogenation, which is considered an intermediate step in the deoxygenation process to produce 2MF (*30–32*). Compared to the NiFe catalyst, the NiCoFe catalyst exhibited a slightly higher selectivity toward 2,3,5-trimethylfuran and 2-acetylfuran, which indicates side reactions such as transalkylation or the production of maltol and isomaltol followed by deoxygenation (*33, 34*). Neither maltol nor isomaltol were identified within the liquid products.

Furan was a suspected product, but it was not possible to distinguish it from other potential products such as acetone. The two compounds are noted as having similar retention indices. The GC–MS spectra showed what looked like a combination of acetone and furan mass spectra.

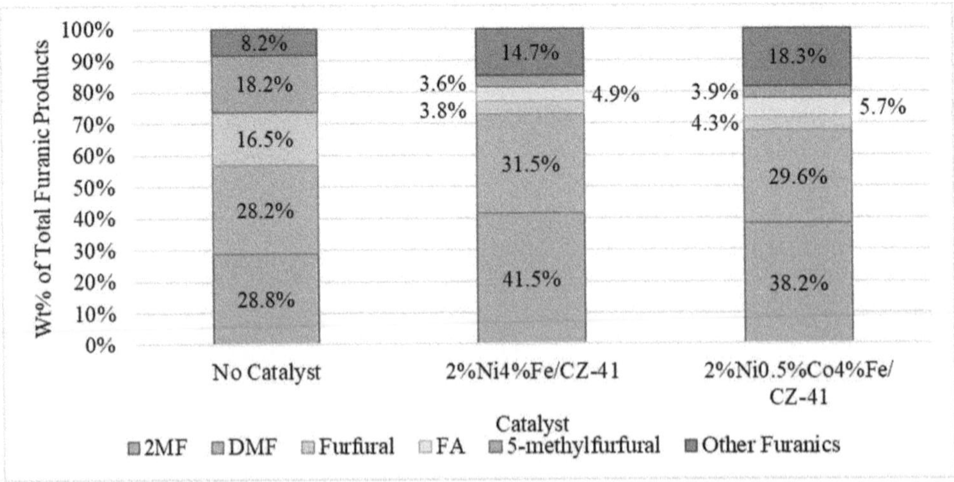

Figure 4. Distribution and relative selectivities of detectable furanic compounds from use of CZ-41–supported catalysts. Reaction temperature, 350 °C; reaction time, 1 h.

Thermogravimetric Analysis

Thermogravimetric analyses were performed to provide a better understanding of the effects of the CZ-41 catalysts on glucose decomposition. As can be seen in Figure 5, the addition of the CZ-41 catalysts, including CZ-41, had a very noticeable effect on glucose decomposition. With the catalysts, weight loss occurred ~50 °C sooner and was much more rapid at ~200 °C. Although the catalysts were added at a 1:5 ratio with the glucose, at temperatures greater than 650 °C, the remaining weight dipped below 16.7%, demonstrating that a weight loss was also associated with the catalyst—potentially the release of oxygen from the catalyst surface as CO_2 at high temperatures. Pairing the TGA instrument with an RGA200 mass spectrometer to monitor outlet gases revealed the release of H_2O and CO_2 during the early stages of weight loss, as well as the release of H_2, CO, and CO_2 at high temperatures (>650 °C).

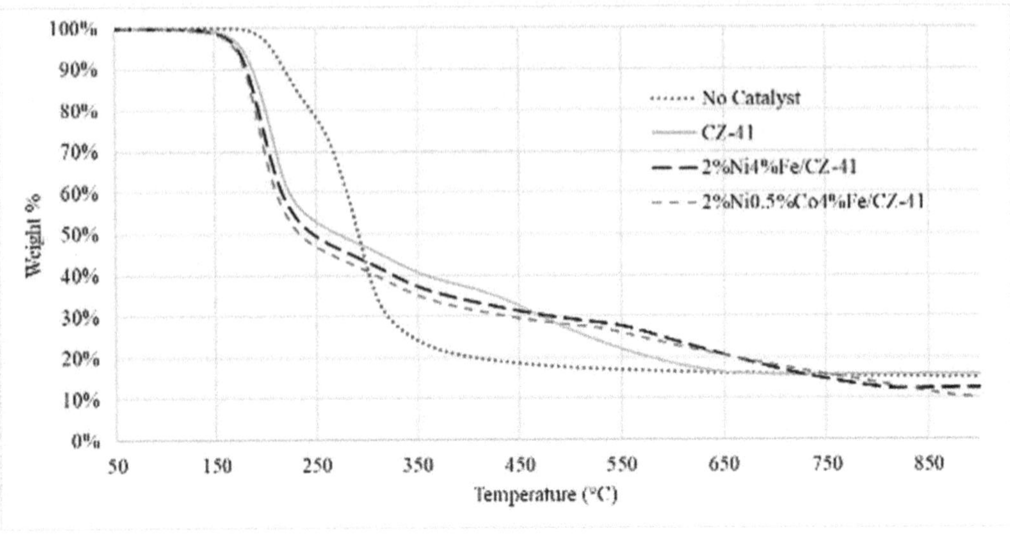

Figure 5. TGA results for glucose with the addition of CZ-41 catalysts at a 5:1 glucose-to-catalyst ratio.

Differential thermal analysis (DTA, or negative weight loss derivative) of the glucose with the catalysts further revealed that the CZ-41 catalysts significantly altered the reaction pathway for glucose decomposition (see Figure 6). Glucose on its own decomposes in two periods, the first being its initial dehydration. Addition of the catalysts caused a more rapid weight loss during the early dehydration stages. Given that such a massive shift also occurred with unpromoted CZ-41, CZ-41 clearly contributes the most. This is likely due to the Lewis acidity of the CZ-41 material, as Lewis acids have been shown to promote significant dehydration of sugars (*21*, *22*).

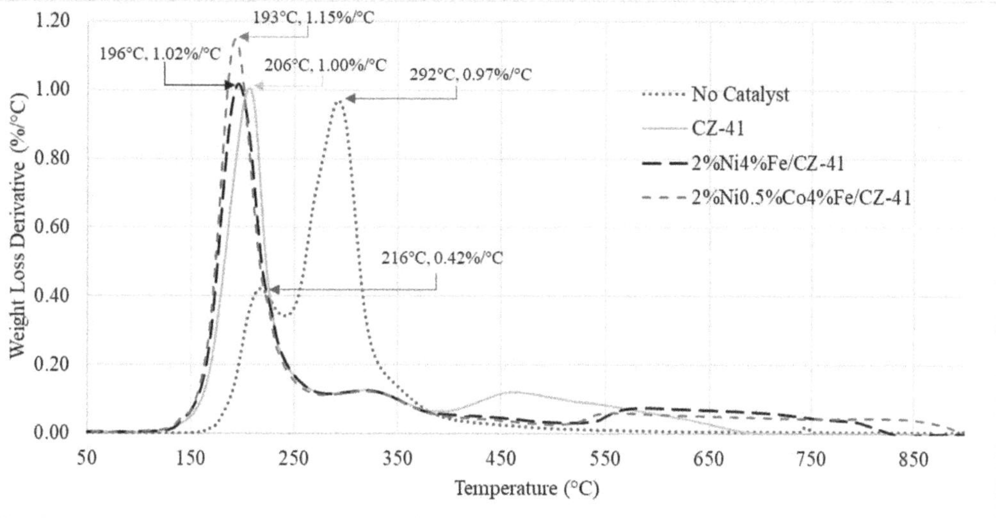

Figure 6. DTA (weight loss derivative from TGA) results for glucose with the addition of CZ-41 catalysts at a 5:1 glucose-to-catalyst ratio.

Again, the 2%Ni0.5%Co4%Fe/CZ-41 catalyst had a greater effect than the 2%Ni4%Fe/CZ-41 catalyst. This could be because of CoFe having a greater interaction with CZ-41 or because of CoFe's own reaction with glucose, as it was previous demonstrated that 0.5%Co4%Fe/SiO_2 also has an effect on glucose decomposition, favoring earlier dehydration (*4*).

Effects of Temperature

A selected catalyst was tested at various temperatures to further optimize the system for liquid product yield and in situ deoxygenation. The effect of temperature is extremely important, as Co and Ni catalysts are known to favor tar cracking and reforming reactions at temperatures greater than 350 °C because of a buildup of steam (*8*, *10*). Steam buildup is a strong possibility in the batch reactor used in this system, especially because the catalysts promote the dehydration of glucose. Although reforming would significantly improve internal H_2 production, it would inadvertently lower the liquid product yields. As such, substantial reforming activity is to be avoided here. The catalyst that was selected was 2%Ni0.5%Co4%Fe/CZ-41, as this catalyst had the greatest liquid product yields with gas and liquid product compositions comparable to those of the 2%Ni4%Fe/CZ-41 catalyst.

Increasing temperatures of 300, 325, and 350 °C, as shown in Figure 7, increased the solids conversion and liquid product yield. This is no surprise, as it was expected that increasing the temperature would help with the decomposition of glucose and also improve tar cracking activity. Conversely, the H/C molar ratio decreased slightly with temperature. In conjunction with the gas results from Table 2, it is clear that increasing temperature helps retain hydrogen in the liquid product.

Evidently, increased temperatures help liberate some hydrogen content from the solids, possibly through tar cracking.

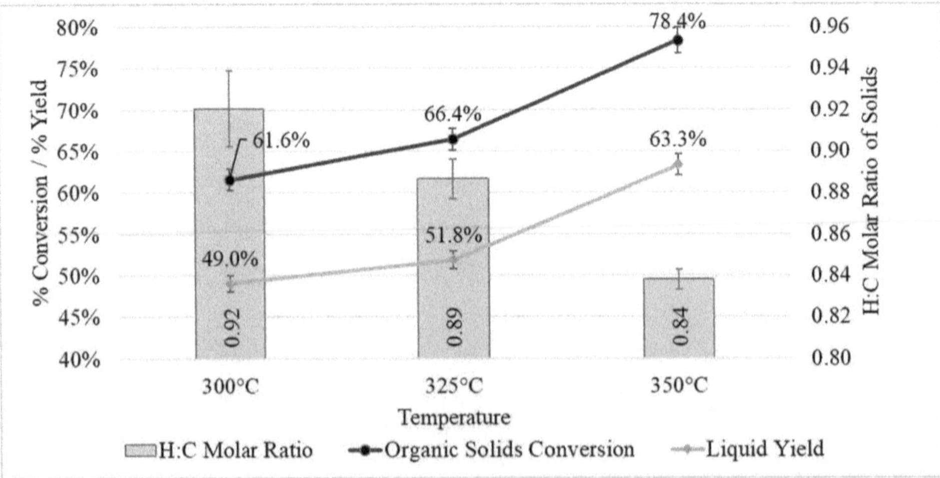

Figure 7. Effects of temperature on organic solids conversion, liquid product yield, and H/C molar ratio of solid residue. Catalyst, 2%Ni0.5%Co4%Fe/CZ-41; reaction time, 1 h.

Table 2. Effects of Temperature on Product Gas Composition[a]

	Product Gas Partial Pressure (kPaA) at 22.5°C			
Temperature (°C)	H_2	CO_2	CO	CH_4
300	9	99	22	5
325	8	112	24	5
350	9	116	30	6

[a] Catalyst, 2%Ni0.5%Co4%Fe/CZ-41; reaction time, 1 h.

Although the outputs of H_2 and CH_4 appeared to remain unchanged with increasing temperature, the CO and CO_2 outputs did increase. The increase in CO_2 and the retention of hydrogen in the liquid products suggest WGSR activity; however, H_2 was rapidly consumed. It is possible that, during the WGSR, H_2 was not released from the catalyst surface but instead remained disassociated on the surface where it could be readily consumed for deoxygenation. This would be beneficial, as it would reduce the mass-transfer steps of H_2 to and from the catalyst surface—as was the intent of developing a bifunctional catalyst.

The lack of difference in CO_2 outputs at 325 and 350 °C suggests that there may not be a significant increase in WGSR activity with increasing temperature. Indeed, the WGSR, as an exothermic reaction, can have reduced conversion at elevated temperatures because of equilibrium. Typical iron-based catalysts for the industrial WGSR tend to have these issues at temperatures exceeding 350 °C (35).

As can be seen in Figure 8, increased temperatures decreased the fractions of furfural and 2-methylfurfural in the furanic products while slightly increasing the selectivity toward DMF, demonstrating that the temperature gains may be beneficial for in situ deoxygenation. However, it is noted that 2-acetylfuran selectivity also increased, which currently has an unknown origin. Because of the higher liquid product yields and decent 2MF and DMF fractions, 350 °C was selected as the most suitable temperature.

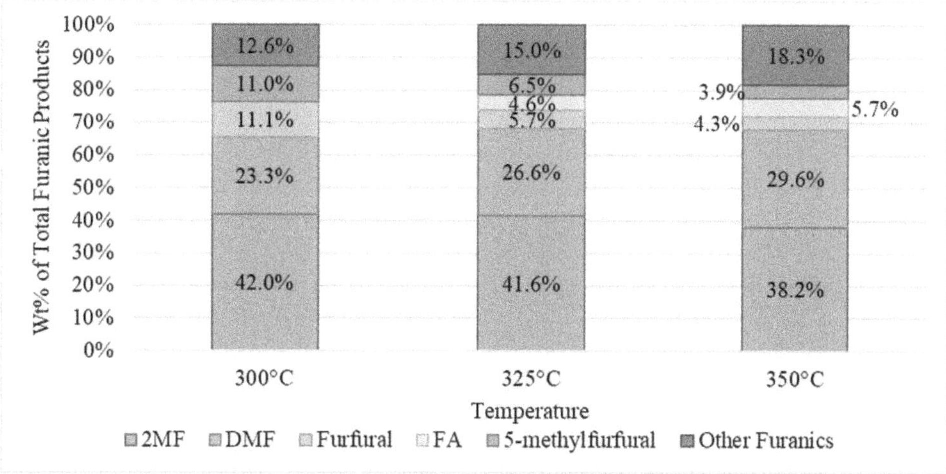

Figure 8. Effects of temperature on total furanic product distribution and relative selectivities. Catalyst, 2%Ni0.5%Co4%Fe/CZ-41; reaction time, 1 h.

Effects of Reaction Time

To further optimize the reaction system, time trials were performed at 350 °C with the 2%Ni0.5%Co4%Fe/CZ-41 catalyst with additional trials at reaction times of 0.5 and 1.5 h. As can be seen in Figure 9, reducing the reaction time from 1 to 0.5 h resulted in a lower organic solids conversion and less liquid product. Increasing the reaction time to 1.5 h caused slight decreases in the organic solids conversion and liquid yield. Among the three time trials, the H/C molar ratio changed insignificantly. Over extended reaction times, it is suspected that side reactions such as oligomerization or reforming may occur, leading to the production of heavy compounds and H_2, respectively.

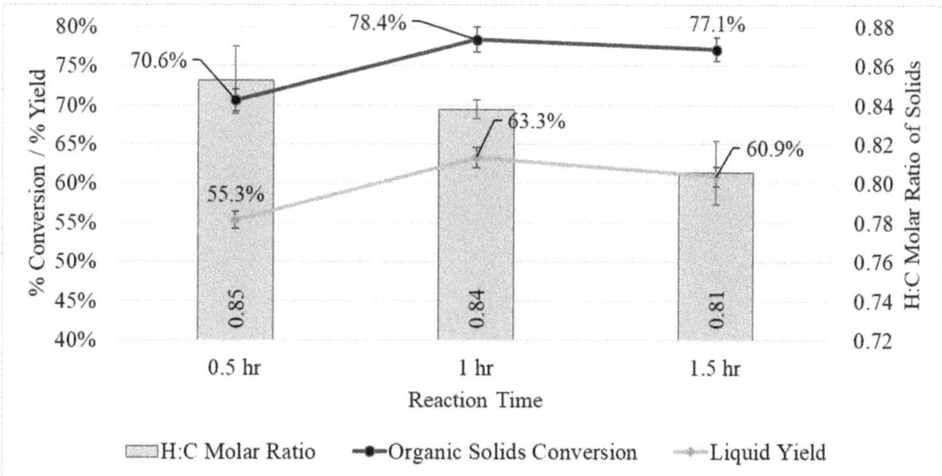

Figure 9. Effects of reaction time on organic solids conversion, liquid product yield, and H/C molar ratio of solid residue. Catalyst, 2%Ni0.5%Co4%Fe/CZ-41; reaction temperature, 350 °C.

Reaction time did not have a significant effect on gas composition aside from CO_2 output, which changed slightly (see Table 3). Significant reforming is not suspected to have occurred as there was no change in either H_2 content or CH_4. As can be seen in Figure 10, the relative selectivities of the furanic compounds did not change much with reaction time: the 2MF and DMF selectivities changed slightly. This independence from reaction time indicates that in situ deoxygenation occurs quickly. Therefore, improving in situ deoxygenation would depend on improving the internal hydrogen production (whether it is gaseous or surface bound). Above all, for the reaction system used herein, a reaction time of 1 h was found to be most suitable because of the resulting higher liquid product yields.

Table 3. Effects of Reaction Time on Product Gas Composition[a]

	Product Gas Partial Pressure (kPaA) at 22.5°C			
Reaction Time (h) at 350 °C	H_2	CO_2	*CO*	CH_4
0.5	8	120	27	6
1	9	116	30	6
1.5	8	128	28	6

[a] Catalyst, 2%Ni0.5%Co4%Fe/CZ-41; reaction temperature, 350 °C.

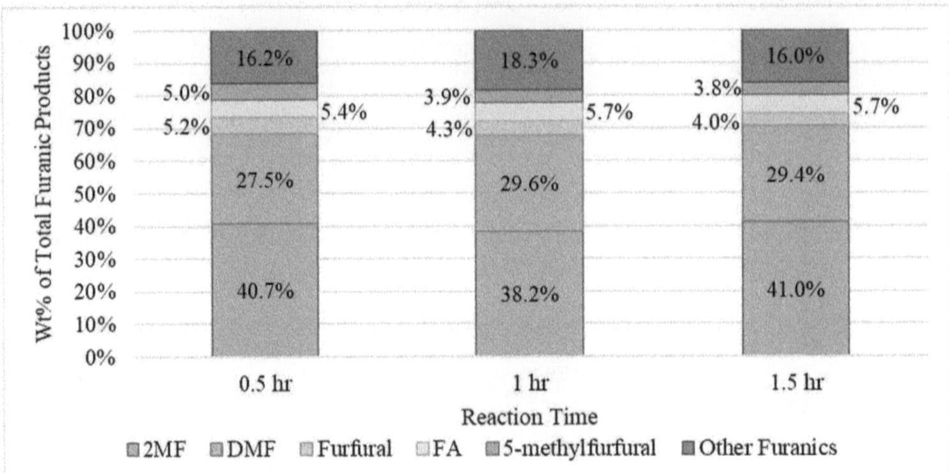

Figure 10. Effects of reaction time on total furanic product distribution and relative selectivities; Catalyst: 2%Ni0.5%Co4%Fe/CZ-41, Reaction temperature: 350 °C.

Cellulose Decomposition

The purpose of this study was to demonstrate the effects that the catalyst can have on real biomass decomposition with internal hydrogen production and deoxygenation by decomposing cellulose with 2%Ni0.5%Co4%Fe/CZ-41.

For these tests (including TGA), cellulose was decomposed under the ideal conditions of 350 °C for 1 h with 1MN as the solvent. Without any catalyst, the organic solid content was reduced by 69%, and the liquid product yield was 54%. Use of 2%Ni0.5%Co4%Fe/CZ-41 increased the conversion of solids to 78% and the liquid product yield to 57%. As can be seen in Table 4, the addition of the catalyst led to a massive change in gas production. The overall H_2 production almost doubled, and the CO output decreased. Meanwhile, addition of the catalyst led to a large increase of CO_2 in the product gas from 87 to 141 kPaA. These results demonstrate the catalyst's ability to produce internal hydrogen by the WGSR that could then be used for deoxygenation. However, the low drop in CO occurred regardless of the high increase in CO_2. This suggests either that CO_2 was produced from an alternative route or that additional CO was produced through the steam reforming of light components whereas H_2 was consumed in the treatment of condensable products.

Table 4. Product Gas Composition from Cellulose Decomposition with and without 2%Ni0.5%Co4%Fe/CZ-41

	Product Gas Partial Pressure (kPaA) at 22.5°C			
	H_2	CO_2	CO	CH_4
None	6	87	70	7
2%Ni0.5%Co4%Fe/CZ-41	11	141	64	7

Surprisingly, on its own, cellulose decomposition produced competitive amounts of 2MF and DMF, likely because of the extended reaction times and the catalytic effects of the reactor wall (see Figure 11). However, oxygenated compounds, furfural, and 5-methylfurfural were also prevalent. Upon addition of the catalyst, the selectivity was shifted toward 2MF at the expense of DMF, possibly as a result of decarbonylation activity. However, as with glucose, the 2%Ni0.5%Co4%Fe/CZ-41 catalyst demonstrated its activity toward transalkylation, as trimethylfuran was a noteworthy product. Undoubtedly, the catalysts affected the decomposition of the cellulose intermediates. For example, addition of the catalyst led to an increase in 2-acetylfuran, which could be generated by transalkylation or an alteration of the decomposition pathways. In addition, when the catalyst was employed, nonaromatic ring products such as cyclopentenone were much more apparent than they were without the catalyst.

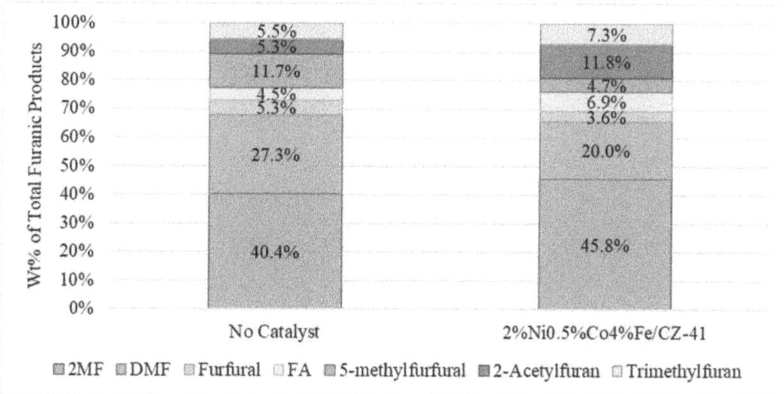

Figure 11. Total furanic product distribution for cellulose decomposition.

Thermogravimetric analyses (and DTA), similar to those of glucose, were performed to further study the effects of 2%Ni0.5%Co4%Fe/CZ-41 on cellulose decomposition. As can be seen in Figures 12 and 13, the catalyst did not play as significant a role in the decomposition of cellulose as it did with glucose. In Figure 12, an offset in final weight after major decomposition is noticed because of the additional weight of the catalyst. As in the glucose experiments, the final weight dropped below 16.7%, representing a loss of weight from the catalyst itself. As can be seen in Figure

13, although the temperature did not decrease, the addition of catalyst slightly decreased the rate of weight loss, suggesting that it contributed to the production of slightly lower amounts of volatile compounds. In earlier tests with glucose, it was noted that the catalyst had significant activity toward the dehydration of glucose, which is considered a product of cellulose decomposition (*2*, *19*, *36*, *37*). Given that the decomposition of cellulose occurs at 340 °C—nearly 50 °C higher than that of glucose—it is suspected that, with or without the catalyst, once glucose is produced, it will readily undergo dehydration. It is not surprising that the catalysts do not have a significant effect on the initial breakdown of cellulose because cellulose has a large polymeric structure, which prevents it from adsorbing onto the surface of the heterogeneous catalyst.

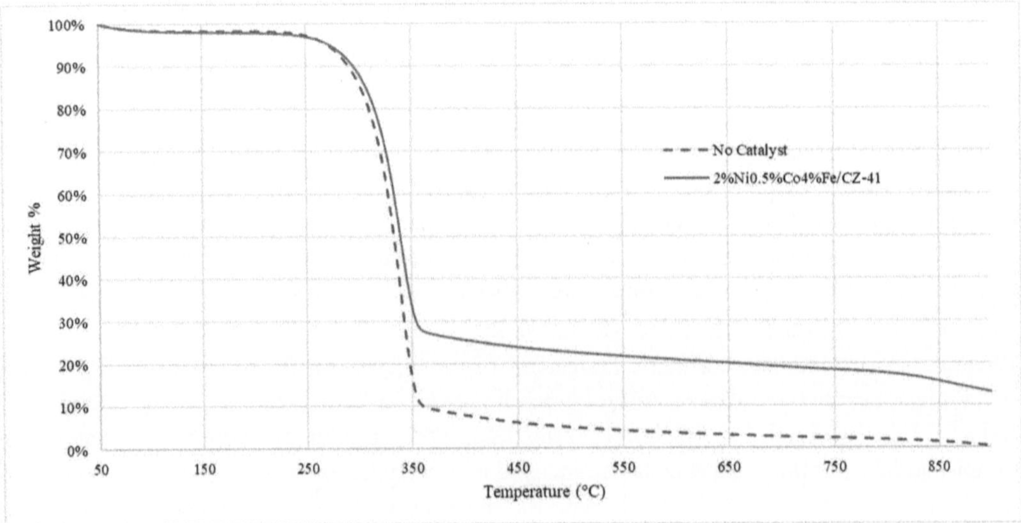

Figure 12. TGA results for cellulose with the addition of 2%Ni0.5%Co4%Fe/CZ-41 catalyst at a 5:1 cellulose-to-catalyst ratio.

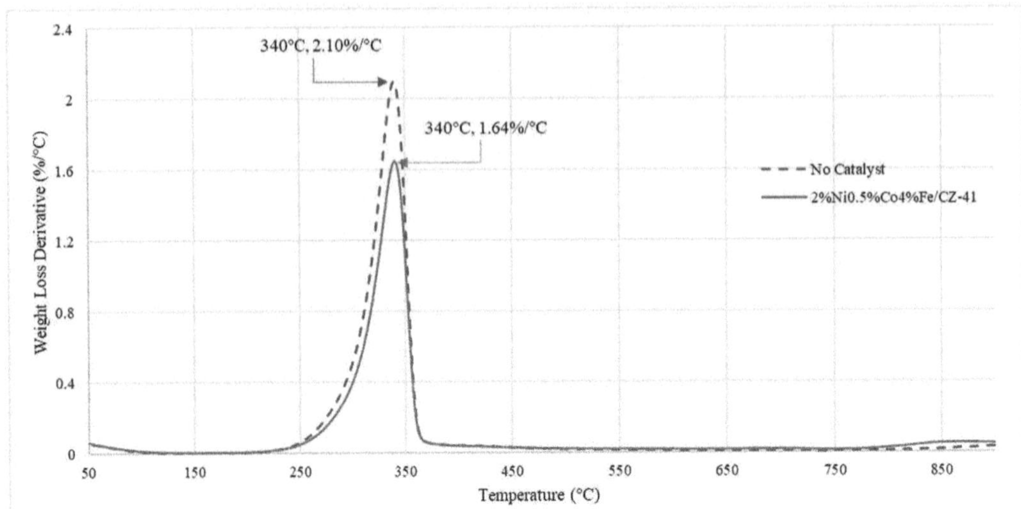

Figure 13. DTA (weight loss derivative from TGA) results of cellulose with the addition of 2%Ni0.5%Co4%Fe/CZ-41 catalyst at a 5:1 cellulose-to-catalyst ratio.

Discussion

For the process of in situ deoxygenation using internally produced hydrogen, reactor design requires attention. Activity tests were performed in batch operation, over long periods of time. During these tests, with glucose, it was found that 350 °C was a suitable reaction temperature. Under these conditions, the highest liquid product yield was achieved with a modest deoxygenation of furanic compounds. Reforming was not significant even when the reaction time was extended. This selection of temperature was further substantiated by TGA tests using cellulose, which decomposed most rapidly at 340 °C. Overall, in situ deoxygenation is highly dependent on the internal hydrogen production, which needs to be improved. For this reaction system, it was assumed that deoxygenation follows traditional reaction routes such as hydrogenation and hydrogenolysis reactions to convert furfural into FA and 2MF, respectively. The difference here is the origin of hydrogen on the catalyst surface. Hydrogen can be produced on the catalyst surface by reactions, such as the WGSR, as opposed to the disassociation of H_2 gas from the bulk phase. One limitation then is the mass transfer of the WGSR reactants to the catalyst surface. Potential solutions include using a more appropriate reactor design and eliminating the use of the solvent.

The solvent, 1MN, was added primarily for heat-transfer purposes. The intent was for the solvent to prevent biomass material from immediately sintering to the wall of the reactor. Using 1MN as a solvent, however, is a hindrance toward the WGSR, as it presents a mass-transfer barrier. With or without a solvent, ensuring adequate contact between the gaseous products and the heterogeneous catalysts needs to be a priority.

Therefore, an ideal reactor type that could be used is a flow reactor in which both the biomass and the catalyst are mobilized. Such a reactor could more appropriately control the temperature, for example, with a cold initial section for the primary decomposition stage and a transition to a hotter section for internal hydrogen production and deoxygenation. This may also allow for better gas/vapor contact with the catalyst and prevent oligomerization and tar formation by reducing the exposure time for intermediates. Without a solvent, high-temperature tests can also be performed without concern for solvent reactions or pressurizing the reactor with an inert gas to prevent solvent evaporation. The testing of high temperatures may be crucial, as real biomass consists of hemicellulose and lignin, with the latter tending to have a higher decomposition temperature than cellulose (*1*). In summary, the proposed reaction system for future work resembles a low-temperature pyrolysis unit in which biomass and a bifunctional catalyst are mobilized for internal hydrogen production and deoxygenation.

Conclusions

Glucose was decomposed in the presence of CZ-41–supported catalysts for the purpose of internal hydrogen production and deoxygenation. Of the two supported catalysts that were used, the trimetallic catalyst, 2%Ni0.5%Co4%Fe/CZ-41, was found to be superior, as it yielded more liquid product and furanic product distribution comparable to that obtained with 2%Ni4%Fe/CZ-41. The addition of 0.5 wt % Co to the Fe phase makes the Fe more reducible because of the stronger interactions with the support, potentially leading to better oxygen transfer and storage capabilities. Of the conditions that were tested, a reaction temperature of 350 °C and a reaction time of 1 h were found to be most suitable for producing a high yield of liquid product. For the decomposition of cellulose, the addition of 2%Ni0.5%Co4%Fe/CZ-41 increased the conversion of organic solids into liquid- and gas-phase products. The catalyst also altered the reaction pathways, as there was evidence of additional transalkylation and dehydration reactions along with higher production of CO_2.

Acknowledgments

The authors thank the Natural Science and Engineering Research Council of Canada (NSERC) and the Canada Foundation for Innovation (CFI) for their financial support.

References

1. Collard, F. X.; Blin, J. A review on pyrolysis of biomass constituents: Mechanisms and composition of the products obtained from the conversion of cellulose, hemicelluloses and lignin. *Renewable Sustainable Energy Rev.* **2014**, *38*, 594–608.
2. Vinu, R.; Broadbelt, L. J. A mechanistic model of fast pyrolysis of glucose-based carbohydrates to predict bio-oil composition. *Energy Environ. Sci.* **2012**, *5*, 9808–9826.
3. Rogers, K. A.; Zheng, Y. Decomposition of glucose with in situ deoxygenation in a low H_2 pressure environment – Part I: Monometallic catalysts. *Appl. Catal. A: Gen.* **2019**, *569*, 75–85.
4. Rogers, K. A.; Zheng, Y. Decomposition of glucose with in situ deoxygenation in a low H_2 pressure environment – Pt. II: Bimetallic catalysts. *Appl. Catal. A: Gen.* **2019**, *578*, 10–19.
5. Florin, N. H.; Harris, A. T. Hydrogen production from biomass coupled with carbon dioxide capture: The implications of thermodynamic equilibrium. *Int. J. Hydrogen Energy* **2007**, *32*, 4119–4134.
6. Widyawati, M.; Church, T. L.; Florin, N. H.; Harris, A. T. Hydrogen synthesis from biomass pyrolysis with in situ carbon dioxide capture using calcium oxide. *Int. J. Hydrogen Energy* **2011**, *36*, 4800–4813.
7. Widyaningrum, R. N.; Church, T. L.; Harris, A. T. Promoting effect of Pd on H_2 production from cellulose pyrolysis over mesocellular-foam-supported Ni catalysts. *Catal. Commun.* **2013**, *35*, 45–50.
8. Widyaningrum, R. N.; Church, T. L.; Zhao, M.; Harris, A. T. Mesocellular-foam-silica-supported Ni catalyst: Effect of pore size on H_2 production from cellulose pyrolysis. *Int. J. Hydrogen Energy* **2012**, *37*, 9590–9601.
9. Zhao, M.; Florin, N. H.; Harris, A. T. The influence of supported Ni catalysts on the product gas distribution and H_2 yield during cellulose pyrolysis. *Appl. Catal. B: Environ.* **2009**, *92*, 185–193.
10. Zhao, M.; Florin, N. H.; Harris, A. T. Mesoporous supported cobalt catalysts for enhanced hydrogen production during cellulose decomposition. *Appl. Catal. B: Environ.* **2010**, *97*, 142–150.
11. Schimming, S. M.; Lamont, O. D.; König, M.; Rogers, A. K.; D'Amico, A. D.; Yung, M. M.; Sievers, C. Hydrodeoxygenation of Guaiacol over Ceria–Zirconia Catalysts. *ChemSusChem* **2015**, *8*, 2073–2083.
12. Shim, J. O.; Jeong, D. W.; Jang, W. J.; Jeon, D. W.; Jeon, B. H.; Cho, S. Y.; Roh, H. S.; Na, J. G.; Ko, C. H.; Oh, Y. K.; Han, S. S. Deoxygenation of oleic acid over $Ce_{(1-x)}Zr_{(x)}O_2$ catalysts in hydrogen environment. *Renewable Energy* **2014**, *65*, 36–40.
13. Teles, C. A.; de Souza, P. M.; Braga, A. H.; Rabelo-Neto, R. C.; Teran, A.; Jacobs, G.; Resasco, D. E.; Noronha, F. B. The role of defect sites and oxophilicity of the support on the phenol hydrodeoxygenation reaction. *Appl. Catal. B: Environ.* **2019**, *249*, 292–305.

14. Chamnankid, B.; Föttinger, K.; Rupprechter, G.; Kongkachuichay, P. Cu/Ni-loaded CeO_2–ZrO_2 catalyst for the water-gas shift reaction: Effects of loaded metals and CeO_2 addition. *Chem. Eng. Technol.* **2014**, *37*, 2129–2134.
15. Watanabe, K.; Miyao, T.; Higashiyama, K.; Yamashita, H.; Watanabe, M. Preparation of a mesoporous ceria–zirconia supported Ni–Fe catalyst for the high temperature water–gas shift reaction. *Catal. Commun.* **2011**, *12*, 976–979.
16. Deshpande, P. A.; Madras, G. Support-dependent activity of noble metal substituted oxide catalysts for the water gas shift reaction. *AIChE J.* **2010**, *56*, 2662–2676.
17. Hilaire, S.; Wang, X.; Luo, T.; Gorte, R. J.; Wagner, J. A comparative study of water-gas-shift reaction over ceria-supported metallic catalysts. *Appl. Catal. A: Gen.* **2004**, *258*, 271–276.
18. Vasconcelos, S. J. S.; Lima, C. L.; Filho, J. M.; Oliveira, A. C.; Barros, E. B.; de Sousa, F. F.; Rocha, M. G. C.; Bargiela, P.; Oliveira, A. C. Activity of nanocasted oxides for gas-phase dehydration of glycerol. *Chem. Eng. J.* **2011**, *168*, 656–664.
19. Mayes, H. B.; Nolte, M. W.; Beckham, G. T.; Shanks, B. H.; Broadbelt, L. J. The Alpha–Bet(a) of Glucose Pyrolysis: Computational and Experimental Investigations of 5-Hydroxymethylfurfural and Levoglucosan Formation Reveal Implications for Cellulose Pyrolysis. *ACS Sustainable Chem. Eng.* **2014**, *2*, 1461–1473.
20. Mayes, H. B.; Nolte, M. W.; Beckham, G. T.; Shanks, B. H.; Broadbelt, L. J. The Alpha–Bet(a) of Salty Glucose Pyrolysis: Computational Investigations Reveal Carbohydrate Pyrolysis Catalytic Action by Sodium Ions. *ACS Catal.* **2015**, *5*, 192–202.
21. Carvalho, W. S.; Cunha, I. F.; Pereira, M. S.; Ataíde, C. H. Thermal decomposition profile and product selectivity of analytical pyrolysis of sweet sorghum bagasse: Effect of addition of inorganic salts. *Ind. Crops Prod.* **2015**, *74*, 372–380.
22. Liu, D.; Yu, Y.; Hayashi, J. I.; Moghtaderi, B.; Wu, H. Contribution of dehydration and depolymerization reactions during the fast pyrolysis of various salt-loaded celluloses at low temperatures. *Fuel* **2014**, *136*, 62–68.
23. Kan, T.; Strezov, V.; Evans, T. Catalytic pyrolysis of coffee grounds using NiCu-impregnated catalysts. *Energy Fuels* **2014**, *28*, 228–235.
24. Ding, K.; He, A.; Zhong, D.; Fan, L.; Liu, S.; Wang, Y.; Liu, Y.; Chen, P.; Lei, H.; Ruan, R. Improving hydrocarbon yield via catalytic fast co-pyrolysis of biomass and plastic over ceria and HZSM-5: An analytical pyrolyzer analysis. *Bioresour. Technol.* **2018**, *268*, 1–8.
25. Jeon, M. J.; Jeon, J. K.; Suh, D. J.; Park, S. H.; Sa, Y. J.; Joo, S. H.; Park, Y. K. Catalytic pyrolysis of biomass components over mesoporous catalysts using Py-GC/MS. *Catal. Today* **2013**, *204*, 170–178.
26. Deng, X.; Chen, K.; Tüysüz, H. Protocol for the Nanocasting Method: Preparation of Ordered Mesoporous Metal Oxides. *Chem. Mater.* **2017**, *29*, 40–52.
27. Rogers, K. *Reducing Bio-Oil Oxygen Content via in-Situ Deoxygenation and Hydrogen Production.* Ph.D. Thesis, University of New Brunswick, Fredericton, NB, Canada, 2020.
28. Tepamatr, P.; Laosiripojana, N.; Charojrochkul, S. Water gas shift reaction over monometallic and bimetallic catalysts supported by mixed oxide materials. *Appl. Catal. A: Gen.* **2016**, *523*, 255–262.

29. Ruppert, A. M.; Niewiadomski, M.; Grams, J.; Kwapiński, W. Optimization of Ni/ZrO_2 catalytic performance in thermochemical cellulose conversion for enhanced hydrogen production. *Appl. Catal. B: Environ.* **2014**, *145*, 85–90.
30. De, S.; Saha, B.; Luque, R. Hydrodeoxygenation processes: Advances on catalytic transformations of biomass-derived platform chemicals into hydrocarbon fuels. *Bioresour. Technol.* **2015**, *178*, 108–118.
31. Mortensen, P. M.; Grunwaldt, J. D.; Jensen, P. A.; Knudsen, K. G.; Jensen, A. D. A review of catalytic upgrading of bio-oil to engine fuels. *Appl. Catal. A: Gen.* **2011**, *407*, 1–19.
32. Nakagawa, Y.; Tamura, M.; Tomishige, K. Catalytic reduction of biomass-derived furanic compounds with hydrogen. *ACS Catal.* **2013**, 3, 2655–2668.
33. Lu, Q.; Zhang, Z. F.; Dong, C. Q.; Zhu, X. F. Catalytic upgrading of biomass fast pyrolysis vapors with nano metal oxides: An analytical Py-GC/MS study. *Energies* **2010**, 3, 1805–1820.
34. Wang, S.; Guo, X.; Liang, T.; Zhou, Y.; Luo, Z. Mechanism research on cellulose pyrolysis by Py-GC/MS and subsequent density functional theory studies. *Bioresour. Technol.* **2012**, *104*, 722–728.
35. Ratnasamy, C.; Wagner, J. P. Water gas shift catalysis. *Catal. Rev.* **2009**, *51*, 325–440.
36. Patwardhan, P. R.; Dalluge, D. L.; Shanks, B. H.; Brown, R. C. Distinguishing primary and secondary reactions of cellulose pyrolysis. *Bioresour. Technol.* **2011**, *102*, 5265–5269.
37. Shen, D. K.; Gu, S. The mechanism for thermal decomposition of cellulose and its main products. *Bioresour. Technol.* **2009**, *100*, 6496–6504.

Chapter 6

Catalytic Oxidative Desulfurization of Liquid Fuels

J. M. Campos-Martin* and M. C. Capel-Sanchez

Sustainable Energy-and-Chemistry Group (EQS), Instituto de Catálisis y Petroleoquímica, CSIC, Marie Curie 2, Cantoblanco, 28049 Madrid, Spain

*Email: jm.campos@csic.es

Environmental concerns have introduced a need to remove sulfur-containing compounds from fuels. As oxidative desulfurization is conducted under very mild reaction conditions, much attention has recently been devoted to this process. In this chapter, developments in selective removal of organosulfur compounds present in liquid fuels via catalytic oxidative desulfurization are reviewed. At the end of the chapter, a brief account of the research directions needed in this field is also included.

Introduction

In recent years, due to the increasing use of transportation vehicles and emission of hazardous materials from exhaust engines, air pollution has dramatically increased, especially in metropolitan areas (*1–3*). Some common hazardous materials exist in the environment as NO_X, SO_X, CO, volatile organic compounds, particulate matter, nitrogen-containing compounds, H_2S, and sulfur-containing compounds. These pollutants arise from burning of fossil fuels, including diesel, gasoline, and jet fuel, widely used in transportation.

Sulfur compounds are one of the most undesirable petroleum contaminants, and a large amount of these compounds can be transferred to diesel oil during the refining process (*4*). The presence of sulfur reduces the performance of fuels and oils, causes corrosion of equipment, decreases the activity of antiknock and antioxidant stability of fuel, and reduces the efficiency of vehicle catalysts, and upon combustion, sulfur compounds are converted to sulfur oxides (SO_X) that contribute to acid rain and environmental pollution (*5*).

Many environmental laws have been adopted and implemented to eliminate sulfur content, and in this regard, many researchers have been investigating various methods for sulfur removal from oil (*6, 7*). The most important and most common industrial process is that of treating the fuel under high temperatures and high pressures with hydrogen. This process is called hydrodesulfurization (HDS) and has received extensive attention since its discovery in the 1930s (*8*). In the HDS process, the hydrocarbon oil is mixed with hydrogen at around 5–160 bar pressure and temperature ranging from 260 to 380 °C. Sulfur-containing compounds are converted into sulfur-free hydrocarbons

© 2021 American Chemical Society

by producing hydrogen sulfide (H_2S) gas, which is then removed by absorption in lean amines accompanied by an amine treatment unit (*9*). This technology is very efficient for mercaptan, thioether, sulfide, disulfide, and thiophene removal, but it has shown limitations of treatment for refractory compounds like alkylated aromatic sulfur compounds such as 4,6-dimethyldibenzothiophene (4,6-DMDBT), benzothiophene (BT), and dibenzothiophene (DBT) (*4, 6*). The amount of sulfur in a distillation fraction increases with an increase in boiling range, and the sulfur compounds become more refractory (*10*). In order to comply with the ultralow sulfur regulations, the refractory part should also be removed. To achieve it, conventional HDS requires more severe conditions, such as higher temperature, pressure, and hydrogen consumption and a long residence time (*6, 7*). The refining cost (financial and environmental) increases as heavier and more sulfur-rich crude oils are being processed. Therefore, alternative desulfurization methods are demanded to achieve fuels with ultralow sulfur content.

Among the alternative processes, the oxidative desulfurization (ODS) process has received much attention because of its efficiency in the removal of sulfur aromatic compounds (*1–3, 5, 11–14*). This desulfurization process includes two stages (Figure 1): (1) sulfur catalytic oxidation in a first step, which changes the nature of the sulfur compounds; and (2) liquid extraction at the end, where oxidized sulfur compounds are removed, taking advantage of the properties of oxidized sulfur compounds being significantly different from that of nonoxidized sulfur compounds.

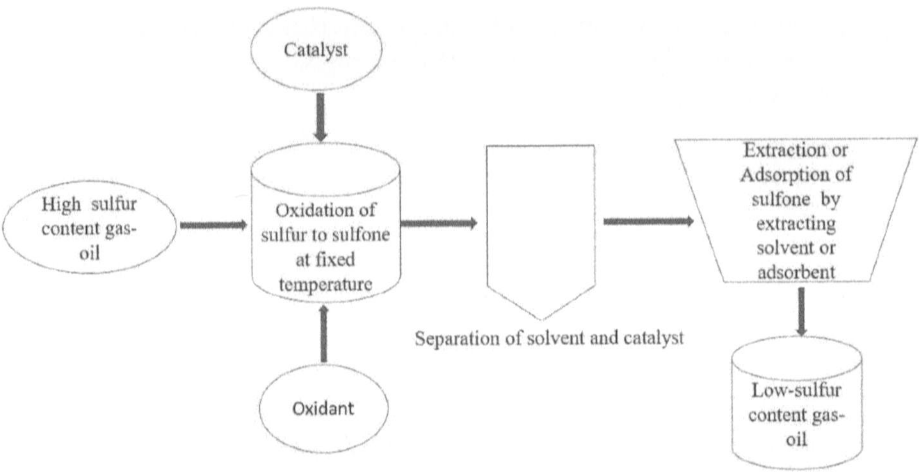

Figure 1. Schematic representation of sulfur removal in the ODS process (14).

It is evident that the greatest advantages of the ODS process are the low reaction temperature and pressure, and hydrogen is not used in the process (*5*). Furthermore, ODS is able to easily remove refractory sulfur compounds by oxidation to the corresponding sulfones or sulfoxides (Figure 2) (*15*). The sulfoxides and sulfones have two properties that are different from the unoxidized sulfur compounds and that facilitate desulfurization. First, the sulfoxides and sulfones are more polar, which increases selectivity during solvent extraction. Although oxidation does not mitigate issues such as solvent loss and energy costs associated with solvent recycling, it improves the extraction selectivity. Second, the C–S bond strength is decreased when the sulfur is oxidized. It is therefore easier to remove the oxidized sulfur by thermal decomposition (*10*).

The oxidant, the oxygen donor, is a very important part in ODS. Different types of oxidants can be used in the ODS process, which must be carefully tailored to the sulfur compounds to prevent the oxidation of olefins or aromatic compounds that are present in the fuel feed. Thus, different types of oxidants are used in the ODS process, such as organic peracid, organic hydroperoxide, H_2O_2, O_2, air, super oxide (O^{2-}), and ozone (*1–3*). Therefore, choosing the right kind of oxidant will depend not only on its oxidizing power and active oxygen but also on its cost, environmental friendliness, and availability.

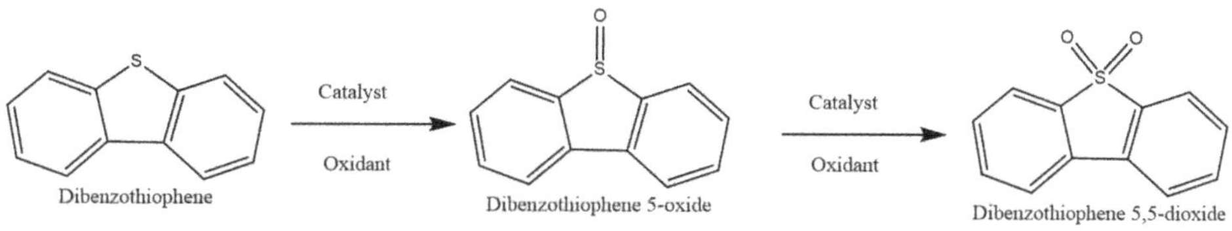

Figure 2. Oxidation desulfurization steps.

In order to conduct ODS in an efficient mode, a catalyst is required in order to employ the oxidant efficiently in the oxidation of the sulfur compounds.

Catalytic Systems Employed in ODS

One of the critical points in the selection of an ODS process is the selection of the catalyst. In this section the main catalytic system studied is described.

Carboxylic Acids

The first approaches to catalysts are carboxylic acids, where among them the formic and acetic acids are considered the most-used catalysts in ODS. The acids form peracids in situ with hydrogen peroxide. The peracids are very powerful oxidizing agents and are able to oxidize sulfur compounds. These compounds are highly reactive and corrosive, and for this reason they have to be produced in situ (*16–21*). The oxidation is carried out according to the following steps (Figure 3): peracids (RCOOOH) are produced by the reaction of hydrogen peroxide and the carboxylic acid. Then the formed peracid reacts with the sulfur compound to give the corresponding sulfone.

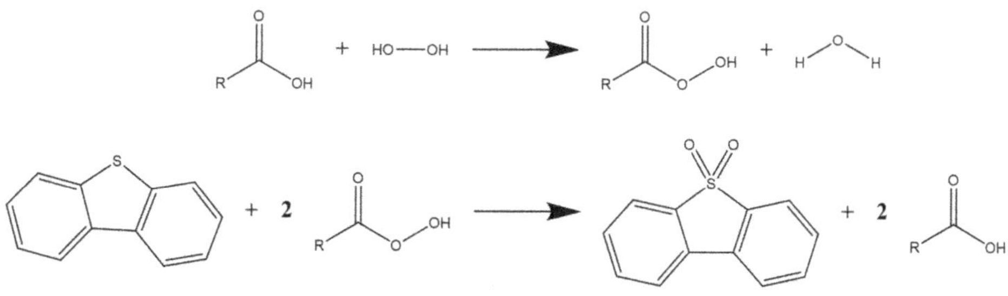

Figure 3. Reaction scheme of ODS with organic peracids.

However, the amount of oxidizing agent employed is very high in comparison to the sulfur removed; in consequence, this procedure is difficult to make feasible economically. For this reason, the addition of catalysts that can oxidize the sulfur compound more efficiently was necessary. The most common option is the use of transition metal salts, and specifically tungsten salts (*22–24*), which are very active and selective in the oxidation of sulfur compounds with peracids. Modified active carbons are also catalysts for this reaction (*25–28*). In the presence of active carbon with acid sites, more sulfur is removed from the fuel than in their absence; it seems that carbon promotes ODS by formation of percarboxylic acid species at defect sites within the carbon basal planes (*28*).

A second option is to use transition metal salts like Sn (*29*), Co (*30*), or Fe (*31*, *32*), which are able to form radicals from peracids (R–COOO·); these radicals are very reactive and have high oxidation capacity. A similar approach has been used with the addition of other substances such as SiO_2 (*33*).

The peracids and their solvents (short-chain carboxylic acids) used in these reactions are highly polar and form two immiscible liquid phases in the reactor. For this reason, there is some mass-transfer limitation between the fuel and the polar phase containing the oxidizing agent. Accordingly, some studies of the use of phase-transfer compounds have been carried out. The presence of the phase-transfer compound facilitates the transfer of products and reagents at the polar–apolar interface, increasing notably the mass transfer across the interphase (*34–36*).

Homogeneous Catalysts

Homogeneous catalysts are able to catalyze the ODS without the use of organic acids as cocatalysts. The main types of homogeneous catalysts reported in the literature are basically transition metal compounds, as well as polyoxometalates. Among salts of transition metals, vanadium, molybdenum, tungsten, niobium, iron, chromium, and zinc (*1–3*, *5*, *11–14*, *37–39*) are used, which form peroxo complexes in reactions with hydrogen peroxide and are also liable to form complexes of different stability with sulfur-containing organic compounds. The first three metals are of considerable interest because their activity is not inferior to that of the acid catalysts.

Heteropolyacids (HPAs) and polyoxometalates (POMs) have been used widely in ODS (*11*, *40*). HPAs are multifunctional heterogeneous acid catalysts, which have been used in developing green technologies and as replacements for conventional liquid homogeneous acid catalysts. They are a special type of POM catalyst and are formed by the condensation of various metallic oxyanions (Mo, W, V, and Nb). In recent times, HPAs have been recognized for their versatile functions (acid–base characteristics and redox power). The first option was their use as an acid. HPA exhibits better activity than common acids (e.g., formic acid) in ODS, and the most-used catalyst is phosphotungstic acid (*41–45*). However, better results are obtained when it acts as a redox system. The ODS activity is revealed by the differences observed when different HPAs are employed, and better results are obtained with POM derivatives with organic cations. These POM derivatives are more efficient because the presence of this organocation (substituted for inorganic cations) favors the lipophilicity of the compound (*46–49*). The resulting compounds are amphiphilic catalysts due to their hydrophilic anion and lipophilic long-chain alkyl cation. Due to the amphiphilicity of POM derivatives, the rate of ODS reactions has been remarkably improved (*50*). The synthesis of these POM derivatives is quite simple, using an ion-exchange substitution process. The selection of the organic cation is a critical issue (*51–53*) because one long alkyl chain enhances the reactivity of the compounds, but a double alkyl chain hinders it because the steric hindrance induced by the double alkyl chains limits the interaction between the active centers and the sulfides (*51*).

Adequate selection of the components that regulate the amphipathic character of the catalysts can allow the formation of catalytic emulsion (water/oil) droplets during ODS reactions (*51*, *54–58*). The formation of emulsion droplets promotes a high dispersion of the POM catalyst throughout the oil phase, increasing ODS activity (Figure 4). These metastable emulsion droplets may be stabilized by the amphiphilic POM catalyst. Emulsions in ODS are made up of three regions: the interior of a droplet (aqueous solvent and anionic part of the amphipathic POM), the continuous phase (oil and cationic part of the amphipathic POM), and the interfacial membrane. The interfacial membrane is a very significant narrow region where the substrate, oxidant, and catalysts interact with each other to facilitate ODS (*51*, *58*). The amphipathic catalysts have the ability to form emulsions that could be easily separated by centrifugation and utilized in the next run without any special treatment (*58*).

The elements of the anionic part of POM can be substituted by transition metals up to lanthanides. These substitutions produce changes in the reactive properties of the POMs (*24, 52, 55, 56, 59–61*). The substitution of tungsten(VI) with vanadium(V) would also result in the generation of more reactive lattice oxygen associated with W–O–V species and seams that can turn the acid-dominant properties of phosphotungstic acid into redox-dominant properties, which is highly desirable for a POM to be a catalyst in the oxidation reactions (*24*). The zinc-substituted polyoxotungstate anion has a high reactivity and stability in the presence of acetonitrile and can be easily recovered from the reaction mixture and reused several times (*52*). However, substitution with other different heteroelements (titanium, manganese, iron, cobalt, nickel, and copper) can produce a loss in the ODS activity, due to the impossibility of forming the active polyperoxometalate in the presence of H_2O_2 (*62*).

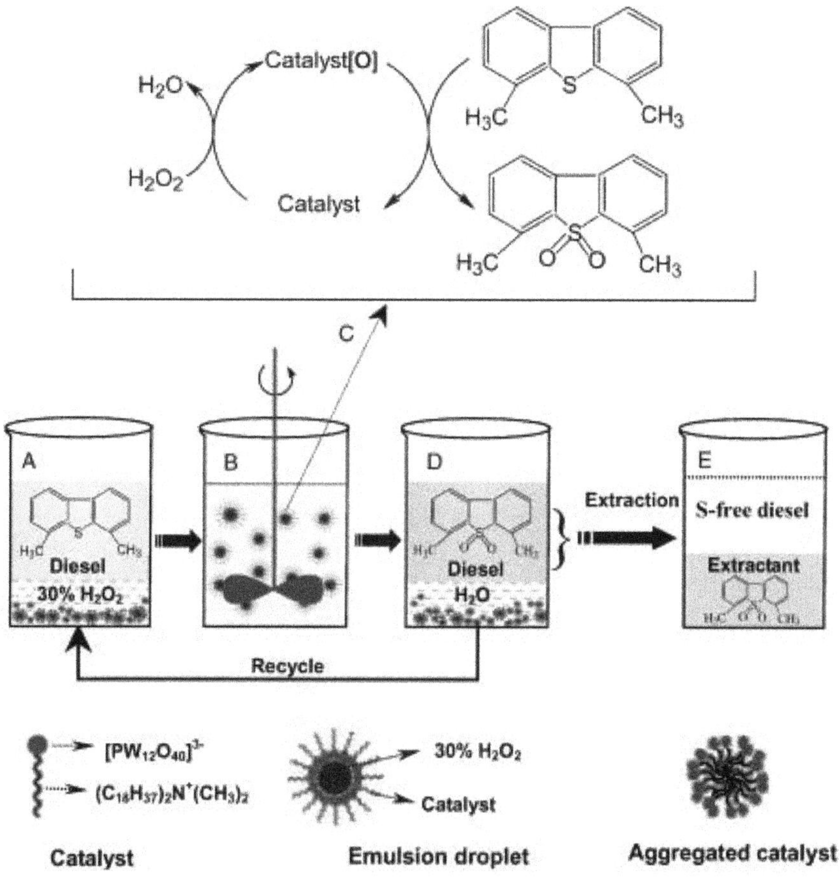

Figure 4. Catalytic oxidation and extraction of sulfur-containing molecules present in real diesel: (A) before oxidation; (B) during oxidation; (C) catalytic oxidation of sulfur-containing molecules in emulsion droplets; (D) after oxidation; (E) extraction with a polar extractant. Reproduced with permission from reference (58). Copyright 2004 John Wiley & Sons.

Ionic Liquids

Ionic liquids (ILs) are able to extract sulfur compounds from the fuels (*63, 64*). This ability can be enhanced by the addition of metal salts: Sn(II), Cu(I), and Ag(I) ions were found to be especially efficient, due to their tendency to form π complexes with thiophene derivatives (*65–67*).

As a general rule, ILs themselves, in the absence of oxidants, fail to provide a high degree of sulfur removal (*68, 69*). For example, the ionic liquid [HMIm]BF_4 is able to remove only 6% of the sulfur, but in the presence of hydrogen peroxide it gave a sulfur removal of 65%, 70%, and 93% at 70, 80, and 90 °C, respectively (*70*), and has proved to be a promising desulfurization method. This effect can be related to the similar polarity between the alkenes, aromatics, and sulfur compounds, and the sulfur removal is improved by increasing the polarity of sulfur-oxidizing sulfur compounds to the corresponding sulfoxides and sulfones. However, when only ionic liquids are employed, a very high H_2O_2/S ratio is necessary to reach these sulfur-removal levels (*71, 72*). This drawback can be avoided by the use of catalysts: Brønsted acids (*73–75*), Lewis acids (*76*), tungsten and molybdenum compounds (*77–80*), vanadium (*81–84*), iron catalyst (*85–87*), or POM (*46, 88–90*) can remove the majority of sulfur with a low ratio of hydrogen peroxide to sulfur. This finding provides evidence in favor of combining catalytic oxidation and extraction. Moreover, POM-based IL can be reused several times after regeneration with negligible loss of efficiency; some studies use immobilized IL on large-surface supports (*91–95*).

The ODS of fuels using ILs has a number of advantages: high energy efficiency, effectiveness in achieving ultralow levels of sulfur, and process simplification by combining extraction and reaction into a "one-pot" procedure. However, challenges associated with this process certainly still remain, including (1) the high cost associated with ILs based on relatively expensive cations (e.g., imidazoliums, pyridiniums) or anions (in particular, fluorinated species like Tf2N–); (2) the relatively high viscosities of ILs, especially amphiphilic (surfactant-inspired) ones capable of performing phase-transfer catalysis; and (3) the poor biodegradability and high eco(cyto)toxicity of some ILs, causing serious consequences if they are released into the environment or come into contact with personnel. Therefore, there is a strong demand to develop new, inexpensive, and environmentally benign ILs as extractants and reaction media for ODS. Although deep eutectic solvents appear promising for addressing criteria (1) and (3), they are likely to be far worse on point (2), not to mention the conspicuous lack of availability of hydrophobic deep eutectic solvent. In any case, (4) there is still plenty of room for improving the efficiency, stability, and reusability and reducing the cost of new catalysts exploited within (or, particularly, as an integral part of) ILs and deep eutectic solvents.

Heterogeneous Catalyst

Heterogeneous catalysts have clear advantages over homogeneous ones in reuse and easy recovery and a better integration in a continuous-flow reaction system. Consequently, efforts have been devoted to the development of processes using highly robust and recyclable catalysts that provide higher atom utilization and minimize pollution levels using greener ingredients.

The first option is to use systems equivalent to the homogeneous counterparts incorporating transition metals in a high oxidation state with Lewis acidity, such as Mo(IV), Ti(IV), V(V), and W(IV). Based on this, several redox catalysts supported on a large-surface carrier have been studied (*1–3, 5, 11–14*). The catalytic activity of any metal oxide in ODS is dependent on the nature of the metal, but also the nature of the support has an influence on the catalyst's performance. For instance, here we show some catalysts based on V (*96–109*), W (*110–120*), and Mo (*121–131*) that have been studied in the last two years.

A variation in the ODS activity is observed using different metals under the same reaction conditions; for this reason, it is critical to choose the oxide of the correct metal for efficient ODS. However, in addition to the metal type, other significant factors govern the overall catalytic activity of metal oxides in ODS. Those significant factors include types of support, textural properties, size and morphology, and oxygen-deficient sites.

The same metal oxide could demonstrate different activities in ODS when it is supported on different supports. The variation of the catalytic activity in ODS is very obvious when using a different support with a different Lewis acidity, showing a higher catalytic activity than metals on a support with a higher Lewis acidity.

The catalytic activity of supported metal oxides can be adjusted by tuning the textural properties of the catalyst. Supports exhibiting high surface area and mesoporosity usually promote the dispersion of metal oxides, in comparison with those exhibiting poor surface area and microporosity, and thus increase the activity of metal oxides in ODS.

A formidable fact related to the catalyst size is that the smaller the size of the catalyst, the higher the catalytic activity is, due to the high ratio of volume to surface area. For ODS, different nanosized metal oxides have been reported as catalysts in the literature so far.

Metal oxides with more oxygen deficiency show a high catalytic activity in the oxidation reactions. Thus, the introduction of more oxygen-deficient sites in metal oxides to boost ODS has become intriguing. Different methods have been followed to introduce more oxygen sites into metal oxide catalysts to improve the ODS activity.

The high activity of catalysts based on POM and the difficulties of recycling homogeneous systems prompt large interest in the heterogenization of POM and HPA by tethering and encapsulation (*94*, *132–139*). The simple deposition of POM or HPA is not an adequate strategy because its high solubility in water produces a severe leaching during the reaction in the presence of hydrogen peroxide/water solutions. A variety of heterogeneous POM (het-POM) catalysts are reported with the advantage of facile separation from the ODS system. Six common preparation methods are available for het-POMs: (1) precipitation with metal ions, (2) hybridization with bulky cations including organic cations, (3) dispersion on supports like SiO_2, (4) encapsulation into porous materials, (5) tethering on supports using spacers (noncovalent interaction), and (6) grafting on the supports (covalent interaction). The nature of the metal ions (precipitation) and bulky cation molecules (hybridization) can remarkably influence the ODS efficiency of het-POMs. Likewise, the size, surface area, and morphology of the supports used to prepare het-POMs can also affect the ODS activity. Therefore, it is advised that, to make an efficient het-POM for ODS, the choice of a suitable metal cation and bulky cation and support with respect to the preparation method is a key factor. During the ODS reaction catalyzed by any type of het-POM, the rapid immersion of het-POMs in liquid oxidants is required for the increased ODS rate because it takes time for any het-POM to construct the catalytically active species through the solid–liquid reaction. In addition to the better separation and reusability, the het-POMs demonstrate higher ODS performance than their corresponding POMs.

In addition, new heterogeneous catalytic systems that have no equivalent in homogeneous systems have been developed, for instance, titanium-based systems, Ti/SiO_2 (*140–142*), TiO_2 (*143–146*), and Ti in the framework of zeolites or mesostructured silica (*147–157*).

Titanium-based systems are broadly classified into four types (conventional, hierarchical, lamellar, and mesoporous) in view of their textural properties, size and shape of the pores, and the location and content of Ti. TS-1 is a microporous solid, and thus it could not exhibit better performance in ODS of bulky sulfur compounds like DBT. This leads to the appearance of new TS-1 ODS catalysts with mesoporosity or bimodal porosity (coexistence of microporosity and mesoporosity). Apart from TS-1, a few other TS titanium-based system catalysts possessing different topologies are reported as catalysts for ODS, such as Ti-β, Ti-MCM-41, Ti-MWW, Ti-SBA-16, Ti-SBA-15, and Ti-SBA-2 (*147–159*). These TSs are synthesized via surfactant-mediated synthesis. The presence of free silanols in titanium-based system catalysts may diminish the lifetime of titanium-based systems during the course of repeated use in ODS since the free silanols tend to adsorb sulfones (the oxidized sulfur compounds), which in turn deactivate the titanium-based system catalysts. However, this problem can be circumvented by silylating the surface silanols with suitable silylating agents.

Finally, some new catalytic systems have been developed based on metal–organic frameworks (MOFs). MOFs are a fascinating class of materials that are suitable as catalysts for the ODS process due to the theoretically infinite number of combinations of polynuclear metal clusters and organic linkers (*160–179*). Based on their characteristics, MOF materials have great potential for application in the ODS process, as either pure MOF catalysts, composites, or MOF-derived materials. Considering the large family of MOFs, their use in ODS applications is still limited, mainly due to their stability and the accessibility of catalytically active sites within the MOFs. The next step in this field will be developing stable MOFs with a sufficient number of accessible catalytically active sites for applications in ODS. Creating defects or mesopores in MOF structures is helpful in increasing the number of available catalytically active sites and their accessibility to reactants. Further efforts should focus on both the stability and functionality of MOFs to obtain improved performance in ODS applications. In conclusion, MOFs, as an emerging class of porous materials, have great potential in the field of desulfurization. (1) MOFs themselves can be efficient catalysts for the ODS process. (2) MOFs can be used as supports or precursors of the heterogeneous catalysts for ODS reactions. (3) For ODS, creating accessible open metal sites to interact with the sulfur-containing compounds is highly important for the performance of MOFs. (4) Using environmentally friendly oxidants (e.g., O_2) and improving the reusability of MOFs are needed for applications in desulfurization. Considering the rapid development of MOF materials, we believe more functionalized MOF materials will be fabricated and utilized in the highly efficient and green desulfurization of fossil fuels.

Main Reaction Conditions That Affect ODS Activity

Not only the catalytic system employed affects the performance of the ODS; some other variables affect the ODS performance and have a significant influence on the activity of ODS catalysts, either directly or indirectly. In this section we describe some of the most important variables.

Temperature

The temperature usually has a dramatic effect on the reaction rate; an increase in the reaction temperature produces an increase in the reaction rate. However, the ODS reactions usually work in a temperature window of 323–373 K. In this range of temperature, the optimum is found to achieve better ODS performance in the studies published. The first reason is that the boiling point of the common solvents or the biphasic system formed (fuel-solvent) falls in this temperature range. But the use of high temperatures can degrade the oxidants, especially when H_2O_2 is utilized, yielding an unproductive consumption of oxidant and reducing the efficiency of the process. With liquid oxidants like H_2O_2, ODS catalysts usually show maximum performance at atmospheric pressure. However, when using gaseous oxidants like O_2, high pressure is sometimes required in order to increase the gas solubility.

Solvent

Solvents are present in the ODS process, especially when hydrogen peroxide is employed as oxidant. Hydrogen peroxide is nonmiscible in a nonpolar medium, so there is always a biphasic reaction system (polar and nonpolar). In consequence, the solvent plays a critical role in the ODS performance. The nonpolar medium (oil phase) is the real liquid fuel, so the influence of solvents on the activity of ODS catalysts is assumed to be related to the polar solvent.

Due to the existence of two phases, transport phenomena are highly important, and the reaction conditions that favor mass transport between phases are very important in the ODS process. During ODS, some exchange of phases can happen: a small part of the polar solvent can be dissolved in the fuel phase, and vice versa. In addition, some sulfones can be extracted from the fuel phase to the polar phase. The catalysts and oxidant (hydrogen peroxide) are in the polar phase and there take part in the oxidation (Figure 5). For this reason, the reaction rate can be limited by the transfer of the sulfur compounds from the oil phase to the polar phase, and the mass-transfer rate depends on the characteristics of the solvent employed and its interactions with the fuel properties.

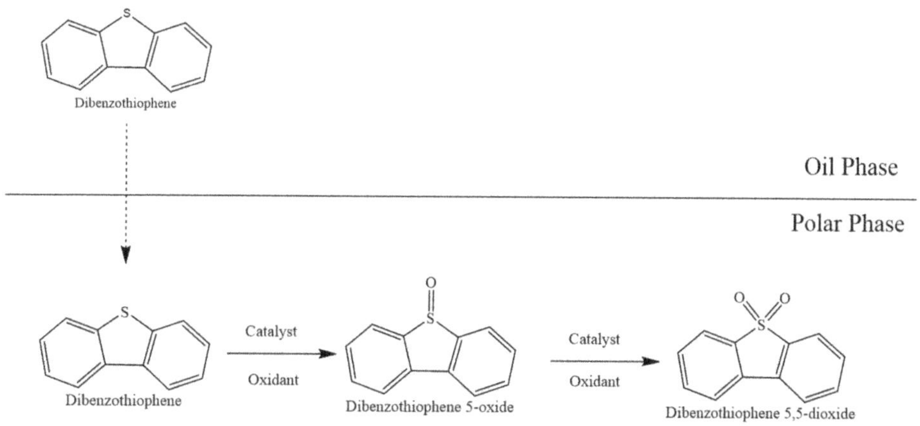

Figure 5. Oxidation of sulfur compounds in a biphase system.

In order to decrease the dependence on the mass transfer of the sulfur compounds, some authors employ different strategies, such as the use of a phase-transfer catalyst (*34, 38, 50, 180*) (Figure 6) or designing amphiphilic catalysts that produce microemulsions of water in oil (*55–58, 181–183*) (see Figure 4). The presence of a phase-transfer catalyst increases the oxidation rate because the catalyst/oxidant can be transferred from the polar to the oil phase and oxidize the sulfur compounds in parallel with the oxidation in the polar phase. This combination enhances the oxidation efficiency.

Based on this assumption, the solvent with the best extraction ability may be the best solvent in the ODS process. However, this is not true, because there is a need for synergy between the solvent and catalyst. In some cases, a good extractant solvent blocks the catalyst, reducing or suppressing the activity (*184*). Catalysts based on oxidant activation by radical formation (HO· and HO_2·), like Fenton systems, work better in aprotic solvents that stabilize the radicals formed, favoring the reaction with sulfur species. In contrast, for catalysts based on the formation of peroxo species (transition metals with high oxidation state or POM systems), the solvents should not occupy the vacant coordination sites of the catalyst, which are essentially allotted to the formation of peroxide species, a critical step in the oxidation of sulfur compounds.

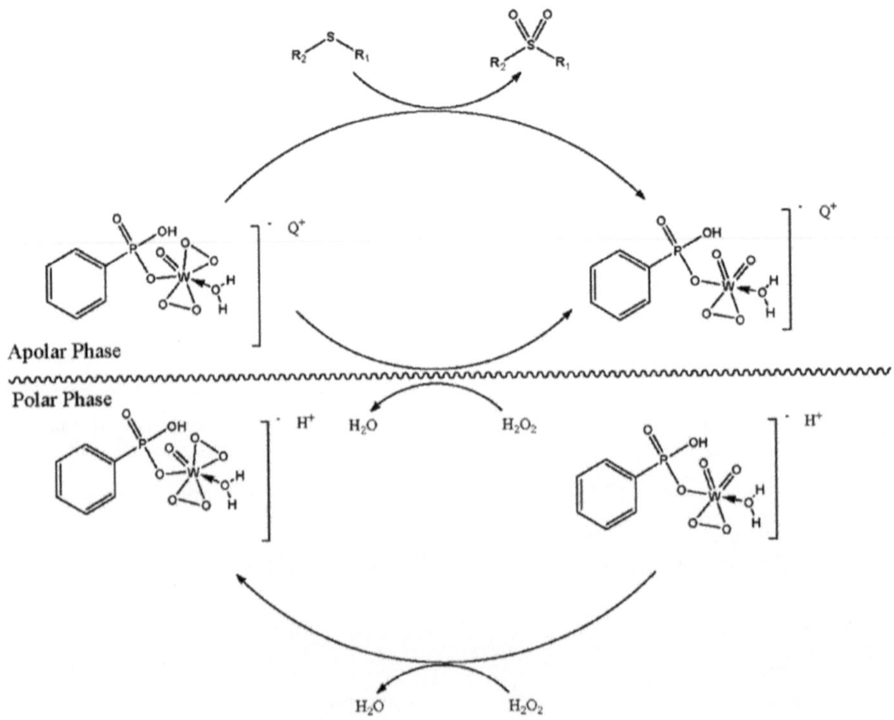

Figure 6. Catalytic cycle of sulfur compound oxidation with phase-transfer catalyst. Reproduced with permission from reference (38). Copyright 2004 Royal Society of Chemistry.

Oxidant

The oxidant, the oxygen donor, is a very important part in ODS. Different types of oxidants can be used in the ODS process, which must be carefully tailored to the sulfur compounds to prevent the oxidation of olefins or aromatic compounds that are present in the fuel feed. Thus, different types of oxidants are used in the ODS process, such as organic peracid, organic hydroperoxide, H_2O_2, oxygen, air, super oxide (O^{2-}), ozone, and others (*1–3, 5, 11–14*). Therefore, choosing the right kind of oxidant will depend not only on its oxidizing power and active oxygen but also on its cost, environmental friendliness, and availability.

The best candidate as an oxidizing agent is hydrogen peroxide. It presents a high amount of active oxygen by mass (47%). It is a commercial product often used at the industrial level and gives only water as a by-product (*185*). The main drawback of using hydrogen peroxide is the need to use a biphasic system, because the hydrogen peroxide is commercially available in a water solution and is immiscible with fuels.

A second family of oxidants that has been studied is based on the use of organic hydroperoxides. The most important advantage of this kind of oxidant is its miscibility with the fuel, which clearly simplifies the reactor engineering and eliminates the mass-transfer problems of a biphasic system. The most widely employed hydroperoxide is *t*-butylhydroperoxide (*12, 45, 186–188*). Other oil-soluble alkyl peroxides have also been tried. Their reactivity in the oxidation of sulfur compounds is as follows: cyclohexanone peroxide > *tert*-amyl hydroperoxide > *tert*-butyl hydroperoxide (*82, 189,*

190). Cyclohexanone peroxide is formed by reaction with hydrogen peroxide (*36*), which makes it less competitive than other hydroperoxides that are produced by hydrocarbon autoxidation.

An important drawback of organic hydroperoxides is the low amount of active oxygen by mass (10–17%). For this reason, the transportation of this compound is very expensive, so the installation of a hydrocarbon oxidation unit to produce hydroperoxides is recommended, with an increase in capital costs. A second major drawback is the waste treatment of the alcohol by-product generated, as well as sulfone waste treatment. When using *tert*-butyl hydroperoxide, the by-product *tert*-butyl alcohol can be used as a potential octane-boosting compound for gasoline.

The use of liquid oxidants has the drawback of the need for large amounts of oxidants, which usually are not present in a refinery and are relatively high in price. When a large amount of sulfur needs to be eliminated, the process can be rentable, but the logistics of the transport of oxidants or the need in comparison with the world market cannot be viable.

For this reason, some approaches have focused on the use of a gaseous oxidant (oxygen or ozone). These oxidants are abundant, available everywhere, and inexpensive (oxygen), but at present the results are far from practical. The reaction puts in contact a highly oxidizing gas (oxygen or ozone) and fuels; this point is a clear safety issue for several petroleum fractions that can produce flammable mixtures. Ozone is a very reactive oxidant that can desulfurize fuels by ODS, but due to its intrinsic reactivity it is also able to oxidize other products present in the fuel, mainly olefins and cycloalkanes. In addition, it can generate radicals in the reaction media, producing the formation of gums by polymerization (*191–195*). The use of oxygen was also studied, but in several cases the use of sacrificial aldehyde makes this procedure impractical at the industrial scale (*196–199*). Some interesting works have been done using the reaction of oxidation with molecular oxygen in the absence of sacrificial compounds (*200–205*). This reaction can be an ideal way to perform the ODS process because the oxidant is present in the air, and consequently there is no need of transport or produce oxidant to the process.

Influence of Feedstock Characteristics and Composition on ODS Activity

The nature of the feedstock employed has a dramatic effect on the ODS performance, because it totally changes the characteristics of the interphase between the polar and nonpolar phase. In addition, the viscosity and density can dramatically affect the mixture of phases. Summing up, the extraction capability of the solvent will be dramatically modified. Another point is the presence of olefins and cycloalkanes in real fuels, which can compete with the oxidation of the sulfur, wasting oxidant. For this reason, the translation of excellent activity results with model fuels to real fuels usually fails. The differences observed between model and real samples in the gasoline or diesel fraction are relatively low, but with heavy oils the differences are dramatic.

The effect of each individual component is very difficult to analyze, because only a few studies have focused on the study of simulated real fuels, and the response of real fuels is too complex to separate the effect of each component.

Some studies showed that the presence of long-chain hydrocarbons reduces the effectiveness of the ODS process. The use of model fuels with three different nonpolar solvents (*n*-hexane, *n*-heptane, and *n*-octane) has shown a faster DBT conversion during ODS when *n*-hexane is employed (*50, 206*).

Secondly, the possible competitive oxidation of olefins and cycloalkanes may be quite problematic and implies the use of greater quantities of oxidant and decreases the quality of the fuel, due to the presence of oxygen. However, the correct selection of catalysts and reaction conditions can minimize the interference of competitive oxidations.

Other compounds like aromatics and polyaromatics present in real fuels can affect the ODS performance. The extraction of sulfur compounds with polar solvents is reduced in the presence of aromatics (*54*), which is the first step in the oxidation. In consequence, this change in the interface between polar and nonpolar solvents has an important effect on the catalytic performance. This effect is negligible in the single-phase oxidation with hydroperoxides (*207*). A similar effect can be attributed to the polyaromatic compounds: their ability to dissolve sulfones can make difficult the desulfurization of the fuel, but when using heterogeneous catalysts, it can help the ODS catalysts to maintain constant activity during the course of ODS (*42, 208*).

Real fuels also contain nitrogen compounds. These compounds are Lewis bases that can be adsorbed on the catalysts, blocking their catalytic activity (*42, 208*). The negative effect of the nitrogen compounds depends on their nature. Carbazone and quinoline have a limited effect (*54*), while indole has shown a negative effect (*42, 96, 207, 208*). The high inhibitor effect of indole can be electron density.

Types of Sulfur Compounds

Sulfur compounds of real fuel feedstocks can be grouped by type, as described in Figure 7. The families of sulfur compounds are present in different fuels depending on their boiling point. Their reactivity in ODS is pretty different from that typical of HDS. As a general rule, the reactivity of organic sulfur compounds in ODS is higher, as high as the electron density on the sulfur atom (*3, 16*). Then the order of expected reactivity of typical sulfur compounds present in diesel fuel is 4,6-DMDBT > MDBT > DBT > BT. However, ODS performance is higher against DBT than 4,6-DMDBT, because the steric hindrance of bulky 4,6-DMDBT prevents the catalytically active sites from freely accessing the sulfur atom (*132, 140*). Following this rule, for DBTs with more alkyl substituents, the rate of catalytic ODS depends on both the number and position of the alkyl substituents present in the alkyl DBTs (*209*). This situation is more complicated in the case of heavy fuel oils, where more complicated structures of polyaromatic compounds are present (*3*).

The catalyst employed has an effect on the reactivity of the sulfur compounds. Heterogeneous catalysts have more problems with bulky molecules with alkyl substituents, while for their homogeneous counterparts the steric hindrance has less of an effect. When catalysts of Fenton type are employed, they activate the oxidant, forming radicals, and the reactivity follows that expected based on the electron density. The previous discussion emphasizes that the relation between the ODS catalysts and the reactivity order of different organic sulfur compounds depends on the electron density and the type and characteristics of the ODS catalyst employed.

Challenges for ODS

ODS has proved that it is possible to remove sulfur compounds from fuels to meet strict regulation limits. Among different chemical oxidants, the best option is hydrogen peroxide. We are waiting for a reliable system than can use oxygen (air). Using this oxidant, the oxidation/extraction is simultaneous and can reach very high levels of sulfur removal. However, in order to make an ODS process more competitive, some improvements are needed: (1) increase the mass transfer in a biphasic system containing an oil fraction and a polar phase, (2) develop applications for heavy fuel fractions, and (3) enhance the posttreatment of the sulfones produced.

Increase the Mass Transfer in the Reactor

The reaction is conducted in the presence of two liquid reaction phases: an oil phase with the sulfur compound and a polar phase that contains the oxidant agent (H_2O_2), which is not soluble in the oil phase. In consequence, the overall sulfur-removal rate is limited by the mass transfer between phases with very active catalysts. For this reason, an increase in the mass transfer clearly improves the overall sulfur-removal rate. Some attempts have been made to employ microstructured reactors (*211*, *212*). Chemical microstructured reactors are devices containing open paths for fluid flow with dimensions in the submillimeter range. Most such reactors are designed with multiple parallel channels with diameters between ten and several hundred micrometers, where the chemical transformation occurs (*213*). This gives a high specific surface area in the range of 10,000 to 50,000 m^2/m^3 (*213*, *214*) and allows for effective mass and heat transfer when compared to more traditional chemical reactors having $\approx$100 m^2/m^3. These reactor systems can improve the mass transfer 100 times.

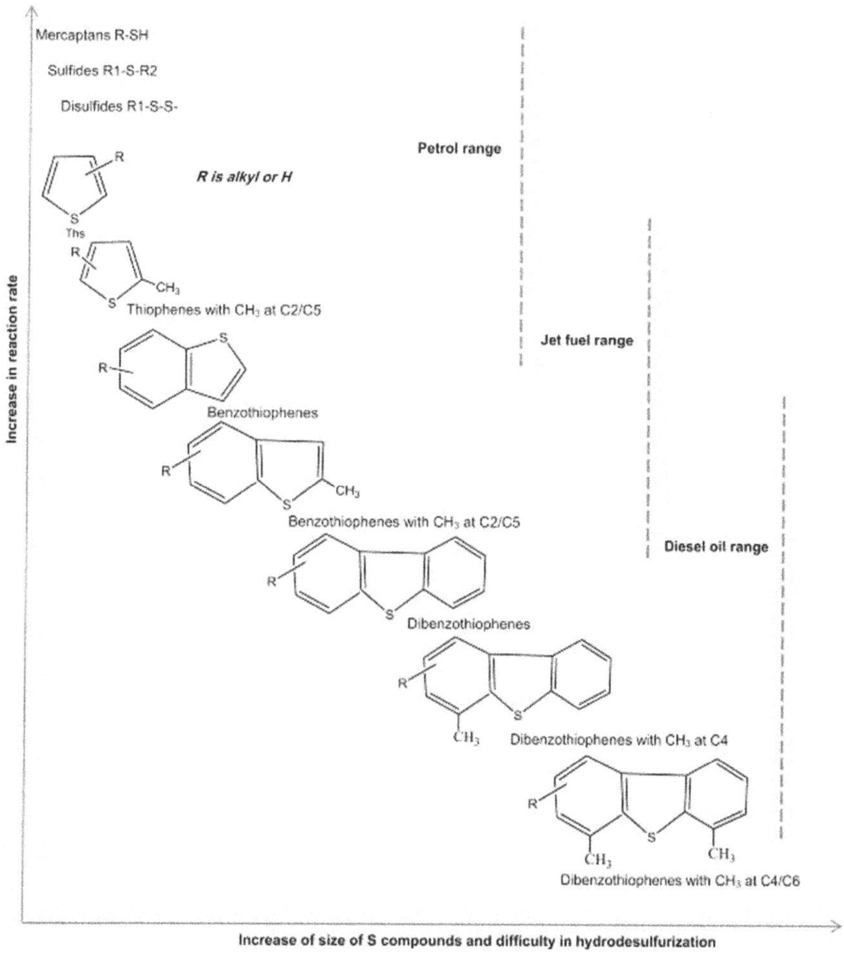

Figure 7. Families of sulfur compounds present in real light fuels. Reproduced with permission from reference (210). Copyright 2004 Elsevier.

Another interesting alternative is the use of ultrasound-assisted oxidative desulfurization (*1, 215*). Ultrasound effects account for the generation of intense convection in the medium in the form of heat and mass transfer and a characteristic improvement in the system. Ultrasound irradiation produces a fine dispersion of both liquid phases (nonpolar and polar) that enhances the mass transfer between the phases that will produce an increase in the reaction rate and consequently the efficiency of the system. But ultrasound irradiation affects not only the mass-transfer rate: the ultrasound irradiation produces a series of microbubbles that collapse very quickly, producing a cavitation phenomenon. The sudden collapse produces localized hot spots, producing a temperature and pressure around 5000 °C and 2000 atm, respectively, with heating and cooling rates above 1000 °C s^{-1}. Under these conditions, highly reactive free radicals are produced that favor oxidation reactions. Although this technology can be considered as an attractive strategy to desulfurize fuels, further development must be done to scale up the process. It still needs to be evaluated more, including the effect of factors like the shape of the reactor, the probe diameter, and liquid heights that affect sonoreactors in general. One must also take into account the electric consumption needed in this treatment.

Application to Heavy Fuel Fractions

The application of ODS to heavy oils is scarce, and only a few studies have been reported to date. In most papers, desulfurization rates are reported after a purification step through extraction or adsorption with no calculation of the actual conversion rate of sulfides to sulfones. Despite major developments in heterogeneous catalysis and its advantageous catalyst recovery, most of the reactions reported were carried out using homogeneous catalysis, mainly to overcome the problem of reaction in the presence of two phases. Heterogeneous catalyst deactivation and reuse also need to be improved before the process can be scaled up.

Enhancement of the Posttreatment of the Sulfones Produced

The concentration of sulfur compounds is very low, but the refinery stream to be treated is huge, so the amount of sulfones produced is large. Some sulfones can be used as chemical intermediates, but in real fuel the variety of sulfur compounds is very high, and thus the number of different sulfones is also high and very difficult to separate. For these reasons, adequate posttreatment of the sulfones produced is very important. The amount of fuel lost in the ODS process depends on the initial sulfur concentration and the amount of sulfur removed. For instance, 1000 ppm of S removed implies a decrease of about 5–6% of fuel mass. For this reason, it is very interesting to recover, at least in part, the hydrocarbon portion of the oxidized sulfur compounds.

The first option is to treat these sulfones to eliminate the sulfur and produce a sulfur-free fuel. The oxidized sulfur-containing hydrocarbon is put in contact with a metal to form a metal-sulfur-containing compound. This process therefore relies on the adsorption of oxidized sulfur compounds from the hydrocarbon using a metal capable of forming a metal sulfide. The metal is selected from the group consisting of Ni, Mo, Co, W, Fe, Zn, V, Cu, Mn, Hg, and their mixtures. This process is distinguished from conventional HDS in that the sulfur is immobilized in the form of a metallic sulfur compound (e.g., a metal sulfide) rather than being converted to hydrogen sulfide. For this reason, the addition of free molecular hydrogen, as is required in HDS, is avoided. The second option is catalytic decomposition on acidic/basic solid catalysts, like zeolites or silica/alumina (*216, 217*), or basic catalysts like MgO or hydrotalcite (*218–221*). This catalytic conversion takes place under relatively mild conditions, without the use of a hydrogen atmosphere. However, these processes need to be improved.

Acknowledgments

We are grateful to many of our colleagues for stimulating discussions and to our research sponsor King Abdullah University of Science and Technology (KAUST) under grant OSR-2019-CPF-4103.2.

References

1. Ja'fari, M.; Ebrahimi, S. L.; Khosravi-Nikou, M. R. Ultrasound-Assisted Oxidative Desulfurization and Denitrogenation of Liquid Hydrocarbon Fuels: A Critical Review. *Ultrason. Sonochem.* **2018**, *40*, 955–968. https://doi.org/10.1016/j.ultsonch.2017.09.002.
2. Rajendran, A.; Cui, T. Y.; Fan, H. X.; Yang, Z. F.; Feng, J.; Li, W. Y. A Comprehensive Review on Oxidative Desulfurization Catalysts Targeting Clean Energy and Environment. *J. Mater. Chem. A* **2020**, *8* (5), 2246–2285. https://doi.org/10.1039/c9ta12555h.
3. Houda, S.; Lancelot, C.; Blanchard, P.; Poinel, L.; Lamonier, C. Oxidative Desulfurization of Heavy Oils with High Sulfur Content: A Review. *Catalysts* **2018**, *8* (9), 344. https://doi.org/10.3390/catal8090344.
4. Duarte, F. A.; Mello, P. D. A.; Bizzi, C. A.; Nunes, M. A. G.; Moreira, E. M.; Alencar, M. S.; Motta, H. N.; Dressler, V. L.; Flores, É. M. M. Sulfur Removal from Hydrotreated Petroleum Fractions Using Ultrasound-Assisted Oxidative Desulfurization Process. *Fuel* **2011**, *90* (6), 2158–2164. https://doi.org/10.1016/j.fuel.2011.01.030.
5. Campos-Martin, J. M.; Capel-Sanchez, M. C.; Perez-Presas, P.; Fierro, J. L. G. Oxidative Processes of Desulfurization of Liquid Fuels. *J. Chem. Technol. Biotechnol.* **2010**, *85* (7), 879–890. https://doi.org/10.1002/jctb.2371.
6. Song, C. An Overview of New Approaches to Deep Desulfurization for Ultra-Clean Gasoline, Diesel Fuel and Jet Fuel. *Catal. Today* **2003**, *86* (1–4), 211–263. https://doi.org/10.1016/S0920-5861(03)00412-7.
7. Babich, I. V.; Moulijn, J. A. Science and Technology of Novel Processes for Deep Desulfurization of Oil Refinery Streams: A Review. *Fuel* **2003**, *82* (6), 607–631. https://doi.org/10.1016/S0016-2361(02)00324-1.
8. Song, C.; Ma, X. New Design Approaches to Ultra-Clean Diesel Fuels by Deep Desulfurization and Deep Dearomatization. *Appl. Catal. B Environ.* **2003**, *41* (1–2), 207–238. https://doi.org/10.1016/S0926-3373(02)00212-6.
9. Shafiq, I.; Shafique, S.; Akhter, P.; Yang, W.; Hussain, M. Recent Developments in Alumina Supported Hydrodesulfurization Catalysts for the Production of Sulfur-Free Refinery Products: A Technical Review. *Catal. Rev.: Sci. Eng.*, in press. https://doi.org/10.1080/01614940.2020.1780824.
10. Kuwahara, H. Desulfurization of Heavy Oil. *Chem. Econ. Eng. Rev.* **1973**, *5* (10), 35–40. https://doi.org/10.1007/s13203-012-0006-6.
11. Ismagilov, Z.; Yashnik, S.; Kerzhentsev, M.; Parmon, V.; Bourane, A.; Al-Shahrani, F. M.; Hajji, A. A.; Koseoglu, O. R. Oxidative Desulfurization of Hydrocarbon Fuels. *Catal. Rev.: Sci. Eng.* **2011**, *53* (3), 199–255. https://doi.org/10.1080/01614940.2011.596426.

12. Jiang, Z.; Lü, H.; Zhang, Y.; Li, C. Oxidative Desulfurization of Fuel Oils. *Cuihua Xuebao/Chinese J. Catal.* **2011**, *32* (5), 707–715. https://doi.org/10.1016/s1872-2067(10)60246-x.
13. Haghighi, M.; Gooneh-Farahani, S. Insights to the Oxidative Desulfurization Process of Fossil Fuels over Organic and Inorganic Heterogeneous Catalysts: Advantages and Issues. *Environ. Sci. Pollut. Res.* **2020**, *27* (32), 39923–39945. https://doi.org/10.1007/s11356-020-10310-4.
14. Hossain, M. N.; Park, H. C.; Choi, H. S. A Comprehensive Review on Catalytic Oxidative Desulfurization of Liquid Fuel Oil. *Catalysts* **2019**, *9* (3), 229. https://doi.org/10.3390/catal9030229.
15. Mjalli, F. S.; Ahmed, O. U.; Al-Wahaibi, T.; Al-Wahaibi, Y.; AlNashef, I. M. Deep Oxidative Desulfurization of Liquid Fuels. *Rev. Chem. Eng.* **2014**, *30* (4), 337–378. https://doi.org/10.1515/revce-2014-0001.
16. Otsuki, S.; Nonaka, T.; Takashima, N.; Qian, W.; Ishihara, A.; Imai, T.; Kabe, T. Oxidative Desulfurization of Light Gas Oil and Vacuum Gas Oil by Oxidation and Solvent Extraction. *Energy Fuels* **2000**, *14* (6), 1232–1239. https://doi.org/10.1021/ef000096i.
17. De Filippis, P.; Scarsella, M. Oxidative Desulfurization: Oxidation Reactivity of Sulfur Compounds in Different Organic Matrixes. *Energy Fuels* **2003**, *17* (6), 1452–1455. https://doi.org/10.1021/ef0202539.
18. Dehkordi, A. M.; Sobati, M. A.; Nazem, M. A. Oxidative Desulfurization of Non-Hydrotreated Kerosene Using Hydrogen Peroxide and Acetic Acid. *Chinese J. Chem. Eng.* **2009**, *17* (5), 869–874. https://doi.org/10.1016/S1004-9541(08)60289-X.
19. Yazu, K.; Makino, M.; Ukegawa, K. Oxidative Desulfurization of Diesel Oil with Hydrogen Peroxide in the Presence of Acid Catalyst in Diesel Oil/Acetic Acid Biphasic System. *Chem. Lett.* **2004**, *33* (10), 1306–1307. https://doi.org/10.1246/cl.2004.1306.
20. Ali, S. H.; Hamad, D. M.; Albusairi, B. H.; Fahim, M. A. Removal of Dibenzothiophenes from Fuels by Oxy-Desulfurization. *Energy Fuels* **2009**, *23* (12), 5986–5994. https://doi.org/10.1021/ef900683d.
21. Farshi, A.; Shiralizadeh, P. Sulfur Reduction of Heavy Fuel Oil by Oxidative Desulfurization (ODS) Method. *Pet. Coal* **2015**, *57* (3), 295–302.
22. Yazu, K.; Furuya, T.; Miki, K.; Ukegawa, K. Tungstophosphoric Acid-Catalyzed Oxidative Desulfurization of Light Oil with Hydrogen Peroxide in a Light Oil/Acetic Acid Biphasic System. *Chem. Lett.* **2003**, *32* (10), 920–921. https://doi.org/10.1246/cl.2003.920.
23. Al-Shahrani, F.; Xiao, T.; Llewellyn, S. A.; Barri, S.; Jiang, Z.; Shi, H.; Martinie, G.; Green, M. L. H. Desulfurization of Diesel via the H_2O_2 Oxidation of Aromatic Sulfides to Sulfones Using a Tungstate Catalyst. *Appl. Catal. B Environ.* **2007**, *73* (3), 311–316. https://doi.org/10.1016/j.apcatb.2006.12.016.
24. Trakarnpruk, W.; Rujiraworawut, K. Oxidative Desulfurization of Gas Oil by Polyoxometalates Catalysts. *Fuel Process. Technol.* **2009**, *90* (3), 411–414. https://doi.org/10.1016/j.fuproc.2008.11.002.
25. Yu, G.; Lu, S.; Chen, H.; Zhu, Z. Diesel Fuel Desulfurization with Hydrogen Peroxide Promoted by Formic Acid and Catalyzed by Activated Carbon. *Carbon N. Y.* **2005**, *43* (11), 2285–2294. https://doi.org/10.1016/j.carbon.2005.04.008.

26. Yu, G.; Lu, S.; Chen, H.; Zhu, Z. Oxidative Desulfurization of Diesel Fuels with Hydrogen Peroxide in the Presence of Activated Carbon and Formic Acid. *Energy Fuels* **2005**, *19* (2), 447–452. https://doi.org/10.1021/ef049760b.
27. Zhou, X. L.; Tan, Q.; Yu, G. X.; Chen, L. F.; Wang, J. A.; Novaro, O. Removal of Dibenzothiophene in Diesel Oil by Oxidation over a Promoted Activated Carbon Catalyst. *Kinet. Catal.* **2009**, *50* (4), 543–549. https://doi.org/10.1134/S0023158409040119.
28. Timko, M. T.; Wang, J. A.; Burgess, J.; Kracke, P.; Gonzalez, L.; Jaye, C.; Fischer, D. A. Roles of Surface Chemistry and Structural Defects of Activated Carbons in the Oxidative Desulfurization of Benzothiophenes. *Fuel* **2016**, *163*, 223–231. https://doi.org/10.1016/j.fuel.2015.09.075.
29. Pandey, S.; Srivastava, V. C. Oxidative-Extractive Desulfurization of Liquid Fuel Using Stannous Chloride-Acetic Acid Mixture as Catalyst. *Pet. Sci. Technol.* **2018**, *36* (1), 40–47. https://doi.org/10.1080/10916466.2017.1403451.
30. Rezvani, M. A.; Shaterian, M.; Akbarzadeh, F.; Khandan, S. Deep Oxidative Desulfurization of Gasoline Induced by PMoCu@$MgCu_2O_4$-PVA Composite as a High-Performance Heterogeneous Nanocatalyst. *Chem. Eng. J.* **2018**, *333*, 537–544. https://doi.org/10.1016/j.cej.2017.09.184.
31. Arellano, U.; Wang, J. A.; Timko, M. T.; Chen, L. F.; Paredes Carrera, S. P.; Asomoza, M.; González Vargas, O. A.; Llanos, M. E. Oxidative Removal of Dibenzothiophene in a Biphasic System Using Sol-Gel $FeTiO_2$ Catalysts and H_2O_2 Promoted with Acetic Acid. *Fuel* **2014**, *126*, 16–25. https://doi.org/10.1016/j.fuel.2014.02.028.
32. de Souza, W. F.; Guimarães, I. R.; Guerreiro, M. C.; Oliveira, L. C. A. Catalytic Oxidation of Sulfur and Nitrogen Compounds from Diesel Fuel. *Appl. Catal. A Gen.* **2009**, *360* (2), 205–209. https://doi.org/10.1016/j.apcata.2009.03.023.
33. Chen, L.; Guo, S.; Zhao, D. Oxidation of Thiophenes over Silica Gel in Hydrogen Peroxide/Formic Acid System. *Chinese J. Chem. Eng.* **2006**, *14* (6), 835–838. https://doi.org/10.1016/S1004-9541(07)60022-6.
34. Zhao, D.; Ren, H.; Wang, J.; Yang, Y.; Zhao, Y. Kinetics and Mechanism of Quaternary Ammonium Salts as Phase-Transfer Catalysts in the Liquid-Liquid Phase for Oxidation of Thiopene. *Energy Fuels* **2007**, *21* (5), 2543–2547. https://doi.org/10.1021/ef070152g.
35. Zhao, D.; Ren, H.; Zhao, Y.; Jia, J. Mechanism and Catalytic Behavior of Quaternary Ammonium Salts in Oxidative Desulfurization. *Pet. Sci. Technol.* **2009**, *27* (12), 1338–1348. https://doi.org/10.1080/10916460802105682.
36. Desai, K.; Bhatt, S.; Dharaskar, S.; Unnarkat, A.; Jampa, S. S.; Khalid, M. Butyl Triphenyl Phosphonium Bromide as an Effective Catalyst for Ultrasound Assisted Oxidative Desulfurization Process. *Mater. Today Proc.* **2019**, *28*, 836–841. https://doi.org/10.1016/j.matpr.2019.12.308.
37. Anisimov, A. V.; Fedorova, E. V.; Lesnugin, A. Z.; Senyavin, V. M.; Aslanov, L. A.; Rybakov, V. B.; Tarakanova, A. V. Vanadium Peroxocomplexes as Oxidation Catalysts of Sulfur Organic Compounds by Hydrogen Peroxide in Bi-Phase Systems. *Catal. Today* **2003**, *78* (1–4 SPEC.), 319–325. https://doi.org/10.1016/S0920-5861(02)00311-5.

38. Campos-Martin, J. M.; Capel-Sanchez, M. C.; Fierro, J. L. G. Highly Efficient Deep Desulfurization of Fuels by Chemical Oxidation. *Green Chem.* **2004**, *6* (11), 557–562. https://doi.org/10.1039/b409882j.
39. Singh, S.; Srivastava, V. C.; Gautam, S. Oxidative-Extractive Desulfurization of Liquid Fuel by Dimethyl Sulfoxide and $ZnCl_2$ Based Ionic Liquid. *Int. J. Chem. React. Eng.* **2016**, *14* (1), 539–545. https://doi.org/10.1515/ijcre-2015-0026.
40. Taghizadeh, M.; Mehrvarz, E.; Taghipour, A. Polyoxometalate as an Effective Catalyst for the Oxidative Desulfurization of Liquid Fuels: A Critical Review. *Rev. Chem. Eng.* **2020**, *36* (7), 831–858. https://doi.org/10.1515/revce-2018-0058.
41. Collins, F. M.; Lucy, A. R.; Sharp, C. Oxidative Desulphurisation of Oils via Hydrogen Peroxide and Heteropolyanion Catalysis. *J. Mol. Catal. A Chem.* **1997**, *117* (1–3), 397–403. https://doi.org/10.1016/S1381-1169(96)00251-8.
42. Yun, G. N.; Lee, Y. K. Beneficial Effects of Polycyclic Aromatics on Oxidative Desulfurization of Light Cycle Oil over Phosphotungstic Acid (PTA) Catalyst. *Fuel Process. Technol.* **2013**, *114*, 1–5. https://doi.org/10.1016/j.fuproc.2013.03.035.
43. Chuang, L. L.; Huang, J. F.; Lo, W. H.; Wei, G. T. Deep Desulfurization of Light Oil through Extraction and Oxidation Processes Using H_2O_2/Tungstophosphoric Acid in Room-Temperature Ionic Liquids. *J. Chinese Chem. Soc.* **2012**, *59* (3), 324–330. https://doi.org/10.1002/jccs.201100672.
44. Liu, G.; Cao, Y.; Jiang, R.; Wang, L.; Zhang, X.; Mi, Z. Oxidative Desulfurization of Jet Fuels and Its Impact on Thermal-Oxidative Stability. *Energy Fuels* **2009**, *23* (12), 5978–5985. https://doi.org/10.1021/ef900669b.
45. Li, J.; Yang, L. N.; Shen, J. One-Step Oxidation-Desulfurization of FCC Gasoline Catalyzed by Tungstophosphoric Acid. *Pet. Sci. Technol.* **2011**, *29* (3), 247–253. https://doi.org/10.1080/10916460903073987.
46. Zhang, B.; Jiang, W.; Fan, X.; Zhang, M.; Wei, Y.; He, J.; Li, C.; Li, H.; Zhu, W. Synthesis of Amphiphilic Peroxophosphomolybdates for Oxidative Desulfurization of Fuels in Ionic Liquids. *Pet. Sci.* **2018**, *15* (4), 890–897. https://doi.org/10.1007/s12182-018-0272-8.
47. Yu, F.; Wang, R. Oxidative Desulfurization of Diesel Using Organic Salt of Polyoxometalate as an Efficient and Recoverable Phase-Transfer Catalyst. *Chem. Lett.* **2014**, *43* (6), 834–836. https://doi.org/10.1246/cl.140132.
48. Zhang, X.; Shi, Y.; Liu, G. Direct Preparation of $[(CH_3)_3NC_{16}H_{33}]_4Mo_8O_{26}$ and Its Catalytic Performance in Oxidative Desulfurization. *Catal. Sci. Technol.* **2016**, *6* (4), 1016–1024. https://doi.org/10.1039/c5cy01439e.
49. Julião, D.; Gomes, A. C.; Cunha-Silva, L.; Valença, R.; Ribeiro, J. C.; Pillinger, M.; de Castro, B.; Gonçalves, I. S.; Balula, S. S. A Sustainable Peroxophosphomolybdate/H_2O_2 System for the Oxidative Removal of Organosulfur Compounds from Simulated and Real High-Sulfur Diesels. *Appl. Catal. A Gen.* **2020**, *589*, 117154. https://doi.org/10.1016/j.apcata.2019.117154.
50. Huang, D.; Zhai, Z.; Lu, Y. C.; Yang, L. M.; Luo, G. S. Optimization of Composition of a Directly Combined Catalyst in Dibenzothiophene Oxidation for Deep Desulfurization. *Ind. Eng. Chem. Res.* **2007**, *46* (5), 1447–1451. https://doi.org/10.1021/ie0611857.

51. Xu, J.; Zhao, S.; Ji, Y.; Song, Y. F. Deep Desulfurization by Amphiphilic Lanthanide-Containing Polyoxometalates in Ionic-Liquid Emulsion Systems under Mild Conditions. *Chem. - Eur. J.* **2013,** *19* (2), 709–715. https://doi.org/10.1002/chem.201202595.
52. Ribeiro, S. O.; Julião, D.; Cunha-Silva, L.; Domingues, V. F.; Valença, R.; Ribeiro, J. C.; De Castro, B.; Balula, S. S. Catalytic Oxidative/Extractive Desulfurization of Model and Untreated Diesel Using Hybrid Based Zinc-Substituted Polyoxometalates. *Fuel* **2016**, *166*, 268–275. https://doi.org/10.1016/j.fuel.2015.10.095.
53. Qiu, J.; Wang, G.; Zeng, D.; Tang, Y.; Wang, M.; Li, Y. Oxidative Desulfurization of Diesel Fuel Using Amphiphilic Quaternary Ammonium Phosphomolybdate Catalysts. *Fuel Process. Technol.* **2009,** *90* (12), 1538–1542. https://doi.org/10.1016/j.fuproc.2009.08.001.
54. Banisharif, F.; Dehghani, M. R.; Capel-Sanchez, M. C.; Campos-Martin, J. M. Extractive-Oxidative Removals of Dibenzothiophene and Quinoline Using Vanadium Substituted Dawson Emulsion Catalyst and Ionic Liquid Based Solvents. *J. Ind. Eng. Chem.* **2017**, *47*, 348–359. https://doi.org/10.1016/j.jiec.2016.12.005.
55. Banisharif, F.; Dehghani, M. R.; Campos-Martin, J. M. Oxidative Desulfurization of Diesel Using Vanadium-Substituted Dawson-Type Emulsion Catalysts. *Energy Fuels* **2017**, *31* (5), 5419–5427. https://doi.org/10.1021/acs.energyfuels.6b02791.
56. Banisharif, F.; Dehghani, M. R.; Parvari, M.; Campos-Martin, J. M. Oxidative-Extractive Desulfurization of Model Diesel Oil by CTA-Type Dawson Type Emulsion Catalysts. In *Catalysis and Reaction Engineering Division 2016 - Core Programming Area at the 2016 AIChE Annual Meeting*; AICHE, New York, NY, USA, 2016; Vol. 2, pp 747–753.
57. Zhang, Y.; Lü, H.; Wang, L.; Zhang, Y.; Liu, P.; Han, H.; Jiang, Z.; Li, C. The Oxidation of Benzothiophene Using the Keggin-Type Lacunary Polytungstophosphate as Catalysts in Emulsion. *J. Mol. Catal. A Chem.* **2010**, *332* (1–2), 59–64. https://doi.org/10.1016/j.molcata.2010.08.021.
58. Li, C.; Jiang, Z.; Gao, J.; Yang, Y.; Wang, S.; Tian, F.; Sun, F.; Sun, X.; Ying, P.; Han, C. Ultra-Deep Desulfurization of Diesel: Oxidation with a Recoverable Catalyst Assembled in Emulsion. *Chem. - Eur. J.* **2004,** *10* (9), 2277–2280. https://doi.org/10.1002/chem.200305679.
59. Komintarachat, C.; Trakarnpruk, W. Oxidative Desulfurization Using Polyoxometalates. *Ind. Eng. Chem. Res.* **2006,** *45* (6), 1853–1856. https://doi.org/10.1021/ie051199x.
60. Dong, X.; Yu, C.; Wang, D.; Zhang, Y.; Wu, P.; Hu, H.; Xue, G. Cu and Fe-Doped Monolacunary Tungstosilicate Catalysts with Efficient Catalytic Activity for Benzyl Alcohol Oxidation and Simulation Gasoline Desulfurization. *Mater. Res. Bull.* **2017**, *85*, 152–160. https://doi.org/10.1016/j.materresbull.2016.09.014.
61. Ribeiro, S. O.; Granadeiro, C. M.; Almeida, P. L.; Pires, J.; Capel-Sanchez, M. C.; Campos-Martin, J. M.; Gago, S.; de Castro, B.; Balula, S. S. Oxidative Desulfurization Strategies Using Keggin-Type Polyoxometalate Catalysts: Biphasic versus Solvent-Free Systems. *Catal. Today* **2019,** *333*, 226–236. https://doi.org/10.1016/j.cattod.2018.10.046.
62. Jin, A. H.; Li, B. S.; Dai, Z. J. Oxidative Desulfurization of Fuel Oil with Hydrogen Peroxide Catalyzed by Keggin-Type Polyoxotungstate in a DC Magnetic Field. *Pet. Sci. Technol.* **2010,** *28* (7), 700–711. https://doi.org/10.1080/10916460902804671.
63. Abro, R.; Abdeltawab, A. A.; Al-Deyab, S. S.; Yu, G.; Qazi, A. B.; Gao, S.; Chen, X. A Review of Extractive Desulfurization of Fuel Oils Using Ionic Liquids. *RSC Adv.* **2014**, *4* (67), 35302–35317. https://doi.org/10.1039/c4ra03478c.

64. Chandran, D.; Khalid, M.; Walvekar, R.; Mubarak, N. M.; Dharaskar, S.; Wong, W. Y.; Gupta, T. C. S. M. Deep Eutectic Solvents for Extraction-Desulphurization: A Review. *J. Mol. Liq.* **2019**, *275*, 312–322. https://doi.org/10.1016/j.molliq.2018.11.051.
65. Li, F. T.; Kou, C. G.; Sun, Z. M.; Hao, Y. J.; Liu, R. H.; Zhao, D. S. Deep Extractive and Oxidative Desulfurization of Dibenzothiophene with $C_5H_9NO{\cdot}SnCl_2$ Coordinated Ionic Liquid. *J. Hazard. Mater.* **2012**, *205–206*, 164–170. https://doi.org/10.1016/j.jhazmat.2011.12.054.
66. Liu, D.; Gui, J.; Song, L.; Zhang, X.; Sun, Z. Deep Desulfurization of Diesel Fuel by Extraction with Task-Specific Ionic Liquids. *Pet. Sci. Technol.* **2008**, *26* (9), 973–982. https://doi.org/10.1080/10916460600695496.
67. Huang, C.; Chen, B.; Zhang, J.; Liu, Z.; Li, Y. Desulfurization of Gasoline by Extraction with New Ionic Liquids. *Energy Fuels* **2004**, *18* (6), 1862–1864. https://doi.org/10.1021/ef049879k.
68. Kulkarni, P. S.; Afonso, C. A. M. Deep Desulfurization of Diesel Fuel Using Ionic Liquids: Current Status and Future Challenges. *Green Chem.* **2010**, *12* (7), 1139–1149. https://doi.org/10.1039/c002113j.
69. Martínez-Palou, R.; Luque, R. Applications of Ionic Liquids in the Removal of Contaminants from Refinery Feedstocks: An Industrial Perspective. *Energy Environ. Sci.* **2014**, *7* (8), 2414–2447. https://doi.org/10.1039/c3ee43837f.
70. Lu, L.; Cheng, S.; Gao, J.; Gao, G.; He, M. Y. Deep Oxidative Desulfurization of Fuels Catalyzed by Ionic Liquid in the Presence of H_2O_2. *Energy Fuels* **2007**, *21* (1), 383–384. https://doi.org/10.1021/ef060345o.
71. Chen, J.; Chen, C.; Zhang, R.; Guo, L.; Hua, L.; Chen, A.; Xiu, Y.; Liu, X.; Hou, Z. Deep Oxidative Desulfurization Catalyzed by an Ionic Liquid-Type Peroxotungsten Catalyst. *RSC Adv.* **2015**, *5* (33), 25904–25910. https://doi.org/10.1039/c5ra00136f.
72. Zhao, H.; Baker, G. A. Oxidative Desulfurization of Fuels Using Ionic Liquids: A Review. *Front. Chem. Sci. Eng.* **2015**, *9* (3), 262–279. https://doi.org/10.1007/s11705-015-1528-0.
73. Yansheng, C.; Changping, L.; Qingzhu, J.; Qingshan, L.; Peifang, Y.; Xiumei, L.; Welz-Biermann, U. Desulfurization by Oxidation Combined with Extraction Using Acidic Room-Temperature Ionic Liquids. *Green Chem.* **2011**, *13* (5), 1224–1229. https://doi.org/10.1039/c0gc00745e.
74. Zhao, D.; Wang, J.; Zhou, E. Oxidative Desulfurization of Diesel Fuel Using a Brønsted Acid Room Temperature Ionic Liquid in the Presence of H_2O_2. *Green Chem.* **2007**, *9* (11), 1219–1222. https://doi.org/10.1039/b706574d.
75. Wang, J.; Zhao, D.; Li, K. Oxidative Desulfurization of Dibenzothiophene Catalyzed by Brønsted Acid Ionic Liquid. *Energy Fuels* **2009**, *23* (8), 3831–3834. https://doi.org/10.1021/ef900251a.
76. Wang, J.; Zhang, L.; Sun, Y.; Jiang, B.; Chen, Y.; Gao, X.; Yang, H. Deep Catalytic Oxidative Desulfurization of Fuels by Novel Lewis Acidic Ionic Liquids. *Fuel Process. Technol.* **2018**, *177*, 81–88. https://doi.org/10.1016/j.fuproc.2018.04.013.
77. Julião, D.; Gomes, A. C.; Pillinger, M.; Valença, R.; Ribeiro, J. C.; Gonçalves, I. S.; Balula, S. S. Desulfurization of Liquid Fuels by Extraction and Sulfoxidation Using H_2O_2 and

[CpMo(CO)$_3$R] as Catalysts. *Appl. Catal. B Environ.* **2018**, *230*, 177–183. https://doi.org/10.1016/j.apcatb.2018.02.036.

78. Zhu, W.; Li, H.; Jiang, X.; Yan, Y.; Lu, J.; Xia, J. Oxidative Desulfurization of Fuels Catalyzed by Peroxotungsten and Peroxomolybdenum Complexes in Ionic Liquids. *Energy Fuels* **2007**, *21* (5), 2514–2516. https://doi.org/10.1021/ef700310r.
79. He, L.; Li, H.; Zhu, W.; Guo, J.; Jiang, X.; Lu, J.; Yan, Y. Deep Oxidative Desulfurization of Fuels Using Peroxophosphomolybdate Catalysts in Ionic Liquids. *Ind. Eng. Chem. Res.* **2008**, *47* (18), 6890–6895. https://doi.org/10.1021/ie800857a.
80. Liu, D.; Gui, J.; Ding, J.; Ma, J.; Lee, J.; Sun, Z. Oxidation of Dibenzothiophene Catalyzed by Na_2WO_4 in a Halogen-Free Ionic Liquid. *React. Kinet. Mech. Catal.* **2011**, *104* (1), 111–123. https://doi.org/10.1007/s11144-011-0347-7.
81. Xu, D.; Zhu, W.; Li, H.; Zhang, J.; Zou, F.; Shi, H.; Yan, Y. Oxidative Desulfurization of Fuels Catalyzed by V_2O_5 in Ionic Liquids at Room Temperature. *Energy Fuels* **2009**, *23* (12), 5929–5933. https://doi.org/10.1021/ef900686q.
82. Chen, Y.; Song, H. Y.; Lu, Y. Z.; Meng, H.; Li, C. X.; Lei, Z. G.; Chen, B. H. Unified Catalytic Oxidation-Adsorption Desulfurization Process Using Cumene Hydroperoxide as Oxidant and Vanadate Based Polyionic Liquid as Catalyst and Sorbent. *Ind. Eng. Chem. Res.* **2016**, *55* (39), 10394–10403. https://doi.org/10.1021/acs.iecr.6b02464.
83. Mesdour, S.; Lekbir, C.; Doumandji, L.; Hamada, B. Microwave-Assisted Extractive Catalytic-Oxidative Desulfurization of Diesel Fuel via a VO(acac)$_2$/Ionic Liquid System with H_2O_2 and H_2SO_4 as Oxidizing Agents. *J. Sulfur Chem.* **2017**, *38* (4), 421–439. https://doi.org/10.1080/17415993.2017.1304550.
84. Campitelli, P.; Aschi, M.; Di Nicola, C.; Marchetti, F.; Pettinari, R.; Crucianelli, M. Ionic Liquids vs Conventional Solvents: A Comparative Study in the Selective Catalytic Oxidations Promoted by Oxovanadium(IV) Complexes. *Appl. Catal. A Gen.* **2020**, *599*, 117622. https://doi.org/10.1016/j.apcata.2020.117622.
85. Zhao, R.; Wang, J.; Zhang, D.; Sun, Y.; Han, B.; Tang, N.; Wang, N.; Li, K. Biomimetic Oxidative Desulfurization of Fuel Oil in Ionic Liquids Catalyzed by Fe(III) Porphyrins. *Appl. Catal. A Gen.* **2017**, *532*, 26–31. https://doi.org/10.1016/j.apcata.2016.12.008.
86. Zhang, J.; Zhu, W.; Li, H.; Jiang, W.; Jiang, Y.; Huang, W.; Yan, Y. Deep Oxidative Desulfurization of Fuels by Fenton-like Reagent in Ionic Liquids. *Green Chem.* **2009**, *11* (11), 1801–1807. https://doi.org/10.1039/b914130h.
87. Li, H.; Zhu, W.; Wang, Y.; Zhang, J.; Lu, J.; Yan, Y. Deep Oxidative Desulfurization of Fuels in Redox Ionic Liquids Based on Iron Chloride. *Green Chem.* **2009**, *11* (6), 810–881. https://doi.org/10.1039/b901127g.
88. Lü, H.; Li, P.; Liu, Y.; Hao, L.; Ren, W.; Zhu, W.; Deng, C.; Yang, F. Synthesis of a Hybrid Anderson-Type Polyoxometalate in Deep Eutectic Solvents (DESs) for Deep Desulphurization of Model Diesel in Ionic Liquids (ILs). *Chem. Eng. J.* **2017**, *313*, 1004–1009. https://doi.org/10.1016/j.cej.2016.10.140.
89. Jiang, W.; Jia, H.; Zheng, Z.; Zhu, L.; Dong, L.; Liu, W.; Zhu, W.; Li, H. Catalytic Oxidative Desulfurization of Fuels in Acidic Deep Eutectic Solvents with $[(C_6H_{13})_3P(C_{14}H_{29})]_3PMo_{12}O_{40}$ as a Catalyst. *Pet. Sci.* **2018**, *15* (4), 841–848. https://doi.org/10.1007/s12182-018-0263-9.

90. Hao, L.; Sun, L.; Su, T.; Hao, D.; Liao, W.; Deng, C.; Ren, W.; Zhang, Y.; Lü, H. Polyoxometalate-Based Ionic Liquid Catalyst with Unprecedented Activity and Selectivity for Oxidative Desulfurization of Diesel in [Omim]BF_4. *Chem. Eng. J.* **2019**, *358*, 419–426. https://doi.org/10.1016/j.cej.2018.10.006.
91. Azimzadeh, H.; Akbari, A.; Omidkhah, M. R. Catalytic Oxidative Desulfurization Performance of Immobilized NMP.$FeCl_3$ Ionic Liquid on γ-Al_2O_3 Support. *Chem. Eng. J.* **2017**, *320*, 189–200. https://doi.org/10.1016/j.cej.2017.03.027.
92. Xia, L.; Zhang, H.; Wei, Z.; Jiang, Y.; Zhang, L.; Zhao, J.; Zhang, J.; Dong, L.; Li, E.; Ruhlmann, L.; Zhang, Q. Catalytic Emulsion Based on Janus Nanosheets for Ultra-Deep Desulfurization. *Chem. - Eur. J.* **2017**, *23* (8), 1920–1929. https://doi.org/10.1002/chem.201604737.
93. Yang, H.; Jiang, B.; Sun, Y.; Zhang, L.; Sun, Z.; Wang, J.; Tantai, X. Polymeric Cation and Isopolyanion Ionic Self-Assembly: Novel Thin-Layer Mesoporous Catalyst for Oxidative Desulfurization. *Chem. Eng. J.* **2017**, *317*, 32–41. https://doi.org/10.1016/j.cej.2017.01.135.
94. Mirante, F.; Dias, L.; Silva, M.; Ribeiro, S. O.; Corvo, M. C.; de Castro, B.; Granadeiro, C. M.; Balula, S. S. Efficient Heterogeneous Polyoxometalate-Hybrid Catalysts for the Oxidative Desulfurization of Fuels. *Catal. Commun.* **2018**, *104*, 1–8. https://doi.org/10.1016/j.catcom.2017.10.006.
95. Zhang, H.; Zhang, Q.; Zhang, L.; Pei, T.; Dong, L.; Zhou, P.; Li, C.; Xia, L. Acidic Polymeric Ionic Liquids Based Reduced Graphene Oxide: An Efficient and Rewriteable Catalyst for Oxidative Desulfurization. *Chem. Eng. J.* **2018**, *334*, 285–295. https://doi.org/10.1016/j.cej.2017.10.042.
96. Shiraishi, Y.; Naito, T.; Hirai, T. Vanadosilicate Molecular Sieve as a Catalyst for Oxidative Desulfurization of Light Oil. *Ind. Eng. Chem. Res.* **2003**, *42* (24), 6034–6039. https://doi.org/10.1021/ie030328b.
97. Caero, L. C.; Hernández, E.; Pedraza, F.; Murrieta, F. Oxidative Desulfurization of Synthetic Diesel Using Supported Catalysts: Part I. Study of the Operation Conditions with a Vanadium Oxide Based Catalyst. *Catal. Today* **2005**, *107–108*, 564–569. https://doi.org/10.1016/j.cattod.2005.07.017.
98. Cedeño Caero, L.; F., J.; A., N.; Gutiérrez-Alejandre, A. Oxidative Desulfurization of Synthetic Diesel Using Supported Catalysts. Part II. Effect of Oxidant and Nitrogen-Compounds on Extraction-Oxidation Process. *Catal. Today* **2006**, *116* (4), 562–568. https://doi.org/10.1016/j.cattod.2006.06.031.
99. Cedeño-Caero, L.; Gomez-Bernal, H.; Fraustro-Cuevas, A.; Guerra-Gomez, H. D.; Cuevas-Garcia, R. Oxidative Desulfurization of Synthetic Diesel Using Supported Catalysts. Part III. Support Effect on Vanadium-Based Catalysts. *Catal. Today* **2008**, *133–135* (1–4), 244–254. https://doi.org/10.1016/j.cattod.2007.12.017.
100. Gregori, F.; Nobili, I.; Bigi, F.; Maggi, R.; Predieri, G.; Sartori, G. Selective Oxidation of Sulfides to Sulfoxides and Sulfones Using 30% Aqueous Hydrogen Peroxide and Silica-Vanadia Catalyst. *J. Mol. Catal. A Chem.* **2008**, *286* (1–2), 124–127. https://doi.org/10.1016/j.molcata.2008.02.004.
101. Gómez-Bernal, H.; Cedeño-Caero, L.; Gutiérrez-Alejandre, A. Liquid Phase Oxidation of Dibenzothiophene with Alumina-Supported Vanadium Oxide Catalysts: An Alternative to

Deep Desulfurization of Diesel. *Catal. Today* **2009**, *142* (3–4), 227–233. https://doi.org/10.1016/j.cattod.2008.08.018.
102. Liu, S.; Zhao, F.; Sun, H.; Liu, X.; Cui, B. Iron Promotion of V-HMS Mesoporous Catalysts for Ultra-Deep Oxidative Desulfurization. *Appl. Organomet. Chem.* **2018**, *32* (2), e4082. https://doi.org/10.1002/aoc.4082.
103. Teimouri, A.; Mahmoudsalehi, M.; Salavati, H. Catalytic Oxidative Desulfurization of Dibenzothiophene Utilizing Molybdenum and Vanadium Oxides Supported on MCM-41. *Int. J. Hydrogen Energy* **2018**, *43* (31), 14816–14833. https://doi.org/10.1016/j.ijhydene.2018.05.102.
104. Fazaeli, R.; Aliyan, H.; Ghanbari, R. Oxidative Desulfurization of Model Diesel Fuel by Vanadium Granular-Shape Mesoporous (V/GMS) Silica System as Nanocatalyst. *Russ. J. Appl. Chem.* **2018**, *91* (9), 1520–1527. https://doi.org/10.1134/S107042721809015X.
105. De Mello, M. I. S.; Sobrinho, E. V.; Teixeira Da Silva, V.; Pergher, S. B. C. Vanadium Incorporation in Montmorillonite Clays for Oxidative Diesel Desulfurization. *Ind. Eng. Chem. Res.* **2018**, *57* (46), 15663–15669. https://doi.org/10.1021/acs.iecr.8b03232.
106. Moslemi, A.; Najafi Chermahini, A.; Najafi Sarpiri, J.; Rezaei, S.; Barati, M. $VOHPO_4.5H_2O$/KIT-6 Composites: Preparation and Their Application in Extractive and Catalytic Oxidation Desulfurization of Benzothiophene and Dibenzothiphene. *J. Taiwan Inst. Chem. Eng.* **2019**, *97*, 237–246. https://doi.org/10.1016/j.jtice.2019.01.030.
107. de Mello, M. I. S.; Sobrinho, E. V.; da Silva, V. L. S. T.; Pergher, S. B. C. V or Mn Zeolite Catalysts for the Oxidative Desulfurization of Diesel Fractions Using Dibenzothiophene as a Probe Molecule: Preliminary Study. *Mol. Catal.* **2020**, *482*, 100495. https://doi.org/10.1016/j.mcat.2018.02.009.
108. Chen, L.; Ren, J. T.; Yuan, Z. Y. Atomic Heterojunction-Induced Electron Interaction in P-Doped g-C_3N_4 Nanosheets Supported V-Based Nanocomposites for Enhanced Oxidative Desulfurization. *Chem. Eng. J.* **2020**, *387*, 124164. https://doi.org/10.1016/j.cej.2020.124164.
109. Lv, G.; Zhang, L.; Zhai, Y.; Zhang, X.; Wang, F.; Jiang, T.; Shen, Y. Preparation of Vanadium-MFI Zeolite for Oxidative Desulfurization with the Aid of Ammonium Carbonate. *Appl. Catal. A Gen.* **2020**, *607*, 117862. https://doi.org/10.1016/j.apcata.2020.117862.
110. Luna, M. L.; Alvarez-Amparán, M. A.; Cedeño-Caero, L. Performance of WO_x–VO_x Based Catalysts for ODS of Dibenzothiophene Compounds. *J. Taiwan Inst. Chem. Eng.* **2019**, *95*, 175–184. https://doi.org/10.1016/j.jtice.2018.06.010.
111. Xun, S.; Hou, C.; Li, H.; He, M.; Ma, R.; Zhang, M.; Zhu, W.; Li, H. Synthesis of WO_3/Mesoporous ZrO_2 Catalyst as a High-Efficiency Catalyst for Catalytic Oxidation of Dibenzothiophene in Diesel. *J. Mater. Sci.* **2018**, *53* (23), 15927–15938. https://doi.org/10.1007/s10853-018-2720-7.
112. Feng, Z.; Zhu, Y.; Zhou, Q.; Wu, Y.; Wu, T. Magnetic WO_3/Fe_3O_4 as Catalyst for Deep Oxidative Desulfurization of Model Oil. *Mater. Sci. Eng. B Solid-State Mater. Adv. Technol.* **2019**, *240*, 85–91. https://doi.org/10.1016/j.mseb.2019.01.009.
113. Quyen, N. D. V.; Tuyen, T. N.; Khieu, D. Q.; Hai, H. V. M.; Tin, D. X.; Itatani, K. Oxidation of Dibenzothiophene Using the Heterogeneous Catalyst of Tungsten-Based Carbon

Nanotubes. *Green Process. Synth.* **2019**, *8* (1), 68–77. https://doi.org/10.1515/gps-2017-0189.

114. Ma, R.; Guo, J.; Wang, D.; He, M.; Xun, S.; Gu, J.; Zhu, W.; Li, H. Preparation of Highly Dispersed WO_3/Few Layer g-C_3N_4 and Its Enhancement of Catalytic Oxidative Desulfurization Activity. *Colloids Surf., A* **2019**, *572*, 250–258. https://doi.org/10.1016/j.colsurfa.2019.04.006.
115. Wang, C.; Li, A.; Xu, J.; Wen, J.; Zhang, H.; Zhang, L. Preparation of WO_3/CNT Catalysts in Presence of Ionic Liquid [C_{16}mim]Cl and Catalytic Efficiency in Oxidative Desulfurization. *J. Chem. Technol. Biotechnol.* **2019**, *94* (10), 3403–3412. https://doi.org/10.1002/jctb.6154.
116. Khan, N. A.; Bhadra, B. N.; Park, S. W.; Han, Y. S.; Jhung, S. H. Tungsten Nitride, Well-Dispersed on Porous Carbon: Remarkable Catalyst, Produced without Addition of Ammonia, for the Oxidative Desulfurization of Liquid Fuel. *Small* **2020**, *16* (12), 1901564. https://doi.org/10.1002/smll.201901564.
117. Wang, B.; Dai, B.; Kang, L.; Zhu, M. Synthesis of Three-Dimensional Ordered Mesoporous W-Doped KIT-6 for Oxidative Desulfurization Catalyst of Fuels. *Fuel* **2020**, *265*, 117029. https://doi.org/10.1016/j.fuel.2020.117029.
118. Thalgaspitiya, W. R. K.; Kankanam Kapuge, T.; He, J.; Rathnayake, D.; Suib, S. L. High Surface Area Mesoporous Tungsten Oxide for Fast, Green Oxidation of Organosulfur Compounds in Crude Oil. *Appl. Mater. Today* **2020**, *19*, 100616. https://doi.org/10.1016/j.apmt.2020.100616.
119. Wang, H.; Tang, M.; Shi, F.; Ding, R.; Wang, L.; Wu, J.; Li, X.; Liu, Z.; Lv, B. Amorphous Cr_2WO_6-Modified WO_3 Nanowires with a Large Specific Surface Area and Rich Lewis Acid Sites: A Highly Efficient Catalyst for Oxidative Desulfurization. *ACS Appl. Mater. Interfaces* **2020**, *12* (34), 38140–38152. https://doi.org/10.1021/acsami.0c10118.
120. Seifikar Gomi, L.; Afsharpour, M.; Lianos, P. Novel Porous SiO_2@SiC Core-Shell Nanospheres Functionalized with an Amino Hybrid of WO_3 as an Oxidative Desulfurization Catalyst. *J. Ind. Eng. Chem.* **2020**, *89*, 448–457. https://doi.org/10.1016/j.jiec.2020.06.019.
121. Haghighi, M.; Gooneh-Farahani, S. Oxidative Desulfurization of Dibenzothiophenes over Metallic and Bimetallic Supported ZSM-11 Catalysts: XLa/YMo-ZSM-11 as an Efficient Bimetallic Catalyst. *Inorg. Chem. Commun.* **2019**, *106*, 61–69. https://doi.org/10.1016/j.inoche.2019.05.007.
122. Mokhtari, B.; Akbari, A.; Omidkhah, M. Superior Deep Desulfurization of Real Diesel over MoO_3/Silica Gel as an Efficient Catalyst for Oxidation of Refractory Compounds. *Energy Fuels* **2019**, 33 (8), 7276–7286. https://doi.org/10.1021/acs.energyfuels.9b01646.
123. Tu, Y.; Li, T.; Yu, G.; Wei, L.; Ta, L.; Zhou, Z.; Ren, Z. Study on Modification and Desulfurization Performance of a Molybdenum-Based Catalyst. *Energy Fuels* **2019**, 33 (9), 8503–8510. https://doi.org/10.1021/acs.energyfuels.9b02132.
124. Ghorbani, N.; Moradi, G. Oxidative Desulfurization of Model and Real Oil Samples Using Mo Supported on Hierarchical Alumina–Silica: Process Optimization by Box–Behnken Experimental Design. *Chinese J. Chem. Eng.* **2019**, 27 (11), 2759–2770. https://doi.org/10.1016/j.cjche.2019.01.037.

125. Zhang, P.; Tong, Y.; Zhu, M.; Dai, B. Oxidative Desulfurization of Dibenzothiophene Catalyzed by Molybdenum Dioxide Immobilized on Zirconia-Modified Silica. *New J. Chem.* **2020,** *44* (9), 3786–3793. https://doi.org/10.1039/c9nj06182g.
126. Kamel, L.; Anbia, M. Composite Beads of Molybdenum Oxide Supported on Textured Silicon as an Oxidative Desulfurization Nanocatalyst. *Silicon* **2020,** *12* (6), 1325–1336. https://doi.org/10.1007/s12633-019-00227-0.
127. Li, S. W.; Wang, W.; Zhao, J. S. Catalytic Oxidation of DBT for Ultra-Deep Desulfurization under MoO_3 Modified Magnetic Catalyst: The Comparison Influence on Various Morphologies of MoO_3. *Appl. Catal. A Gen.* **2020,** *602*, 117671. https://doi.org/10.1016/j.apcata.2020.117671.
128. Said, S.; Abdelrahman, A. A. Atomic Layer Deposition of MoO_3 on Mesoporous γ-Al_2O_3 Prepared by Sol–Gel Method as Efficient Catalyst for Oxidative Desulfurization of Refractory Dibenzothiophene Compound. *J. Sol-Gel Sci. Technol.* **2020,** *95* (2), 308–320. https://doi.org/10.1007/s10971-020-05332-w.
129. Yang, G.; Yang, H.; Zhang, X.; lqbal, K.; Feng, F.; Ma, J.; Qin, J.; Yuan, F.; Cai, Y.; Ma, J. Surfactant-Free Self-Assembly to the Synthesis of MoO_3 Nanoparticles on Mesoporous SiO_2 to Form MoO_3/SiO_2 Nanosphere Networks with Excellent Oxidative Desulfurization Catalytic Performance. *J. Hazard. Mater.* **2020,** *397*, 122654. https://doi.org/10.1016/j.jhazmat.2020.122654.
130. Wang, C.; Miao, Q.; Huang, X.; Li, J.; Duan, Y.; Yan, L.; Jiang, Y.; Lu, S. Fabrication of Various Morphological Forms of a g-C_3N_4-Supported MoO_3 Catalyst for the Oxidative Desulfurization of Dibenzothiophene. *New J. Chem.* **2020,** *44* (43), 18745–18755. https://doi.org/10.1039/d0nj04105j.
131. Bibak, F.; Moradi, G. Oxidative Desulfurization of Model Oil and Oil Cuts with MoO_3/SBA-15: Experimental Design and Optimization by Box–Behnken Method. *React. Kinet. Mech. Catal.* **2020,** *131* (2), 935–951. https://doi.org/10.1007/s11144-020-01852-2.
132. Capel-Sanchez, M. C.; Perez-Presas, P.; Campos-Martin, J. M.; Fierro, J. L. G. Highly Efficient Deep Desulfurization of Fuels by Chemical Oxidation. *Catal. Today* **2010,** *157* (1–4), 390–396. https://doi.org/10.1016/j.cattod.2010.01.047.
133. Banisharif, F.; Dehghani, M. R.; Capel-Sanchez, M. C.; Campos-Martin, J. M. Highly Catalytic Oxidative Desulfurization and Denitrogenation of Diesel Using Anchored-Silica-Gel Vanadium-Substituted Dawson-Type Polyoxometalate. *Catal. Today* **2019,** *333*, 219–225. https://doi.org/10.1016/j.cattod.2018.07.009.
134. Ribeiro, S. O.; Granadeiro, C. M.; Corvo, M. C.; Pires, J.; Campos-Martin, J. M.; de Castro, B.; Balula, S. S. Mesoporous Silica vs. Organosilica Composites to Desulfurize Diesel. *Front. Chem.* **2019,** *7*, 756. https://doi.org/10.3389/fchem.2019.00756.
135. Ribeiro, S. O.; Granadeiro, C. M.; Almeida, P. L.; Pires, J.; Valença, R.; Campos-Martin, J. M.; Ribeiro, J. C.; De Castro, B.; Balula, S. S. Effective Zinc-Substituted Keggin Composite to Catalyze the Removal of Sulfur from Real Diesels under a Solvent-Free System. *Ind. Eng. Chem. Res.* **2019,** *58* (40), 18540–18549. https://doi.org/10.1021/acs.iecr.9b02340.
136. Wang, C.; Li, A. R.; Ma, Y. L. Phosphomolybdic Acid Niched in the Metal-Organic Framework UiO-66 with Defects: An Efficient and Stable Catalyst for Oxidative Desulfurization. *Fuel Process. Technol.* **2021,** *212*, 106629. https://doi.org/10.1016/j.fuproc.2020.106629.

137. Lu, Y.; Yue, C.; Liu, B.; Zhang, M.; Li, Y.; Yang, W.; Lin, Y.; Pan, Y.; Sun, D.; Liu, Y. The Encapsulation of POM Clusters into MIL-101(Cr) at Molecular Level: LaW10O36@MIL-101(Cr), an Efficient Catalyst for Oxidative Desulfurization. *Microporous Mesoporous Mater.* **2021**, *311*, 110694. https://doi.org/10.1016/j.micromeso.2020.110694.

138. Estephane, G.; Lancelot, C.; Blanchard, P.; Dufaud, V.; Chambrey, S.; Nuns, N.; Toufaily, J.; Hamiye, T.; Lamonier, C. W-SBA Based Materials as Efficient Catalysts for the ODS of Model and Real Feeds: Improvement of Their Lifetime through Active Phase Encapsulation. *Appl. Catal. A Gen.* **2019**, *571*, 42–50. https://doi.org/10.1016/j.apcata.2018.12.007.

139. Wang, P.; Wang, P.; Jiang, L.; Zou, X.; Tan, H.; Zhang, P.; Li, J.; Liu, B.; Zhu, G. Confining Polyoxometalate Clusters into Porous Aromatic Framework Materials for Catalytic Desulfurization of Dibenzothiophene. *ACS Appl. Mater. Interfaces* **2020**, *12* (23), 25910–25919. https://doi.org/10.1021/acsami.0c05392.

140. Capel-Sanchez, M. C.; Campos-Martin, J. M.; Fierro, J. L. G. Removal of Refractory Organosulfur Compounds via Oxidation with Hydrogen Peroxide on Amorphous Ti/SiO_2 Catalysts. *Energy Environ. Sci.* **2010**, *3* (3), 328–333. https://doi.org/10.1039/b923795j.

141. Cruz, P.; Granados, E. A.; Fajardo, M.; del Hierro, I.; Pérez, Y. Heterogeneous Oxidative Desulfurization Catalysed by Titanium Grafted Mesoporous Silica Nanoparticles Containing Tethered Hydrophobic Ionic Liquid: A Dual Activation Mechanism. *Appl. Catal. A Gen.* **2019**, *587*, 117241. https://doi.org/10.1016/j.apcata.2019.117241.

142. Ghahramaninezhad, M.; Ahmadpour, A. A New Simple Protocol for the Synthesis of Nanohybrid Catalyst for Oxidative Desulfurization of Dibenzothiophene. *Environ. Sci. Pollut. Res.* **2020**, *27* (4), 4104–4114. https://doi.org/10.1007/s11356-019-07048-z.

143. Rivoira, L. P.; Martínez, M. L.; Falcón, H.; Beltramone, A. R.; Campos-Martin, J. M.; Fierro, J. L. G.; Tartaj, P. Probing the Catalytic Activity of Sulfate-Derived Pristine and Post-Treated Porous TiO_2(101) Anatase Mesocrystals by the Oxidative Desulfurization of Dibenzothiophenes. *ACS Omega* **2017**, *2* (5), 2351–2359. https://doi.org/10.1021/acsomega.7b00307.

144. Zuo, M.; Huang, X.; Li, J.; Chang, Q.; Duan, Y.; Yan, L.; Xiao, Z.; Mei, S.; Lu, S.; Yao, Y. Oxidative Desulfurization in Diesel: Via a Titanium Dioxide Triggered Thermocatalytic Mechanism. *Catal. Sci. Technol.* **2019**, *9* (11), 2923–2930. https://doi.org/10.1039/c9cy00298g.

145. Yao, Y.; Zuo, M.; Shao, P.; Huang, X.; Li, J.; Duan, Y.; Zhou, H.; Yan, L.; Lu, S. Oxidative Desulfurization of 4,6-Dimethyldibenzothiophene over Short Titanate Nanotubes: A Non-Classical Shape Selective Catalysis. *J. Porous Mater.* **2020**, *27* (2), 331–338. https://doi.org/10.1007/s10934-019-00809-8.

146. He, M.; Zhu, W.; Xun, S.; Ti, Q.; Jiao, Z.; Wu, L.; Chen, L.; Zhu, L.; Li, H. Dispersing TiO_2 Nanoparticles on Graphite Carbon for an Enhanced Catalytic Oxidative Desulfurization Performance. *Ind. Eng. Chem. Res.* **2020**, *59* (41), 18471–18479. https://doi.org/10.1021/acs.iecr.0c03202.

147. Wei, X.; Wang, Y. J.; Ren, T. Q.; Wang, H. Y.; Wei, M. The Synthesis of Nano-Sized TS-1 Zeolites under Rotational Crystallisation Conditions Can Inhibit Anatase Formation. *RSC Adv.* **2019**, *10* (2), 1015–1020. https://doi.org/10.1039/c9ra09273k.

148. Shi, D.; Xu, L.; Chen, P.; Ma, T.; Lin, C.; Wang, X.; Xu, D.; Sun, J. Hydroxyl Free Radical Route to the Stable Siliceous Ti-UTL with Extra-Large Pores for Oxidative Desulfurization. *Chem. Commun.* **2019,** *55* (10), 1390–1393. https://doi.org/10.1039/c8cc09225g.
149. Du, Q.; Guo, Y.; Wu, P.; Liu, H.; Chen, Y. Facile Synthesis of Hierarchical TS-1 Zeolite without Using Mesopore Templates and Its Application in Deep Oxidative Desulfurization. *Microporous Mesoporous Mater.* **2019,** *275,* 61–68. https://doi.org/10.1016/j.micromeso.2018.08.018.
150. Pei, X.; Liu, X.; Liu, X.; Shan, J.; Fu, H.; Xie, Y.; Yan, X.; Meng, X.; Zheng, Y.; Li, G.; Wang, Q.; Li, H. Synthesis of Hierarchical Titanium Silicalite-1 Using a Carbon-Silica-Titania Composite from Xerogel Mild Carbonization. *Catalysts* **2019,** *9* (8), 672. https://doi.org/10.3390/catal9080672.
151. Wang, H.; Shi, C.; Chen, S.; Chen, R.; Sun, P.; Chen, T. Hierarchically Mesoporous Titanosilicate Single-Crystalline Nanospheres for Room Temperature Oxidative-Adsorptive Desulfurization. *ACS Appl. Nano Mater.* **2019,** 2 (10), 6602–6610. https://doi.org/10.1021/acsanm.9b01496.
152. Bian, H.; Zhang, H.; Li, D.; Duan, Z.; Zhang, H.; Zhang, S.; Xu, B. Insight into the Oxidative Desulfurization Mechanism of Aromatic Sulfur Compounds over Ti-MWW Zeolite: A Computational Study. *Microporous Mesoporous Mater.* **2020,** *294,* 109837. https://doi.org/10.1016/j.micromeso.2019.109837.
153. Hulea, V.; Fajula, F.; Bousquet, J. Mild Oxidation with H_2O_2 over Ti-Containing Molecular Sieves - A Very Efficient Method for Removing Aromatic Sulfur Compounds from Fuels. *J. Catal.* **2001,** *198* (2), 179–186. https://doi.org/10.1006/jcat.2000.3149.
154. Kong, L.; Li, G.; Wang, X. Mild Oxidation of Thiophene over TS-1/H_2O_2. *Catal. Today* **2004,** *93–95,* 341–345. https://doi.org/10.1016/j.cattod.2004.06.016.
155. Kong, L.; Li, G.; Wang, X. Kinetics and Mechanism of Liquid-Phase Oxidation of Thiophene over TS-1 Using H_2O_2 under Mild Conditions. *Catal. Letters* **2004,** *92* (3–4), 163–167. https://doi.org/10.1023/b:catl.0000014340.08449.d0.
156. Cheng, S.; Liu, Y.; Gao, J.; Wang, L.; Liu, X.; Gao, G.; Wu, P.; He, M. Catalytic Oxidation of Benzothiophene and Dibenzothiophene in Model Light Oil Ti-MWW. *Chinese J. Catal.* **2006,** *27* (7), 547–549. https://doi.org/10.1016/S1872-2067(06)60031-4.
157. Wang, Y.; Li, G.; Wang, X.; Jin, C. Oxidative Desulphurization of 4,6-Dimethyldibenzothiophene with Hydrogen Peroxide over Ti-HMS. *Energy Fuels* **2007,** *21* (3), 1415–1419. https://doi.org/10.1021/ef070006a.
158. Jin, C.; Li, G.; Wang, X.; Wang, Y.; Zhao, L.; Sun, D. A Titanium Containing Micro/Mesoporous Composite and Its Catalytic Performance in Oxidative Desulfurization. *Microporous Mesoporous Mater.* **2008,** *111* (1–3), 236–242. https://doi.org/10.1016/j.micromeso.2007.07.037.
159. Napanang, T.; Sooknoi, T. Oxidative Extraction of Thiophene from N-Dodecane over TS-1 in Continuous Process: A Model for Non-Severe Sulfur Removal from Liquid Fuels. *Catal. Commun.* **2009,** *11* (1), 1–6. https://doi.org/10.1016/j.catcom.2009.07.022.
160. Ribeiro, S.; Barbosa, A. D. S.; Gomes, A. C.; Pillinger, M.; Gonçalves, I. S.; Cunha-Silva, L.; Balula, S. S. Catalytic Oxidative Desulfurization Systems Based on Keggin Phosphotungstate and Metal-Organic Framework MIL-101. *Fuel Process. Technol.* **2013,** *116,* 350–357. https://doi.org/10.1016/j.fuproc.2013.07.011.

161. Afzali, N.; Kardanpour, R.; Zadehahmadi, F.; Tangestaninejad, S.; Moghadam, M.; Mirkhani, V.; Mechler, A.; Mohammadpoor-Baltork, I.; Bahadori, M. Molybdenum(VI)-Functionalized UiO-66 Provides an Efficient Heterogeneous Nanocatalyst in Oxidation Reactions. *Appl. Organomet. Chem.* **2019**, *33* (11) https://doi.org/10.1002/aoc.5225.
162. Piscopo, C. G.; Voellinger, L.; Schwarzer, M.; Polyzoidis, A.; Bošković, D.; Loebbecke, S. Continuous Flow Desulfurization of a Model Fuel Catalysed by Titanium Functionalized UiO-66. *ChemistrySelect* **2019**, *4* (9), 2806–2809. https://doi.org/10.1002/slct.201900342.
163. Smolders, S.; Willhammar, T.; Krajnc, A.; Sentosun, K.; Wharmby, M. T.; Lomachenko, K. A.; Bals, S.; Mali, G.; Roeffaers, M. B. J.; De Vos, D. E.; Bueken, B. A Titanium(IV)-Based Metal–Organic Framework Featuring Defect-Rich Ti-O Sheets as an Oxidative Desulfurization Catalyst. *Angew. Chem., Int. Ed.* **2019**, *58* (27), 9160–9165. https://doi.org/10.1002/anie.201904347.
164. Li, X.; Zhang, L.; Sun, Y. Titanium-Modified Mil-101(Cr) Derived Titanium-Chromium-Oxide as Highly Efficient Oxidative Desulfurization Catalyst. *Catalysts* **2020**, *10* (9), 1–10. https://doi.org/10.3390/catal10091091.
165. Liao, X.; Zhang, M.; Wang, X.; Zhou, Y.; Wang, F.; Chen, L.; Yao, Y.; Lu, S. Vacuum Degassed Treatment for Fabricating Quasi-MIL-125(Ti) with Enhanced Catalytic Oxidative Desulfurization Activity. *Microporous Mesoporous Mater.* **2020**, *308*, 110529. https://doi.org/10.1016/j.micromeso.2020.110529.
166. Ye, G.; Gu, Y.; Zhou, W.; Xu, W.; Sun, Y. Synthesis of Defect-Rich Titanium Terephthalate with the Assistance of Acetic Acid for Room-Temperature Oxidative Desulfurization of Fuel Oil. *ACS Catal.* **2020**, *10* (3), 2384–2394. https://doi.org/10.1021/acscatal.9b04937.
167. Zhang, Y.; Li, G.; Kong, L.; Lu, H. Deep Oxidative Desulfurization Catalyzed by Ti-Based Metal-Organic Frameworks. *Fuel* **2018**, *219*, 103–110. https://doi.org/10.1016/j.fuel.2018.01.050.
168. Ye, G.; Sun, Y.; Zhang, D.; Zhou, W.; Lancelot, C.; Rives, A.; Lamonier, C.; Xu, W. Hierarchical Porous Titanium Terephthalate Based Material with Highly Active Sites for Deep Oxidative Desulfurization. *Microporous Mesoporous Mater.* **2018**, *270*, 241–247. https://doi.org/10.1016/j.micromeso.2018.05.025.
169. Abazari, R.; Sanati, S.; Morsali, A.; Slawin, A.; Carpenter-Warren, C. L. Dual-Purpose 3D Pillared Metal-Organic Framework with Excellent Properties for Catalysis of Oxidative Desulfurization and Energy Storage in Asymmetric Supercapacitor. *ACS Appl. Mater. Interfaces* **2019**, *11* (16), 14759–14773. https://doi.org/10.1021/acsami.9b00415.
170. Zheng, H. Q.; Zeng, Y. N.; Chen, J.; Lin, R. G.; Zhuang, W. E.; Cao, R.; Lin, Z. J. Zr-Based Metal-Organic Frameworks with Intrinsic Peroxidase-Like Activity for Ultradeep Oxidative Desulfurization: Mechanism of H_2O_2 Decomposition. *Inorg. Chem.* **2019**, *58* (10), 6983–6992. https://doi.org/10.1021/acs.inorgchem.9b00604.
171. Li, X.; Gu, Y.; Chu, H.; Ye, G.; Zhou, W.; Xu, W.; Sun, Y. MFM-300(V) as an Active Heterogeneous Catalyst for Deep Desulfurization of Fuel Oil by Aerobic Oxidation. *Appl. Catal. A Gen.* **2019**, *584*, 117152. https://doi.org/10.1016/j.apcata.2019.117152.
172. Viana, A. M.; Ribeiro, S. O.; de Castro, B.; Balula, S. S.; Cunha-Silva, L. Influence of UiO-66(Zr) Preparation Strategies in Its Catalytic Efficiency for Desulfurization Process. *Materials (Basel)* **2019**, *12* (18), 3009. https://doi.org/10.3390/ma12183009.

173. Bhadra, B. N.; Jhung, S. H. Oxidative Desulfurization and Denitrogenation of Fuels Using Metal-Organic Framework-Based/-Derived Catalysts. *Appl. Catal. B Environ.* **2019**, *259*, 118021. https://doi.org/10.1016/j.apcatb.2019.118021.
174. Gu, Y.; Ye, G.; Xu, W.; Zhou, W.; Sun, Y. Creation of Active Sites in MOF-808(Zr) by a Facile Route for Oxidative Desulfurization of Model Diesel Oil. *ChemistrySelect* **2020**, *5* (1), 244–251. https://doi.org/10.1002/slct.201903376.
175. Hao, L.; Hurlock, M. J.; Ding, G.; Zhang, Q. Metal–Organic Frameworks Towards Desulfurization of Fuels. *Top. Curr. Chem.* **2020**, *378* (1), 17. https://doi.org/10.1007/s41061-020-0280-1.
176. Liao, X.; Wei, W.; Zhou, Y.; Zhang, M.; Cai, Y.; Liu, H.; Yao, Y.; Lu, S.; Hao, Q. A Ti-Based Bi-MOF for the Tandem Reaction of H_2O_2 Generation and Catalytic Oxidative Desulfurization. *Catal. Sci. Technol.* **2020**, *10* (4), 1015–1022. https://doi.org/10.1039/c9cy02318f.
177. Liu, S.; Li, X.; Zhang, H. Synergistic Effects of MOF-76 on Layered Double Hydroxides with Superior Activity for Extractive Catalytic Oxidative Desulfurization. *New J. Chem.* **2020**, *44* (16), 6269–6276. https://doi.org/10.1039/d0nj00612b.
178. Piscopo, C. G.; Granadeiro, C. M.; Balula, S. S.; Bošković, D. Metal-Organic Framework-Based Catalysts for Oxidative Desulfurization. *ChemCatChem* **2020**, *12* (19), 4721–4731. https://doi.org/10.1002/cctc.202000688.
179. Hao, L.; Stoian, S. A.; Weddle, L. R.; Zhang, Q. Zr-Based MOFs for Oxidative Desulfurization: What Matters? *Green Chem.* **2020**, *22* (19), 6351–6356. https://doi.org/10.1039/d0gc01999b.
180. Ge, J.; Zhou, Y.; Yang, Y.; Gao, Q.; Wu, X.; Sun, Q. The Deep Oxidative Desulfurization of Fuels Using $Na_2WO_4.2H_2O$ in Acidic Ionic Liquids. *Pet. Sci. Technol.* **2013**, *31* (21), 2280–2286. https://doi.org/10.1080/10916466.2011.567203.
181. Banisharif, F.; Dehghani, M. R.; Capel-Sánchez, M. C.; Campos-Martin, J. M. Desulfurization of Fuel by Extraction and Catalytic Oxidation Using a Vanadium Substituted Dawson-Type Emulsion Catalyst. *Ind. Eng. Chem. Res.* **2017**, *56* (14), 3839–3852. https://doi.org/10.1021/acs.iecr.7b00089.
182. Lü, H.; Gao, J.; Jiang, Z.; Jing, F.; Yang, Y.; Wang, G.; Li, C. Ultra-Deep Desulfurization of Diesel by Selective Oxidation with $[C_{18}H_{37}N(CH_3)_3]_4[H_2NaPW_{10}O_{36}]$ Catalyst Assembled in Emulsion Droplets. *J. Catal.* **2006**, *239* (2), 369–375. https://doi.org/10.1016/j.jcat.2006.01.025.
183. Gao, J.; Wang, S.; Jiang, Z.; Lu, H.; Yang, Y.; Jing, F.; Li, C. Deep Desulfurization from Fuel Oil via Selective Oxidation Using an Amphiphilic Peroxotungsten Catalyst Assembled in Emulsion Droplets. *J. Mol. Catal. A Chem.* **2006**, *258* (1–2), 261–266. https://doi.org/10.1016/j.molcata.2006.05.058.
184. Tian, Y.; Yao, Y.; Zhi, Y.; Yan, L.; Lu, S. Combined Extraction-Oxidation System for Oxidative Desulfurization (ODS) of a Model Fuel. *Energy Fuels* **2015**, *29* (2), 618–625. https://doi.org/10.1021/ef502396b.
185. Campos-Martin, J. M.; Blanco-Brieva, G.; Fierro, J. L. G. Hydrogen Peroxide Synthesis: An Outlook beyond the Anthraquinone Process. *Angew. Chem., Int. Ed.* **2006**, *45* (42), 6962–6984. https://doi.org/10.1002/anie.200503779.

186. Chica, A.; Corma, A.; Dómine, M. E. Catalytic Oxidative Desulfurization (ODS) of Diesel Fuel on a Continuous Fixed-Bed Reactor. *J. Catal.* **2006**, *242* (2), 299–308. https://doi.org/10.1016/j.jcat.2006.06.013.
187. Chica, A.; Gatti, G.; Moden, B.; Marchese, L.; Iglesia, E. Selective Catalytic Oxidation of Organosulfur Compounds with Tert-Butyl Hydroperoxide. *Chem. - Eur. J.* **2006**, *12* (7), 1960–1967. https://doi.org/10.1002/chem.200500858.
188. Ishihara, A.; Wang, D.; Dumeignil, F.; Amano, H.; Qian, E. W.; Kabe, T. Oxidative Desulfurization and Denitrogenation of a Light Gas Oil Using an Oxidation/Adsorption Continuous Flow Process. *Appl. Catal. A Gen.* **2005**, *279* (1–2), 279–287. https://doi.org/10.1016/j.apcata.2004.10.037.
189. Zhou, X.; Gai, H.; Wang, J.; Zhang, S.; Yang, J.; Zhang, S. Oxidation of Benzothiophenes Using Tert-Amyl Hydroperoxide. *Chinese J. Chem. Eng.* **2009**, *17* (2), 189–194. https://doi.org/10.1016/S1004-9541(08)60192-5.
190. Anisimov, A. V.; Tarakanova, A. V. Oxidative Desulfurization of Hydrocarbon Raw Materials. *Russ. J. Gen. Chem.* **2009**, *79* (6), 1264–1273. https://doi.org/10.1134/S1070363209060437.
191. Wang, J.; Zhao, D.; Li, K. Oxidative Desulfurization of Dibenzothiophene Using Ozone and Hydrogen Peroxide in Ionic Liquid. *Energy Fuels* **2010**, *24* (4), 2527–2529. https://doi.org/10.1021/ef901324p.
192. Ma, C.; Dai, B.; Liu, P.; Zhou, N.; Shi, A.; Ban, L.; Chen, H. Deep Oxidative Desulfurization of Model Fuel Using Ozone Generated by Dielectric Barrier Discharge Plasma Combined with Ionic Liquid Extraction. *J. Ind. Eng. Chem.* **2014**, *20* (5), 2769–2774. https://doi.org/10.1016/j.jiec.2013.11.005.
193. Otsuki, S.; Nonaka, T.; Qian, W.; Ishihara, A.; Kabe, T. Oxidative Desulfurization of Middle Distillate Using Ozone. *Sekiyu Gakkaishi (J. Japan Pet. Inst.)* **1999**, *42* (5), 315–320. https://doi.org/10.1627/jpi1958.42.315.
194. Ma, C.; Chen, D.; Liu, F.; Sun, X.; Xiao, F.; Dai, B. Oxidative Desulfurization of a Model Fuel Using Ozone Oxidation Generated by Dielectric Barrier Discharge Plasma Combined with $Co_3O_4/\gamma\text{-}Al_2O_3$ Catalysis. *RSC Adv.* **2015**, *5* (117), 96945–96952. https://doi.org/10.1039/c5ra16277g.
195. Akopyan, A. V.; Grigoriev, D. A.; Polikarpova, P. L.; Eseva, E. A.; Litvinova, V. V.; Anisimov, A. V. Ozone-Assisted Oxidative Desulfurization of Light Oil Fractions. *Pet. Chem.* **2017**, *57* (10), 904–907. https://doi.org/10.1134/S0965544117100024.
196. Murata, S.; Murata, K.; Kidena, K.; Nomura, M. A Novel Oxidative Desulfurization System for Diesel Fuels with Molecular Oxygen in the Presence of Cobalt Catalysts and Aldehydes. *Energy Fuels* **2004**, *18* (1), 116–121. https://doi.org/10.1021/ef034001z.
197. Lü, H.; Gao, J.; Jiang, Z.; Yang, Y.; Song, B.; Li, C. Oxidative Desulfurization of Dibenzothiophene with Molecular Oxygen Using Emulsion Catalysis. *Chem. Commun.* **2007** (2), 150–152. https://doi.org/10.1039/b610504a.
198. Jiang, C.; Wang, J.; Wang, S.; Guan, H. yu; Wang, X.; Huo, M. Oxidative Desulfurization of Dibenzothiophene with Dioxygen and Reverse Micellar Peroxotitanium under Mild Conditions. *Appl. Catal. B Environ.* **2011**, *106* (3–4), 343–349. https://doi.org/10.1016/j.apcatb.2011.05.038.

199. Zhu, W.; Wang, C.; Li, H.; Wu, P.; Xun, S.; Jiang, W.; Chen, Z.; Zhao, Z.; Li, H. One-Pot Extraction Combined with Metal-Free Photochemical Aerobic Oxidative Desulfurization in Deep Eutectic Solvent. *Green Chem.* **2015**, *17* (4), 2464–2472. https://doi.org/10.1039/c4gc02425g.

200. Bertleff, B.; Goebel, R.; Claußnitzer, J.; Korth, W.; Skiborowski, M.; Wasserscheid, P.; Jess, A.; Albert, J. Investigations on Catalyst Stability and Product Isolation in the Extractive Oxidative Desulfurization of Fuels Using Polyoxometalates and Molecular Oxygen. *ChemCatChem* **2018**, *10* (20), 4602–4609. https://doi.org/10.1002/cctc.201801081.

201. Bertleff, B.; Haider, M. S.; Claußnitzer, J.; Korth, W.; Wasserscheid, P.; Jess, A.; Albert, J. Extractive Catalytic Oxidative Denitrogenation of Fuels and Their Promoting Effect for Desulfurization Catalyzed by Vanadium Substituted Heteropolyacids and Molecular Oxygen. *Energy Fuels* **2020**, *34* (7), 8099–8109. https://doi.org/10.1021/acs.energyfuels.0c00864.

202. Lü, H.; Zhang, Y.; Jiang, Z.; Li, C. Aerobic Oxidative Desulfurization of Benzothiophene, Dibenzothiophene and 4,6-Dimethyldibenzothiophene Using an Anderson-Type Catalyst $[(C_{18}H_{37})_2N(CH_3)_2]_5[IMo_6O_{24}]$. *Green Chem.* **2010**, *12* (11), 1954–1958. https://doi.org/10.1039/c0gc00271b.

203. Ma, X.; Zhou, A.; Song, C. A Novel Method for Oxidative Desulfurization of Liquid Hydrocarbon Fuels Based on Catalytic Oxidation Using Molecular Oxygen Coupled with Selective Adsorption. *Catal. Today* **2007**, *123* (1–4), 276–284. https://doi.org/10.1016/j.cattod.2007.02.036.

204. Zhang, W.; Zhang, H.; Xiao, J.; Zhao, Z.; Yu, M.; Li, Z. Carbon Nanotube Catalysts for Oxidative Desulfurization of a Model Diesel Fuel Using Molecular Oxygen. *Green Chem.* **2014**, *16* (1), 211–220. https://doi.org/10.1039/c3gc41106k.

205. Tang, N.; Jiang, Z.; Li, C. Oxidation of Refractory Sulfur-Containing Compounds with Molecular Oxygen Catalyzed by Vanadoperiodate. *Green Chem.* **2015**, *17* (2), 817–820. https://doi.org/10.1039/c4gc01790k.

206. Hassan, S. I.; Sif El-Din, O. I.; Tawfik, S. M.; Abd El-Aty, D. M. Solvent Extraction of Oxidized Diesel Fuel: Phase Equilibrium. *Fuel Process. Technol.* **2013**, *106*, 127–132. https://doi.org/10.1016/j.fuproc.2012.07.012.

207. Wu, L.; Miao, G.; Dai, X.; Dong, L.; Li, Z.; Xiao, J. Ultra-Deep Desulfurization of Real Diesel Using Two-Layer Silica Gels under Mild Conditions. *Energy Fuels* **2019**, *33* (8), 7287–7296. https://doi.org/10.1021/acs.energyfuels.9b01896.

208. Cho, K. S.; Lee, Y. K. Effects of Nitrogen Compounds, Aromatics, and Aprotic Solvents on the Oxidative Desulfurization (ODS) of Light Cycle Oil over Ti-SBA-15 Catalyst. *Appl. Catal. B Environ.* **2014**, *147*, 35–42. https://doi.org/10.1016/j.apcatb.2013.08.017.

209. Safa, M. A.; Bouresli, R.; Al-Majren, R.; Al-Shamary, T.; Ma, X. Oxidative Desulfurization Kinetics of Refractory Sulfur Compounds in Hydrotreated Middle Distillates. *Fuel* **2019**, *239*, 24–31. https://doi.org/10.1016/j.fuel.2018.10.141.

210. Speight J. G., El-Gendy N. S. Biocatalytic Desulfurization. In *Introduction to Petroleum Biotechnology*; Speight, J. G., El-Gendy, N. S. , Eds.; Gulf Professional Publishing: Cambridge, MA, USA, 2018; Chapter 6, pp 165–227. https://doi.org/10.1016/B978-0-12-805151-1.00006-0.

211. Hu, S. Z.; Li, F. Y.; Fan, Z. P. Oxidative Desulfurization of Dibenzothiophene in a Micro-Structured Reactor. *Asian J. Chem.* **2012**, *24* (3), 1199–1202.

212. Huang, D.; Lu, Y. C.; Wang, Y. J.; Yang, L.; Luo, G. S. Intensification of Catalytic Oxidation with a T-Junction Microchannel Reactor for Deep Desulfurization. *Ind. Eng. Chem. Res.* **2008**, *47* (11), 3870–3875. https://doi.org/10.1021/ie701781r.

213. Jähnisch, K.; Hessel, V.; Löwe, H.; Baerns, M. Chemistry in Microstructured Reactors. *Angew. Chem. Int. Ed.* **2004**, *43* (4), 406–446. https://doi.org/10.1002/anie.200300577.

214. Kiwi-Minsker, L.; Renken, A. Microstructured Reactors for Catalytic Reactions. *Catal. Today* **2005**, *110* (1–2), 2–14. https://doi.org/10.1016/j.cattod.2005.09.011.

215. Cravotto, G.; Cintas, P. Power Ultrasound in Organic Synthesis: Moving Cavitational Chemistry from Academia to Innovative and Large-Scale Applications. *Chem. Soc. Rev.* **2006**, 35 (2), 180–196. https://doi.org/10.1039/b503848k.

216. Kim, H.; Jeong, K. E.; Jeong, S. Y.; Park, Y. K.; Kim, D. H.; Jeon, J. K. Utilization of a By-Product Produced from Oxidative Desulfurization Process over Cs-Mesoporous Silica Catalysts. *J. Nanosci. Nanotechnol.* **2011**, *11* (2), 1706–1709. https://doi.org/10.1166/jnn.2011.3333.

217. Ko, C. H.; Han, S. S.; Lim, S. M.; Park, J.; Kim, J.; Ryu, J. W.; Chu, Y. H.; Kim, Y. W.; Park, S. Y.; Yi, K. B.; Park, J. H.; Kim, J. N. R Energy-Efficient Sulfone Separation Process for the Production of Ultralow Sulfur Diesel by Two-Step Adsorption. *Energy Fuels* **2012**, *26* (4), 2168–2174. https://doi.org/10.1021/ef201964v.

218. Sundararaman, R.; Song, C. Catalytic Decomposition of Benzothiophenic and Dibenzothiophenic Sulfones over MgO-Based Catalysts. *Appl. Catal. B Environ.* **2014**, *148–149*, 80–90. https://doi.org/10.1016/j.apcatb.2013.10.033.

219. You, N.; Kim, M. J.; Jeong, K. E.; Jeong, S. Y.; Park, Y. K.; Jeon, J. K. Catalytic Removal of Sulfur Dioxide from Dibenzothiophene Sulfone over Mg-Al Mixed Oxides Supported on Mesoporous Silica. *J. Nanosci. Nanotechnol.* **2010**, *10* (5), 3663–3666. https://doi.org/10.1166/jnn.2010.2313.

220. Park, Y. K.; Kim, S. Y.; Kim, H. J.; Jung, K. Y.; Jeong, K.-E.; Jeong, S.-Y.; Jeon, J.-K. Removal of Sulfur Dioxide from Dibenzothiophene Sulfone over Mg-Based Oxide Catalysts Prepared by Spray Pyrolysis. *Korean J. Chem. Eng.* **2010**, 27 (2), 459–464. https://doi.org/10.1007/s11814-010-0086-x.

221. Kim, M. J.; Kim, H.; Jeong, K. E.; Jeong, S. Y.; Park, Y. K.; Jeon, J. K. Catalytic Decomposition of Dibenzothiophene Sulfones over Layered Double Hydroxide Catalysts. *J. Ind. Eng. Chem.* **2010**, *16* (4), 539–545. https://doi.org/10.1016/j.jiec.2010.03.026.

Chapter 7

Performance of Low-Cost Carbon-Based Adsorbent on Desulfurization of Heavy Gas Oil

Biswajit Saha, Sundaramurthy Vedachalam, Alivia Mukherjee, and Ajay K. Dalai*

Catalysis and Chemical Reaction Engineering Laboratories, University of Saskatchewan, Saskatoon, SK S7N 5A9, Canada

***Email: ajay.dalai@usask.ca**

This chapter discusses the performance of low-cost activated carbon for adsorptive desulfurization of heavy gas oil. The activated carbon adsorbent was prepared from exhausted coffee residue by steam activation. Fourier-transform infrared spectroscopy revealed that steam activated carbon contained aromatic C=C bonds and acidic functional groups on the surface. Adsorptive desulfurization on steam activated carbon involved the acid–base and π–π stacking interaction mechanisms. The sulfur compounds adsorption on steam activated carbon obeyed pseudo–first order kinetics. Mechanism insights revealed that surface adsorption and intraparticle diffusion controlled the rate of adsorption, and adsorption of sulfur compounds on activated carbon involved physisorption through electrostatic forces.

Introduction

According to the State of Global Air/2019 report (*1*), air pollution is the fifth-highest mortality risk factor globally. Poor air quality is the leading cause of cardiovascular and respiratory diseases and contributed to almost 4.9 million deaths in 2017. The main cause of air pollution is the emission of SO_x, NO_x, CO, CO_2, and so on into the environment due to modern globalization, growth in industrialization, and increase in the use of motor vehicles (*1*). Based on the U.S. Energy Information Administration's (EIA) annual energy outlook, Canada ranks third with 880,000 barrels of motor gasoline consumed per day (*2*). Organic sulfur compounds present in petroleum fuels pollute the environment by releasing sulfur and nitrogen oxides, which are produced by the combustion of sulfur and nitrogen compounds in internal combustion engines. Nitrogen oxides lead to global warming, whereas sulfur dioxide causes acid rain, decreases plants' growth by interfering with photosynthesis, and increases the acidity of soils, thus reducing soil fertility (*3*). Sulfur dioxide also causes health issues such as reduction of lung function and asthma. According to a 2020 survey of the Canadian council of ministers of the environment, around 14,000–15,000 deaths in Canada are due to air

© 2021 American Chemical Society

pollution every year (3). Therefore, it is essential to remove sulfur from petroleum fuels in a sustainable and environmentally friendly manner.

Hydrodesulfurization (HDS) is the only commercial process widely used in refineries worldwide to remove sulfur compounds from petroleum fluids. This process typically operates at high temperatures between 360 and 420 °C, and high H_2 pressures in the range of 8–15 MPa over γ-Al_2O_3–supported Ni- or Co-promoted Mo or W catalysts. This process requires hydrogen, which is expensive (*4*). In this regard, different nonconventional desulfurization approaches based on oxidation, adsorption, extraction, membrane-based separation, and microorganisms have been explored. These alternative desulfurization processes take place at mild temperature and pressure conditions and do not require costly hydrogen gas as a reactant.

One of the promising alternatives to HDS, adsorptive desulfurization (ADS), shows good prospects to produce ultra-low sulfur fuel. Adsorption is a mass transfer process where molecules in the free phase bind onto the surface of an adsorbent via intermolecular forces (5). This process occurs at a relatively lower temperature and pressure than HDS. Research has reported on ADS using various adsorbents like metal oxide, zeolite, mesoporous material, metal-organic framework, clay material, industrial waste, and carbon-based material (*6*). Among them, carbon-based adsorbents, mainly activated carbon, attract researchers' attention due to their low cost, high surface area (500–3000 m^2/g), high porosity, and adjustable pore size. It can be prepared from low-cost, readily available precursors, such as rice husks, bamboo, scrap tires, wood, fruit stone waste, waste plastic, and coconut shells (7). The adsorption of sulfur compounds depends on the nature and number of oxygen functional groups (such as carboxylic acid, pyrone, lactone, and phenolic group) and active metals on the surface of activated carbon (*6, 8–12*). Table 1 summarizes some results of acid- or metal-modified activated carbons for dibenzothiophene (DBT) adsorption, where activated carbons showed remarkable improvement due to the existence of surface oxygen functional groups, metal active sites, and Lewis acid sites.

Table 1. Adsorption Efficiency of Acid- and Metal-Modified Activated Carbons for DBT Adsorption

Adsorbent	*Feed Sulfur Amount (ppm)*	*Desulfurization (%)*	*Adsorption Capacity (mg/g)*	*References*
HCl-treated activated carbon	1000	98.5	18.4	(*8*)
Sn/activated carbon	1000	99.4	18.6	(*9*)
Al_2O_3/activated carbon	100	95.0	9.5	(*10*)
$ZnCl_2$/activated carbon	500	92.6	11.6	(*11*)
Ni/activated carbon	59	96.0	2.8	(*12*)

Saleh et al. (*13*) studied manganese oxide–incorporated activated carbon (MnO/AC) for ADS of DBT. It showed promising activity due to π-complexation. The ADS activities of mesoporous carbon and Cu_2O/mesoporous carbon for benzothiophene were studied by Jiang et al. (*14*). The adsorption efficiency of Cu-modified mesoporous carbon was found to be better due to the presence of a strong interaction between the sulfur atom and the hydroxyl and carboxyl groups as well as

π-complexation between Cu_2O species and sulfur. Similarly, the strong interaction between Lewis acidic sites of Zr and Lewis basic sites of sulfur promoted the adsorption capacity of ZrO_2/AC by 17% against pristine activated carbon (*15*). Most ADS studies on activated carbon were conducted with model sulfur compounds. Studies on activated carbon for desulfurization of real fuel are limited. Recently, Zubaidi et al. (*16*) used granular activated carbon for desulfurization of diesel fuel. Shah et al. (*9*) reported Sn-impregnated activated carbon for ADS of real feedstocks. It was capable of removing 81% and 72% of sulfur from kerosene and diesel range fuel, respectively. In this chapter, desulfurization of heavy gas oil feedstock by adsorption on low-cost activated carbon is studied. The adsorption kinetics and mechanism as well as sulfur molecules' diffusion into adsorbent pores are discussed in detail.

Materials and Methods

Preparation of Activated Carbon

The exhausted coffee residue (ECR) was upgraded by a two-step physical activation process using steam. ECR was collected from a local café in Saskatoon, Canada. ECR was first washed using distilled water to remove impurities. After washing, ECR was air-dried for 12 h at 110 °C. The two-step physical activation of ECR was carried out in a lab-scale fixed bed reactor. The detailed procedure for carbonization and activation was described elsewhere (*17*). The biochar was prepared from ECR by slow pyrolysis at 400 °C with nitrogen gas. The prepared biochar was then steam activated at 800 °C using nitrogen as a carrier gas. The yield of this process was around 50 wt%. The steam activated carbon was stored in desiccators for further analysis.

Characterization of Activated Carbon

Adsorbent materials were characterized by a Micrometrics ASAP 2020 analyzer to determine the specific surface area, pore volume, and average pore diameter. Temperature programmed desorption of ammonia (NH_3-TPD) analysis was carried out with an AutoChem HP chemisorption analyzer from Micromeritics. The sample (70 mg) was loaded into a u-shaped tube and degassed at 300 °C. The degassed sample was purged with 15% NH_3 in He at 100 °C. The TPD profile was obtained by heating the sample from 100 to 800 °C at a ramp rate of 10 °C/min. A VERTEX 70 Fourier-transform infrared spectroscopy (FTIR) instrument equipped with a diamond attenuated total reflectance (ATR)was used to characterize the functional groups of steam activated carbon. The spectra were recorded in the range of 400–4000 cm^{-1} wavenumbers, with a nominal resolution of 4 cm^{-1} and 128 scans.

ADS

A typical adsorption experiment was conducted using 10 mL of feedstock and 0.5 gm of adsorbent in a glass vial placed on a hot plate and stirred at 35 °C. The adsorption experiments were performed in a batch setup for 24 h. The liquid product was separated from the adsorbent by vacuum filtration. Mild hydrotreated bitumen-derived heavy gas oil was used as a feedstock. The total sulfur content of the feedstock and desulfurized oils was determined using an Antek N/S analyzer as per ASTM D4629. The sulfur removal efficiency (% R) and sulfur adsorption per unit mass of the adsorbent (mg/g) were calculated using eqs 1 and 2:

$$\% R = (C_i\text{-}C_t)/C_i \text{ ....(1)}$$

$$q_t = [(C_i\text{-}C_t)V]/W \text{ ....(2)}$$

All terms are stated in the nomenclature section.

Results and Discussion

Activated carbon was produced by steam activation of ECR biochar. Brunauer–Emmett–Teller surface area, pore volume, and pore diameter of activated carbon were found to be 725 m^2/g, 0.5 cm^3/g, and 4.2 nm, respectively, as determined by Brunauer–Emmett–Teller–N_2 physisorption. The ADS of mild hydrotreated heavy gas oil on steam activated carbon was carried out at 35 °C in a batch mode for 24 h. The external mass transfer in a stirred batch reactor depends on the agitation speed. In this study, a noticeable enhancement in the S adsorption capacity occurred when the agitation speed increased from 100 to 300 rpm, then the S adsorption capacity approached a plateau at 300 and 500 rpm (figure not shown). This demonstrates the absence of external mass transfer resistance when the agitation speed is >300 rpm. In this study, all experiments kept the agitation speed at 400 rpm to prevent external diffusional resistance. The hydrotreated gas oil contained 0.73 and 0.17 wt% sulfur and nitrogen, respectively. Activated carbon showed an adsorption capacity of 38.4 mg of sulfur per gram of adsorbent. Along with sulfur, this material also removed 25% of nitrogen from heavy gas oil, which corresponds to an adsorption capacity of 8.4 mg/g. Like HDS, ADS improves the quality of fuel by selectively removing sulfur and nitrogen compounds. However, the product yield is expected to be lower in ADS compared with HDS since full molecules (C, H, N, S) are removed in ADS, whereas only S and N atoms are removed in HDS, leaving carbon and hydrogen atoms in the final fuel product. Nevertheless, as a low-cost technology, ADS might offset any loss in the product yield.

Mechanism of Sulfur Adsorption on Activated Carbon

The activated carbon used in this study contained a certain degree of surface chemical heterogeneity, which was evidenced by FTIR. The spectrum is shown in Figure 1.

The broad weak absorbance band between 3600 and 3200 cm^{-1} corresponds to O-H vibrations of hydroxyl groups of water, alcohols, and phenols and N-H stretching vibrations of the amino group (*18, 19*). The bands between 3000 and 2800 cm^{-1} are associated with C-H vibrations of aliphatic CH, CH_2, and CH_3 groups (*18*). The bands in the 1620–1400 cm^{-1} region can be assigned to C=C vibrations of aromatic rings and C=O vibrations of carboxylic acid and ketone functional groups (*19*). The prominent peak at about 1000 cm^{-1} represents the C-O stretching vibration of hydroxyl, ester, or ether functional groups on the surface of activated carbon. The FTIR data evidence the presence of aromatic C=C bonds and organic molecules containing hydroxyl and carboxylic acid functional groups on the surface of activated carbon.

The NH_3-TPD profile shown in Figure 2 evidences that the activated carbon consists of medium and strong acidic sites with a total acidity of 0.43 mmol/g on its surface.

The surface acidity of activated carbon mainly originated from acidic oxygen functional groups (OH, COOH). Organic sulfur compounds present in heavy gas oil are benzothiophenes, DBTs, and benzonaphthothiophenes. These compounds are aromatic and contain basic sulfur atoms. Acid sites in activated carbon can attract the basic sulfur compounds present in heavy gas oil (*20, 21*) through the acid–base interaction mechanism shown in Figure 3.

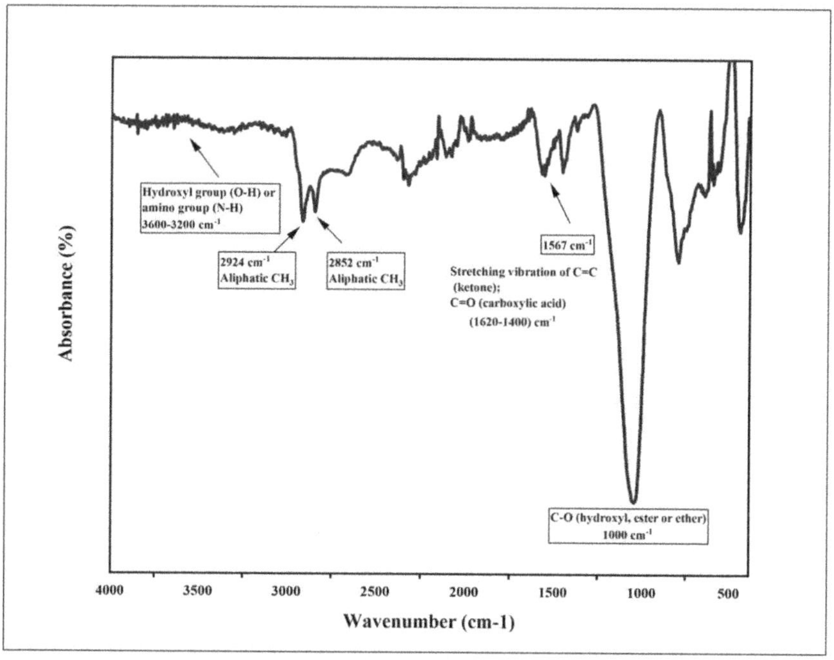

Figure 1. FTIR spectrum of steam activated carbon from ECR.

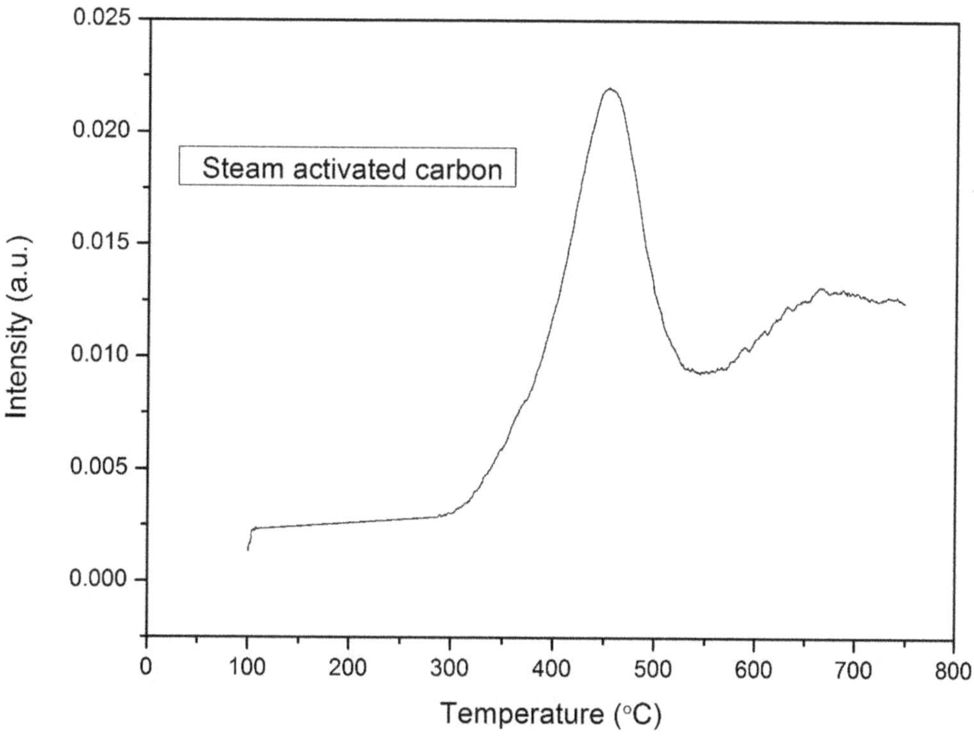

Figure 2. NH_3-TPD profile of steam activated carbon.

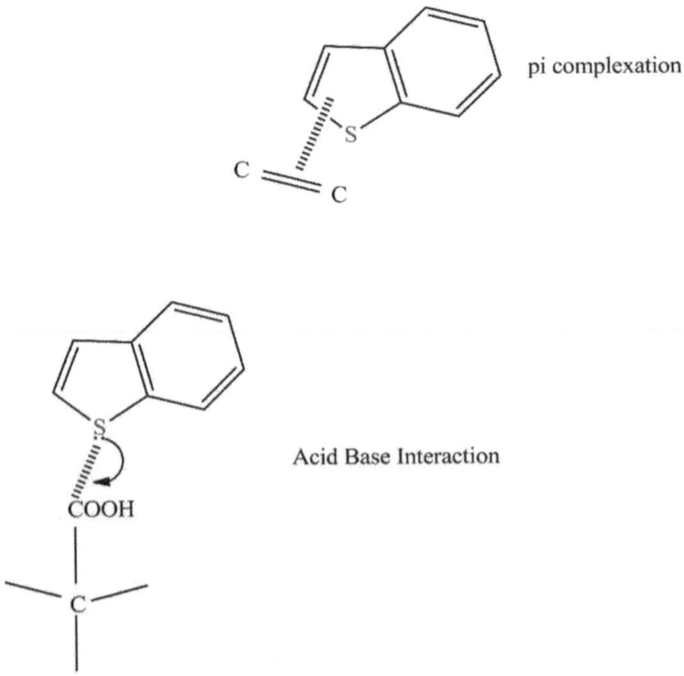

Figure 3. Mechanism of ADS on activated carbon.

The sulfur atom of the thiophene ring contains two lone pairs of electrons and one of those pairs is utilized to retain the aromaticity. The other is free and increases the interaction between the adsorbent acid sites and the sulfur compounds. The mechanism of sulfur compound adsorption on activated carbon may also involve a π–π stacking interaction between the delocalized π electrons of sulfur in the thiophene ring and C=C and C=O functional groups present on the activated carbon surface (22, 23), forming an electron donor–acceptor complex. The FTIR study confirmed the presence of C=C and C=O functional groups on activated carbon.

Heavy gas oil contains several saturated, naphthenic, and aromatic molecules along with aromatic sulfur and nitrogen compounds. Contrary to sulfur compounds, saturated and naphthenic compounds do not have basic groups and π electrons to be adsorbed on activated carbon. The selective adsorption of sulfur compounds in the presence of aromatics is challenging as they can be adsorbed on activated carbon through π–π interaction. However, an enhanced selectivity of activated carbon to sulfur compounds is expected due to the presence of acid–base interaction along with π–π interaction. Most of the nitrogen-containing compounds in heavy gas oil exist as polycyclic aromatic nitrogen derivatives. Like sulfur compounds, nitrogen compounds in heavy gas oil are basic and thus they are adsorbed via acid–base and π–π interactions. As discussed earlier, activated carbon prepared in this study adsorbed 25% of nitrogen from heavy gas oil, which is equivalent to the nitrogen adsorption capacity of 8.4 mg/g. The ADS mechanisms emphasized the importance of the high specific surface area and acidic functional groups (active sites of adsorbent), which make activated carbon a better adsorbent compared with other adsorbents reported for ADS of real feedstocks (Table 2).

Table 2. Comparison of Activated Carbon with Other Reported Adsorbents for ADS of Real Feedstock

Feedstock	*Adsorbent*	*S Adsorption Capacity (mg/g)*	*Reference*
4-Methyldibenzothiophene (4-MDBT) in ultra-low sulfur (ULS) diesel (300 ppm of S)	Pd/MCM-41	2.3	(24)
CN-II diesel (952.9 ppm of S)	$Ag\text{-}CeO_X/TiO_2\text{-}SiO_2$	5.4	(25)
DBT in diesel (800 ppm of S)	Ni-CNF/activated carbon	8.1	(26)
Jet propellant-8 (JP-8) fuel (2230 ppm of S)	Cu-Na/Y	36.0	(27)
Gasoline (76 ppm of S)	Cu/Y	23.0	(28)
Diesel (49 ppm of S)	Cu/Y	36.0	(28)
Heavy gas oil (7300 ppm of S)	Activated carbon	38.4	present

Kinetic of ADS on Steam Activated Carbon

The kinetics of adsorption can provide a better understanding of the mechanisms of the mass transfer of the adsorbate to the liquid–solid interface and the interactions between the adsorbate and the adsorbent. The typical steps of adsorption are as follows: (1) external diffusion, which involves the transfer of adsorbate to the external surface of adsorbent due to the concentration difference driving force, (2) internal diffusion, which involves diffusion of adsorbate into pores of adsorbent, and (3) adsorption of the adsorbate's molecules through interaction with active sites of adsorbent (29). In this study, the typical empirical pseudo–first order and pseudo–second order rate laws were investigated for ADS on steam activated carbon:

$$ln(q_e - q_t) = lnq_e - K_1 t \qquad (3)$$

$$\frac{t}{q_t} = \frac{1}{K_2 {q_e}^2} + \frac{t}{q_e} \qquad (4)$$

where q_e and q_t are the amounts of sulfur compounds adsorbed at equilibrium (q_e) and at various times (q_t) calculated in mg/g and K_1 and K_2 are pseudo–first order and pseudo–second order rate constants, respectively. Eqs 3 and 4 represent the kinetic model equation of pseudo–first order and pseudo–second order kinetics, respectively (30). The linear plots of these equations are shown in Figure 4. The values of K_1, K_2, q_e, and C were determined for ADS of heavy gas oil on activated carbon and given in Table 3. The experimental data show the best fit with the pseudo–first order model with a R^2 value of 0.96. The calculated q_e value of the pseudo–first order model is close to the experimental q_e value of 40.8 mg/g.

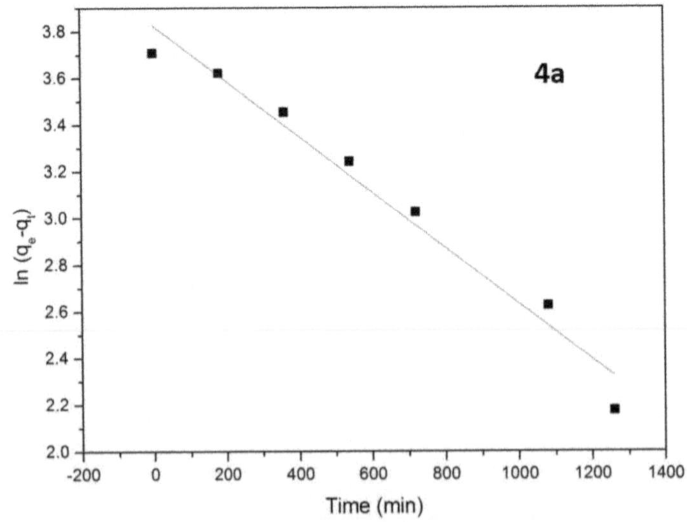

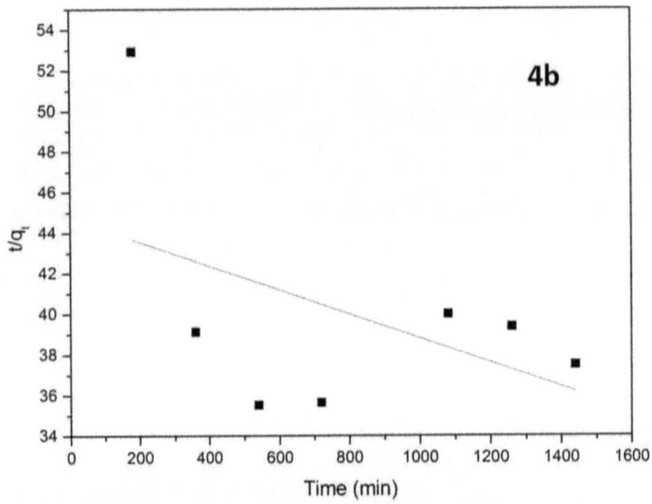

Figure 4. (a) Pseudo-first order and (b) pseudo-second order models for ADS of heavy gas oil. Adsorbent amount: 50 g/L; Temperature: 35 °C; Sulfur concentration: 7300 ppm.

Table 3. Values of Adsorption Kinetic Parameters

Kinetic Model	*Parameters*	*Value*	R^2
Pseudo-first order	K_1(min^{-1})	1.19×10^{-3}	0.96
	q_e (mg/g)	45.60	
Pseudo-second order	K_2(g/mg.min)	7.86×10^{-7}	0.06
	q_e (mg/g)	−168.63	
Intraparticle diffusion	K_{id} (mg/g.min$^{0.5}$)	1.37	0.98
	C (mg/g)	−6.25	

Adsorption Diffusion Mechanism on Activated Carbon

To understand the mechanism of diffusion, the adsorption kinetic data was further analyzed by an intraparticle diffusion model as given in eq 5:

$$q_t = K_{id}t^{\frac{1}{2}} + C \ldots\ldots\ldots\ldots\ldots\ldots\ldots\ldots (5)$$

where K_{id} is the intraparticle diffusion rate constant (mg. g^{-1} $min^{-0.5}$) and C is a constant that provides the boundary layer thickness (mg/g). As per this model, if the plot of q_t versus $t^{0.5}$ provides a straight line, then adsorption involves intraparticle diffusion, and if the data show multilinear plots, then more than one step controls adsorption of sulfur compounds on activated carbon. Besides, if the straight line goes through the origin, the intraparticle diffusion is the sole rate-controlling step. If not, some other mechanisms may coexist. In the present study, as shown in Figure 5, the linear plot does not pass through the origin; therefore, intraparticle diffusion is not the sole mechanism that controls the adsorption of sulfur compounds on activated carbon.

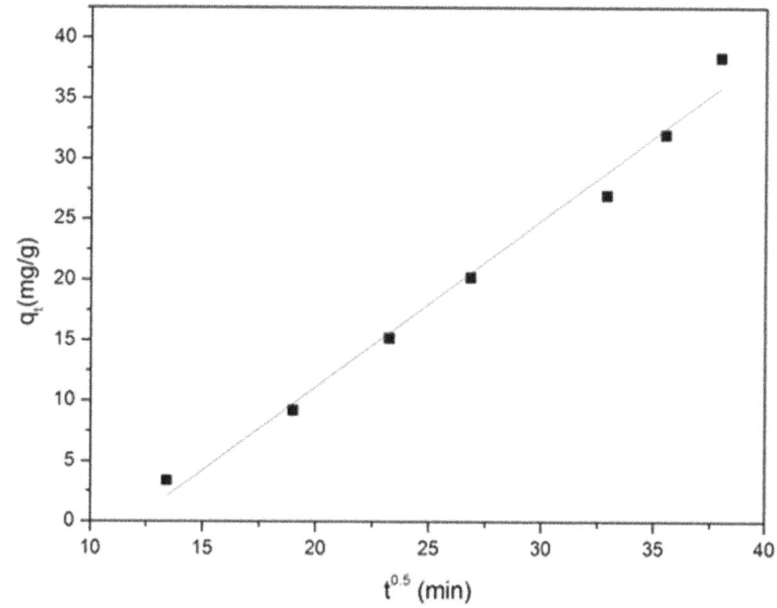

Figure 5. Intraparticle diffusion kinetic model for ADS of heavy gas oil on activated carbon. Adsorbent amount: 50 g/L; Temperature: 35 °C; Sulfur concentration: 7300 ppm.

The following mechanism is suggested for ADS of heavy gas oil over activated carbon. Adsorption of sulfur compounds on activated carbon is influenced by its textural characteristics. At the initial stage of adsorption, there is a possibility that sulfur molecules interact with the functional groups on the external surface of activated carbon (external surface adsorption). At the start, ADS is controlled by surface adsorption due to the higher concentration gradient between the adsorbate and the adsorbent. After the saturation point, sulfur molecules start to enter the pores of activated carbon and then adsorb on the interior surface (pores), where intraparticle diffusion is the rate-limiting step. Although adsorption on activated carbon is affected by intraparticle diffusion and surface adsorption, intraparticle diffusion is the main rate-controlling factor for adsorption of sulfur molecules on activated carbon. This result agrees with a previous report (*31*).

Since the adsorption kinetic was controlled by intraparticle diffusion, the kinetic rate constants were determined at different temperatures (35, 50, 65, and 80 °C) using the intraparticle diffusion control model. The adsorption plots at different temperatures are shown in Figure 6(a). Table 4 shows the values of rate constants at different temperatures.

Table 4. Values of the Rate Constant at Different Temperatures

Temperature (K)	*Rate Constant (K_{id}) (mg/g.min$^{0.5}$)*
308	1.373
323	1.225
338	1.083
353	1.027

The activation energy was calculated to gain more insight into the adsorption mechanism. Using the Arrhenius eq 6, the activation energy was calculated in the temperature range of 208–353 K:

$$ln\ K_{id} = ln\ k_0 - \frac{E_a}{RT} \ldots\ldots\ldots (6)$$

where k_0 (g mmol^{-1} min^{-1}) is the temperature-independent factor, E_a (kJ/mol) is the activation energy of the adsorption process, T is the absolute temperature in Kelvin, and R is the universal gas constant.

Figure 6(b) shows the Arrhenius plot from which the E_a of −6.06 kJmol^{-1} was calculated. It is known that the E_a of <40 kJ/mol corresponds to physical adsorption, while the E_a of >40 kJ/mol corresponds to chemical adsorption (32, 33). The calculated low activation energy of the present study suggests that ADS on activated carbon occurs via the physical adsorption pathway, involving weak van der Waals forces and hydrogen bonds.

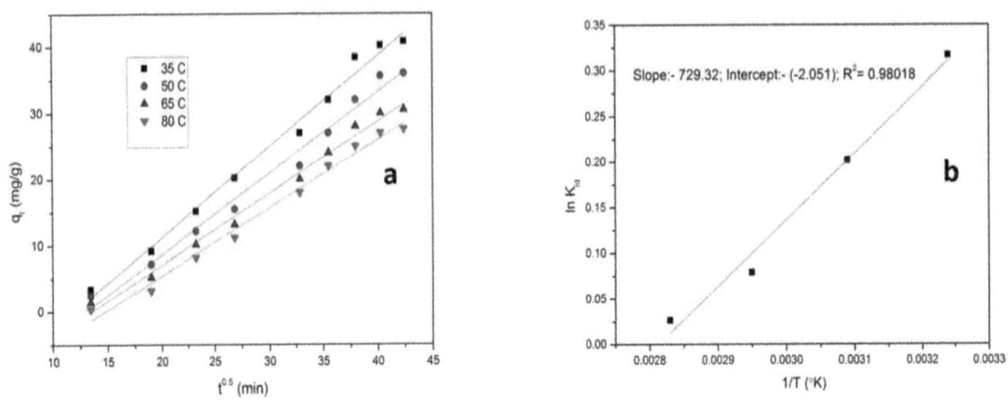

Figure 6. (a) Rate constants at different temperatures and (b) Arrhenius plot of ADS of heavy gas oil.

Conclusions

ADS of heavy gas oil was performed on steam activated carbon. The pseudo–first order adsorption model fit best to ADS of heavy gas oil on activated carbon. The diffusion model analysis implies that ADS of heavy gas oil on activated carbon involves pore diffusion. The low activation energy of ADS of heavy gas oil indicates that adsorption on activated carbon is physical and involves mainly electrostatic interactions. Overall, the kinetics data suggest that intraparticle diffusion is the main rate-controlling factor for adsorption of sulfur molecules on activated carbon.

Nomenclature

q_t = Adsorption capacity at time t (mg/g)

q_e = Adsorption capacity at equilibrium (mg/g)

C_e = Residual concentration of sulfur compound in model fuel at equilibrium (mg/L)

C_i = Initial concentration of sulfur compound in model fuel (mg/L)

C_t = Residual concentration of sulfur compound in model fuel at time t (mg/L)

V = Volume of model fuel (L)

W = Mass of adsorbent (gm)

K_1 = Pseudo–first order rate constant (min^{-1})

K_2 = Pseudo–second order rate constant (g/mg.min)

K_{id} = Intraparticle diffusion model rate constant ($mg/g.min^{0.5}$)

C = Intercept (mg/g)

k_0 = Pre-exponential factor

E_a = activation energy (kJ/mol)

R = Universal gas constant (8.314 J/(mol.K))

References

1. *State of Global Air/2019, A Special Report on Global Exposure to Air Pollution and Its Disease Burden*, Health Effects Institute, Boston, ISSN 2578-6873. http://www.stateofglobalair.org/sites/default/files/soga_2019_report.pdf (accessed Oct 8, 2020).
2. *Gasoline Consumption by Country, around the World*, TheGlobalEconomy.com. https://www.theglobaleconomy.com/rankings/gasoline_consumption/ (accessed Oct 10, 2020).
3. *Guidance Document on Achievement Determination for Canadian Ambient Air Quality Standards for Sulphur Dioxide*. https://www.ccme.ca/en/res/gdadforcaaqsforsulphurdioxide_en1.0.pdf (accessed Oct 20, 2020).
4. AL-Hammadi, S. A.; Al-Amer, A. M.; Saleh, T. A. Alumina-Carbon Nanofiber Composite as a Support for MoCo Catalysts in Hydrodesulfurization Reactions. *Chem. Eng. J.* **2018**, *345*, 242–251.
5. Adekanmi, A. A.; Folorunsho, A. Comparative Analysis of Adsorptive Desulfurization of Crude Oil by Manganese Dioxide and Zinc Oxide. *Res. J. Chem. Sci.* **2012**, *2*, 14–20.
6. Lee, K. X.; Valla, J. A. Adsorptive Desulfurization of Liquid Hydrocarbons Using Zeolite-Based Sorbents: A Comprehensive Review. *React. Chem. Eng.* **2019**, *4*, 1357–1386.

7. Saleem, J.; Bin Shahid, U.; Hijab, M.; Mackey, H.; McKay, G. Production and Applications of Activated Carbons as Adsorbents from Olive Stones. *Biomass Convers. Biorefinery* **2019**, *9*, 775–802.
8. Shah, S. S.; Ahmad, I.; Ahmad, W.; Ishaq, M.; Gul, K.; Khan, R.; Khan, H. J. Study on Adsorptive Capability of Acid Activated Charcoal for Desulphurization of Model and Commercial Fuel Oil Samples. *Environ. Chem. Eng.* **2018**, *6*, 4037–4043.
9. Shah, S. S.; Ahmad, I.; Ahmad, W. Adsorptive Desulphurization Study of Liquid Fuels using Tin (Sn) Impregnated Activated Charcoal. *J. Hazard. Mater.* **2016**, *304*, 205–213.
10. Ganiyu, S. A.; Alhooshani, K.; Sulaiman, K. O.; Qamaruddin, M.; Bakare, I. A.; Tanimu, A.; Saleh, T. A. Influence of Aluminium Impregnation on Activated Carbon for Enhanced Desulfurization Of DBT at Ambient Temperature: Role of Surface Acidity and Textural Properties. *Chem. Eng. J.* **2016**, *303*, 489–500.
11. Alhamed, Y. A.; Bamufleh, H. S. Sulfur Removal from Model Diesel Fuel Using Granular Activated Carbon From Dates' Stones Activated by $ZnCl_2$. *Fuel* **2009**, *88*, 87–94.
12. Danmaliki, G. I.; Saleh, T. A.; Shamsuddeen, A. A. Response Surface Methodology Optimization of Adsorptive Desulfurization on Nickel/Activated Carbon. *Chem. Eng. J.* **2017**, *313*, 993–1003.
13. Saleh, T. A.; Sulaiman, K. O.; AL-Hammadi, S. A.; Dafalla, H.; Danmaliki, G. I. Adsorptive Desulfurization of Thiophene, Benzothiophene and Dibenzothiophene Over Activated Carbon Manganese Oxide Nanocomposite: With Column System Evaluation. *J. Clean. Prod.* **2017**, *154*, 401–412.
14. Jiang, X.; Xu, W.; Liu, W.; Yue, M.; Zhu, Y.; Yang, M. Facile Preparation of Cuprous Oxide Decorated Mesoporous Carbon by One-step Reductive Decomposition for Deep Desulfurization. *Fuel* **2019**, *241*, 777–785.
15. Xiong, L.; Yan, X. M.; Mei, P. Synthesis and Characterization of a ZrO_2/AC Composite as a Novel Adsorbent for Dibenzothiophene. *Adsorpt. Sci. Technol.* **2010**, *28*, 341–350.
16. Al Zubaidi, I.; Darwish, N. N.; El Sayed, Y.; Shareefdeen, Z.; Sara, Z. Adsorptive Desulfurization of Commercial Diesel Oil Using Granular Activated Charcoal. *Int. J. Adv. Chem. Eng. Biol. Sci.* **2015**, *2*, 15–18.
17. Azargohar, R.; Dalai, A. K. Production of Activated Carbon from Luscar Char: Experimental and Modeling Studies. *Microporous Mesoporous Mater.* **2005**, *85*, 219–225.
18. Hao, Z.; Wang, C.; Yan, Z.; Jiang, H.; Xu, H. Magnetic Particles Modification of Coconut Shell-Derived Activated Carbon and Biochar for Effective Removal of Phenol from Water. *Chemosphere* **2018**, *211*, 962–969.
19. Liang, J.; Xu, X.; Qamar Zaman, W.; Hu, X.; Zhao, L.; Qiu, H.; Cao, X. Different Mechanisms Between Biochar and Activated Carbon for the Persulfate Catalytic Degradation of Sulfamethoxazole: Roles of Radicals in Solution or Solid Phase. *Chem. Eng. J.* **2019**, *375*, 121908.
20. Li, J. J.; Zhang, H.; Tang, X. D.; Lu, H. Adsorptive Desulfurization of Dibenzothiophene Over Lignin-Derived Biochar by One-step Modification with Potassium Hydrogen Phthalate. *RSC Adv.* **2016**, *6*, 100352–100360.

21. Yu, M.; Zhang, N.; Fan, L.; Zhang, C.; He, X.; Zheng, M.; Li, Z. Removal of Organic Sulfur Compounds from Diesel by Adsorption on Carbon Materials. *Rev. Chem. Eng.* **2015**, *31*, 27–43.
22. Mao, L. S.; Wang, Y. L.; Hu, X. C. - Stacking Interactions in the Peridinin-Chlorophy II-Protein of Amphidinium Carterae. *J. Phys. Chem. B* **2003**, *107*, 3963–3971.
23. Qu, D.; Feng, X.; Li, N.; Ma, X.; Shanga, C.; Chen, X. D. Adsorption of Heterocyclic Sulfur and Nitrogen Compounds in Liquid Hydrocarbons on Activated Carbons Modified by Oxidation: Capacity, Selectivity and Mechanism. *RSC Adv.* **2016**, *6*, 41982–41990.
24. Teymouri, M.; Samadi-Maybodi, A.; Vahid, A.; Miranbeigi, A. Adsorptive Desulfurization of Low Sulfur Diesel Fuel Using Palladium Containing Mesoporous Silica Synthesized Via a Novel In-situ Approach. *Fuel Process. Technol.* **2013**, *116*, 257–264.
25. Xu, C. Z.; Zheng, M. Q.; Chen, K.; Hu, H.; Chen, X. H. CeO_x Doping on a TiO_2-SiO_2 Supporter Enhances Ag Based Adsorptive Desulfurization for Diesel. *J. Fuel Chem. Technol.* **2016**, *44*, 943–953.
26. Prajapati, Y. N.; Verma, N. Adsorptive Desulfurization of Diesel Oil Using Nickel Nanoparticle-Doped Activated Carbon Beads with/without Carbon Nanofibers: Effects of Adsorbate Size and Adsorbent Texture. *Fuel* **2017**, *189*, 186–194.
27. Dias Da Silva, P.; Samaniego Andrade, S. K.; Zygourakis, K.; Wong, M. S. Adsorptive Desulfurization of Liquid Fuels at Elevated Temperatures Using Metal Exchanged Zeolite Y. *Ind. Eng. Chem. Res.* **2019**, *58*, 19623–19632.
28. Lin, K. S.; Adhikari, A. K.; Wang, Z. P. Desulfurization of Gasoline and Diesel by Adsorption with Cu (I) – Y Zeolite. *J. Chem. Eng. Japan* **2009**, *42*, 168–175.
29. Wang, J.; Guo, X. Adsorption Kinetic Models: Physical Meaning, Applications and Solving Method. *J. Hazard. Mater.* **2020**, *390*, 122156.
30. Jha, D.; Haider, M. B.; Kumar, R.; Shim, W. G.; Sivagnanam, B. M. Batch and Continuous Adsorptive Desulfurization of Model Diesel Fuels Using Graphene Nanoplatelets. *J. Chem. Eng. Data.* **2020**, *65*, 2120–2132.
31. Liu, Y.; Liu, X.; Dong, W.; Zhang, L.; Kong, Q.; Wang, W. Efficient Adsorption of Sulfamethazine onto Modified Activated Carbon: A Plausible Adsorption Mechanism. *Sci. Rep.* **2017**, *7*, 12437.
32. Mahmoud, M. A. Kinetics and Thermodynamics of Aluminum Oxide Nanopowder as Adsorbent for Fe (III) from Aqueous Solution. *Beni-Suef University J Basic Appl. Sci.* **2015**, *4*, 142–149.
33. Karaoglu, M. H.; Zor, S.; Ugurlu, M. Biosorption of Cr(III) from Solutions Using Vineyard Pruning Waste. *Chem. Eng. J.* **2010**, *159*, 98–106.

Chapter 8

Heavy Oil Upgrading by Oxidative Ring-Opening of Polycyclic Aromatic Hydrocarbons

Natalia Montoya Sánchez and Arno de Klerk*

Department of Chemical and Materials Engineering, University of Alberta, Edmonton, Alberta T6G 1H9, Canada

***Email: deklerk@ualberta.ca**

Heavy oil and bitumen, which are feedstocks rich in polycyclic aromatic hydrocarbons (PAHs), are upgraded and refined to lighter products mainly by thermal conversion and hydroprocessing. An alternative process involving oxidative ring opening to convert PAHs to products with fewer aromatic rings and a slightly increased hydrogen-to-carbon ratio appears possible. The steps in such a process are (1) oxidation of the PAHs to quinonoids, (2) ring-opening to dicarboxylic acids, and (3) decarboxylation. This chapter critically evaluates the oxidation of PAHs, highlighting how differences in their structural motifs (i.e., linear and angular ring-configurations) affect the oxidation chemistry. Although several gaps in fundamental descriptions for oxidative ring-opening by autoxidation and catalytic oxidation were identified, the reaction pathway leading to aromatic ring-opening is possible. The investigation brought insights regarding the conversion chemistry. The way in which oxygen adds to the PAHs in the preoxidative step is dependent on the ring configuration. The steps following the addition of oxygen to the aromatic structure involved not only the formation of quinonoids but also the formation of anhydride species, with at least one reversible pathway proceeding through the anhydride (i.e., elimination of oxygen from the anhydride to produce the quinonoid and oxygen addition to the quinonoid to produce the anhydride). This reaction is more easily accomplished by the oxidation of outer rings in PAHs and internal rings in PAHs with an angular ring configuration. Quinonoid and anhydride species can also be produced by selective catalytic oxidation of the PAHs. Irrespective of whether autoxidation or catalytic oxidation is performed, ring-opening proceeds through the oxidative formation and hydrolysis of anhydride species, which leads to the formation of carboxylic acid products. Aromatic ring-opening requires both oxidation and hydrolysis.

© 2021 American Chemical Society

1. Introduction

When we talk about heavy oil and bitumen upgrading, one of the aspects that immediately comes to mind is that we are dealing with complex materials. These materials are rich in hydrogen-deficient and bulky molecules, that is, polycyclic aromatic hydrocarbons (PAHs). PAHs require strong conditions to react in order to yield lighter products. PAHs also contain molecules that are responsible for some of the observed undesired macroscopic properties that make processing of heavy feedstocks so challenging, for example, high viscosity and high coking tendency.

Existing technologies for processing of heavy feedstocks, and so processing of PAHs, make use of thermal conversion and hydrogen addition (*1–3*). Thermal conversion allows for disproportionation or "redistribution" of the available hydrogen to give fractions that are richer or poorer in hydrogen. Thermal conversion is the heart of industrially successful visbreaking and coking processes.

Interestingly, the chemistry involved in hydrogen disproportionation does not affect the core of the PAHs, which retain the same skeletal configuration. The transfer of hydrogen and methyl groups can cause the saturation of aromatic rings. Intermediate free-radical products can also participate in free-radical addition reactions to form even heavier products. The products that become more hydrogen enriched are more susceptible to cracking to lighter products, whereas the products that become more hydrogen deficient are less susceptible to cracking. Without an external source of hydrogen, there is a finite capacity for thermal cracking of PAHs facilitated by hydrogen disproportionation.

Hydrogen-addition technologies allow for a higher yield of lighter products. However, there is a significant cost associated with the production and use of large amounts of high-pressure hydrogen. The hydrogen consumption in residue hydroconversion processes is of the order of 2 wt % of the vacuum residue feed material (*3*).

Existing processes for heavy oil and bitumen upgrading are mature technologies. To move beyond process improvement of mature technologies, it is necessary to develop new approaches to the upgrading of heavy feedstocks. In other words, we need to *rethink* the way in which key molecules such as the PAHs can be converted. The question is, what can we do differently to selectively break aromatic C–C bonds to obtain simpler products? Of equal importance, what can we do differently that does not involve hydrogen addition and that is able to retain the hydrogen already available in PAHs?

One possible and potentially elegant solution could be the use of oxygen (in air) as a changing agent to break aromatic C–C bonds and produce ring-opened hydrocarbon products. This conversion chemistry is called oxidative ring-opening of polycyclic aromatic hydrocarbons.

Conceptually, the process pathway for oxidative ring-opening of polycyclic aromatic hydrocarbons to produce products with fewer aromatic rings and a slightly increased hydrogen-to-carbon ratio comprises three steps. These three process steps are as follows (Figure 1):

- (*i*) Oxidation of polycyclic aromatic hydrocarbons to produce quinonoids, which are di-ketones.
- (*ii*) Further oxidation of the quinonoids to produce aromatic carboxylic acid species, which concomitantly lead to the opening of the aromatic ring.
- (*iii*) Decarboxylation of the aromatic carboxylic acids produced by ring-opening to eliminate CO_2 and yield the ring-opened hydrocarbon product with slightly higher hydrogen-to-carbon ratio than the polycyclic aromatic hydrocarbon feedstock.

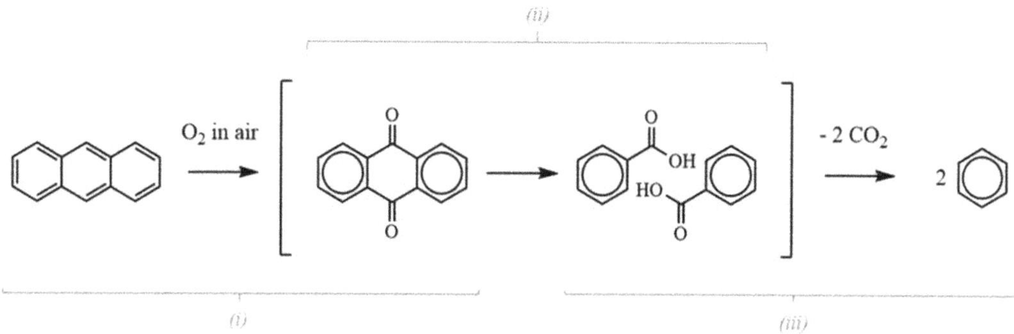

Figure 1. Oxidative ring-opening of anthracene.

The high-level description of the steps leading to oxidative ring-opening of PAHs hides the complexity associated with each step. The purpose of this chapter is to critically look at the oxidation of PAHs and how structural motifs in heavy oils and bitumen will affect the oxidation. In terms of scope, this means that only steps (*i*) and (*ii*) will be discussed and not step (*iii*). The discussion is also limited to oxidation using oxygen (in air) as oxidant. Conversion of aromatic materials such as heavy oils and bitumen calls for an approach in which cheap oxidants are more likely to make economic sense.

2. Structural Motifs in Bitumen and Heavy Oil and Their Implications for Oxidation

Crude oils can be separated into different solubility classes by paraffin precipitation of asphaltenes, followed by column chromatography. In this way, the gross composition of crude oils can be defined by the content of *saturated hydrocarbons* (paraffins and naphthenes), *aromatic hydrocarbons* (aromatics and naphthenoaromatics), and *resins and asphaltenes* (high-molecular-weight polycyclic fraction and material enriched in S-, N-, and O-containing compounds) (*4*). The relative proportion of each one determines how light or how heavy an oil is.

In the case of bitumen and heavy oils, bacterial degradation has largely removed paraffins along with the light boiling fractions; therefore, these oils have a lower concentration of paraffinic groups, many of which are actually present as side chains in larger and complex molecules (*1*). Similarly, biodegradation of monocyclic (single-ring) aromatic compounds, which are common in lighter oils, has left behind a significant fraction of the less-soluble polycyclic (multiring) aromatic compounds. These are typically substituted with alkyl groups or with bridges to other aromatics (*1*). Naphthenoaromatics, including one or several condensed aromatic rings fused with naphthenic rings and paraffinic chains, are significant constituents of the high boiling fractions (*4*). Table 1 illustrates different compound classes and functional groups found in bitumen and heavy oils (*5*).

Despite the considerable molecular diversity found in bitumen and heavy oils, this discussion is limited to polycyclic aromatic hydrocarbons. When looking at polycyclic aromatic structures, some general motifs can be identified. In reality, these motifs exist in complex structures, but from a reactivity point of view, their chemistry remains mostly unaffected. Three exceptions can have an effect on reactivity: (*i*) the presence of pendant groups that cause considerable steric hindrance, (*ii*) the presence of pendant groups that are very strong activating or deactivating groups, and (*iii*) polycyclic structures that have an inherent free-radical nature.

The most common polycyclic structures found in heavy oil are polycyclic aromatic hydrocarbons and polycyclic aromatic hydrocarbons in association with heterocycles. In heavy-petroleum-derived oil, the most common heterocyclic heteroatom is sulfur, with nitrogen usually present at lower levels

(6). Oxygen, when present in the heavy oil, is not often present in a heterocycle (7). These statements are not applicable to all heavy oils irrespective of their origin, and exceptions to these generalizations can be found. Coal-derived oils tend to have higher heterocyclic nitrogen content (8), whereas biomass-derived oils tend to have some heterocyclic oxygen content (9, 10).

Table 1. Compound Classes and Functional Groups Found in Bitumen and Heavy Oils

Compound Class	*Examples of Compounds or Functional Groups*
Alkanes (paraffins)	*n*-Paraffins or *i*-paraffins (typically carbon number <19)
Cycloalkanes (naphthenes)	Decalin, cheilanthane, sterane, gammacerane
Aromatics	Naphthalene, anthracene, phenanthrene, perylene, fluorantracene, chrysene, picene, pyrene
Naphtheno-aromatics	Tetralin, fluorene, 9,10-dihydrophenanthrene
S-containing compounds	Thiophene, benzothiophene, dibenzothiophene, sulfide, thiol, sulfoxide, sulfone, sulfonic acids
N-containing compounds	Pyrrole, pyridine, carbazole, quinoline
O-containing compounds	Furan, benzofuran, fluorenone, carboxylic acids, alcohol, ketone, ether
Metal chelates	Nickel and vanadyl porphyrins

The true diversity of compounds transcends a simple classification into molecules with only a single functional group. Detailed analyses of heavy oil fractions (11) indicate that heteroatoms occur in different combinations, and it is likely that compounds with mixed heterocycles are present in oil. The polycyclic aromatic hydrocarbons present in oil further consist of mixtures of five-membered and six-membered rings.

In this respect, a rather simplified view of polycyclic aromatic oxidation is presented in this chapter, which is limited in scope to two structural motifs of PAHs: (*i*) linear and (*ii*) angular. Aromatic compounds consisting of three fused rings are employed to illustrate the concept, namely, anthracene as representative of a linear structure, and phenanthrene as representative of an angular structure. This selection will be explained in the next section. Although oxidation of PAHs in association with heterocyclic compounds is outside the scope of this chapter, discussion of the bulk autoxidation of aromatic N-, S-, and O-heterocyclic compounds in aromatic-rich heavy oils can be found in references (12–14).

2.1. Reactivity of Aromatics through the Aromatic Sextet

The relative stability of aromatic compounds compared to cyclic aliphatic compounds and conjugated polyenes has been a topic of significant scientific discussion. Two valuable contributions to this discussion are the ones by Hückel and Clar. On one hand, Hückel's rule is employed to determine whether a cyclic structure has aromaticity. According to Hückel's rule, a cyclic structure is aromatic when the number of π-electrons is equal to $4n + 2$, when n is zero or an integer. Hückel's rule successfully explained the stability of benzene compared to other monocyclic conjugated polyenes (15). On the other hand, Clar's π-sextet rule, based on electron density, successfully explained the stability of polycyclic systems (16, 17). As we will see later, Clar's rule is a valuable tool to rationalize the reactivity of polycyclic aromatics toward oxidation. For that reason, a brief description of Clar's rule is included here.

The working principle of Clar's rule relies on the concept of the aromatic sextet. The aromatic sextet or π-sextet refers to the six π-electrons localized in a single benzene ring, which is separated from other rings by formal carbon–carbon single bonds (*18*). It is noted by a circle. A complete explanation on how to properly draw formulas of polycyclic aromatic hydrocarbons using the circle notation is found in reference (*19*).

Clar's rule classifies aromatic rings into four groups: rings having localized π-sextets, migrating π-sextets, localized double bonds, and empty rings (Figure 2). Aromatic rings having localized π-sextets, also known as benzenoid rings, exhibit benzenoid-like stability (*16*, *17*); they are the most aromatic centers and, therefore, the more stable rings in polycyclic aromatic hydrocarbons (*18*). *p*-Terphenyl is an example of a PAH displaying localized π-sextets (Figure 2a).

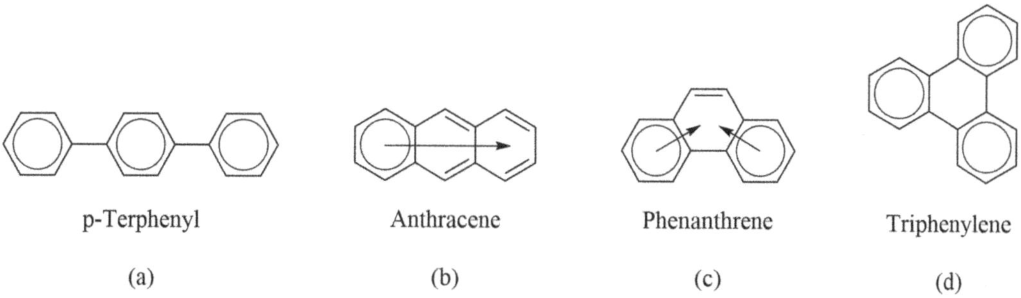

Figure 2. Examples of the different types of aromatic rings according to Clar's rule.

According to Clar's rule, the Kekulé resonance structure displaying the largest number of disjoint aromatic sextets, that is, benzenoid-like rings, determines the properties of a given polycyclic aromatic hydrocarbon (*18*). In this way, if different Clar representations of a PAH are possible, the representation with the greater number of π-sextets would explain its observed characteristics, due to the greater stability offered by the increased number of sextets (*20*). Furthermore, PAHs with a higher fraction of aromatic sextet carbons are more stable than PAHs with a higher fraction of isolated double bonds.

In some PAHs, resonance makes the rings equal in terms of electron density, so it is not possible to assign the π-sextet to any ring alone. Instead, the π-sextet migrates between the rings. Anthracene is a clear example of a PAH having a migrating π-sextet (Figure 2b). In linear acenes, the aromatic character of a molecule is diluted with the increase in the number of aromatic rings (*16*, *17*). However, even in this case, it is possible for the anthracene structure to have higher stability by forming two π-sextets on the outer two rings, thereby giving the carbons on the 9,10-positions in the central ring some free-radical character. On the other hand, the presence of localized π-sextets in PAHs with fused angular rings results in the formation of true double bonds for the same reason. In the case of phenanthrene, the outer rings exhibit a higher local aromaticity compared to the central ring; the latter exhibits a double bond, which is as reactive as any olefinic bond ((*17*); Figure 2c). Aromatic rings without π-electrons, known as empty rings, are found in fully benzenoid structures (*19*). The central ring in triphenylene is an example of an empty ring (Figure 2d).

The use of anthracene as representative of linear PAH structures and of phenanthrene as representative of angular PAH structures therefore captures the essence of the influence of ring configuration on reactions, that is, free-radical versus double-bond character.

2.2. Anthracene

The reactivity of anthracene, the first of the structural motifs analyzed in this chapter, can be rationalized using Clar's rule. Due to its linear configuration, a migrating π-sextet is shared between the three aromatic rings, making their aromatic character comparable. According to the literature, the similar aromaticity of the anthracene rings has been confirmed using several indicators (*21*, *22*). Despite this, anthracene (and acenes in general) undergoes a significant number of reactions across the 9,10-positions of the central ring. By doing so, two π-sextets are formed in the final product, and Clar's rule shows its usefulness in predicting and explaining the reactivity of the 9,10-positions.

The reactivity of the *para* position, which is caused by the localization of the π-electrons to form two π-sextets, is important in the addition reactions of acenes (*16*). In this way, the central ring of anthracene is protonated, adds bromine, and undergoes a Diels–Alder reaction readily (*18*). Similarly, in the case of autoxidation, oxygen preferentially attacks the 9,10-positions in the anthracene molecule, forming a resonance structure with two π-sextets in the two outer rings and a diradical in the 9,10-positions. The formation of anthraquinone by autoxidation is a key step in one of the major processes for the industrial production of hydrogen peroxide (*23*). Considering that oxidation is at the core of the oxidative ring-opening concept, additional details on the autoxidation of anthracene will be presented in Section 3.

In a reductive environment, anthracene is readily hydrogenated to 9,10-dihydroanthracene (*24*). The central-ring hydrocracking of anthracene is capable of producing more monocyclic products at low hydrogen consumption than the terminal-ring hydrocracking (*25*). During hydroprocessing, opening of the central ring of anthracene represents an efficient route leading to the formation of more valuable intermediates with lower hydrogen consumption (*24*). Similarly, the ability of anthracene to transfer hydrogen on the 9,10-positions, that is, from the central ring, is observed during thermal conversion (*26*).

2.3. Phenanthrene

Clar's rule can also be applied to inspect the reactivity of phenanthrene, which is the second structural motif of interest for this chapter. In this case, two resonance structures are possible. The first one displays a π-sextet in the central ring and two double bonds in the outer rings. The second one exhibits π-sextets in the outer rings and a double bond in the central ring. The latter, which has a greater number of aromatic sextets, is the preferred one. The higher aromaticity of the outer rings has been observed by different methods (*21*, *22*). From a reactivity point of view, the presence of the π-sextets in the outer rings of phenanthrene makes the associated carbon atoms rather stable, whereas the double bond in the central ring is quite reactive. Thus, it is not surprising that most reactions in phenanthrene occur at the 9- and 10-positions (*18*, *27*).

Industrially, the oxidation of phenanthrene is the basis for the production of 9,10-phenanthrenequinone. This compound is used as an intermediate in the production of dyes, as well as in the synthesis of herbicides and pharmaceuticals (*27*). Processes using different oxidizing agents have been developed. Some examples are liquid-phase oxidation using chromates, catalytic oxidation with oxygen, and with *tert*-butyl hydroperoxide (*27*). Similar to anthracene, during the autoxidation of phenanthrene, oxygen preferentially attacks the carbons in the 9,10-positions, in which oxygen is added to a carbon–carbon double bond. Section 3 presents details on this type of chemistry. Hydrogenation in the presence of Raney nickel catalysts produces 9,10-di-, tetra-, octa-, or perhydrophenanthrene (*27*).

Compared to anthracene, phenanthrene is known to be more stable by about 17–33 kJ/mol (4–8 kcal/mol) (*28*). This difference in thermochemical stability is related to differences in their aromaticity. An analysis of the structures in view of Clar's formalism shows that phenanthrene has two aromatic sextets.

3. Autoxidation and Free-Radical Chemistry

The chemistry of the liquid-phase autoxidation (oxidation with air) of hydrocarbons has been extensively studied and documented (*29*).

It is generally accepted that in the absence of a catalyst, autoxidation proceeds via a free-radical mechanism involving the typical steps of initiation, propagation, and termination. Autoxidation is characterized by the presence of an induction period, autocatalytic propagation, and zero-order kinetics with respect to oxygen, irrespective of the hydrocarbon (*30*). The induction period refers to the time required for sufficient free radicals to form so that the oxidation reaction can proceed at a significant rate. The induction time depends on the compound classes (*31*), as well as experimental conditions, such as temperature and gas–liquid interfacial area per unit volume liquid (*32*). In saturated hydrocarbons, two slow steps contribute to the induction period of autoxidation. First, there is the abstraction of hydrogen by oxygen to form a free-radical hydrocarbon intermediate; once this reaction has taken place, further oxidation to produce a hydroperoxide is comparatively fast. Second, there is the decomposition of the hydroperoxide, which can be accelerated (*33*) and determines the rate of subsequent oxidation.

The chemistry of the initiation step of aromatic oxidation is discussed in more detail, because there are differences compared to the autoxidation of saturated hydrocarbons. The possibility of hydrogen abstraction by radical species in aromatic compounds without alkyl chains is small because of the strong nature of the sp^2 C–H bonds (*34*). The dissociation energies of the sp^2 aromatic C–H bonds and sp^3 aromatic C–CR_2–H are 469 and 377 kJ/mol (112 and 90 kcal/mol), respectively (*35, 36*). This added "difficulty" in the abstraction of hydrogen, paired with the electronic configuration of aromatic rings, translates into a different sequence in the initiation of the oxidation of aromatic hydrocarbons compared to that of saturated hydrocarbons. The oxidation behavior of anthracene and phenanthrene is used to illustrate the point and should be viewed as representative of structures in PAHs.

3.1. Preoxidative Oxygen Addition to Polycyclic Aromatics

During autoxidation of anthracene, the resonance structure of the molecule is disturbed. The two π-electrons at the 9,10-positions become localized, and the oxygen is added to these nonadjacent carbons through a *para* attack ((*37*); Figure 3a). According to Clar's formalism, the formation of two π-sextets in the outer rings of the oxidized product favors the transannular addition of oxygen. Furthermore, this initial addition of oxygen causes the carbons in the 9,10-positions to change from sp^2 to sp^3 hybridized carbons. It is only after this "preoxidation" step takes place that the hydrogen in the sp^3 carbon is abstracted or transferred to continue the oxidation reaction (*14*). It is also this "preoxidation" step that facilitates the formation of hydrogen peroxide from the oxidation of 9,10-dihydroxyanthracene to anthraquinone.

Autoxidation of phenanthrene exhibits similar characteristics. In this case, oxygen directly adds to one of the carbons of the double bond, the carbons in the 9,10-positions. By doing so, a diradical intermediate is initially formed: a free radical is located on one of the carbons involved in the double bond, while another radical is located on the second oxygen of the oxygen molecule (O_2) that was added (Figure 3b). In other words, the addition of oxygen to the double bond (olefinic bond) forms a C–O bond with one of the carbons of the C=C group and a carbon-centered radical on the other carbon. The hydrogen in these carbons, which are now aliphatic carbons, can be abstracted or transferred for the oxidation reaction to take place.

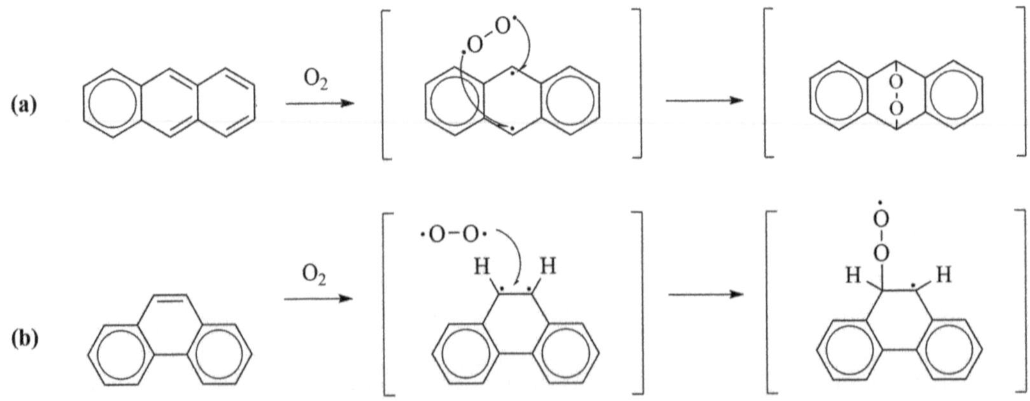

Figure 3. Preoxidative oxygen addition of (a) anthracene and (b) phenanthrene.

Detailed literature on the autoxidation of sp^3 C–H bonds is readily available. However, little has been published on the oxidation of aromatic sp^2 C–H bonds (*30*). This disparity in the existing information could lead us to assume that both oxidation chemistries follow the same steps. However, the use of Clar's rule highlighted two important and potentially overlooked aspects that are different in the "preoxidative" behavior during aromatic autoxidation. First, the interaction of oxygen with aromatic hydrocarbons depends on their structure. As a result, oxidation chemistries that are fundamentally different are possible, here transannular oxygen addition and oxygen addition to a carbon–carbon double bond. Second, hydrogen abstraction is not necessary for the preoxidative step in aromatic oxidation. The oxygen initially acts as the electron acceptor, while the aromatic molecule acts as the electron donor (*14, 37*).

Understanding the fundamental differences in autoxidation chemistry of the linear and angular structural motifs is relevant to the development of the oxidative ring-opening of polycyclic aromatics as a process.

3.2. Oxidation to Quinonoids

The first step in the oxidative ring-opening process targets the oxidation of PAHs to quinonoid species. In the particular case of anthracene and phenanthrene, this translates into the production of anthraquinone and phenanthrenequinone, respectively.

Autoxidation of anthracene to anthraquinone looks as though it is a rather straightforward reaction. Oxygen adds through a *para* attack to the 9,10-carbons of anthracene, leading to the formation of the carbonyl groups. This type of *para* attack has also been suggested as a pathway to explain the formation of maleic anhydride from benzene (*38*). It is less clear whether the main reaction pathway proceeds through hydroperoxide decomposition, leading to stepwise oxidation, or whether the main reaction pathway involves $2O_2$ with concerted elimination of hydrogen as hydrogen peroxide. Both pathways are potentially active. Intermediate products such as anthrone and oxanthrone have been detected in the oxidation of anthracene (*39*). This could have been viewed as evidence of stepwise oxidation, but anthraquinone is not immune to thermal decomposition, and compounds such as anthrone were positively identified in the product of anthraquinone pyrolysis at 380 °C (*40*).

Autoxidation of phenanthrene to phenanthrenequinone, on the other hand, seems to be a more complex reaction. Although no detailed description of the steps following the initial addition of oxygen to the double bond was found in the literature, a plausible mechanism is presented here (Figure 4). Oxygen is very reactive to free-radical species once they are formed. It is therefore possible that once oxygen successfully attacks phenanthrene, forming a carbon-centered free radical, another oxygen molecule would attack the carbon-centered free radical. There does not appear to be any evidence that phenanthrene oxidation leads to alcohols or oxidation only on the one carbon. Thus, the hydrogen must be unavailable for free-radical transfer to species of the form RO•. Furthermore, the presence of diphenic anhydride was repeatedly detected during oxidation of phenanthrene with a metal oxide catalyst (*40*). There is also an evident analogy with the oxidation of cyclic ketones by peracids that leads to oxygen insertion (*41*), and evidence of such oxygen insertion was also found in petroleum-derived materials (*42*). Viewed together, these observations suggest that the pathway of oxidation involves $2O_2$. No direct proof for the reaction pathway in Figure 4 is provided here; nevertheless, it is able to successfully explain the products, as well as the absence of some products.

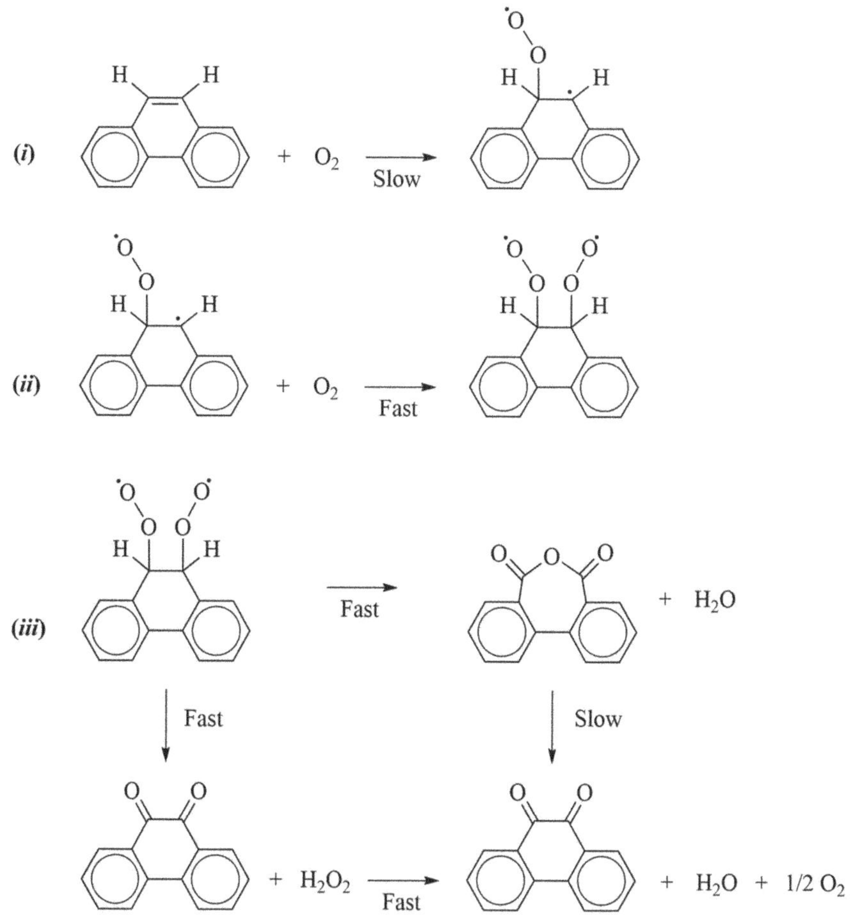

Figure 4. Plausible mechanism for the formation of 9,10-phenanthrenequinone by autoxidation of phenanthrene at elevated temperature. Only hydrogen atoms at the 9,10-positions are explicitly shown.

3.3. Oxidative Ring-Opening

Oxidative ring-opening involves the scission of a C–C bond. The product from oxidative ring-opening of a quinonoid is the corresponding dicarboxylic acid. There are a few observations that can immediately be made in this respect. First, the hydrogen that was lost during the oxidation of the PAHs to produce the quinonoids is returned to the molecule, but in association with oxygen. Second, the C–C bond that must be broken in angular PAHs is located between two aliphatic carbons, whereas the C–C bond that must be broken in linear aromatics is located between an aliphatic and an aromatic carbon, which is a stronger bond. Third, forming the dicarboxylic acid from the quinonoid of an angular structure requires scission of only one C–C bond, whereas the same reaction in a linear structure requires scission of two C–C bonds.

It is unrealistic to expect that aromatic ring-opening will be a selective autoxidation pathway. Some side reactions that do not lead to ring-opening will occur in parallel with ring-opening. Thermal conversion leading to free-radical addition and total combustion will obscure the conversion chemistry (*14*).

The importance of water in oxidative degradation is recognized (*42*), and water is likely one potential source of hydrogen for oxidative ring-opening. In carboxylic acid pyrolysis (*43*), but in the reverse, the formation of an anhydride or ester as an intermediate seems important (Figure 5). In the case of angular PAHs, such as phenanthrene, the anhydride is also an intermediate of oxidation to phenanthrenequinone, as shown in step (*iii*) of Figure 4. If elimination of oxygen from the anhydride does not take place, then hydrolysis of diphenic anhydride will form the ring-opened dicarboxylic acid species, biphenyl-2,2′-dicarboxylic. The reverse reaction, the dehydration of the dicarboxylic acid species to produce the anhydride, was observed in the pyrolysis product from biphenyl-2,2′-dicarboxylic (*44*).

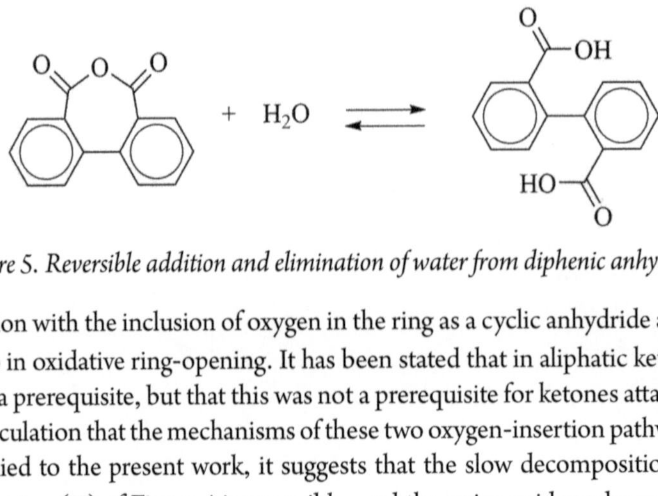

Figure 5. Reversible addition and elimination of water from diphenic anhydride.

Ring expansion with the inclusion of oxygen in the ring as a cyclic anhydride appears to be a key intermediate step in oxidative ring-opening. It has been stated that in aliphatic ketones the presence of α-hydrogen is a prerequisite, but that this was not a prerequisite for ketones attached to aromatics, leading to the speculation that the mechanisms of these two oxygen-insertion pathways were different (*41*). When applied to the present work, it suggests that the slow decomposition of an anhydride to a quinonoid in step (*iii*) of Figure 4 is reversible, and the quinonoid can be oxidized to form the anhydride.

It has been reported that oxygen exchange in quinonoids due to reaction with water is considerable. The carbonyl group reacts with water to form di-alcohol species. However, the reaction is reversible and favors the carbonyl. The dynamic oxygen exchange with water can be observed through oxygen-isotope labeling (*45*).

Further evidence that ring-opening requires oxidation of the quinonoid to the anhydride is found in the observation that anthraquinone decomposition around 450 °C was "not materially affected" by the presence of steam (*46*). Both anthraquinone and phenanthrenequinone were quite stable and could be vaporized without decomposition at around 350–360 °C (*46*). The inference is that for water to react with these quinonoids, they have to be oxidized. It is important to recognize that under different conditions, such as hydrothermal conditions, water affects the chemistry that takes place. While anthraquinone seems stable when reacted with water at supercritical conditions (no oxidant present), apart from oxygen exchange already mentioned (*45*), phenanthrenequinone readily reacts to form a series of ring-closed products, associated with CO elimination with a concomitant ring contraction (*47*). Another study showed that the oxidation of phenanthrenequinone under hydrothermal conditions, and using H_2O_2 as oxidant, led to ring-opening by forming a di-aldehyde, which is further oxidized to aromatic acids (*48*). The same study showed that hydrothermal conditions lead to the complete conversion of phenanthrene to carbon oxides and water.

Hydrolysis of the anhydride in the case of angular PAHs therefore appears to be the pathway by which hydrogen is restored to the oxidized species during oxidation. In the case of linear PAHs, hydrolysis of an ester-type bond also appears to be involved in some way. These observations do not rule out hydrogen-transfer reactions (*26*). In fact, the formation of products with higher molecular mass and carbonaceous appearance was observed during the pyrolysis of quinonoids (*40*).

The mechanistic steps that are involved in the attack of the quinonoid by oxygen were not clear from the literature, despite empirical evidence that it happens. The formation of diphenic anhydride by oxidation of phenanthrenequinone requires insertion of oxygen between two aliphatic carbons; it appears to be easier than the corresponding oxidation of anthraquinone. It was reported that "anthraquinone is a relatively inert compound exhibiting stability towards oxidation" (*49*). In fact, there is no simple pathway by which anthraquinone can form an anhydride in the central ring. Yet, diphenic anhydride was positively identified as a decomposition product of anthraquinone (*40*), which suggests that this could be a bimolecular reaction that also involves exchange of phenyl groups.

In terms of selectivity, generating an anhydride as an intermediate from linear PAH oxidation, or even performing the oxidative ring-opening following a different pathway that is yet not identified, is more difficult and less selective than generating an anhydride from an angular PAH. The formation of quinonoids and anhydrides can be achieved by autoxidation with reasonable selectivity. However, ring-opening does not take place by oxidation only. Both oxidation and hydrolysis are required for ring-opening.

Making a C–C bond seems to be easier and is better documented than breaking it. In the liquid phase, free-radical addition takes place even at thermal conditions normally associated with cracking. Using thermal conversion to achieve a balance between oxidation and hydrolysis that leads to a high selectivity to ring-opening looks as though it may be an unrealistic target.

4. Catalytic Oxidation

Autoxidation of PAHs provides a suitable pathway for production of quinonoid and anhydride species; however, further oxidation leads to undesirable side products, such as ring-closed products and heavier products (*50*). Some of these side products may also be formed during the autoxidation of the PAHs to quinonoids. One potential way to limit side-product formation is to perform the oxidation by catalysis. Catalysts are able to influence the relative rates of reaction, accelerating the elementary steps that lead to the desired products and interfering with the ones that form undesired products. In this way, the choice of catalyst, reaction conditions, and reaction engineering (e.g., heat management and reactor type) has a significant impact on oxidation selectivity and production rate (*51*).

Applied to the oxidative ring-opening of PAHs, oxidation catalysis was of interest because it could potentially achieve good selectivity to anhydrides or enable oxidation with ring-opening to carboxylic acids without the elimination of water.

4.1. Catalytic Oxidation of Aromatics

Most of the knowledge on catalytic oxidation of aromatic compounds originates from industrially successful processes, which received a lot of attention in the first half of the 20th century. These processes include (*i*) oxidation of benzene to maleic anhydride, (*ii*) oxidation of naphthalene to phthalic anhydride, and (*iii*) oxidation of anthracene and phenanthrene to anthraquinone and phenanthrenequinone,. In all cases, vapor-phase oxidation of the corresponding aromatic feedstock results in the production of anhydride and quinonoid compounds that are high-value intermediates in the synthesis of polymers and other chemicals. A review on the topic can be found in reference (*14*). Note that the products from these catalytic oxidation reactions (anhydrides or quinonoids) correspond to the compound classes that are intermediates in oxidative ring-opening of PAHs.

The major products from vapor-phase catalytic oxidation of benzene, naphthalene, anthracene, and phenanthrene are summarized in Table 2. There is an increase in the complexity of the product spectrum with an increase in the number of rings in these aromatic hydrocarbons. Oxidation to the anhydride takes place more readily when a –(CH)=(CH)– substructure is present. Anhydrides are major products from outer-ring oxidation or internal-ring oxidation of an angular PAH, such as phenanthrene. The central ring of anthracene, which does not have a –(CH)=(CH)– substructure, does not lead to appreciable anhydride formation.

The products listed in Table 2 resemble the product description of autoxidation. However, it was early realized that the color change of catalysts that took place during the oxidation of aromatics was associated with a change in the oxidation state of the catalytic materials (*46*). Catalytic oxidation was a redox process, and metal oxides, selected mainly from the transition metals, dominate the catalyst formulations employed for aromatic oxidation. The different oxidation states available in these materials make it possible to control the selectivity during oxidation (*52*).

For benzene oxidation, catalysts are based on mixtures of V_2O_5 and MoO_3 (*53*). Moreover, oxides or salts of elements such as Ti, Ni, Cu, Ag, P, Bi, and W are often used as promoters (*54*). Naphthalene is more readily oxidized than benzene. Catalysts based on vanadium oxides are quite active and can cause overoxidation of naphthalene to maleic anhydride and carbon oxides (*46*). Therefore, the addition of less-active oxides is used to regulate vanadium oxide activity. Mixtures of V_2O_5 and K_2SO_4 supported on SiO_2 have been used at commercial scale for naphthalene oxidation, achieving selectivities close to 90% toward phthalic anhydride (*55*). Anthracene and phenanthrene are more readily oxidized than naphthalene, and catalyst activity must also be moderated to limit overoxidation. Industrially, oxidation of anthracene using catalysts consisting of V_2O_5, K_2SO_4, and Fe_2O_3 on pumice has yielded 95–97 wt % of 9,10-anthraquinone (*56*). Oxidation of the outer rings of anthracene takes place to a minor extent and results in products such as 1,4-anthraquinone, 2,3-naphthalic anhydride, and pyromellitic anhydride (*57*). Reaction of phenanthrene over V_2O_5 supported on SiO_2, with and without K_2SO_4, results in oxidation products of the inner and outer rings ((*58*); Table 2).

Table 2. Reaction Products from Vapor-Phase Catalytic Oxidation of Polycyclic Aromatic Hydrocarbons

Aromatic compound	Products of selective oxidation	Other oxidation products	Decomposition products
 Benzene	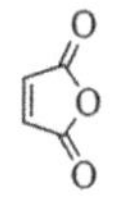 Maleic Anhydride		$CO + CO_2$
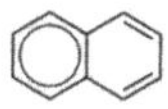 Naphthalene	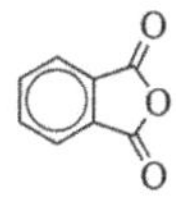 Phthalic Anhydride	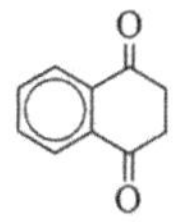 1,4-naphthenequinone	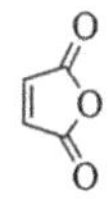 Maleic Anhydride $CO + CO_2$
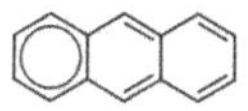 Anthracene	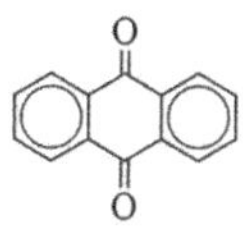 9,10-anthraquinone	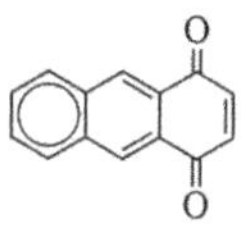 1,4-antraquinone 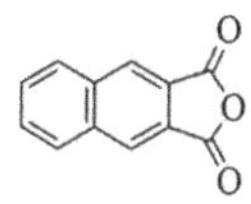 2,3-naphthalic anhydride 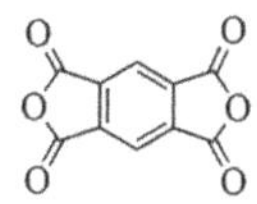 pyromellitic anhydride	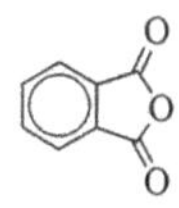 Phthalic Anhydride 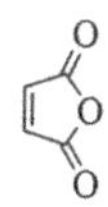 Maleic Anhydride $CO + CO_2$
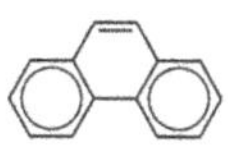 Phenanthrene	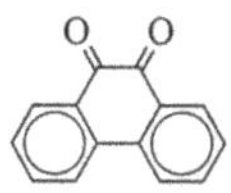 9,10-phenanthrenequinone	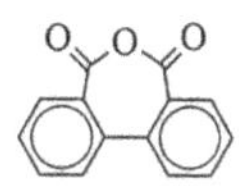 Diphenic anhydride 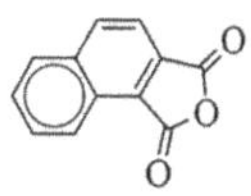 1,2-naphthalic anhydride 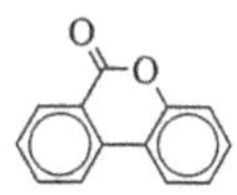 2-hydroxydiphenyl-2'carboxylic acid lactone 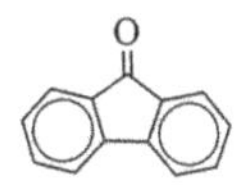 Fluorenone	

4.2. Mars–Van Krevelen Mechanism for Oxidation

When considering the role of a catalyst in the oxidation of PAHs, it is possible to conceive of different ways by which the oxidation can take place (Figure 6).

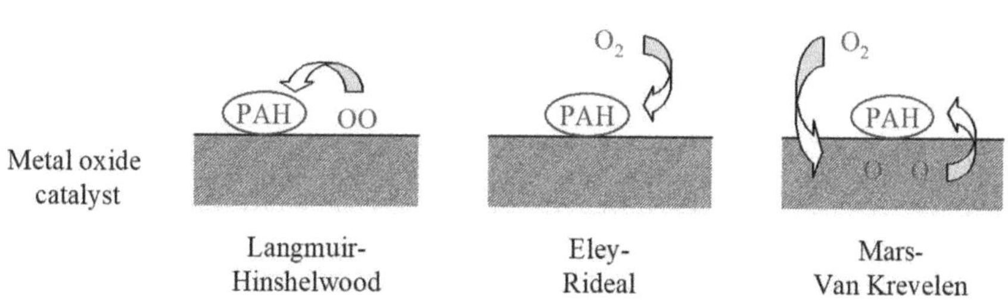

Figure 6. Potential catalytic oxidation pathways for the oxidation of PAHs by oxygen.

First, adsorption of both the PAH and O_2 on the catalyst in a way that facilitates attack of the adsorbed PAH by the adsorbed O_2 is possible, as in the Langmuir–Hinshelwood description. Second, PAH can adsorb on the catalyst in a way that facilitates attack from oxygen from the gas phase (the Eley–Rideal description). Finally, PAH can adsorb on the catalyst and lead to a redox reaction with the catalyst, which is reoxidized by O_2 as a separate step (the Mars–Van Krevelen description).

There is considerable experimental evidence that selective catalytic oxidation over metal oxides takes place by a redox reaction and that the dominant mechanism corresponds to the Mars–Van Krevelen description (originally in reference (59)). However, as Kung (52) points out, in practice, there may be more than one source of oxygen (different oxygen species may be present at the surface and lattice of the catalyst), each playing a role during the reaction. In addition, the presence of water vapor further complicates the description. A high-level explanation in terms of only the Mars–Van Krevelen mechanism hides the complexity of oxidation over metal oxides and the nature of the oxygen-containing species that is responsible for the oxidation.

A comprehensive discussion of the different types of oxygen species that could be present in a metal oxide catalyst can be found in the book by Bielański and Haber (53). They state that for selective oxidation to occur, as opposed to total oxidation, the oxygen must be present as the oxide ion (O^{2-}). If this description is accepted, it implies that the PAH must be activated by the catalyst for oxidation to proceed. The activation of the PAH must occur in such a way that the catalyst induces a partial positive charge on the carbons it is interacting with to provide a driving force for the reaction with O^{2-}, because O^{2-} species do not have oxidizing power.

The Mars–Van Krevelen mechanism considers two independent steps. In the first step, the PAH reacts with oxygen from the metal oxide lattice to give the oxidized product. Desorption of the oxidized product leaves an "oxygen vacancy" in the metal oxide catalyst, and it leaves the active site in a reduced state at the surface of the catalyst. In the second step, gas-phase oxygen adsorbs and reoxidizes the reduced catalyst.

The same occurs during oxidation of already-oxidized PAHs. Using phenanthrenequinone as an example, Figure 7 shows how an anhydride (diphenic anhydride) may be formed by the Mars–Van Krevelen type of oxidation. Although not indicated, it is likely that the oxygen in phenanthrenequinone also interacts with the metal ions to assist with stabilizing the intermediate.

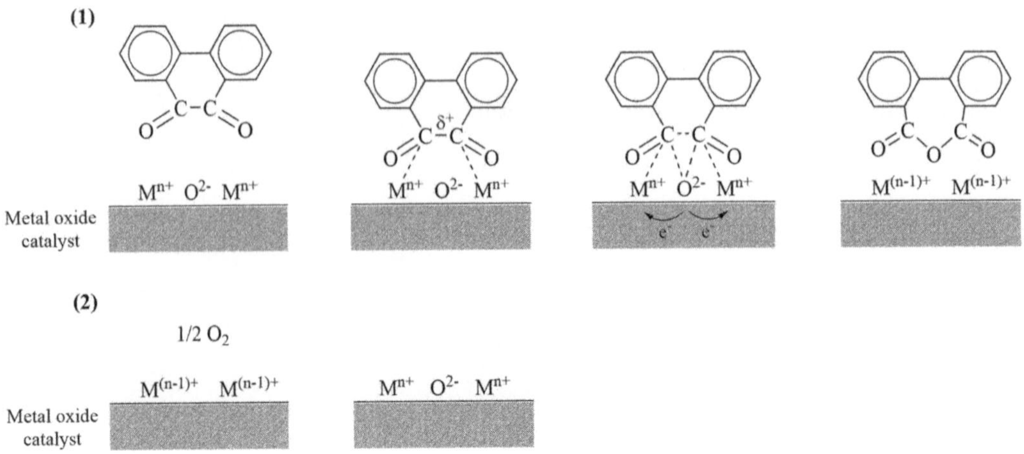

Figure 7. Simplified scheme for the formation of an anhydride according to the Mars–Van Krevelen mechanism.

An important aspect of aromatic oxidation by the Mars–Van Krevelen mechanism, which is not elucidated, is how hydrogen migrates during the reaction when a PAH is oxidized to produce a quinonoid. This was conveniently sidestepped in Figure 7 by showing only the oxidation of the quinonoid to the anhydride. According to the Mars–Van Krevelen description, the reaction should proceed by single oxygen (O^{2-}) addition. A few potential avenues by which the oxidation proceeds with respect to the fate of the hydrogen in the PAH can be conceived, but which pathways are followed is not clear from the literature: (*i*) Epoxide forms on the outer ring in angular PAHs, which does not require hydrogen migration. At the same time, epoxide formation is difficult to imagine as a pathway for internal-ring oxidation in linear PAHs. (*ii*) Hydrogen transfers from the carbon to which oxygen is added, to the oxygen that was added, to form an alcohol. (*iii*) Intramolecular transfer of the hydrogen to another carbon forms a $-CH_2-$ group. (*iv*) Hydrogen transfers from the carbon to which oxygen is added to the catalyst.

It should also be clear that further oxidation of anthraquinone to add oxygen is not easily explained by the mechanistic steps described.

The example in Figure 7 also hides the complexity of the metal oxide lattice, as well as the different steps that are involved in going from gas-phase O_2 to lattice O^{2-} species. This process has been described in terms of multiple steps that include O_2 adsorption, dissociation of the adsorbed O_2, and reactions leading to electron transfer to the oxygen that takes place at high temperature (*53*). In the process of oxygen incorporation into the oxide structure, various types of electrophilic activated oxygen species, such as O^- (oxide) and O_2^- (superoxide), form before the oxygen is incorporated into the oxide lattice in the form of nucleophilic oxygen, that is, O^{2-} species (*60*). O_2^- and O^- species are electron-deficient species that become strongly electrophilic reactants. They may abstract hydrogen from a hydrocarbon molecule to form alkyl radicals, which in turn might start a chain reaction leading to total combustion. The type of catalysts required for total combustion of PAHs in exhaust systems of combustion engines and furnaces is therefore different from that required for selective partial oxidation.

The two-step Mars–Van Krevelen mechanism seems to explain the selectivity achieved during oxidation using metal oxide catalysts. Moreover, it is actually beneficial that the steps take place separately. When both steps are conducted together, selectivity will be degraded unless reoxidation of the catalyst and formation of the O^{2-} are fast. More specifically, the formation of O^{2-} must be sufficiently fast to limit the presence of other oxygen species that could lead to destructive oxidation reactions.

4.3. Catalytic Oxidative Ring-Opening

As stated before, oxidation catalysis was of interest because it was considered likely to achieve high selectivity to anhydrides, or even assist oxidation with ring-opening to carboxylic acids without the elimination of water. Oxidation catalysis by metal oxides appears useful in producing quinonoid and anhydride species; however, it seems unlikely that ring-opening of PAHs can rely on selective catalytic oxidation alone. In other words, once anhydride species are formed, further oxidation seems more likely to yield ring-closed products. Oxidation of diphenic anhydride illustrates the point. If the addition of another oxygen to diphenic anhydride were possible, the subsequent removal of CO_2 would produce 3,4-benzocoumarin, an undesirable ring-closed product.

In this way, it is important to remember that selective oxidation is only part of the chemistry required for oxidative ring-opening of PAHs. Hydrolysis also plays a role. Oxidation of PAHs to quinonoids or anhydrides involves elimination of water as a reaction product (Figure 4). In the stoichiometry, it can be seen that subsequent ring-opening of the quinonoid or anhydride would require a source of hydrogen. One potential source of hydrogen is the water that was eliminated during oxidation (e.g., Figure 5). The same stoichiometric constraint found in autoxidation applies to oxidative ring-opening over a catalyst.

For catalytic oxidation to be successful at ring-opening, the water produced during oxidation should not be lost. This is possible if the hydrogen that is eliminated during oxidation is retained on the catalyst surface in the form of hydroxyl groups. The importance of water during catalysis over metal oxides is recognized (52) and it affects the catalysis. Metal oxides can catalyze hydration, dehydration, and hydrolysis. However, oxidation and hydrolysis take place by different pathways, and metal oxides that perform well for selective oxidation of PAHs by redox catalysis are not necessarily efficient catalysts for hydrolysis by acid–base catalysis. Furthermore, ester hydrolysis by an acid is a reversible reaction, whereas hydrolysis by a base is practically irreversible (*61*).

This highlights an opportunity for further investigation. Catalytic ring-opening of aromatics with metal oxides is not a straightforward reaction. It needs to accommodate selective oxidation as well as the role of water in forming carboxylic acid species. Only after decarboxylation of these carboxylic acid species can nonoxygenated ring-opened products be obtained.

5. Perspective on PAH Oxidation and Oxidative Ring-Opening

Conversion of PAHs plays a significant role in the upgrading of bitumen and heavy oil. As explained in the introduction, these bulky, hydrogen-deficient, and refractory molecules require strong conditions to yield lighter products, which is why thermal conversion and hydrogen addition are the typical technologies employed. Conceptually, there are alternatives, and Figure 1 showed how it might be possible to make use of oxidative ring-opening to convert PAHs. The steps involved in oxidative ring-opening were explored, leaning on what is known from the literature, as well as on insights from our own work. Oxidative ring-opening of PAHs is, as the saying goes, *easier said than done*.

Oxidation to obtain anhydrides and quinonoids as intermediates or intermediate products is key to successful oxidative ring-opening of PAHs. Autoxidation and catalytic oxidation offer pathways with different requirements to produce such species. Autoxidation requires the carbon atoms to assume a free-radical nature, which also causes a selectivity loss due to undesired reactions, such as ring closure, free-radical addition, and total combustion. Catalytic oxidation with metal oxides, on the other hand, works on the principle of a redox process in which no free radicals but charged species are involved. The surface chemistry of these charged species is complex, but it is also effective. In fact, it is what made vapor-phase selective oxidation of aromatics (accompanied by the cleavage of ring carbon bonds) an industrial success. It is also the type of chemistry that could aid the selective oxidation of PAHs.

Ring-opening to obtain carboxylic acid intermediates requires hydrolysis. The hydrolysis of anhydrides directly yields dicarboxylic acids; however, in the case of quinonoids, oxidation to the anhydride is required before the hydrolysis. In other words, oxidative ring-opening involves oxidation, but it is not purely an oxidative process. In fact, one could argue that the current name of the conversion strategy, *oxidative ring-opening*, does not yet reflect this subtle distinction.

To gain some perspective on the current state of knowledge and the prognosis for the application of oxidative ring-opening of PAHs, the following observations are made:

1. Oxidative ring-opening is a *selectivity quest*. Although the chemistry involved in each step, in principle, appears simple and widely studied, throughout our investigation both theoretical and experimental findings repeatedly indicated the opposite. The steps were neither simple nor sufficiently studied to be fully understood. There are nuances in the way that oxidation takes place that ultimately translate into a loss of selectivity.
2. Oxidation of linear and angular PAHs follows different reaction pathways that affect oxidation selectivity. There is a higher probability of producing ring-opened products from angular PAHs than from linear PAHs. Heavy oil and bitumen contain a combination of both types of structures.
3. Both autoxidation and catalytic oxidation have the ability to produce quinonoids and anhydrides from PAHs. However, catalytic oxidation has the ability to do so in a more selective way. Production of these species is useful, to a certain extent, but it does not equate to opening the aromatic ring, as would be the case in the formation of carboxylic acid species.
4. Ring-opening by the formation of the anhydride and hydrolysis of the anhydride to yield a dicarboxylic acid requires both *oxidation and hydrolysis* to proceed.

The discussion presented in this chapter simplified the conversion chemistry by looking at PAHs in isolation. In practice, purified PAHs are only found in petrochemical production, not in bulk petroleum refining. Heavy oils consist of larger fused ring systems, in which heteroatoms and coordinating metals are present. It is expected that these differences in the feedstock will lead to additional reactions, such as heavier addition product formation. It is also expected that once a source of hydrogen, like water, is integrated as part of the process, such reactions could be moderated. The importance of water in oxidative degradation is recognized. Issues related to the use of catalysts with heavy oil were not discussed either, although it is known from petroleum-refining practice that catalysts are readily fouled and deactivated by heavy petroleum fractions.

Despite the selectivity benefit conferred by catalysis, there are still many practical engineering challenges to overcome to apply oxidation catalysis to heavy oil and bitumen. The overall process in Figure 1 requires the combination of three conversion chemistries: oxidation, hydrolysis, and decarboxylation. We need to consider that these chemistries might require different reaction conditions, namely temperature and catalysts, so the reaction engineering has to accommodate all of them. One could envision a process where catalytic oxidation is used to selectively produce anhydride species as a primary product, which could be further converted and ring-opened after proper separation. Nevertheless, even if these three processes are achieved with acceptable selectivity, the high reaction temperatures derived from the stability of PAHs will make it impossible to completely avoid the contribution of free-radical chemistry and total combustion.

Progress in this field is hampered by lacking descriptions of some oxidation reactions, despite remarkable empirical industrial success in applying aromatic oxidation. Although oxidation is one of the major chemistries used in the petrochemical industry (not a small industry), it was surprising that fundamental aspects of oxidation were not properly developed. For instance, the formation of quinonoid and carboxylic acid species by oxidation of an aromatic substrate was not fully explained from a mechanistic point of view. These are important steps in processes such as fuel combustion and destruction of soot particulate matter from combustion. To this can be added some "enigmatic" observations, such as the formation of diphenic anhydride during oxidation of anthracene, which suggested some annular rearrangement, and that is clearly important in the oxidation of linear PAHs. In the case of autoxidation, the literature offered no significant comments regarding the initial addition of oxygen to aromatic hydrocarbons, or on the mechanistic transition from the preoxidative step of aromatics to aromatic oxidation. Oxygen addition to aromatics in the preoxidative step does not require hydrogen abstraction, unlike oxidation of aliphatic hydrocarbons.

The discussion presented in this chapter is in a way pointing to several gaps in a fundamental understanding of PAH oxidation, but it is also contributed to narrow them.

6. Conclusions

Oxidative ring-opening of polycyclic aromatic hydrocarbons (PAHs) was explored as a potential reaction pathway for upgrading of heavy oil and bitumen, while recognizing that it is also an important reaction step in combustion and treatment after combustion to limit PAH emissions and soot. Although there are still several gaps in our fundamental understanding of oxidative ring-opening of aromatics, the reaction is possible, and some conclusions relating to the conversion chemistry were as follows:

1. Autoxidation of PAHs proceeds through a preoxidative step that involves O_2 addition without hydrogen abstraction. The way in which O_2 is added to the aromatic structure depends on the ring configuration of the PAH.
2. In autoxidation, the steps following aromatic preoxidation have not yet been fully elucidated. It appears that addition of another O_2 molecule to the preoxidized aromatic takes place, which is then followed by rearrangement and elimination of hydrogen, as hydrogen peroxide or water. An anhydride forms as an intermediate product in at least one of the potential pathways.
3. When a quinonoid is formed by autoxidation, the oxidation step leading to the anhydride appears to be a reversible reaction. Elimination of oxygen from the anhydride to produce the quinonoid and oxygen addition to the quinonoid to produce the anhydride are possible.
4. Selective catalytic oxidation of PAHs to produce quinonoids and anhydrides is a redox reaction. Charged species instead of free-radical species offer a more selective pathway to obtain these compounds.
5. The individual steps involved in O^{2-} addition to explain selective catalytic oxidation of PAHs were incompletely described, also with respect to the migration of hydrogen.
6. Irrespective of whether autoxidation or catalytic oxidation is performed, ring-opening of PAHs proceeds through the formation and hydrolysis of anhydrides to yield carboxylic acid species. A combination of oxidation and hydrolysis chemistries is required.
7. Anhydride formation by oxidation of outer rings or internal rings of angular PAHs is more easily accomplished than anhydride formation of internal rings of linear PAHs.

References

1. Gray, M. R. *Upgrading Oils and Bitumen and Heavy Oil*; University of Alberta Press: Edmonton, AB, Canada, 2015.
2. Ramirez-Corredores, M. M. *The Science and Technology of Unconventional Oils*; Academic Press: London, 2017.
3. Kaiser, M. J.; De Klerk, A.; Gary, J. H.; Handwerk, G. E. *Petroleum Refining: Technology, Economics, and Markets*, 6 ed.; CRC Press: Boca Raton, FL, 2020.

4. Tissot, B. P.; Welte, D. H. *Petroleum Formation and Occurrence*; Springer-Verlag: New York, 1978.
5. Strausz, O. P.; Lown, E. M. *The Chemistry of Alberta Oil Sands, Bitumens and Heavy Oils*; Alberta Energy Research Institute: Calgary, AB, 2003.
6. Ancheyta, J.; Speight, J. G. *Hydroprocessing of Heavy Oils and Residua*; CRC Press: Boca Raton, FL, 2007.
7. Fahim, M. A.; Alsahhaf, A. A.; Elkilani, A. *Fundamentals of Petroleum Refining*; Elsevier: Amsterdam, 2010.
8. Van Krevelen, D. W. *Coal: Typology, Physics, Chemistry, Constitution*, 3 ed.; Elsevier: Amsterdam, 1993.
9. Torri, I. D. V.; Paasikallio, V.; Faccini, C. S.; Huff, R.; Caramao, E. B.; Sacon, V.; Oasmaa, A.; Zini, C. A. Bio-oil production of softwood and hardwood forest industry residues through fast and intermediate pyrolysis and its chromatographic characterization. *Bioresource Technol.* **2016**, *200*, 680–690.
10. Lyu, G.; Wu, S.; Zhang, H. Estimation and comparison of bio-oil components from different pyrolysis conditions. *Front. Energy Res.* **2015**, 3, 1–11.
11. Purcell, J. M.; Juyal, P.; Kim, D-G.; Rodgers, R. P.; Hendrickson, C. L.; Marshall, A. G. Sulfur speciation in petroleum: Atmospheric pressure photoionization or chemical derivatization and electrospray ionization Fourier transform ion cyclotron resonance mass spectrometry. *Energy Fuels* **2007**, *21*, 2869–2874.
12. Prado, G. H. C.; Rao, Y.; De Klerk, A. Nitrogen removal: A review. *Energy Fuels* **2017**, *31*, 14–36.
13. Javadli, R.; De Klerk, A. Desulfurization of heavy oil. *Appl. Petrochem. Res.* **2012**, *1*, 3–19.
14. Montoya Sánchez, N.; De Klerk, A. Autoxidation of aromatics. *Appl. Petrochem. Res.* **2018**, *8*, 55–78.
15. Waack, R. The stability of the "Aromatic Sextet". *J. Chem.* **1962**, *39*, 469–472.
16. Clar, E. *Polycyclic Hydrocarbons*, Vol 1; Academic Press: London, 1964.
17. Clar, E. *The Aromatic Sextet*; John Wiley & Sons: London, 1972.
18. Solà, M. Forty years of Clar's aromatic π-sextet rule. *Front. Chem.* **2013**, *1*, 1–8.
19. Balaban, A. T. Clar formulas: how to draw and how not to draw formulas of polycyclic aromatic hydrocarbons. *Polycycl. Aromat. Compd.* **2004**, *24*, 83–89.
20. Ruiz-Morales, Y.; Mullins, O. Polycyclic aromatic hydrocarbons of asphaltenes analyzed by molecular orbital calculations with optical spectroscopy. *Energy Fuels* **2007**, *21*, 256–265.
21. Cyranski, M. C.; Stepién, B. T.; Krygowski, T. M. Global and local aromaticity of linear and angular polyacenes. *Tetrahedron* **2000**, *56*, 9663–9967.
22. Schleyer, P. V. R.; Manoharan, M.; Jiao, H. J.; Stahl, F. The acenes: is there a relationship between aromatic stabilization and reactivity? *Org. Lett.* **2001**, *3*, 3643–3646.
23. Kirchner, J. R. Hydrogen Peroxide. In *Kirk–Othmer Encyclopedia of Chemical Technology*, Vol. 13, 3 ed.; Wiley: New York, 1981; pp 12–38.
24. Fan, H.; Han, B.; Jiang, T.; Guo, J.; Wang, Q.; Cheng, Y.; Wu, S. Hydrocracking of anthracene to ethyl biphenyl promoted by coupling supercritical water and cracking catalysts. *Chem. Cat. Chem.* **2011**, *3*, 1474–1479.

25. Collin, G.; Höke, H.; Talbiersky, J. Anthracene. In *Ullman's Encyclopedia of Industrial Chemistry*; Wiley-VCH Verlag GmbH & Co: Weinheim, Germany, 2012.
26. Rüchardt, C.; Gerst, M.; Ebenhoch, J. Uncatalyzed transfer hydrogenation and transfer hydrogenolysis: Two novel types of hydrogen-transfer reactions. *Angew. Chem., Int. Ed.* **1997**, *36*, 1406–1430.
27. Schmidt, R.; Griesbaum, K.; Behr, A.; Biedenkapp, D.; Voges, H. W.; Garbe, D.; Paetz, C.; Collin, G.; Mayer, D.; Höke, H. Hydrocarbons. In *Ullman's Encyclopedia of Industrial Chemistry*; Wiley-VCH Verlag GmbH & Co: Weinheim, Germany, 2012.
28. Poater, J.; Duran, M.; Solà, M. Aromaticity determines the relative stability of kinked vs. straight topologies in polycyclic aromatic hydrocarbons. *Front. Chem.* **2018**, *561* (6), 1–14.
29. Emanuel, N. M.; Denisov, E. T.; Maizus, Z. K. *Liquid-Phase Oxidation of Hydrocarbons*; Plenum Press: New York, 1967; translated by B. J. Hazzard.
30. Suresh, A. K.; Sharma, M. M.; Shridhar, T. Engineering aspects of industrial liquid-phase air oxidation of hydrocarbons. *Ind. Eng. Chem. Res.* **2000**, *39*, 3958–3997.
31. Scott, G. *Atmospheric Oxidation and Antioxidants*; Elsevier: Amsterdam, 1965.
32. Siddiquee, M. N.; Wu, Y.; De Klerk, A.; Nazemifard, N. The impact of microfluidic reactor configuration on hydrodynamics, conversion and selectivity during indan oxidation. *J. Flow Chem.* **2020**, *10*, 647–660.
33. Sheldon. R. A.; Kochi, J. K. *Metal-Catalyzed Oxidations of Organic Compounds*; Academic Press: New York, 1981.
34. Kholdeeva, O. A. Selective oxidation of aromatic rings. In *Arene Chemistry: Reaction Mechanisms and Methods for Aromatic Compounds*; Mortier, J., Ed.; John Wiley & Sons: Hoboken, NJ, 2016; pp 365–398.
35. Blanksby, S. J.; Ellison, G. B. Bond dissociation energies of organic molecules. *Acc. Chem. Res.* **2003**, *36*, 255–263.
36. Warren, J. J.; Tronic, T. A.; Mayer, J. M. Thermochemistry of proton-coupled electron transfer reagents and its implications. *Chem. Rev.* **2010**, *110*, 6961–7001.
37. Tipson, R. S. *Oxidation of Polycyclic Aromatic Hydrocarbons. A Review of the Literature*; National Bureau of Standards Monograph 87; U.S. Government Printing Office: Washington, DC, 1965.
38. Shtern, V. Y. *The Gas Phase Oxidation of Hydrocarbons*; Pergamon Press: New York, 1964.
39. David, B.; Boule, P. Phototransformation of PAHs Adsorbed on SiO_2 in Aqueous Suspension. In *Polycyclic Aromatic Compounds: Synthesis, Properties, Analytical Measurements, Occurrence and Biological Effects*; Garrigues, P.; Lamotte, M., Eds.; Gordon and Breach Science Publishers: Yverdon, 1993; pp 127–133.
40. Montoya Sánchez, N.; De Klerk, A. *Unpublished results*.
41. Lawesson, S-O.; Schroll, G. Peracids and Peresters. In *The Chemistry of the Carboxylic Acids and Esters*; Patai, S., Ed.; Interscience: London, 1969; pp 669–703.
42. Montoya Sánchez, N.; De Klerk, A. Low-temperature oxidative asphaltenes liquefaction for petrochemicals: fact or fiction? *Appl. Petrochem. Res.* **2016**, *6*, 97–106.
43. Fuson, R. C. Formation of Aldehydes and Ketones from Carboxylic Acids and Their Derivatives. In *The Chemistry of the Carbonyl Group*; Patai, S., Ed.; Interscience: London, 1966; pp 211–232.

44. Montoya Sánchez, N.; De Klerk, A. Oxidative ring-opening of aromatics: Decomposition of biphenyl carboxylic acids and zinc biphenyl carboxylates. *Energy Fuels* **2015**, *29*, 7910–7922.
45. Samuel, D.; Silver, B. L. Oxygen isotope exchange reactions of organic compounds. *Adv. Phys. Org. Chem.* **1965**, 3, 123–186.
46. Marek, L. F.; Hahn, D. A. *The Catalytic Oxidation of Organic Compounds in the Vapor Phase*; ACS Monograph Series 61; Chemical Catalog Company: New York, 1932.
47. Katritzky, A. R.; Ignatchenko, E. S.; Allin, S. M.; Barcock, R. A. Aqueous high-temperature chemistry of carbo- and heterocycles. 29. Reactions of aryl hydrocarbons, aryl N-oxides, and aryl carbonyl compounds in supercritical water at 460 °C. *Energy Fuels* **1997**, *11*, 160–173.
48. Onwudili, J. A.; Williams, P. T. Reaction mechanisms for the decomposition of phenanthrene and naphthalene under hydrothermal conditions. *J. Supercrit. Fluids* **2007**, *39*, 399–408.
49. Cofrancesco, A. J. Anthraquinone. In *Kirk-Othmer Encyclopedia of Chemical Technology*; John Wiley & Sons, 2000.
50. Montoya Sánchez, N.; De Klerk, A. Oxidative ring-opening over metal oxides. *Prepr. Pap.-Am. Chem. Soc., Div. Energy Fuels* **2014**, *59* (2), 558–561.
51. Lücke, B.; Martin, A. The Future Role of Aromatics in Refining and Petrochemistry. In *Proc. DGMK-Conf. Petrochem.*, Erlangen, Germany, 1999; p 139.
52. Kung, H. H. *Transition Metal Oxides*; Studies in Surface Science and Catalysis 45; Elsevier: Amsterdam, 1989; pp 169–199.
53. Bielański, A.; Haber, J. *Oxygen in Catalysis*; Marcel Dekker: New York, 1991.
54. Trivedi, B. C.; Culbertson, B. M. *Maleic Anhydride*; Plenum Press: New York, 1982.
55. Dálessandro, A. F.; Farkas, J. The kinetics of the catalytic oxidation of naphthalene. *Coll. Sci.* **1956**, *11*, 653–670.
56. Dixon, J. K.; Longfield, J. E. Hydrocarbon Oxidation. In *Catalysis: Oxidation, Hydration, Dehydration and Cracking Catalysts*, Vol. VII; Emmett, P. H., Ed.; Reinhold: New York, 1960; pp 183–280.
57. Brückner, A.; Baerns, M. Selective gas-phase oxidation of polycyclic aromatic hydrocarbons on vanadium oxide-based catalysts. *Appl. Catal. A* **1997**, *157*, 311–334.
58. Brooks, J. D. The catalytic oxidation of phenanthrene. *J. Appl. Chem.* **1955**, *5*, 250–260.
59. Mars, P.; Van Krevelen, D. W. Oxidations carried out by means of vanadium oxide catalysts. *Chem. Eng. Sci. Special Suppl.* **1954**, 3, 41–59.
60. Centi, G.; Cavani, F.; Trifiro, F. *Selective Oxidation by Heterogeneous Catalysis*; Kluwer Academic/Plenum Publishers: New York, 2001.
61. Elam, E. U. Esters, organic. In *Kirk–Othmer Encyclopedia of Chemical Technology*, Vol. 9, 3 ed.; Wiley: New York, 1981; pp 311–337.

Chapter 9

Cracking of Heavy Oil over a Catalyst Synthesized from Fly Ash

D. Venkata Ramakrishna,[1] Lakshmana Rao Jeeru,[2] and Narayan C. Pradhan[*,1]

[1]Department of Chemical Engineering, Indian Institute of Technology, Kharagpur 721 302, India

[2]School of Petroleum Technology, Pandit Deendayal Petroleum University, Gandhinagar 382007, India

[*]Email: ncp@che.iitkgp.ac.in

Fly ash containing silica and alumina as the major constituents is a cheap starting material for synthesizing zeolites, which are useful as solid acid catalysts. In the present work, fly ash containing 30% Al_2O_3 and 55% SiO_2 was converted to X-type zeolite using alkali fusion followed by hydrothermal treatment. The synthesized zeolite was characterized for its composition, surface area, and acidity. Cracking of vacuum gas oil (VGO) was carried out over the synthesized zeolite at an atmospheric pressure in the temperature range of 400–450°C. The catalyst-to-oil ratio (w/w) was maintained in the range of 0.2–0.5. The VGO conversion was found to increase with temperature. The gasoline yield initially increased with temperature and then decreased. A maximum gasoline yield of 24.5% was obtained at 425°C. The yield of gaseous products increased with temperature, whereas that of heavy gas oil decreased with temperature. The effects of space-time on VGO conversion and on the yields of different products were studied. More gas and coke were produced at a higher space-time. The performance of the catalyst synthesized from fly ash compared well with the commercial catalysts used in Indian refineries. Kinetic modeling of the cracking process was performed using a lump parameter model and the kinetic constants as well as the deactivation constant were estimated. Activation energies for each reaction were determined and compared with the literature values. The results are of immense importance as fly ash, a solid waste material, could be used to synthesize a valuable catalyst for catalytic cracking practiced in refineries worldwide.

© 2021 American Chemical Society

1. Introduction

During the past 60 years, a group of crystalline aluminosilicates has attracted broad interest among scientists and catalyst manufacturers. These materials, called zeolites, represent a series of solids with properties that are highly desirable in catalyst formulations (*1*). They have high adsorption capacities, high surface areas, and can be made to contain acidic sites of varying strengths.

Despite a large and growing number of distinctive structures and compositions, only a handful of zeolites have achieved commercial importance. These zeolites have proven their importance in petroleum (hydrocarbon) processing and there is a considerable indication that their role will continue to expand worldwide as an effective utilization of an energy resource base. Catalytic cracking is one of the most important petroleum refining processes practiced throughout the world. The first cracking catalysts used on a large scale commercially were synthetic alumina and magnesia (*2*). In the early 1980s, zeolites with rare earth ions replacing sodium ions were added to the amorphous matrix to improve catalyst activity and selectivity (*3–5*). The new catalyst had remarkably higher activity, much better selectivity for gasoline, and higher hydrothermal stability than the amorphous catalysts used at that time. The influence of the zeolite on the yield and selectivity of the cracking catalyst has been reviewed by Magee and Blazek (*6*). Their results can be summarized in part by noting that an increase in the amount of the zeolite present in the catalyst increases the yield of gasoline and light cycle oil, while coke and dry gas makeup are decreased at any given level of conversion. Simultaneously, aromatic content and the octane number of the gasoline fraction increase.

The early workers in the area of catalytic cracking took up the study of pure component cracking intending to establish the processes and mechanisms involved. The field was reactivated later by numerous authors (*7–10*),who have approached the topic with modern analytical, computational, and theoretical knowledge, giving due importance to thermal reactions and catalyst decay. The difficulty in understanding the cracking mechanism of gas oil lies in the involvement of a large number of molecules in the feedstock and products, which react through numerous pathways with at least 10^4 rate constants. From studies on gas oil cracking (*11–14*), it is clear that the assessment of rate constants experimentally for each reactant and intermediate is unrealistic and impractical. Some level and type of lumping is required to study this type of reaction. In principle, the information required for each reaction in cracking of gas oil may be generated through pure component and simple mixture cracking (*15*). Paraffins are generally used to achieve this goal. From the beginning of catalytic cracking operations, cracking of lower hydrocarbons drew the interest of investigators, mainly for academic reasons. Many homologous series of paraffinic compounds are admirably suited to systematic studies of the effects of molecular structure on the cracking behavior of catalysts. In view of this, cracking of *n*-paraffins has been extensively investigated on a wide variety of catalysts (*8*, *12*, *16*). These studies help discern the mechanism of gas oil cracking.

Prior investigations of kinetics in catalytic cracking reactions have focused on both the reactions of complex molecules in the feedstocks and the reactions of single compounds. Model compound studies focus on determining reaction mechanisms and rate constants, while work on actual feedstocks is directed toward empirically predicting product distribution through either rate equations (*17*) or the development of a lumped kinetic model (*18*). Both models have their advantages and limitations. However, the optimal and flexible operation of catalytic cracking is most important and can be achieved through the development of fundamental kinetic models. Mohammad et al. (*19*) experimentally determined the deactivation constant of a fluid catalytic cracking (FCC) process. Kinetic modeling was performed using a four-lump model and an experimentally determined deactivation constant. They reported significant improvement in predicted product yields. Activation energies for each reaction were reported and compared with literature values.

Ojha et al. (*20*) developed a method to prepare an X-type zeolite from fly ash and found optimum conditions for the best zeolite product. They have studied fusion temperature, fusion time, aging time, and hydrothermal treatment as well as the effects of alkali-to-fly ash ratio, curing temperature and curing time, and acid (HCl) treatment on the formation of an X-type zeolite.

Ojha and Pradhan (*21*) studied the catalytic activity of treated fly ash in the cracking of a heavy oil fraction obtained from North Gujarat (India) heavy crude oil and have reported a similar conversion (>55%) as that obtained from commercial 13X zeolite. From X-ray diffraction (XRD) and scanning electron microscopic (SEM) studies, they have confirmed the formation of X-type zeolite from fly ash from alkali fusion followed by hydrothermal treatment.

It is thus evident that no information has been provided in the literature so far on the use of nonconventional catalysts, such as treated fly ash, apart from one study (*21*). Fly ash containing silica and alumina as the major constituents is a cheap starting material for synthesizing zeolites, which are useful as solid acid catalysts. The exploration of the catalytic activity of waste material fly ash via conversion to zeolite for a commercially important reaction was therefore thought to be desirable. A further objective of this study was to develop a suitable kinetic model of the reaction system catalyzed by zeolite synthesized from fly ash.

2. Experimental

2.1. Materials

The main raw material, coal fly ash samples, was collected from the electrostatic precipitator of Kolaghat Thermal Power Station (West Bengal, India). Commercial 13X zeolite was obtained from Sisco Research Laboratories Pvt. Ltd. (Mumbai, India). Sodium hydroxide was procured from Glaxo Laboratories Ltd (India).

Commercial FCC catalysts were obtained from Bharat Petroleum Corporation Ltd. (BPCL, Mumbai, India) and Indian Oil Corporation Ltd. (IOCL, Haldia Refinery, Haldia, India). Vacuum gas oil (VGO) and actual FCC feed oil were also obtained from IOCL (Haldia Refinery, Haldia, India).

2.2. Synthesis of Zeolite

The fly ash samples were first screened through a Tyler sieve of 80-mesh size to eliminate larger particles. The chemical composition and surface area of the fly ash samples are presented in Table 1. The unburnt carbon (4–6%), along with other volatile materials present in fly ash, was removed by calcination at 800°C (±10°C) for 2 h. The fly ash sample was further treated with hydrochloric acid to increase its activity in zeolite formation. The acid treatment helps to dealuminate the fly ash and remove iron, thereby increasing the activity, thermal stability, and acidity of the zeolite for catalytic applications (*12*). The zeolite catalyst was then synthesized from the acid-treated fly ash using a hydrothermal treatment method (*20*). The synthesis conditions of the catalyst used in the present study were as follows: NaOH:fly ash (w/w), 1.3; fusion temperature, 550 °C; aging time, 18 h; and hydrothermal treatment temperature, 90 °C. The zeolite synthesized under these conditions was designated as ZOP-31.

2.3. Modification of the Synthesized Zeolite

The synthesized zeolite was modified by exchanging its Na^+ ions with H^+ (HZOP-31) as the H-form of the zeolite was found to be more active than the Na-form. For this, the freshly prepared zeolite was refluxed with a 5% NH_4NO_3 solution for 18h in three steps, each with a 6 h duration. In between two refluxes, the catalyst was filtered and calcined at 350 °C for deammoniation to yield the H-form of the zeolite. The XRD pattern of the H-form of zeolite matched exactly with that of the parent zeolite, indicating no structural change during ion exchange.

2.4. Characterization

2.4.1. XRD

The X-ray (powder) diffraction pattern of any crystalline material is the fingerprint of its structures. XRD patterns of different fly ash samples and synthetic zeolitic materials were obtained using a Philips X-ray diffractometer (Philips BW 1710). Operating conditions involved the use of Co Kα radiation at 4kV and 30mA. The samples were scanned from 10–50° (2θ, where θ is the angle of diffraction). Various crystalline phases present in the samples were identified with the help of Joint Committee on Powder Diffraction Standards (1967) files for inorganic compounds.

2.4.2. Microanalysis by SEM

The bulk composition was also estimated from scanning electron microscope SEM/EDXS (JEOL, JSM 5800) via an indirect method. The elemental composition of the samples was first determined from the SEM/EDXS and from these data, oxide percentages were calculated.

2.4.3. Determination of the Acidity of the Catalysts

The acidity (both Brønsted and Lewis) of the synthesized catalyst as well as the commercial catalysts, namely BPCL and Haldia FCC catalysts, was determined by the following established methods (22, 23).

2.4.4. Protonic/Brønsted Acidity

KI (0.025 M), KIO_3 (0.025 M), and sodium thiosulphate (0.025 M) were added (20 mL each) to 0.5 g of the catalyst sample in a conical flask. The catalyst was allowed to soak in the mixture for 48 h. The excess of thiosulfate was then titrated against anI_2 solution (0.008 M) using a starch indicator. At the end, the pale yellow solution turned blue. The protonic acidity of the catalyst was calculated from the volume of standard thiosulfate solution consumed by the liberated iodine. The thiosulfate solution was earlier standardized against a standard potassium dichromate solution. Protonic acidity was expressed as meq/g of catalyst.

2.4.5. Lewis Acidity

To determine Lewis acidity, 0.5 g of the catalyst sample was soaked in 25 mL of n-butyl amine (0.025 M)for 24 h. A supernatant liquid (10 mL)was pipetted out and titrated against trichloroacetic acid using 0.1% neutral red as the indicator. Trichloroacetic acid was standardized earlier using a standard n-butyl amine solution with neutral red as the indicator. All solutions were prepared in benzene. At the end, the green-colored solution turned a pink-red color.

2.5. Experimental Setup and Procedure

The catalytic cracking experiments were carried out in a stainless steel tubular reactor (2.5 cm in diameter and 30 cm in length). A schematic diagram of the experimental setup is shown in Figure 1. Before each experimental run, 10–15 g of catalyst was positioned at the center of the reactor and supported by inert ceramic rings on both sides. The ceramic packing in the upstream also helps in the uniform distribution of the feed on the catalyst bed. The temperature of the reactor bed was determined by a thermocouple placed in a thermowell extended from the top to the center of the reactor bed. A preheater was used to vaporize the feed before entering into the reactor. The feed was introduced into the preheater by a metering pump. Nitrogen, the inert carrier gas, was mixed with the feed in the preheater to dilute the feed. The catalyst was activated in situ for 2 h at 350°C in an atmosphere of nitrogen before the experimental runs began. Each experiment was conducted for 60–300 s to study the catalyst's activity and its time-on-stream behavior. The products then enter a stainless steel condenser maintained at 0°C by means of methanol cryostat and eventually enter a gas separator. Product liquid samples were collected in a glass tube immersed in an ice bath. The gaseous products were separated from the liquid product by a separating bubbler and the volume of the gas produced was measured with a gas flow meter. The nitrogen gas flow rate was maintained at 3.0×10^{-6} m^3/s. The heavy oil was first heated in an oil preheater to a temperature of 100°C so that it could be pumped easily.

The reactor was filled with the catalyst (10–15 g) and operated at a temperature between 400–450°Cand atmospheric pressure. The preheater was maintained at a temperature of 400–450°C so that the feed was not only completely vaporized under the inert nitrogen pressure, but it entered the reactor at the same temperature as that of the reactor, maintaining isothermal conditions within the reactor. The total feed processed was between 30–50 g giving a catalyst-to-oil ratio of 0.2–0.5 and a weight hourly space-velocity (WHSV) in the range of 50–105 h^{-1}. The experiments were conducted with VGO as the feed and with three catalysts, namely HZOP-31 (fly ash–based), BPCL, and IOCL cracking catalysts.VGO conversion was calculated by subtracting the unconverted VGO (heavy oil) from the VGO processed and dividing by the VGO processed.

$$VGO\ Conversion = \frac{VGO\ Processed\,(Wt) - Unconverted\ Heavy\ Oil\ Collected\,(Wt)}{VGO\ Processed\,(Wt)}$$

2.6. Product Analysis

2.6.1. ASTM Distillation

The liquid product was characterized by ASTM distillation (ASTM D86-78). The liquid product collected in a glass tube was transferred to a distillation flask and the product was separated into different fractions, namely gasoline (30–200°C), light gas oil (LGO, 200–300 °C), heavy gas oil (HGO, 300–360 °C), and unconverted heavy oil fraction. The characterization was done in terms of distillation cutoffs so the results could be directly used for the estimation of parameters of the lumped kinetic model. The weight of each collected product was measured and reported as the weight fraction of liquid distillate as well as overall weight fraction. All of the fractions, including coke and gas, were reported as weight fractions. The feed distillation range was also determined by ASTM distillation and the distillation curve is shown in Figure 2.

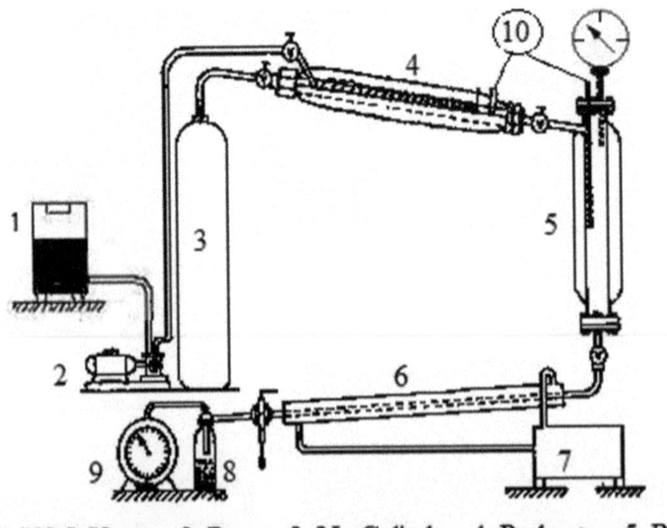

1: VGO Heater, 2: Pump, 3: N_2 Cylinder, 4: Preheater, 5: Reactor, 6: Condenser, 7: Cryostat, 8: Liquid collection bottle, 9: Gas flow meter, 10: Temperature controller

Figure 1. Schematic diagram of the experimental setup.

2.6.2. Coke on Catalyst

The coke on the catalyst was analyzed by the gravimetric method. After the experimental run, the catalyst was purged for 15 min with nitrogen gas to desorb any vapors on the catalyst surface and then the coked catalyst samples were dried at 110°C. Then, 3 g of catalyst was weighed in a silica crucible and ignited in a furnace at 550°C for 3 h, which completely burned the coke on the catalyst. By measuring the initial and final weight of the crucible, the coke on the catalyst was determined. The coke was measured for all catalysts and all experimental runs at various temperatures, space-times, and catalyst-to-oil ratios.

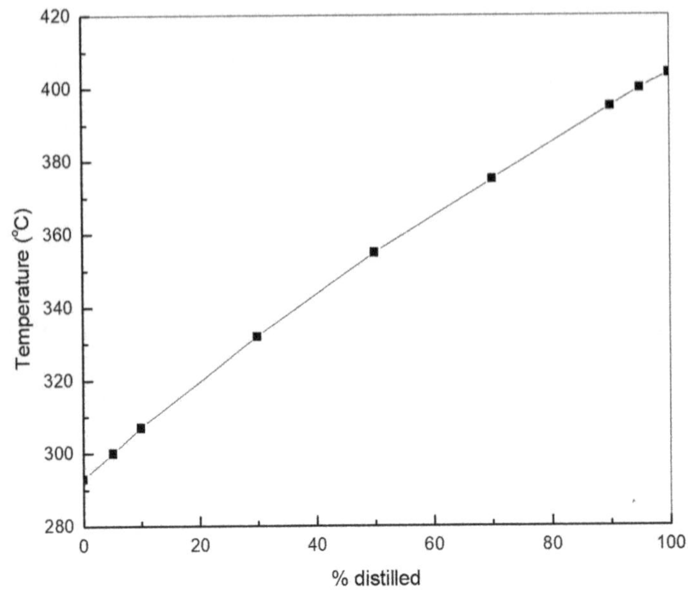

Figure 2. ASTM distillation curve of feed VGO.

3. Results and Discussion

The fly ash of Kolaghat Thermal Power Station, West Bengal, India was converted to zeolite using the method described earlier. The XRD patterns of the synthesized zeolite were found to match exactly with that of the X-type zeolite (*24*), although the relative intensities of the main peaks did not exactly match with the standard one (as in the XRD data file). The change in peak intensity may be due to the change in cation in the crystal structure or the preferred orientation of larger crystals in the X-ray sample holder (*24*). The formation of different crystalline phases on treatments may be due to the variations in chemical composition, nature, and properties between ash samples (*25*). The compositions of the treated fly ash (zeolite) and commercial zeolite, obtained from the micro elemental analysis of SEMs, are shown in Table 1.

Table 1. Compositions of Fly Ash, Synthesized Zeolite, and Commercial 13X Zeolite

Composition (wt %)	*Kolaghat Thermal Power Station Fly Ash*	*Fly Ash–Based Zeolite (ZOP-31)*	*Commercial 13X Zeolite*
SiO_2	55.2	48.0	48.3
Al_2O_3	30.0	27.2	31.8
Fe_2O_3	4.6	3.6	3.2
Na_2O	2.1	15.6	15.6
CaO	0.8	1.5	0.4
K_2O	1.4	0.7	0.7
TiO_2	2.7	2.1	0.0
MgO	1.9	0.5	0.0
BaO	1.3	0.8	0.0

The acidity values of the H-form of synthesized zeolite and other commercial cracking catalysts were determined and are reported in Table 2. From the acidity values, it is evident that the synthesized catalyst acidity value is comparable with that of the commercial catalysts. It could, therefore, be used as a cracking catalyst.

Table 2. Acidity of Various Catalysts

Acidity (meq/g of catalyst)	*Synthesized Catalyst (H-form)*	*BPCL Catalyst*	*IOCL Catalyst*
Lewis Acidity	0.284	0.293	0.366
Brønsted Acidity	0.213	0.243	0.274

Cracking experiments were carried out over the catalyst synthesized from fly ash as well as over two other commercial cracking catalysts with VGO as the feed material. The ASTM distillation characteristics of the feed VGO are presented in Figure 2. Effects of various parameters on the cracking activity and selectivity of different products were studied as reported in the following sections.

3.1. Effect of Temperature

Experiments were carried out with the synthesized catalyst in the temperature range of 400–450°C. Since only fresh catalysts were used, experiments were not carried out above 450°C as it produced a high yield of gaseous products. As can be seen from Figure 3, the increase in temperature increases the overall conversion of VGO. The yield of gas, LGO, and coke products was also increased (Figure 4), and the HGO yield decreased with an increase in temperature due to an increase in secondary cracking. Due to increased cracking, coke formation on the catalyst also increased proportionately with temperature. The effect of the increase in temperature on gasoline yield with the increase in residence time is shown in Figure 5. It was observed that the gasoline yields decreased at higher space-times with an increase in temperature due to over-cracking.

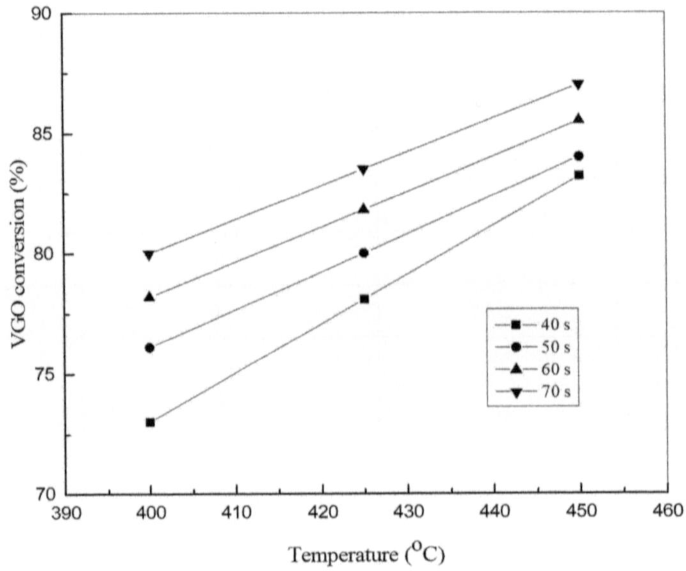

Figure 3. VGO conversion versus temperature at different space-times.

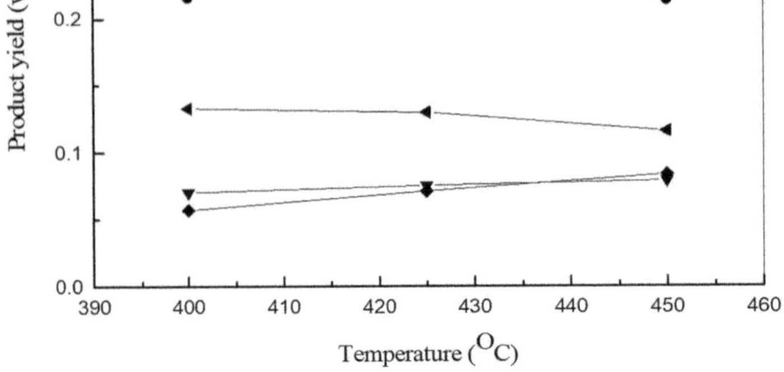

Figure 4. Product yield versus temperature at a space-time of 50 s.

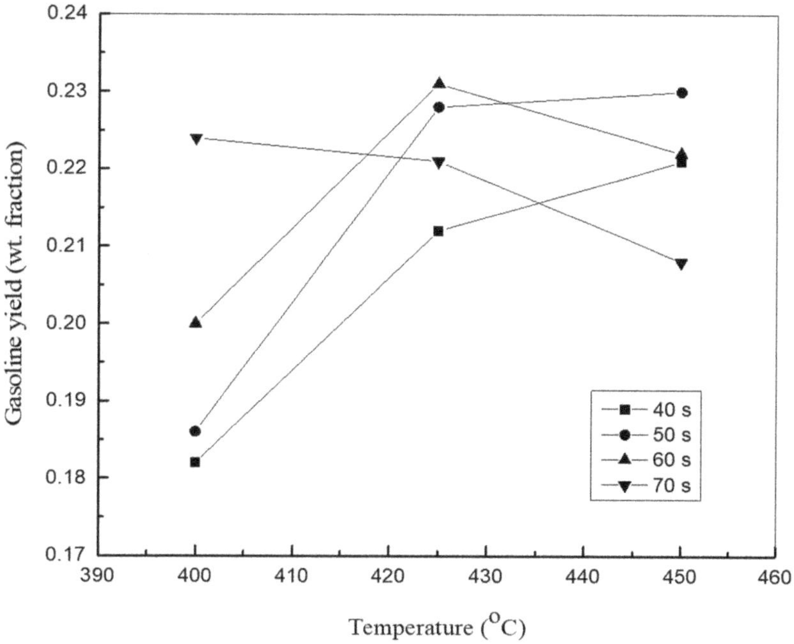

Figure 5. Gasoline yield versus temperature at different space-times.

3.2. Effect of Space-Time

The experiments were conducted for different space-times (in the range of 35–70 s) by varying the feed flow rate and catalyst loading. As can be seen from Figure 6, the increase in space-time increased the VGO conversion. It is evident from Figure 7 that at each temperature, the gasoline fraction attained a maximum value at some space-time and then declined. With an increase in space-time, the conversion of VGO increases, thereby increasing the gasoline in the product, but with an increase in space-time, the cracking of gasoline to gas and coke also increases (i.e., over-cracking occurs). That is why the gasoline content in the product starts falling after a certain space-time.

It was observed that the gas yield increased with an increase in space-time due to an increase in overall conversion and an increase in secondary cracking (i.e., cracking of the products, Figure 8). The effect of an increase in space-time on the yields of LGO, HGO, and coke is shown in Figure 9. It was also observed that the coking of the catalyst increased with an increase in space-time, as can be seen from Figure 10, and increased with temperature for a given space-time. Due to an increase in VGO cracking with an increase in space-time, more coke is deposited, thereby increasing the coke on the catalyst, which again rapidly increases with an increase in temperature due to an increase in secondary cracking reactions (i.e., LGO, gasoline to gas, and coke).

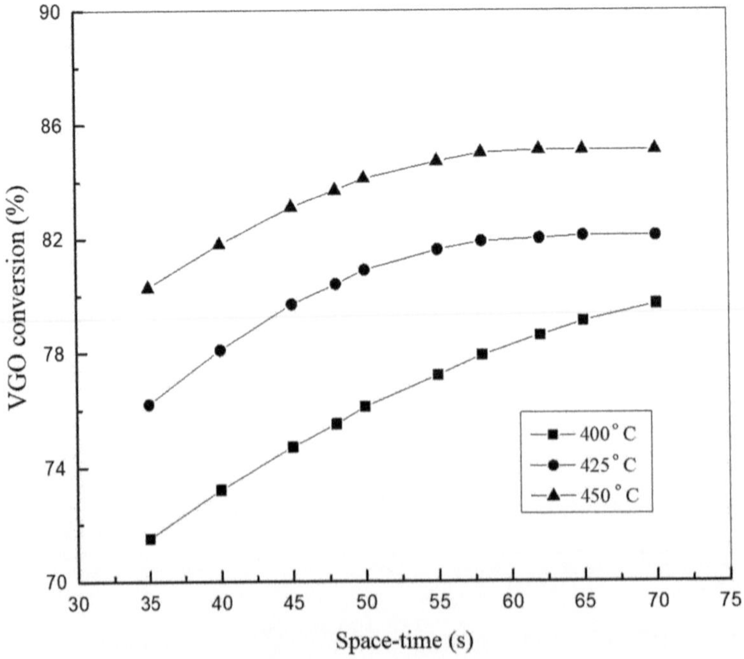

Figure 6. Effect of space-time on VGO conversion at different temperatures.

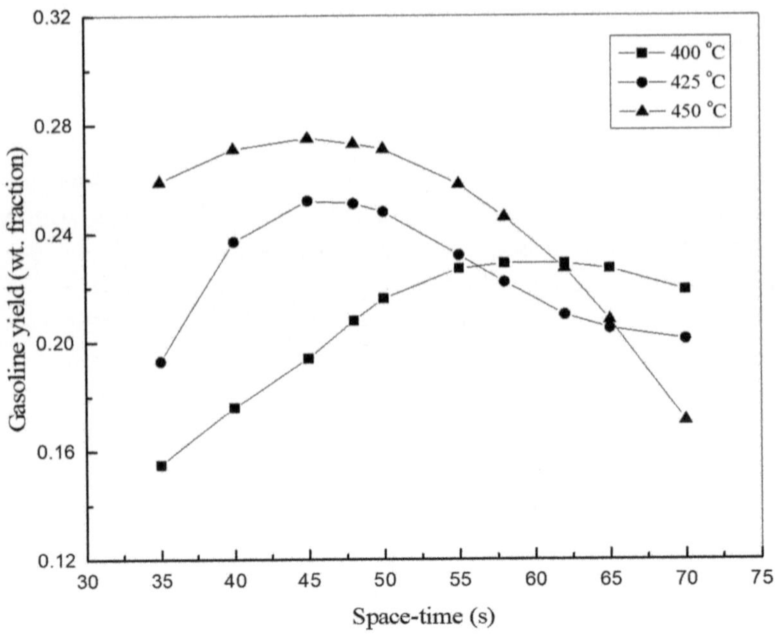

Figure 7. Effect of space-time on gasoline yield at different temperatures.

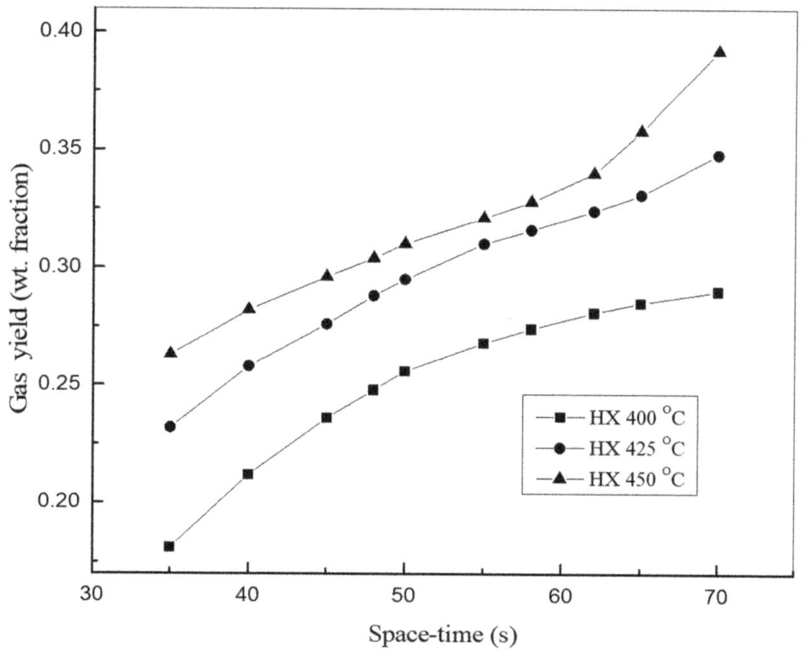

Figure 8. Effect of space-time on gas yield at different temperatures.

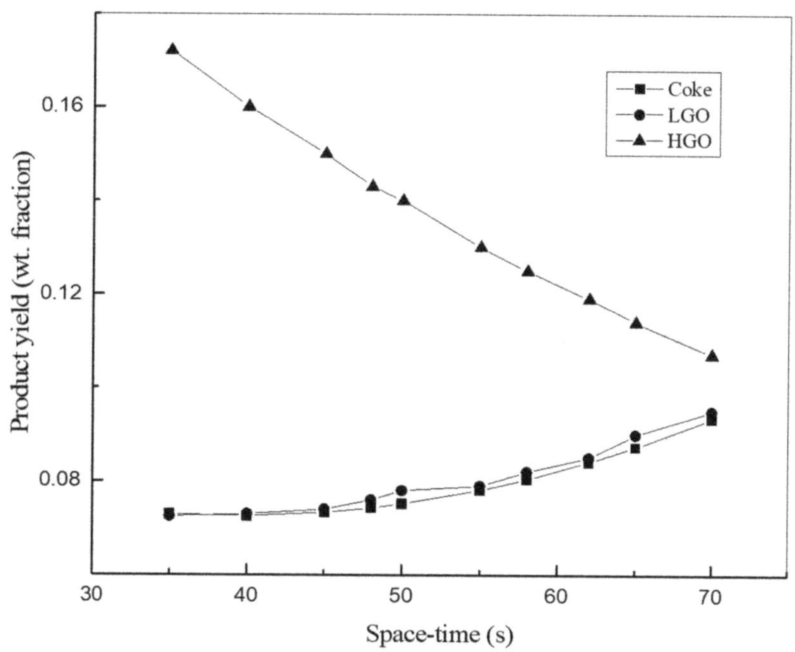

Figure 9. Effect of space-time on product yield at 425 °C.

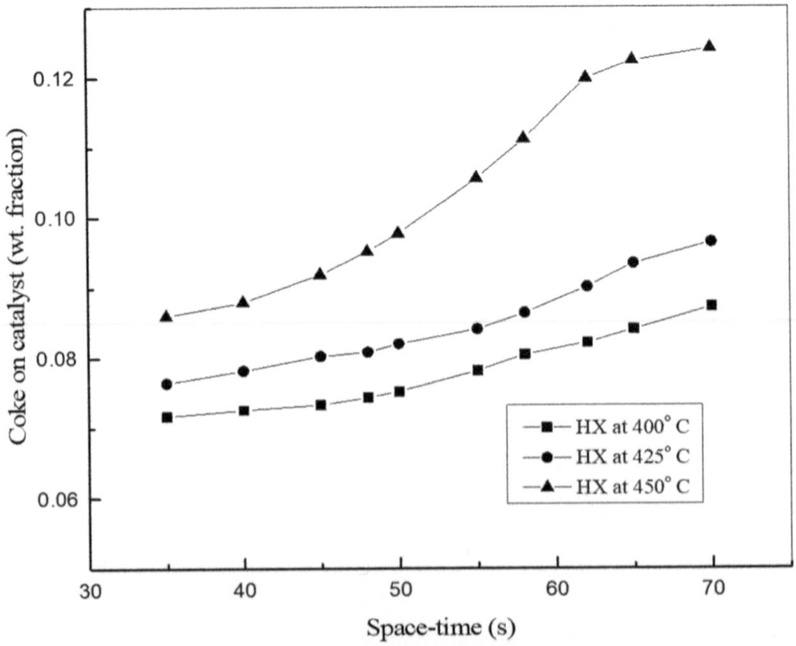

Figure 10. Effect of space-time on coke on catalyst at different temperatures.

3.3. Comparison of the Performance of Synthesized Zeolite with Commercial FCC Catalysts

The cracking performance of the synthesized zeolite was compared with commercial catalysts. Commercial FCC catalyst samples were obtained from BPCL (Mumbai, India) and IOCL (Haldia Refinery, Haldia, India). The experiments were conducted with the synthesized catalyst and the commercial FCC catalysts under otherwise identical conditions. The performance of the synthesized catalyst was observed to be comparable with that of the BPCL and IOCL catalysts. In fact, the performance of the synthesized catalyst at higher space-times was observed to be better than that of the commercial FCC catalysts, as shown in Figure 11. The gasoline selectivity was observed to be high at low space-times for the synthesized catalyst (Figure 12) in comparison with the commercial catalyst. The opposite trend was observed at higher space-times (Figure 13), probably due to the rapid coking of the synthesized catalyst with an increase in space-time.

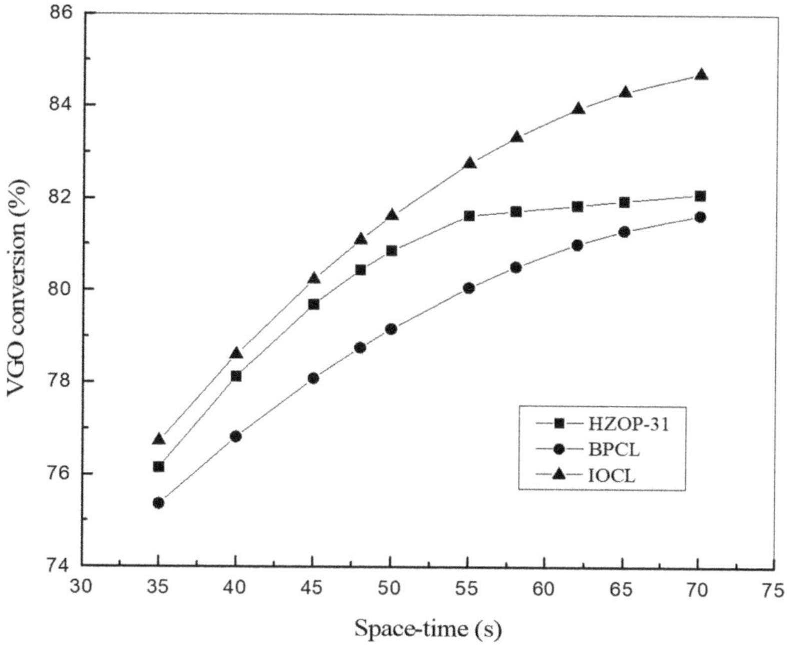

Figure 11. Effect of space-time on VGO conversion with different catalysts at 425 °C.

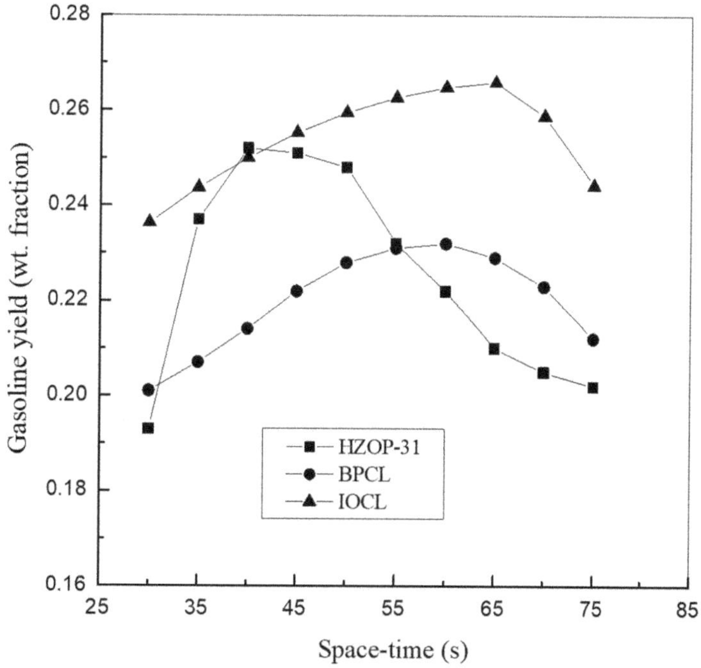

Figure 12. Effect of space-time on gasoline yield with different catalysts.

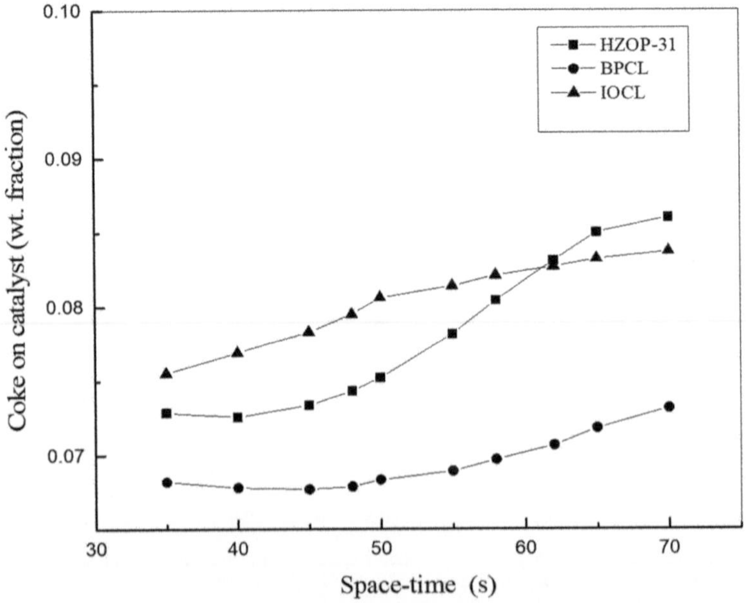

Figure 13. Effect of space-time on coke on catalysts with different catalysts at 425°C.

4. Kinetic Modeling

The reaction scheme for a four-lumped kinetic model is shown in Scheme 1.

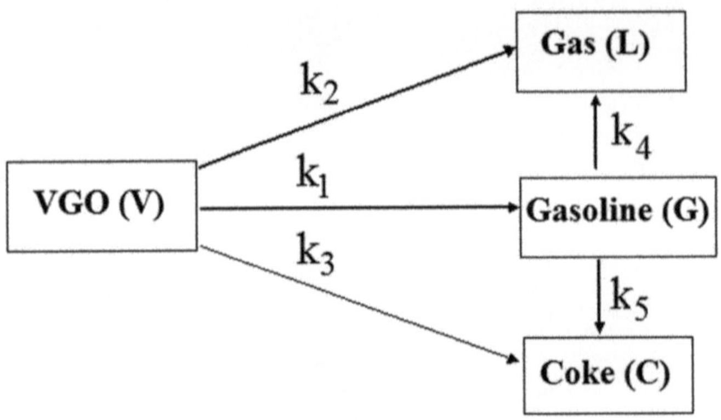

Scheme 1. Reaction scheme for a four-lumped model.

4.1. Model Formulation

Based on this scheme, the model equations for VGO cracking, and gasoline, gas, and coke formation are given as follows (*19*):

$$\frac{dV}{d\tau} = -(K_1 + K_2 + K_3) \times V^2 \times \phi \quad \text{-------- (1)}$$

$$\frac{dG}{d\tau} = (K_1 \times V^2 - (K_4 + K_5) \times G) \times \phi \quad \text{-------- (2)}$$

$$\frac{dL}{d\tau} = (K_2 \times V^2 + K_4 \times G) \times \phi \qquad \text{-------- (3)}$$

$$\frac{dC}{d\tau} = (K_3 \times V^2 + K_5 \times G) \times \phi \qquad \text{-------- (4)}$$

$$\phi = \exp(-\alpha \times C) \qquad \text{-------- (5)}$$

Where V = weight fraction of VGO, G = weight fraction of gasoline, L = weight fraction of gas, C = weight fraction of coke, Φ = catalyst deactivation term, α = deactivation constant, and τ = space-time. For the sake of simplifying the parameter estimation, gasoline and LGO fractions were taken together as gasoline and HGO, and VGO lump was taken as unconverted VGO, giving a total of 4 unknown lumped variables, involving 6 unknown parameters and 10 experimental points fitted to each of the variables. A nonlinear regression method was applied for parameter estimation. Since the feed was pure VGO and contained none of the product fractions, the following initial conditions were applied.

Initial conditions: V = 1.00, G = 0.00, L = 0.00, and C = 0.00.

Assumed initial six parameters (guess parameters): $k_1 = 1.00$, $k_2 = 0.10$, $k_3 = 0.10$, $k_4 = 0.01$, $k_5 = 0.01$, and α = 2.00

4.2. Kinetic Parameter Estimation

Kinetic parameters were estimated from the model equations using a lumped parameter estimation method through a multiple nonlinear regression technique (*26, 27*). The following objective function (**ψ**) was used during regression analysis:

$$\psi = \sum_{j=1}^{v} wj\,(Y_j^* - Y_j)'\,(Y_j^* - Y_j) \qquad \text{--- (6)}$$

$$\Delta \mathbf{b} = \left| \lambda \boldsymbol{I} + \sum_{j=1}^{v} w_j (\boldsymbol{J}_J{}' \boldsymbol{J}_J) \right|^{-1} * \left| \sum_{j=1}^{v} w_j \boldsymbol{J}_J{}' (y_j^* - y_j) \right| \qquad \text{--- (7)}$$

$$b^{(m+1)} = b^{(m)} + \Delta \mathbf{b} \qquad \text{--- (8)}$$

Using the Marquardt algorithm, the kinetic parameters of the model were obtained. From Equation 7, the increment in the initial guess parameters was obtained. Using Equation 8, the next updated parameter vector was obtained. By iterative steps, the objective function (Equation 6) was minimized until either **ψ** did not change appreciably or **Δb** became very small.

From the estimated kinetic parameters, the predicted yields (weight fraction) versus the experimental yields were plotted. The plotted results for the synthesized catalyst and the BPCL and IOCL catalysts at 425°C are shown in Figures 14–16. A statistical analysis was performed to correlate the predicted and experimental values and thereby validate the model and the estimated parameters. From the predicted parameters shown in Table 3, the Arrhenius plots were drawn, as shown in Figure 17. The activation energies obtained from the Arrhenius plots were compared with values from the literature. As seen in Table 4, the estimated kinetic and other parameters of the model compare well with those reported in the literature.

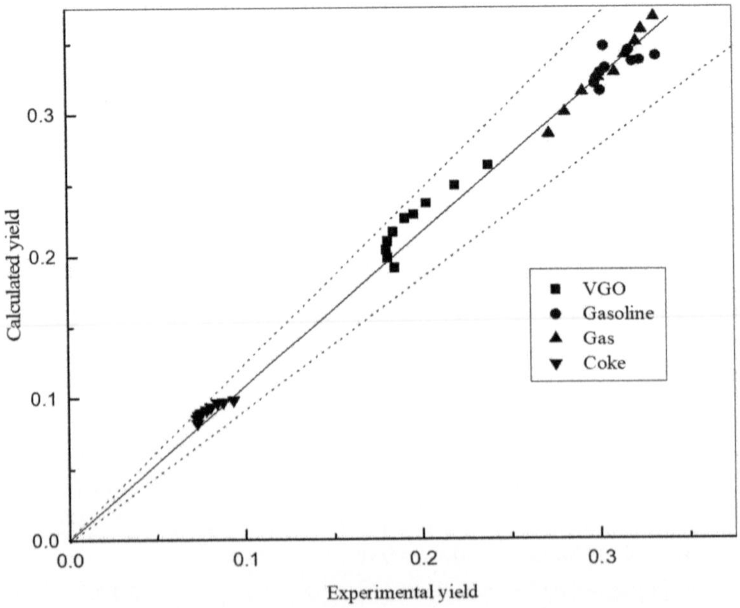

Figure 14. Calculated yields versus experimental yields of products with the synthesized catalyst (HZOP-31) at 425°C and VGO feed. (The dotted lines represent 10% deviation in either direction.)

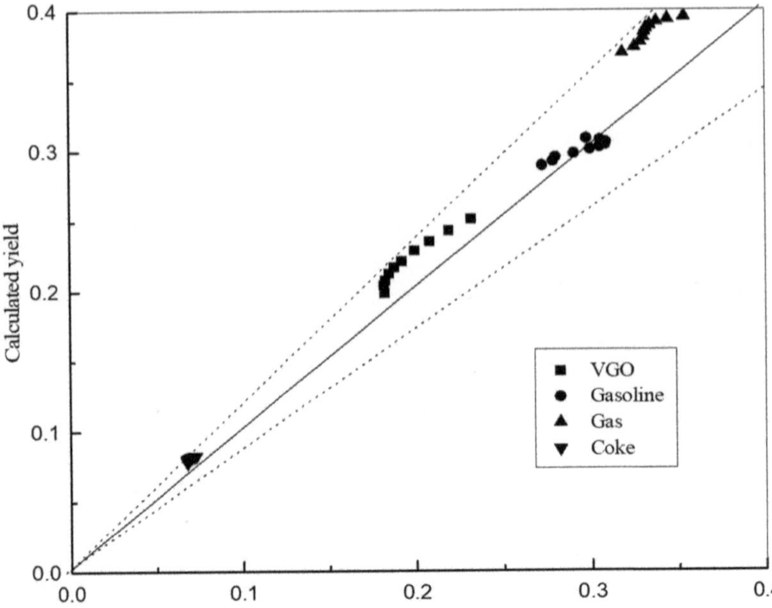

Figure 15. Calculated yields versus experimental yields of products with the BPCL catalyst at 425°C and VGO feed. (The dotted lines represent 10% deviation in either direction.)

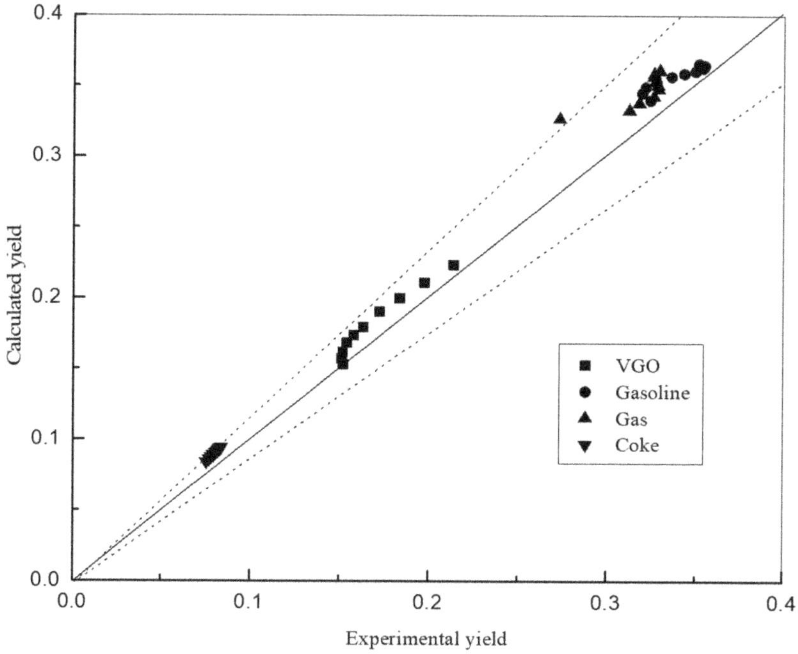

Figure 16. Calculated yields versus experimental yields of products with the IOCL catalyst at 425°C and FCC feed. (The dotted lines represent 10% deviation in either direction.)

Table 3. Estimated Parameters for Different Sets of Experimental Runs[a]

Catalyst	*Temperature (°C)*	*Estimated Parameters*					
		k_1 (s^{-1})	k_2 (s^{-1})	k_3 (s^{-1})	k_4 (s^{-1})	k_5 (s^{-1})	α
HZOP-31	400	1.66	1.02	0.40	0.29	0.017	1.02
HZOP-31	425	2.50	1.40	0.56	0.62	0.042	1.86
HZOP-31	450	3.45	2.24	0.91	0.82	0.076	2.51
BPCL	425	1.65	1.82	0.47	0.0045	0.021	2.01
IOCL	425	2.14	2.00	0.61	0.0087	0.0005	1.91

[a] VGO as feed.

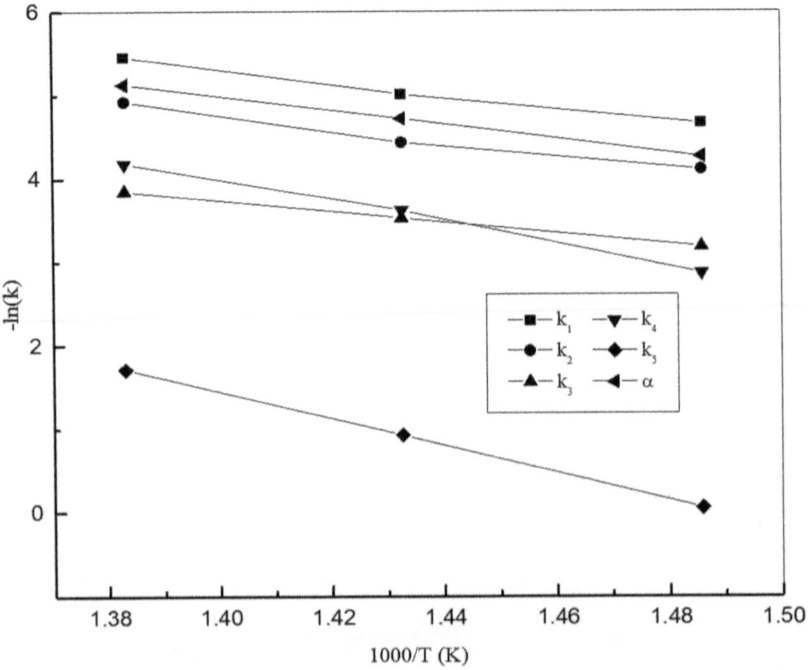

Figure 17. Arrhenius plot for the kinetic constants.

Table 4. Calculated Activation Energies and Literature Values

Kinetic Constant	*Activation Energy (kcal/mol)*	
	Synthesized Catalyst (HZOP-31)	*Literature*
k_1 (VGO to gasoline)	14.24	*10–36*
k_2 (VGO to gas)	15.30	*12–21*
k_3 (VGO to coke)	16.06	*6–18*
k_4 (gasoline to gas)	20.32	*10–19*
k_5 (gasoline to coke)	28.60	*16–27*
α (deactivation constant)	15.54	*10–90*

5. Conclusions

From the results, it can be concluded that the performance of the synthesized catalyst as a cracking catalyst is comparable with that of the commercial catalysts in the temperature range studied with higher conversion and gasoline selectivity. The cracking process could be suitably represented by a four-lump parameters model. The estimated kinetic and other parameters of the model compare well with those reported in the literature. The cracking of gasoline to coke was observed to be insignificant owing to the very low value of k_5. Thus, it can be concluded that the synthesized zeolite can be a promising FCC catalyst due to its comparable activity and less expensive cost. The present work therefore proves that the synthesized catalyst not only provides a solution to the present-day environmental problem but it can also improve the economics of a refinery. The utility of the synthesized catalyst needs to be studied in detail in a pilot plant or the synthesized zeolite can be used in mini-refineries to test its utility as an FCC catalyst.

Nomenclature

α	deactivation constant
BPCL	Bharat Petroleum Corporation Limited
C	coke weight fraction
FCC	fluid catalytic cracker
G	gasoline weight fraction
HGO	heavy gas oil
IOCL	Indian Oil Corporation Limited
J	Jacobian matrix
k_1	kinetic rate constant for VGO to gasoline (s^{-1})
k_2	kinetic rate constant for VGO to gas (s^{-1})
k_3	kinetic rate constant for VGO to coke (s^{-1})
k_4	kinetic rate constant for gasoline to gas (s^{-1})
k_5	kinetic rate constant for gasoline to coke (s^{-1})
LGO	light gas oil
SEM	scanning electronic microscope
V	VGO weight fraction
VGO	vacuum gas oil
XRD	X-ray diffraction
$\boldsymbol{Y_J}$	calculated weight fraction vector
$\boldsymbol{Y_j^*}$	experimental result vector
τ	space-time(s)
Φ	catalyst decay term
$\mathbf{b}$	parameter vector
$\Delta\mathbf{b}$	correction vector
$\boldsymbol{\psi}$	objective function
$b^{(m+1)}$	new updated parameter vector
$b^{(m)}$	old parameter vector

Acknowledgments

Narayan C. Pradhan acknowledges the partial financial support of this work from the Ministry of Human Resource Development, Government of India (Grant No. F26-1/2002 TS.V). The authors are also thankful to BPCL for supplying the FCC catalyst and IOCL (Haldia Refinery) for supplying both the FCC catalyst and cat cracker feedstocks.

References

1. Plank, C. J. The Invention of Zeolite Cracking Catalysts—A Personal Viewpoint. In *Heterogeneous Catalysis*; Davis, B. H., Hettinger, W. P., Eds.; ACS Symposium Series 222; American Chemical Society: Washington, DC, 1983; pp 253-271.
2. Gladrow, G. M.; Krebs, R. W.; Kimberlin, C. N. Reactions of hydrocarbons over cracking catalysts. *Ind. Eng. Chem.* **1953**, *45*, 142–147.
3. Elliot, K. M.; Eastwood, S. C. Durabead-5-a new cracking catalyst. *Oil Gas J.* **1962**, *60*, 142–148.
4. Eastwood, S. C.; Drew, R. D.; Hartzell, F. D. What Mobil learned about Durabead-5. *Oil Gas J.* **1962**, *60*, 152–158.
5. Johning, C. E.; Martin, H. Z.; Campbell, D. L. The development of fluid catalytic cracking. *ACS Symposium Series* **1983**, *222*, 273–291.
6. Magee, J. S.; Blazek, J. J. Preparation and performance of zeolite cracking catalysts. In *Zeolite Chemistry and Catalysis*; Rabo, J. A. Ed.; ACS Monograph, American Chemical Society: Washington, DC. 1976; Vol. 171, p 615.
7. Jecobs, P. A. *Carboniogenic Activity of Zeolites*; Elsevier: Amsterdam, 1977.
8. Corma, A.; Agudo, A.; Nebot, L. I.; Thomas, F. A. Kinetic study of the cracking, isomerization and disproportionation of n-heptane on a chromium exchanged Y-zeolite. *J. Catal.* **1982**, *77*, 159–168.
9. Le Normand, F.; Fajula, F. Degradation mechanism of 3-methyl pentane on a supported superacid catalyst studied by the ^{13}C isotopic tracer technique. In *Studies on Surface Science and Catalysis (Catalysis by Acids and Bases)*; Imelik, B.; Naccache, C.; Coudurier, G.; Ben Taarit, Y.; Vedtine, J. C., Eds.; Elsevier: Amsterdam, 1985; Vol. 20, pp 325-334.
10. Corma, A.; Monton, J. B.; Orchilles, A. V. Influence of acid strength distribution on the cracking selectivity of zeolite Y catalysts. *Ind. Eng. Chem. Prod. Res. Dev.* **1984**, *23*, 404–409.
11. Jacob, S. M.; Gross, B.; Voltz, S. E.; Weekman, V. W., Jr. A lumping and reaction scheme for catalytic cracking. *AIChE J.* **1976**, *22*, 701–720.
12. Wojciechowski, B. D., Corma, A. *Catalytic Cracking: Catalyst, Chemistry and Kinetics*; Marcel Dekker, Inc.: New York, 1986.
13. Pitault, I.; Nevicato, D.; Forissier, M.; Bermard, J. R. Kinetic model based on a molecular description for catalytic cracking of vacuum gas oil. *Chem. Eng. Sci.* **1994**, *49*, 4249–4262.
14. Gianetto, A.; Faraq, H.; Blasetti, A.; de Lasa, H. I. Fluid catalytic cracking catalyst for reformulated gasolines. Kinetic modeling. *Ind. Eng. Chem. Res.* **1994**, *33*, 3053–3062.
15. Watson, B. A.; Michael, T. K.; Robert, H. H. Mechanistic modeling of n-heptane cracking on HZSM-5. *Ind. Eng. Chem. Res.* **1996**, *35*, 1506–1516.
16. Blosma, E.; Martens, J. A.; Jecobs, P. A. Reaction mechanisms of isomerization and cracking of heptane on Pd/H-beta zeolite. *J. Catal.* **1995**, *155*, 141–147.
17. Blanding, F. H. Reaction rates in catalytic cracking of petroleum. *Ind. Eng. Chem. Res.* **1953**, *38*, 1186–1197.
18. Weekman, V. W., Jr. *AIChE Monograph Series* **1979**, *75*, 3–29.

19. Mohammad, A. A. H.; Siddiqui, M. A. B.; Ino, T.; Aitani, A. M. Experimental determination of high-severity fluidized catalytic cracking (hs-fcc) deactivation constant. *Appl. Catal. A: Gen.* **2002**, *237*, 71–80.
20. Ojha, K.; Pradhan, N. C.; Samanta, A. N. Zeolite from fly ash: Synthesis and characterization. *Bull. Mater. Sci.* **2004**, *27*, 555–564.
21. Ojha, K.; Pradhan, N. C. Treated fly ash: A potential catalyst for catalytic cracking. *Indian J. Eng. Mater. Sci.* **2001**, *8*, 100–103.
22. Jeffery, G. H.; Bassett, J.; Mendham, J.; Denney. R. C.; *Vogel's Textbook of Quantitative Chemical Analysis*, 5th ed.; Longman Group UK Limited: UK, 1989, pp 384-399.
23. Barman, S. *Alkylation of Aromatics with Alcohols and Acetone over Modified Zeolites*. Ph.D. Thesis, Indian Institute of Technology, Kharagpur, India, 2006, pp 45–46.
24. Szostak, R. *Molecular Sieves: Principles of Synthesis and Identification*; Van Nostrand Reinhold: New York, 1976.
25. Henmi, T. Increase in cation exchange capacity of coal fly ash by alkali treatment. *Clay Sci.* **1987**, *6*, 277–282.
26. Bard, Y. *Nonlinear Parameter Estimation*; Academic Press: New York, 1974.
27. Constantinides, A.; Mostoufi, N. *Numerical Methods for Chemical Engineers with MATLAB Applications*; Prentice Hall International: New Jersey, 1995.

Chapter 10

Fundamental Aspects and Applications of Ultrasonically Induced Cavitation in Heavy Fuel Oil with a Focus on Deasphalting, Emulsions, and Oxidative Desulfurization

Paolo Guida,[1] Abdul Gani Abdul Jameel,[2] Saumitra Saxena,[*,1] and William L. Roberts[1]

[1]Clean Combustion Research Center (CCRC), King Abdullah University of Science and Technology (KAUST), Thuwal 23055, Kingdom of Saudi Arabia

[2]Department of Chemical Engineering, King Fahd University of Petroleum & Minerals, Dhahran 31261, Kingdom of Saudi Arabia

[*]Email: saumitra.saxena@kaust.edu.sa

The combustion of hydrocarbons will continue to feed the planet's growing demand for mobility and power generation over the next several decades, shifting to lower-value, more-difficult-to-burn fuels while at the same time meeting more stringent emissions regulations. These lower-value fuels include heavy fuel oils and vacuum residuals, which are difficult to burn cleanly due to the presence of asphaltenes, the insoluble fractions with exceptionally high molecular weight that are found in high concentrations in crude oils. In particular, heavy fuel oils (HFO) are widely used in marine and power-generation sectors, and the International Maritime Organization's (IMO2020) promulgation has redistributed the HFO demand and pushed the world's economy into a new paradigm. We seek solutions for such a complex oil industry paradigm by utilizing some state-of-the-art technologies like ultrasonically induced cavitation (UIC). In the current chapter, we have discussed a roadmap for use of "bottom-of-barrel fuel" with high asphaltene content via UIC-based fuel upgrading, desulfurization, and direct use (emulsions). We expect that a strategy of using UIC for asphaltene modification and water-in-HFO-enabled microexplosions will significantly impact the combustion of HFO. Furthermore, ultrasonic-assisted oxidative desulfurization can be utilized to remove undesired sulfur to meet marine or power sector requirements. Deasphalting, emulsions, and desulfurization solutions could be applied in a multiplicity of combustion-driven energy conversion platforms, including compression ignition engines, gas turbines, and boilers.

© 2021 American Chemical Society

1. Introduction

Despite impressive annual growth rates for renewables and hydrogen as energy vectors, a transition period spanning through 2050 and longer will be needed for the complete transformation of the primary energy mix. Fossil-based resources may continue to dominate energy production needs even as the world shifts to cleaner and low-carbon sources (*1*). Thus, the most crucial aspect of reducing global carbon- and combustion-generated air-pollutant emissions will be to improve the carbon utilization efficiency of fossil fuel applications, from resource recovery to end use. However, the availability of "easily extractable" and "low carbon intensity (*2*)" light and sweet crude oil is expected to decrease drastically (*3, 4*) as rampant exploitation to meet burgeoning global petroleum demands has led to progressively depleted oil wells worldwide. Therefore, heavy or sour crude oil and the "bottom of the barrel" products—like heavy fuel oils (HFOs), vacuum residue oils (VROs), pitch, asphalts, and petcoke—could be increasingly utilized (*5*) in combustion applications due to their copious availability and low cost. These so-called "opportunity fuels (*6*)" (in petroleum industry parlance) are very difficult to burn and produce heavy pollution in the aftermath of the burning. Several strategies have been proposed in the past to use heavy fuels as a source of energy. The most direct usage consists of burning fuels in suitable combustion devices. Another option is to perform gasification to obtain syngas, which is a high-value, cleanly combustible, and multi-industry chemical feedstock. The so-called brown and blue hydrogens (*7*) are important products of heavy fuel and vacuum residue gasification.

The combustion or gasification of the HFOs and the subsequent particulate matter (soot, cenosphere, and ash) formation processes are mostly driven by the asphaltenic content (up to 15 wt %) (*8–11*). In industrial burners, asphaltenes in spray droplets are slow to react until the late combustion stages. Much of the energetic content of asphaltenes is wasted in the formation of carbonaceous cenospheres and their inefficient burnout. The delayed reaction affects the heat-release distribution and the heat-transfer profile within the combustion zone and subsequent heat-transfer events, significantly limiting energy-conversion efficiencies for useful power generation. Asphaltenes are defined as the fraction of petroleum that is insoluble in nonpolar aliphatic hydrocarbon solvents (e.g., *n*-pentane and *n*-heptane). They make up the heaviest category among the SARA (saturates, aromatics, resins, and asphaltenes) fractions. They have highly refractory heteroatoms of sulfur and nitrogen and organometallic porphyrins that are laden with trace metals like vanadium and nickel. In countries where very heavy crude, oil sands, and shale-derived oils are in abundance, asphaltenes are considered untapped fuels. If asphaltenes could be burned more effectively, a quantum leap in combustion efficiency gain would be possible.

Asphaltene breakup, heteroatom removal, and viscosity reduction are critical areas for upgrading and utilization of HFOs (*5*). Asphaltene can be efficiently burned in water-in-fuel emulsions that offer tremendous benefits of fuel savings and clean burning during combustion processes. There is evidence that the problematic structure of asphaltenes can be modified by the intense temperatures and pressures associated with the collapse of microbubbles during cavitation (*12*). This acoustically generated cavitation is both a physical and chemical effect; the enhancement or alteration of chemical reactions using ultrasound, in particular, is known as sonochemistry (*13–15*). A few recent review articles have summarized the previous studies on heavy oil upgrading using cavitation methods (*16, 17*) and industrial developments (*18*). Our fundamental understanding of cavitation and its impact on asphaltene/fuel upgrading, desulfurization, and emulsification processes has not yet been sufficiently developed. This chapter describes the potential role played by ultrasonically induced cavitation during selected aspects of fuel treatment and direct usage. HFO has been studied here as the representative "opportunity fuel" since it is heavily utilized in the marine and power-generation sectors. A brief overview of the economic drivers and industrial advancements in the study of cavitation is provided. Based on our research at King Abdullah University of Science and Technology (KAUST) and a review of the literature, we endeavor to provide a path forward for HFO upgrading through water-in-HFO emulsions, oxidative desulfurization, and combustion. To that end, we discuss salient aspects of ultrasonic cavitation with a focus on the following themes or applications:

- Detailed characterization of HFO and asphaltenes
- Water-in-HFO emulsions from ultrasonically induced cavitation (UIC)
- Oxidative desulfurization (ODS) of HFO

To achieve some of the stated goals of the study, we utilized state-of-the-art analytical (Fourier transform ion cyclotron resonance mass spectrometry, FT-ICR-MS; nuclear magnetic resonance, NMR), imaging (cryogenic electron microscopy), and high-speed camera visualization techniques.

Environmental and Economic Drivers

SO_x emissions from ships have been recognized as a significant threat to the global environment because they are highly destabilizing to many extremely vulnerable ecosystems such as the polar region and near-Arctic sea, leading to the International Maritime Organization's IMO2020 mandate of a sulfur cap on marine fuels (*19*). HFO (S > 3.5%) is now restricted in marine transport since the IMO2020 regulations came into effect on January 1, 2020 (*20*). Stringent regulations have been imposed on Emission Controlled Areas. Consequently, HFO demand is plummeting globally, bringing down its price substantially. Diesel or marine fuel oil or other low-sulfur fuel alternatives can meet the displaced demand for HFO in the near term. But these fuels are highly processed, creating a significant carbon footprint, and they will lead to a net loss of export revenue for oil-exporting nations like Saudi Arabia (*21*, *22*). With the current price differential between low-sulfur fuel oil and HFO reaching $200 per ton (*23*), the price of marine transport could skyrocket globally and severely affect international trade. Alternatives like liquefied natural gas are seen as future marine fuels but with a decade or more to wait because of a lack of a sufficient bunkering infrastructure. IMO's 2050 strategy to reduce CO_2 emissions by 50% by 2050 (from base year 2008) adds further strategic challenges to marine transport fuel options (*24*). Liquefied natural gas's supply chain and current engine technologies are prone to fugitive methane emissions. An ARAMCO-led study (*25*) estimates that liquefied natural gas is noncompetitive for climate impact in marine applications based on a 20-year global warming potential. Hence, an HFO with sufficiently reduced sulfur content meeting IMO2020 specifications (S < 0.5% for non-Emission Controlled Area and S < 0.1% for Emission Controlled Area) can potentially play a dominant role in marine transport for decades to come. It is important to note, however, that even low-sulfur HFO operations will require on-board carbon capture to be fully compliant with IMO2050 (*24*, *26*).

There are projections of an up to 3 million barrels per day (50%) drop in demand for high-sulfur HFO in 2020 (*22*). The displaced HFO from the marine industry is likely to be absorbed by the power-generation sector in countries like Saudi Arabia, United Arab Emirates, Iraq, Lebanon, Bangladesh, Sri Lanka, South Korea, Cuba, and Pakistan. Currently, 132 GW of HFO-fired power plants are operational worldwide. Out of these, only 11 GW have any kind of scrubber to control SO_x emissions (*22*). With an 88 GW total generating capacity, Saudi Arabia is perhaps the biggest consumer of HFO in the power and desalination sectors. The country's energy mix includes HFO, crude oil, and diesel-fired in respective shares of 16.5%, 24.2%, and 8.6% (27). In the aftermath of IMO2020, crude and HFO may displace diesel as preferred fuels for power generation (*21*). An estimated 10 GW of new HFO-fired capacity is slated to be online in 2020 alone (22). One of the most critical challenges in burning HFO in boilers is its sulfur content (*28*). Current capital expenditure (CAPEX) and operational expenditure (OPEX) intensive flue gas desulfurization (FGD) units are insufficient in controlling SO_x emissions to the levels prescribed by global regulatory bodies. A power plant's overall efficiency is also affected by cold-end corrosion issues caused by condensation of sulfuric acid in its exhaust systems. Considering the developments mentioned above, desulfurization and clean burning of HFO in steam boilers are essential developments. Indeed, using bottom-of-the-barrel or opportunity fuels like HFO with upgrading and pretreatments like emulsification and desulfurization will make the sustainable utilization of petroleum resources possible in hard-to-abate sectors like power and marine transport.

Cavitation-Based HFO Emulsions

Cavitation-based processes have been used in many engineering applications in the petroleum industry since the 1950s. The earliest application was related to enhanced oil recovery, which used the ultrasonic wave-powered sonocapillary effect for improved oil percolation in capillaries and in debris destruction by sound-induced mechanical vibrations inside a well. The extraction of oil from oil shale and tar sands is currently enabled by major installations of cavitation-based units. Several other processes that heavily utilize cavitation-based methods in downstream refineries, such as desulfurization, removal of trace metals, and breaking the naturally formed emulsions of crude oils, have been used.

Cavitation involves the creation, growth, and eventual collapse of microbubbles. The underlying physics of cavitation is as follows (*18, 29*). When the bubbles collapse under the influence of external parameters like varying pressure, they produce violent implosions. These implosions create shock waves and very high temperatures (~5000 K) and pressures (hundreds of bar) in the vicinity of the site of the microbubbles. Cavitation can lead to the formation of hydroxyl and hydroperoxyl radicals by the splitting of water molecules under some conditions (*13, 14, 30*). These implosions are the foundation of sonochemistry or of the physiochemical transformations that are observed during a cavitation phenomenon. Cavitation can be produced either by ultrasonic (*13, 17*) or hydrodynamic methods (*30, 31*). In the ultrasonic method, a high-intensity sound wave is irradiated on the liquid, and microbubbles form and grow to an unstable size and then collapse to produce implosions. In the hydrodynamic technique, the passage of a liquid through a constriction such as an orifice plate throttling valve, or venturi, is responsible for the creation of microbubbles based on Bernoulli's principle. These bubbles later collapse, leading to sonochemical effects similar to that observed by ultrasonic means. Both methods are used in the petroleum industry for selected applications, and each has distinct advantages and drawbacks (*18*).

Several combustion strategies have been proposed to improve HFO combustion, including combustion-enhancing additives (*32, 33*), staged combustion with natural gas reburning (*34*), oxycombustion (*35*), gasification (*36*), chemical looping (*37*), and water-in-HFO emulsions (*38–45*). Among these, the emulsification of HFO with a certain amount of water has been one of the most-researched methods. Spray combustion involves fuel droplet breakup, vaporization, mixing with air, and subsequent burning. The advantage of burning HFO emulsion consists in the occurrence of microexplosions. The high volatility differential between water (T_b ~ 373 K) and HFOs (T_b ~ 600–700 K) triggers the formation of water bubbles within the fuel droplets when these are exposed to high temperature. The bubbles eventually overcome the surface tension of the droplet, resulting in multiple ejections that create secondary smaller droplets, resulting in the so-called "microexplosion" phenomenon (*38, 42, 43, 46–48*). Water-in-liquid fuel emulsions provide therefore a dramatic improvement in the droplet-break stage (*40, 49*). The surface area of smaller droplets thus formed is several orders of magnitude larger than what is possible to achieve using traditional atomizers. Subsequent mixing with air is intense and uniform, enabling efficient burning, lower temperatures, and less NO_X and particulate matter formation. Emulsion formation by cavitation also leads to breaking of the large asphaltene clusters, which may speed up the burning of these difficult-to-burn fuel components. In addition, the emulsions so produced will also aid in ODS of HFO (*50*). Water-in-liquid emulsions could be applied in multiple combustion platforms such as compression ignition engines, gas turbines, and boilers.

Industrial Advancements in Cavitation-Based Emulsification and ODS

Until a few years ago, emulsions were not an industrially preferred solution for HFO burning. Some of the reasons included the high cost of emulsifiers or surfactants and the huge water storage and desalting (removal of hot corrosion-causing sodium and potassium trace metals) infrastructure and expenses. However, in recent years, commercially available emulsion technologies have seen exponential growth due to the development of cavitation-based devices (*18*), which can enable in situ or in-line-prepared emulsions without the need for long-term storage (*46*). There has been a great focus on scaling up cavitation technologies to make water-in-fuel emulsions using cavitation techniques. This industrial focus is driven by the need to improve spray combustion in practical combustors, involving hard-to-atomize viscous fuels like heavy distillates, crude oils, biofuels, and HFOs. Another important application is ODS, which tailgates on the microemulsions from cavitation- and sonochemistry-induced oxidation of organosulfur species.

In 2000, Arisdyne Systems Inc. (USA) patented a large-scale hydrodynamic cavitation-based device (called a Controlled Cavitation Device) for emulsion preparation (*51*). This device has no moving parts. It harnesses cavitation by using a liquid that moves via a flow-through compartment with an upstream portion and a downstream portion, where the cross-sectional area of the flow-through space increases incrementally in the direction of the fluid flow. The arrangement of several baffles creates a cavitation phenomenon and fuel–water mixing. The manufacturers claimed that their hydrodynamic cavitation is superior to the ultrasonic method, which is violent and hard to control. Alternative Petroleum Technologies (USA) uses a proprietary shear-based mixing of fuel and water with surfactants to create stable emulsions of HFO and other fuels. At the heart of their emulsion technology is a mixing device with a constant flow area, and it includes a high shearing apparatus for creating uniformly distributed water in fuel droplets (*52, 53*). Their emulsion units come with ~18,000 gallons per hour flow capacity, and they claim that they can produce highly reduced NO_X and particulate matter (PM) emissions (*54*). NoNox Ltd. (USA) has a patented hydrodynamic cavitation-based surfactant-free water-in-HFO emulsion preparation method. In a boiler operation with water-in-HFO in-line-prepared emulsion, they claim a reduction of 5–15% in fuel consumption, 25–50% in NO_X, 5–15% in CO_2, and 60–90% in PM. NoNOx units are available in 25–50 gallons per hour flow rates (*55*).

Several other companies are working on ultrasonics-based cavitation methods to manufacture industry-scale emulsifying and ODS units. Powerful shearing and mixing from the cavitation effects allow for effective oxygen transfer from the oxidant to sulfur-containing and other fuel compounds. The process is claimed to rupture the complex asphaltenic structures and break bonds in complex hydrocarbons, leading to overall upgrading of the fuel oil. Technoveritas (Portugal) makes commercial units by the name ENEREMULSION (*56*) for in-line ultrasound emulsification systems (*7*). Technoveritas has conducted large-scale testing of their commercial units with Sulzer/Wartzilla 12ZAV40 marine engines with diesel and HFO as fuels, claiming PM reductions up to 80%. Extensive literature (*18, 29, 57, 58*) suggests that ultrasonic water decomposition leads to various compounds like peroxides. These compounds and derived free radicals, like H_2O_2 (*57*), act as strong oxidizers and aid in combustion and ODS.

The manufacturers of hydrodynamic cavitation-based devices claim that these devices have a better chance of scaling up to industrial scale and being lower in cost compared with ultrasonic-based devices (*18*). However, several advancements in high-power ultrasonic transducer designs have made ultrasonic-based devices more cost competitive. Germany's Hielscher company successfully produced and operated a high-power cylindrical transducer in 2007. China developed first-generation piezoelectric ceramic material-based transducers in 2006, which were modified to give rise to the second generation of such transducers in 2013. This technology was later utilized in oil fields in China for enhanced oil-recovery applications. Some other recent advances include developing new lithium niobate crystals that have a 10^5 times greater mechanical quality factor than conventional piezoelectric transducers. These lithium niobate–based ultrasonic methods are being used in mainstream industrial applications and are bringing highly upgraded ultrasonic-based cavitation technologies to the fore. Hielscher's industrial ultrasound processor, UIP16000, is claimed to be the world's most powerful, designed to work in a cluster of several of these units (*59*). Four of these units can process up to 13,000 gallons per minute of liquid. If we consider a typical 400 MW boiler-based power plant, the mass flow rate of HFO could be in the vicinity of 75,000 to 100,000 gallons per hour. As is evident, the Hielscher unit is closing the gap for direct application in a large-scale power plant. The above list of companies is not exhaustive, and this brief overview was intended only as a glance at the latest developments in cavitation-based emulsification from the industrial world. *The challenge with all of the above-mentioned industrial units for emulsification is that the exact scientific details are either obscure due to the Edisonian approach of the development or are protected as trade secrets.*

2. Heavy Fuel Oil Deasphalting

Physiochemical Properties

Although UIC can effectively reduce the size of asphaltene aggregates, it cannot substantially cut the presence of asphaltene molecules. However, the process of completely deasphalting HFOs is a commercially viable solution to improve the performance of the fuel.

HFO accounts for a significant portion of extracted petroleum, also known as residual fuel oil or bunker fuel in the marine industry. HFO is generally commercialized as products of blending of vacuum residues and diesel or other lighter cuts. HFO has a tar-like appearance characterized by high viscosity and dark color. Because of the extremely variable composition of HFOs, they are distinguished by their viscosity, which has to be a maximum of 700 cSt at 50 °C according to the ISO 8217 standard, or they have a maximum density of 1010 kg/m^3 according to the ISO 8217 standard. A specification of sulfur content is provided to standardize the use of HFOs. As described below, their chemical composition is incredibly complex. The boiling range of HFO spans from 180 to 1000 °C (data obtained from extrapolated normal boiling point curves). HFO has an energy content comparable to distillate fuels, but its high viscosity necessitates heating before spray combustion inside a boiler or internal combustion engine.

HFOs are traditionally separated into four fractions: saturates, aromatics, resins, and asphaltenes (*60–62*). The SARA approach is based on differences in the solubility and polarity of these chemical classes. The saturate fraction mostly contains aliphatic compounds, while the aromatic one consists of aromatic groups with aliphatic side chains. Resin is a class characterized by heavier and more polar molecules than saturates and aromatics. Asphaltenes are hard-to-burn heavy polycyclic aromatic compounds with embedded heteroatoms, which lead to the formation of large cenospheres during inefficient burning. Asphaltenes are insoluble in alkyl solvents and are the most complex fraction. Sulfur is present mainly in thiophene and sulfidic structures. The most abundant form in which nitrogen is found is derivatives of pyrrole and pyridine. Organometallic porphyrins are also present in significant quantities. Oxygen is individuated in carboxylic, phenolic, and ester functional groups (*63–65*). HFO can contain asphaltenes (up to 15 wt %) (*66*), carbon residue, trace metals like vanadium (*67*), and fuel-bound nitrogen and sulfur (*60, 68, 69*). The sulfur content (2–5%) leads to the formation of harmful SO_x emissions. With the use of increasingly lower grades of HFO, the problems engendered by higher asphaltene and sulfur contents are expected to grow in intensity.

Deasphalting HFO

We evaluated and quantified the impact of asphaltene content on the physical properties of HFO in several studies. The properties of HFO samples, as received from Saudi Arabian power plants, can vary significantly in terms of heteroatoms and trace metals. In a previous study by our group (*70*), an HFO sample (named HFO^a) was deasphalted, resulting in deasphalted oil (DAO^a). The deasphalted oil is the portion of the HFO remaining after the asphaltenes are removed by precipitation.

The elemental analysis of DAO^a is compared with that of pure HFO^a in Table 1 (*70*). All measurements were conducted at KAUST in accordance with standard methods for HFO samples of Saudi Arabian origin. The elemental analysis was conducted using a CHNS/O analyzer (Flash 200 Thermo Scientific, USA), whereas the trace metal analysis was conducted using inductively coupled plasma atomic emission spectroscopy (ICP-OES, Agilent 5110). The viscosity of HFO emulsions at different temperatures was determined by an electromagnetically spinning viscometer (EMS-1000, KEM Kyoto Electronic Manufacturing, Japan). The density of HFO was measured by an electronic densitometer (JN-300S/E, Shanghai Jenner Industrial Co., China). The heating values of HFO and DAO were determined by a Parr Instrument 6400 Automatic Isoperibol Calorimeter.

The asphaltene separation process followed the ASTM D3279 standard. We used a three-step process, beginning with the transfer of approximately 1–1.3 g of HFO into a 250 ml flask. This was followed by adding 100 ml of the solvent *n*-heptane per 1 g of HFO that was originally added to the flask. The mixture was heated for 30 min using a reflux condenser. Next, the contents were allowed to pass through double filter pads and then into a filter flask. The filtered asphaltene particles were then kept inside a furnace maintained at a constant temperature of 110 °C for 30 min to remove any solvent impurities. The formed precipitate was then extracted by pouring toluene on the filter pads. The toluene was then evaporated from the asphaltene sample using a rotary evaporator device. Figure 1 provides a pictorial representation of solvent deasphalting of HFO. An image of solid asphaltene appearing as hard lumps extracted from the HFO sample is provided in Figure 2.

Table 1. Physiochemical Properties of HFO[a] and DAO[a] of Saudi Arabian Origin. Reproduced with permission from reference (70). Copyright 2019 Elsevier.

Properties	*Method*	*Units*	*HFO[a]*	*DAO[a]*
Density @ 288 K	ASTM D4052-11	kg/m^3	970.5	930.4
Specific gravity @ 288 K	ASTM D4052-11	–	0.971	0.931
Kinematic viscosity @ 298 K	ASTM D445-12	10^{-6} m^2/s	617.7	113.1
Carbon	EPA 440.0	mass %	85.0	84.5
Hydrogen	EPA 440.0	mass %	10.8	11.3
Sulfur	ASTM D4294-10	mass %	3.3	3.1
Nitrogen	EPA 440.0	mass %	0.24	0.26
Oxygen	EPA 440.0	mass %	0.03	0.05
Asphaltene content	IP 143	mass %	8.2	-*
Vanadium	IP 501-05	mg/kg	18.0	9.8
Nickel	IP 501-05	mg/kg	11.0	2.0
Sodium	IP 501-05	mg/kg	33.4	23.6
Zinc	IP 501-05	mg/kg	<1.0	<1.0
Lead	IP 501-05M	mg/kg	<1.0	<1.0
Potassium	IP 501-05M	mg/kg	<1.0	<1.0
Higher heating value	ASTM D4868	MJ/kg	42.5	43.4
Lower heating value	ASTM D4868	MJ/kg	40.2	41.1

* Relatively very low and unquantifiable by standard methods

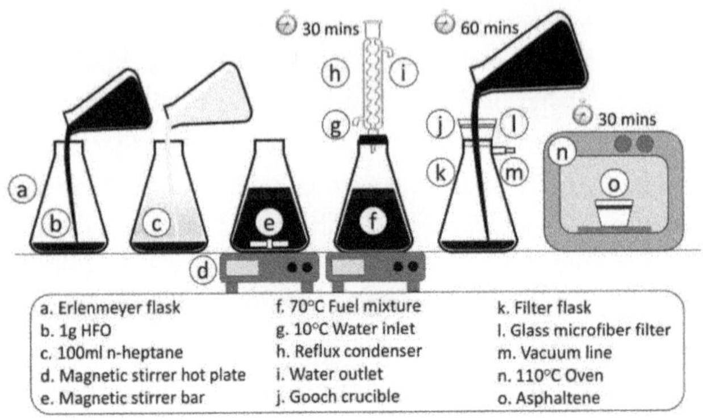

Figure 1. Overview of solvent deasphalting of HFO. Adapted with permission from reference (70). Copyright 2019 Elsevier.

Figure 2. Extracted asphaltenes.

A different sample called HFO[b] was deasphalted more recently, using the same ASTM D3279 method, and used for more detailed analysis. Table 2 reports the elemental composition of HFO[b] and DAO[b]. The table also reports compositional data (CHNSO, trace metals) of HFO[b] and DAO[b] when the asphaltene content of these samples is varied. The HFO[b] sample differs from the HFO[a] described in Table 1 in its sulfur and trace metals content. We quantified the effect of asphaltenes in HFO[b] of Saudi Arabian origin on the elemental composition, viscosity, and density. We used ICP-OES to further analyze the composition of the asphaltene. In comparison to HFO[a], the asphaltenes in HFO[b] have lower hydrogen content (8%) and a higher content of sulfur, nitrogen, nickel, vanadium, and magnesium. The elemental analysis of DAO[b] and extracted asphaltene is compared with that of pure HFO[b]. The extracted asphaltene was used to prepare the HFO[b] samples with 4 and 16 wt % asphaltene.

Asphaltenes have high concentrations of sulfur, nitrogen, and trace metals. The net quantity of asphaltene in pure HFO[b] is nearly 8%; thus, deasphalting only partially removed the refractory sulfur and trace metals from the HFO[b]. Table 3 reports the physical properties of HFO[b] and DAO[b]. The viscosity of DAO[b] is nearly 10 and 8 times lower than the pure fuel at 25 and 40 °C, respectively (see Table 3). The variation in viscosity for HFO[b] samples with 0, 4, 8, 16, and 100% asphaltene is plotted in Figure 3. As is evident, asphaltenes impart high viscosities to heavy fuels, enabling tremendous transport and flow processes. At low temperatures, the viscosity rise is almost asymptotic. The transport problems related to high viscosity can be overcome either by operating at temperatures higher than the ambient temperature, by removing asphaltenes, or by modifying their structure. Hence, the asphaltene needs to be either broken down (i.e., upgraded) or removed. Asphaltenes extracted from heavy oils like HFO can be used as asphalt for construction of roads, production of carbon black (*71*), and as surfactants (*5*), and they can also be used in gasifiers as fuel for syngas production.

Table 2. Effect of Asphaltene Content on Elemental Composition and Trace Metals of HFO[b]

Fuel	*Net Asphaltene Content in Sample, %*	*N, wt %*	*C, wt %*	*H, wt %*	*S, wt %*	*O, wt %*	*K, ppm*	*Na, ppm*	*Ni, ppm*	*V, ppm*
(DAO[b])	0	0.21	83.22	10.81	3.14	2.62	20.8	65.2	17.5	33.3
DAO[b]+additional asphaltene	4	0.23	83.19	10.58	3.32	2.68	–	–	–	–
HFO[b]	8	0.28	84.86	10.65	3.53	0.69	10.7	29.5	24.5	54.0
HFO[b]+additional asphaltene	16	0.28	84.75	10.31	3.88	0.78	–	–	–	–
Asphaltene	100	0.88	82.77	6.53	7.76	2.05	133.1	450.3	254.4	759.9

Table 3. Comparison of Density and Viscosity of HFO[b] and DAO[b] at Selected Temperatures

Fuel	*Density @ 25 °C, gm/ml*	*Dynamic Viscosity @ 25 °C, mPa·s*	*Dynamic Viscosity @ 40 °C, mPa·s*
Pure HFO[b]	0.9668	2194	648.2
DAO[b]	0.929	231	95.9
Asphaltene	1.771	–	–

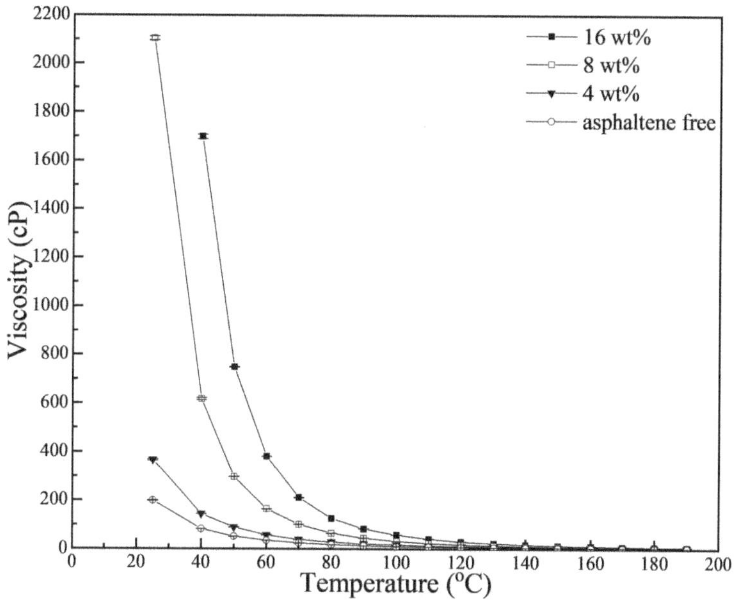

Figure 3. Variation of the viscosity of HFO[b] with the asphaltene content.

Molecular-Level Characterization

Many experimental techniques, such as NMR, Fourier transform infrared spectroscopy (FTIR), X-ray photoelectron spectroscopy, and X-ray absorption near-edge structure, have been widely employed as powerful characterization tools to identify and quantify the elemental composition and chemical state of the elements present in HFO and asphaltenes. State-of-the-art analytical techniques like NMR spectroscopy and FT-ICR mass spectrometry are the preferred techniques when analyzing residual petroleum fuels like HFO or vacuum residues as these complex fuels possess millions of individual molecules of varying sizes and functional groups with the presence of hetero atoms like O, N, and S (*72*). Gas chromatography–mass spectrometry, high-performance liquid chromatography, and other conventional analytical techniques are not suitable for the characterization of petroleum residual fuels like HFO and asphaltenes. Molecular characterization of HFO by our group at KAUST (*66, 73, 74*) has revealed valuable information on the molecular structure and complexity of asphaltenes.

FT-ICR Mass Spectrometry

FT-ICR mass spectrometry has the ability to assign unique elemental formulas to resolved ions, even those ions that have very minor differences in mass at the order of an electron (*75*). First, the sample has to be ionized with a suitable technique that is determined based on the chemical characteristics of the sample. In the present work, HFO, asphaltenes, and DAO samples were ionized by two known techniques: positive ion atmospheric pressure photoionization (APPI) and electrospray ionization (ESI). Both of these ionization techniques have been successfully demonstrated to ionize a number of heavy petroleum fuels. ESI is usually applied for ionizing heteroatom-containing polar molecules ($N_nO_oS_s$ species), which are notorious for causing catalyst fouling, corrosion, and deposit formation. APPI is efficient in ionizing the bulk nonpolar aromatic species present in HFO, like benzo- and dibenzothiophenes, polyaromatic hydrocarbons, and cycloalkanes, and it is therefore suitable for analyzing the composition of petroleum samples. Many researchers have used APPI along with FT-ICR-MS to study the molecular characteristics of several asphaltene samples. The FT-ICR-MS technique, in fact, gives an important, although hardly quantitative, insight into the structure of hydrocarbon mixtures. Marshall's group (*76, 77*) studied a number of crude oil samples using APPI. FT-ICR-MS has been used to analyze a number of other samples, like base oils (*78*), HFO (*79*), fuel oil derived DAO (*80*), vacuum residue derived DAO (*81*), vacuum residues (*82*), furnace oil (*83*), coal liquefaction products (*84*), lubricants (*85*), tyre pyrolysis oils (*86, 87*), and biomass (*88, 89*), besides crude oils (*76, 90–92*) and asphaltenes (*93–96*).

The experimental procedure used to analyze samples using APPI and ESI FT-ICR mass spectrometry is discussed in detail elsewhere (*70*). A Bruker Daltonics FT-ICR mass spectrometer (9.4 T Solarix XR, Bremen, Germany) was used for the analysis. Toluene was used to dilute the fuel samples as it can dissolve the asphaltenes present in the sample and keep them in solution. This was followed by direct injection into the APPI source. For ESI ionization, the samples were dissolved in methanol at a concentration of 100 pm. The aqueous samples were diluted 50 times with toluene APPI and with methanol for ESI. For both APPI and ESI, the ionization was done in the positive mode with a mass range of 154–1200 m/z. For APPI, the nitrogen-nebulizer gas pressure was 2.5 bar, and the vaporizer temperature was set to 400 °C, along with a dry temperature of 280 °C. A krypton light with ionization energy of 10.6 eV was the APPI source. The samples underwent soft ionization into molecular ions, with no fragmentation of the parent molecules. For ESI, a pressure of 0.5 bar was used for the nebulizer, and the dry temperature applied was 180 °C. The ion accumulation time for the APPI hexapole was 0.01 s, and for ESI, the time was 0.07 s. The spectra for both ionization methods were acquired with 300 scans with transient length 4.47 and time-domain size of 8 megapoints.

The FT-ICR mass spectra obtained using APPI and ESI ionization for HFO^a, asphaltenes, and DAO^a samples from one of our previous studies (*70*) are presented in Figure 4. The mass-to-charge ratio (m/z) for all three samples ranged from 154 to 1200. The average molecular weight (AMW) of HFO^a was reduced from 565 m/z to 531 m/z on account of asphaltene removal. This suggests that the asphaltene content of HFO^a has a higher percentage of molecules that are bigger in size as compared to HFO^a, which resulted in a lower AMW for the DAO^a sample. As discussed before, the present HFO^a sample had an asphaltene content of 8%. The AMW of all three samples was calculated by multiplying the masses of the ions with their corresponding signal intensity, followed by the summation of the products. The obtained number was then divided by the total intensity of the observed ions.

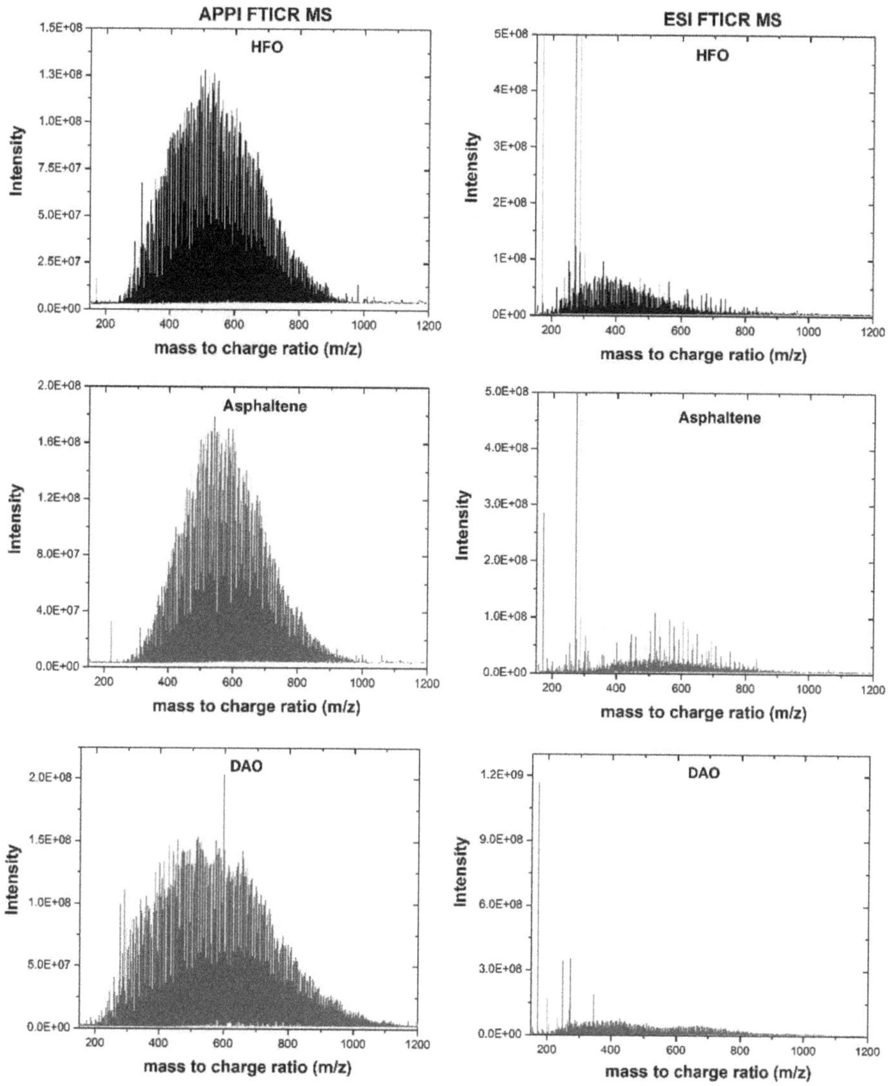

Figure 4. APPI and ESI FT-ICR mass spectra of HFO[a], asphaltenes, and DAO[a].

The distributions of S_1 and S_2 species in HFO[a] and DAO[a] as a function of the double-bond equivalent (DBE) and the carbon number are presented in Figure 5. The size of the bubbles indicates the abundance of the resolved species. The degree of unsaturation (aromaticity) present in the sample with the variation in molecular size can be understood from the DBE versus carbon number plots. Paraffins with a zero DBE and carbon numbers less than 15 are not observed in any of the three samples because they are not present in the samples or they could not be ionized. The latter possibility could be higher due to the stabilized paraffin ions that are difficult to ionize. The DBE plots of S_1 and S_2 species are presented as these classes are the most prevalent in sulfur-laden residual fuels like HFO[a], asphaltenes, and DAO[a]. The S_1 class indicates that the core structure of these fuels can be in the form of thiophenic functional groups or thiolic groups; the former seems to be more probable based on NMR results of petroleum fuels discussed in the literature. The DBE of the polycyclic aromatic sulfur heterocycle species in the samples falls in the vicinity of 7–12 for molecules having carbon numbers between 30 and 50. These molecules are more likely to be benzothiophenes (DBE 6), dibenzothiophenes (DBE 9), and benzonaphthothiophenes (DBE 12) with either short or long paraffinic chains. The DBE plots also indicate the prevalence of tetra and penta-aromatics in HFO[a], asphaltenes, and DAO[a] due to the presence of molecules with high carbon numbers (around 80) and large DBE values of around 20.

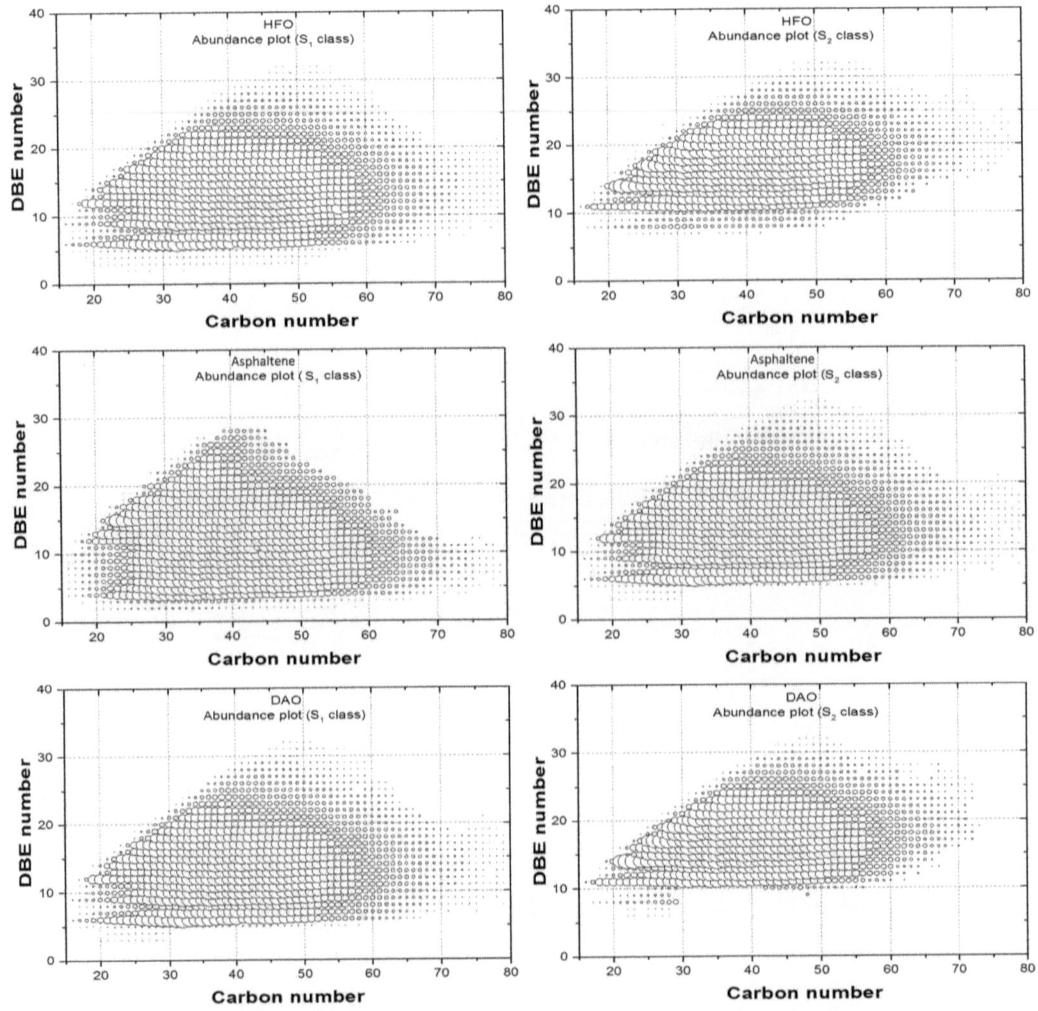

Figure 5. Carbon number vs. DBE plot for S_1 and S_2 species present in HFO[a], asphaltenes, and DAO[a].

The molecular species in HFO[a], asphaltene, and DAO[a] can be disassembled into their underlying classes based on the presence of S, N, and O heteroatoms. They are shown in Figure 6. HC stands for pure hydrocarbon molecules present in the sample. Hydrocarbon molecules containing one, two, and three sulfur atoms are indicated as S_1, S_2, and S_3, respectively. Similarly, N, O, and NS stand for hydrocarbon species with one nitrogen atom, one oxygen atom, and one nitrogen and sulfur atom each, respectively. The ions presented in Figure 6 are a sum of both protonated ions and radical cations, which are both observed in the APPI ionization mode. The species distributions in HFO[a] and DAO[a] are similar with minor differences between the HC, S_1, and S_2 classes. HFO[a] has a higher percentage of molecules with sulfur atoms at 62.3% compared with DAO[a], which has sulfur atoms in 60.8% of its molecules. This suggests that the asphaltenes removed from HFO[a] have a slightly higher sulfur distribution than HFO[a]. The molecular classes observed with APPI indicate the abundance of sulfur atoms distributed in the volumes of HFO[a] and DAO[a], which suggests that desulfurization techniques applied to these fuels will lead to significant structural modifications. Asphaltenes, on the other hand, are slightly lower in HC species than HFO, whereas the S_1 and S_2 class species are higher in asphaltenes compared to both HFO[a] and DAO[a]. The distribution of the species discussed in Figure 6 is with respect to abundance % (in terms of number of species resolved), whereas the elemental analysis in Tables 1 and 2 is in terms of mass %. The numbers of N species observed in DAO (5.4%) and asphaltenes (5%) are both greater than in HFO (4.6%), which may appear to be slightly analogous. This is mostly due to the uneven APPI ionization of the sample that may not accurately reflect the species distribution actually present in the sample. Also, the data reported in Figure 6 correspond to only the resolved ions whose signal-to-noise ratio is greater than 5. However, the errors are minor, and the FT-ICR data are fairly accurate and give a good overview of the molecular diversity present in the sample.

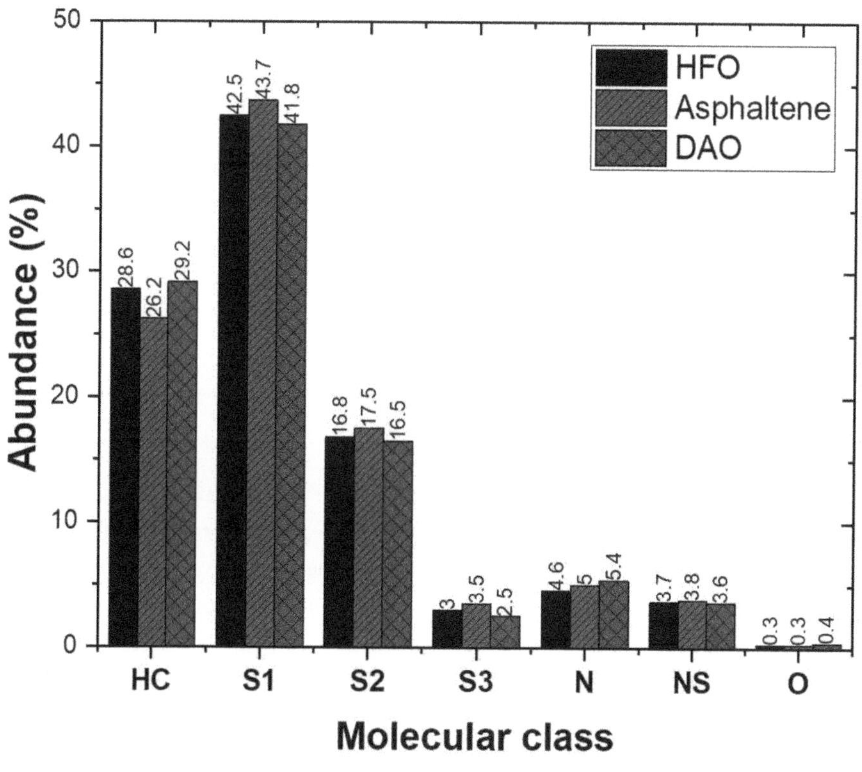

Figure 6. Molecular distribution of species present in HFO[a] and DAO[a].

^{1}H NMR Spectroscopy

High-resolution NMR spectroscopy has the inherent ability to provide qualitative and even quantitative structural information, such as the various hydrocarbon functional groups present in fuels (*97*). The molecular structure is identified by studying the NMR spectra, which are presented in the form of chemical shifts, which in turn are diagnostic of the molecular structure. NMR spectroscopy is ideal for analyzing hydrogen and carbon functional groups, as each of these groups appears distinctly on the spectra. For example, aromatic groups are always observed around 6.7–8 ppm on a ^{1}H NMR spectra. Spectral processing techniques can also remove the overlap seen in some functional groups (like naphthenic CH_2 and paraffinic CH groups) and make NMR convenient for quantifying the groups present. One of the advantages of ^{1}H NMR spectroscopy is that the measurement can be performed in a few minutes, making it ideal for high-throughput operations. Also, the spectra are reproducible almost every time, provided the experimental conditions are the same. The structural information obtained from ^{1}H and ^{13}C NMR can also be used to predict the combustion properties and sooting propensities of hydrocarbon fuels. NMR-derived functional groups have been widely used to predict the combustion properties of fuels, such as octane and cetane numbers. NMR spectroscopy has been applied to study the molecular structure of various fuels like gasoline (*98–100*), diesel (*101*), shale oils (*102, 103*), HFO (*79, 104*), coal-derived liquids (*105*), base oils (*97*), tyre pyrolysis oils (*86, 87*), vacuum residues (*106*), asphalts (*107*), and crude oil (*108*).

HFO[a], asphaltenes, and DAO[a] samples were analyzed by ^{1}H NMR spectroscopy using a Bruker 700 AVANAC III spectrometer. In preparation, 6100 µl of each of the three samples was transferred to 5 mm NMR tubes, and then 1600 µl of $CDCl_3$ was added to the NMR tubes. The ^{1}H NMR spectra were recorded with a recycle delay time of 5 s at a temperature of 298 K. A total of 128 scans were collected using a standard one-dimensional 90° pulse sequence. The detailed measurement procedure and the experimental conditions used were described previously (*70*). The spectra were collected and analyzed using the Bruker Topspin 2.1 software (and MestreNova NMR software was used to process the results). The ^{1}H NMR spectra of the HFO[a], asphaltene, and DAO[a] samples are presented in Figure 7.

Comparisons of various ^{1}H types in the samples are presented in Figure 8. The paraffinic CH_3 and CH_2 groups are slightly higher in DAO[a] (73.4 mol %) compared to HFO[a] (65.5 mol %) as the hydrogen content in DAO[a] is also slightly higher as observed from the elemental analysis of the samples. The aromaticity (number of aromatic carbons divided by total number of carbons) of HFO and DAO is 0.38 and 0.33, respectively, which indicates the remaining carbon atoms are present mostly as substituted alkyl side chains. Also, the average aromatic ring number of HFO and DAO is 3.2 and 2.9, respectively, which indicate the possibility of polycondensed aromatic rings (*70*). This is the reason for the low content of aromatic hydrogen atoms in HFO and DAO. Asphaltenes have lower paraffinic content than both HFO[a] and DAO[a]. The aromatic content in asphaltenes is, however, greater than that observed in HFO[a]. Since asphaltenes have been shown to possess a higher polycyclic aromatic sulfur heterocycle content than the parent source fuel, the polyaromatic content in HFO[a] at 4.5% is greater than that in DAO[a] at 2.9%. The other species distributions in HFO[a] and DAO[a] are comparatively similar to each other, with only minor variations, whereas olefin species are absent in both fuels.

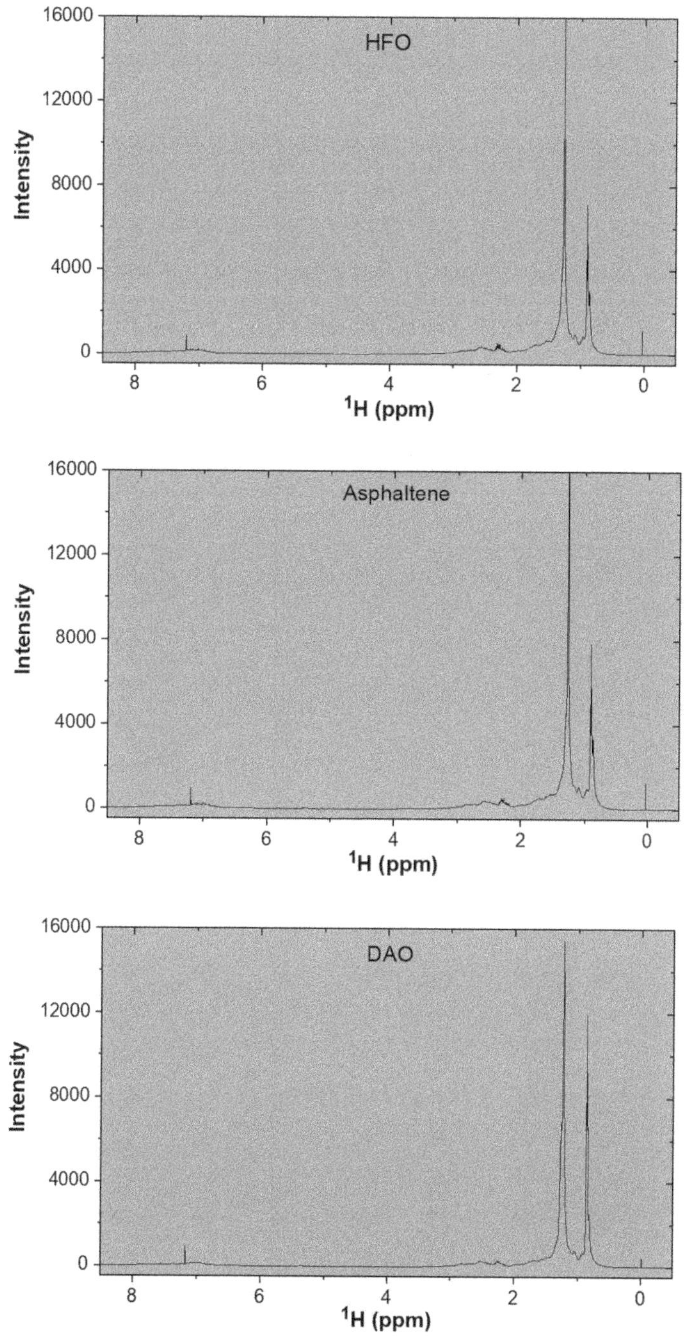

Figure 7. ^1H *NMR spectra of HFO*[a]*, asphaltenes, and DAO*[a]*. Reproduced with permission from reference (70). Copyright 2019 Elsevier.*

Combining the data from ^1H NMR spectra with data from other analytical techniques like FT-ICR-MS and elemental analysis (CHNSO) enables us to calculate the average molecular parameters (AMP) of the fuel sample. The AMP method is based on quantifying various functional groups present in the fuel by imagining a hypothetical average molecule designed from the elemental distribution and AMW obtained from mass spectrometric methods like FT-ICR. An example of an AMP is the number of paraffinic hydrogens (or carbons) present in the average molecule. The advantage of this method is that it can help us to visualize the structure of a complex fuel like HFO or DAO and also to estimate the fuel properties by using various quantitative structure–property relationship techniques. The average molecular formulas of HFO[a] and DAO[a] calculated using the CHNS and AMW data are $C_{40}H_{60.5}O_{0.01}N_{0.1}S_{0.58}$ and $C_{37.4}H_{60}O_{0.01}N_{0.1}S_{0.52}$, respectively. The AMPs calculated for HFO[a], asphaltenes, and DAO[a] are presented in Figure 9. The numbers of hydrogen and carbon atoms in HFO[a] are 61.5 and 40, respectively. HFO[a] has 4.7 paraffinic CH_3 groups and 15.3 paraffinic CH_2 groups in its average molecule, whereas DAO[a] has 4.3 paraffinic CH_3 groups and 14.1 paraffinic CH_2 groups. Asphaltenes have a slightly lower content of paraffinic CH_2 and CH_3 groups compared to both HFO[a] and DAO[a], whereas the proportion of monoaromatic and polyaromatic groups in asphaltene is higher. The number of aromatic atoms is higher in HFO[a] compared to DAO[a], whereas the heteroatoms, namely S, N, and O, are present in a negligible amount in both HFO[a] and DAO[a].

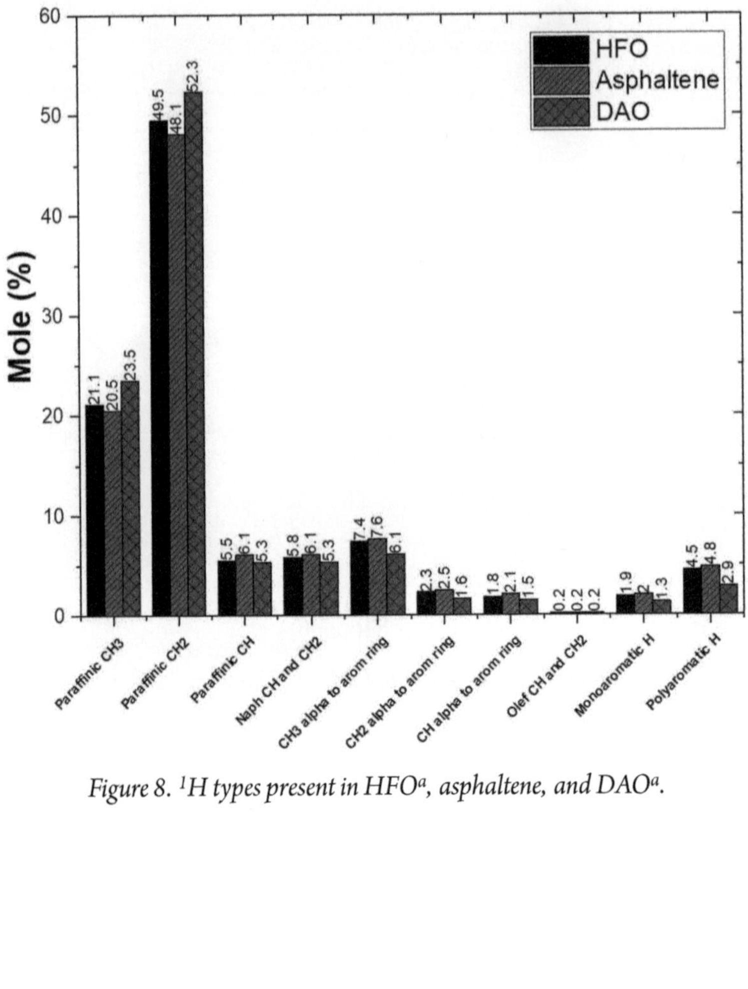

Figure 8. [1]H types present in HFO[a], asphaltene, and DAO[a].

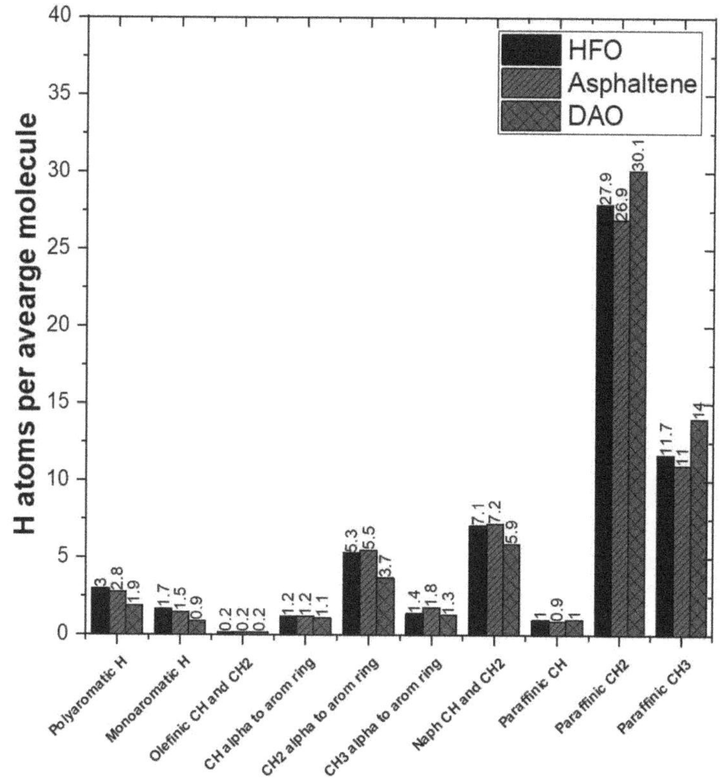

Figure 9. AMPs of HFO[a], asphaltene, and DAO[a] obtained by combining the results from ^{1}H NMR, elemental analysis, and FT-ICR-MS.

3. Ultrasonically Induced Cavitation and Water in HFO Emulsions

Fundamental Aspects

UIC is a phenomenon whereby vapor-filled cavities are formed by the application of a fluctuating pressure field within a liquid medium. This is expected to occur in areas of the liquid domain where the pressure drops below the liquid's vapor pressure. However, recent studies (*109–111*) demonstrated that a pressure below the saturation pressure is not enough to trigger the formation of bubbles. In fact, very high stress is required to pull the liquid apart and finally generate a cavity. Classical nucleation theory does not hold true in the case of homogeneous nucleation induced by pressure gradients. The controlling mechanism of bubble formation in still liquids is expected to be heterogeneous nucleation, with the bubbles nucleating on the sonotrode surface and eventually migrating to the bulk following the velocity streamlines generated by the sonotrode motion (*112*).

However, nucleation can also be favored by solid particles dispersed in a bulk liquid that may lower the energy required for the bubbles to nucleate. It is worth noting that asphaltene clusters may serve this purpose if sonication takes place in HFOs. After nucleation, the bubble lifetime will be affected by the interaction with the oscillating pressure field. At high-pressure amplitudes, bubble motion is strongly nonlinear.

When the acoustic pressure wave enters the rarefaction phase, the bubble's volume grows. Subsequently, the acoustic pressure wave enters its compression phase, and the drop will continue the expansion phase. However, this system configuration is energetically unfavorable; indeed, bubble growth is quickly arrested. The bubble now begins a rapid collapse. The velocity of this collapse in its last stages can be enormous and, in some cases, can be several times the speed of sound in gas at ambient temperature and pressure. The violent implosion of an acoustically driven bubble leads to the generation of high intracavity temperatures and pressures. In these hot spots, temperatures of thousands of kelvin are expected with strains of up to 100 bar.

The dynamics of such bubbles is usually modeled by adopting the Rayleigh–Plesset equation (see eq (1)). The Rayleigh–Plesset equation directly derives from the continuity equation applied to a bubble immersed in an incompressible, viscous Newtonian fluid:

$$R\left(\frac{\mathrm{d}^2 R}{\mathrm{d}t^2}\right)+\frac{3}{2}\left(\frac{\mathrm{d}R}{\mathrm{d}t}\right)^2+\frac{4\nu_L}{R}\frac{\mathrm{d}R}{\mathrm{d}t}+\frac{2\sigma}{\rho_L R}+\frac{\Delta P(t)}{\rho_L}=0 \quad (1)$$

where R is the radius of the bubble, ν_L is the kinematic viscosity, ΔP is the pressure difference between the external and internal pressure, and σ is the surface tension coefficient. No mass transfer between the bubble and the bulk is considered in this analysis. Contributions of viscosity and surface tension to the Rayleigh–Plesset equation are usually neglected. However, this assumption becomes incorrect when dealing with heavy fuels. The assumption of polytropic transformation leads to the following relation for temperature:

$$T_b(t)=T_b^0 \frac{R(t)}{R_0}^{3(1-\gamma)}\left(\frac{2\sigma}{R_0 P_0}+1\right) \quad (2)$$

where γ is the polytropic exponent and the 0 index indicates the initial conditions. Figures 10 and 11 report the results of the numerical solutions of eq (1) and eq (2) using the physical properties of HFO at 50 °C and the pressure profile reported in eq (3):

$$p = p_a sin(\omega t) \quad (3)$$

where p_α is the pressure amplitude and ω is the oscillation frequency.

Although the results were obtained from a simplistic analysis, they are representative of the physics that takes place in UIC-driven phenomena. HFOs are generally a blend of diesel and vacuum residue in a proportion of 40:60 by mass. Diesel will eventually be the fraction that will evaporate once negative pressure is established in the liquid.

Figure 10, in particular, depicts the radius fluctuations of a bubble immersed in a pressure field generated by oscillations of the sonotrode. The sonotrode vibrates at 20 kHz, and the derived pressure field is supposed to have the same periodicity. Collapse occurs when the pressure peaks as the expansion step and short oscillations occur when the pressure moves to negative values. When the bubble collapses, the droplet's temperature increases dramatically, reaching values above 2000 K (see Figure 11), which makes the occurrence of radical formation possible in the hotspot region also with the eventual migration of the radicals to the liquid phase.

The consequences of these extreme conditions are chemical reactions (sonochemistry) and intense physical forces that can be used for various processing applications and for promoting chemical reactions. These effects include radical generation, light emission sonoluminescence, shock waves, microjets, shear forces, and turbulence.

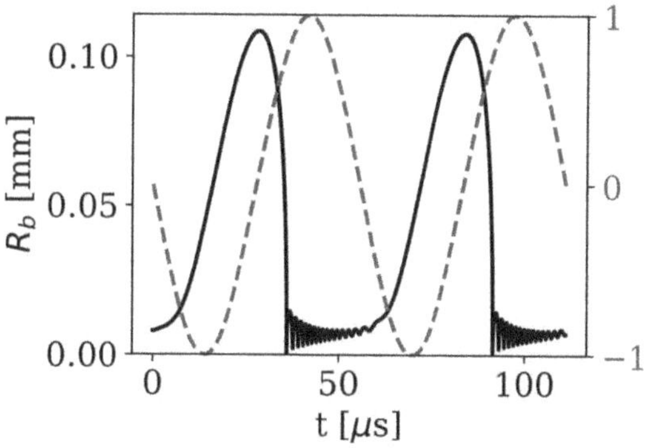

Figure 10. HFO bubble oscillations under an oscillating pressure field. The frequency of the oscillating field is 20 kHz. The vaporizing bubble is supposed to have the same thermodynamic properties as diesel fuel.

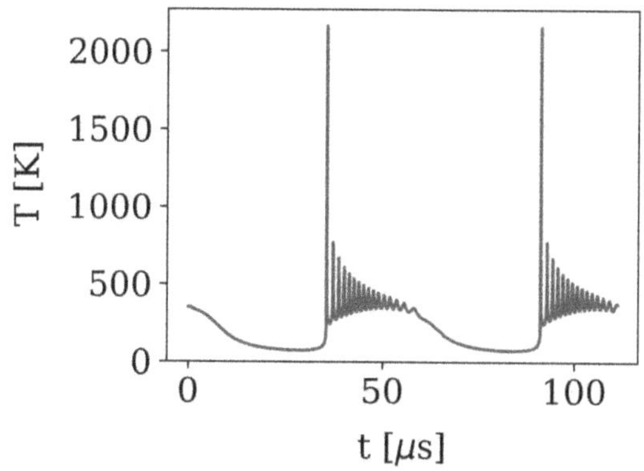

Figure 11. Temperature profile of the oscillating bubble. The temperature fluctuates with the radius of the droplet.

Direct Upgrade of HFOs through Breakage of Asphaltene Clusters and Cracking of Hydrocarbons

As described above, UIC induces local hot spots and, in general, increases the temperature of the mixture. When an ultrasonic source is applied to water, decomposition takes place (Figure 12) (*113, 114*). The water decomposition results in the formation of H· and OH· upon collapse. However, when water is not present in the mixture, some authors suggested other mechanisms would take place, such as the Rice mechanism in the presence of hydrocarbons.

Two effects play a role in changing the structure of the liquid. The first effect is a physical effect: jet formation, which follows bubble collapse and may cause breakage of the asphaltene clusters. The sizes of asphaltene clusters and bubbles are on the same order of magnitude. The second is a chemical effect. A series of reactions takes place as a consequence of radical formation in the local hotspot and because of the increased temperature. These reactions can eventually occur in the gas or liquid phase. Gas-phase reactions take advantage of the high number of radicals made available by extremely high temperatures, as well as of the large pressure gradients generated by bubble oscillations. Liquid-phase reactions occur at lower temperatures because the intense mixing of the strong temperature gradients is promptly distributed within the bulk phase. Typical temperatures range from 60 to 150 °C.

By looking at the system from a macroscopic perspective, a straightforward energy balance allows us to estimate how much fuel can potentially be converted into smaller molecules in a UIC system. We are supposed to break a C–C bond per molecule to form a smaller component using

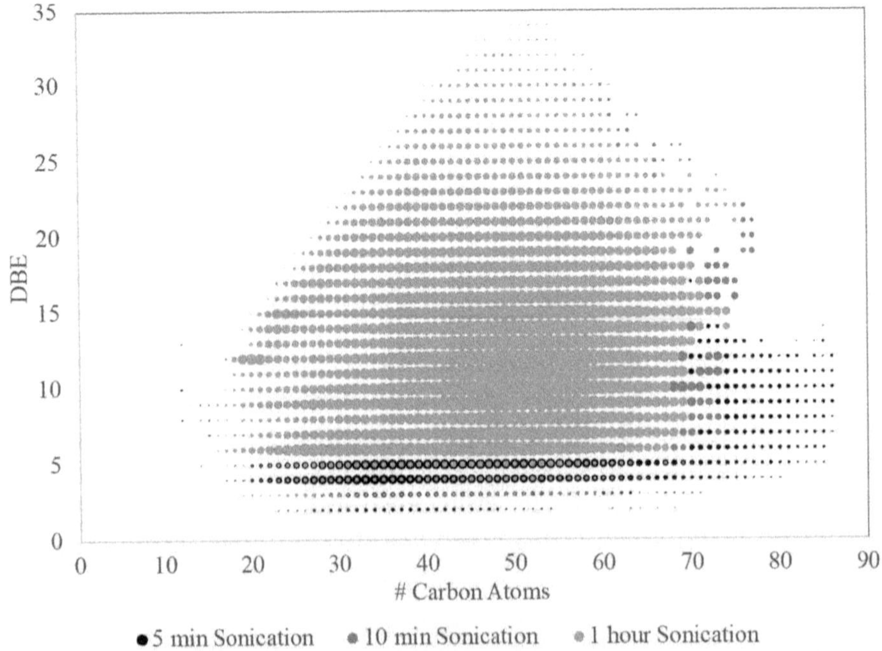

Figure 14. Sulfur species distribution as a function of DBE and the number of carbon atoms at different sonication times.

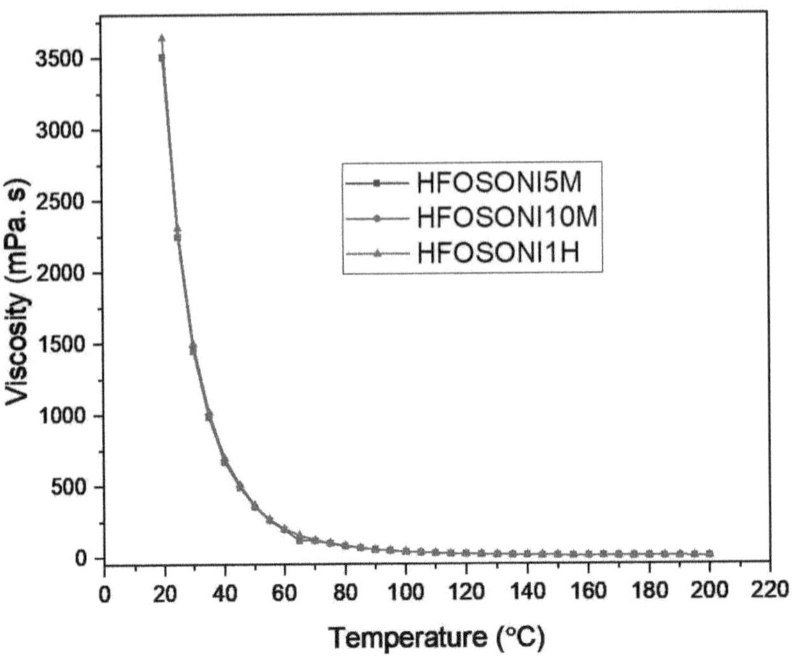

Figure 15. Measured viscosities of sonicated HFO samples at 5, 10, and 60 min.

Water-in-Fuel Emulsions

Forming emulsions from HFOs serves two important purposes in combustion applications:

- Enhancing secondary atomization and thereby increasing carbon efficiency.
- Reducing the production of pollutants, such as NO_x, SO_x, and PM, through the presence of water in the combustion chamber.

The quality of the emulsion is mainly a function of three parameters: water content, the size distribution of the droplets, and the stability of the emulsion. Whereas the water content and size distribution of the droplets directly influence the combustion behavior, the emulsion's stability is relevant only when long-term storage is required. In in-line emulsions, the emulsion's desired stability is necessary only for a short time (on the order of hours or days). Some stability-related concerns can, therefore, be relaxed when emulsions are prepared for immediate burning. However, some stability in emulsions, even those prepared for immediate burning, is essential for maintaining the physical properties of the emulsified fuel until burning. Advanced techniques like sonication can provide small-enough size distributions of droplets for the required stability without the addition of any surfactants. Therefore, the goal for characterization involves determining the optimal water content and droplet size distribution for the emulsions under study.

The formation of stable water-in-HFO emulsions without surfactants is possible because of two factors. The first is the *high asphaltene and resin content of HFO*. As noted above, asphaltenes are a class of hydrocarbons with heteroatoms like S, N, and O defined only by their precipitation in nonpolar solvents like pentane, hexane, or heptane. They are also natural emulsifiers. The resins in HFO are smaller in size (as indicated by the number of carbon atoms) than asphaltenes, but they have a similar structure. Both are polar compounds as characterized by their solubility criteria. In crude and derived fuel oils, asphaltenes are suspended as micelles stabilized in the solution by resins (*120, 121*). Variations in pressure, temperature, and composition of the oils can cause the asphaltene molecules to form aggregates of different mean molecular weights, which can result in precipitation, deposition, and filter clogging. The dominant molecular and colloidal structure of asphaltenes has been categorized in the Yen–Mullins model (*122, 123*). Asphaltenes migrate to the oil–water interface and self-assemble to form a rugged but elastic and stable film around the oil droplet. Resins act as solvents to asphaltenes and provide added stability. In this study, we evaluated the role played by asphaltenes in the stability of water-in-HFO emulsions. Figure 16 shows the typical structures of asphaltenes and resins.

The second factor is the *small droplet size distribution*: emulsions with small droplets are usually more stable than those with large droplets because smaller droplets can resist flocculation, creaming, and coalescence (*124*). Droplet breakup is an important process that determines the size distribution of droplets in an emulsion. Droplets can be broken by applying energy to overcome the pressure difference between the convex and concave sides of the interface, known as the Laplace pressure. This can be accomplished by using high shearing forces or fluctuating velocities and pressures. Typical methods include mechanical homogenization at high rotational speeds or ultrasonics. Interfacial tension can be lowered by adding surfactants, which reduces the energy needed to break the droplets. Ultrasonic homogenization has the potential to produce very tiny droplets in common oil–water mixtures.

We have characterized emulsions as created by UIC with two modern techniques: cryogenic microscopy and high-speed visualization.

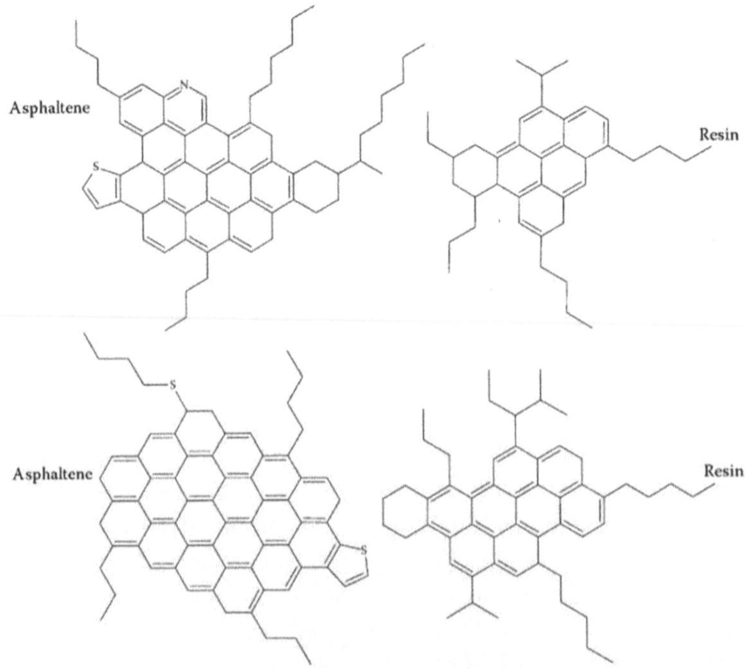

Figure 16. Asphaltenes and resins.

Cryogenic Electron Microscopy

We have utilized highly sophisticated imaging techniques such as cryogenic scanning electron microscopy (cryo-SEM) and cryogenic transmission electron microscopy (cryo-TEM) for emulsion characterization (*125*). A deeper understanding of the role of asphaltenes in the emulsification phenomenon has thereby evolved. The capability to determine the exact droplet size distribution can be critical in testing the emulsified fuel combustion characteristics because earlier work (*42, 126, 127*) suggests it is a crucial parameter for optimizing combustion performance. We have demonstrated accurate measurements of emulsion size distribution via cryo-EM imaging, as explained below.

The specimen preparation for both cryo-SEM and cryo-TEM was described previously by our group (*125*). In Figure 17, we present images of UIC-created water-in-HFO emulsions from both optical and cryo-electron microscopy. The image shown in Figure 17(C) was acquired with an optical microscope with a large field of view. It shows the dispersion of water droplets in a water/oil (W/O) emulsion droplet. The size of the water droplets is found to be in the range of a few micrometers or less. Figure 17(B) was acquired using the cryo-SEM technique. Before cryo-SEM analysis, W/O emulsion droplets were frozen and then transferred to the SEM chamber held at cryoconditions. Before cryo-SEM imaging, the samples were warmed up to -139 °C to sublime the water from the W/O emulsion droplets. This step is essential to having successful cryo-SEM imaging because the presence of water in the specimens degrades the quality of the acquired images. Otherwise, the interpretation of images becomes difficult if not impossible. In this way, the image shown in Figure 17(B) is a result of sublimation. The volumes that were once the water droplets within the emulsified sample are observed as void volumes from which information about the water droplet size distribution in the original W/O emulsion can be obtained.

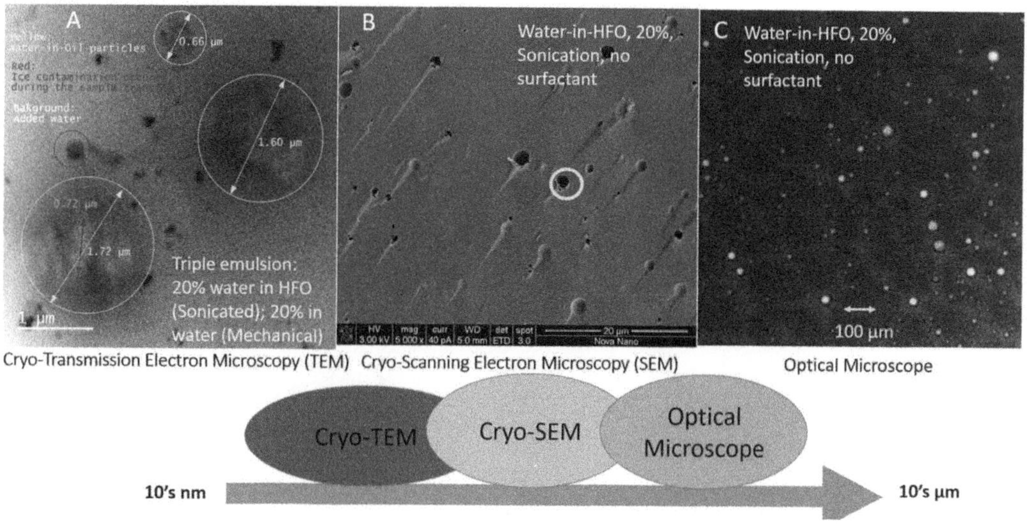

Figure 17. Water-in-HFO emulsions analyzed with cryo-SEM and cryo-TEM.

Nevertheless, the image shows the droplets' W/O droplet size and the interconnectivity of oil and water. The size of W/O droplets is similar to that of direct optical microscopic resolution. Even so, this type of analysis gives improved contrast over direct optical microscopic observations while permitting resolution of water droplet sizes (void sizes) to 100 nm, far smaller than allowed by direct optical microscopy. The image shown in Figure 17(A) is a cryo-TEM image of the W/O specimen. The specimen for cryo-TEM is very different from that for cryo-SEM. For cryo-SEM, the W/O emulsion was dissolved into the oil, while for cryo-TEM analysis, the W/O emulsion was dissolved into water. This is necessary because oil is too viscous to produce a specimen that can be conducive to cryo-TEM analysis. The image in Figure 17(A) shows the presence of the W/O emulsion in water. Lighter regions in the emulsion droplets are actually water that was trapped by the oil network. The cryo-TEM results confirm the cryo-SEM observations and show the presence of water in the W/O emulsion droplets in the native state. One disadvantage of cryo-TEM observations is that they capture only small droplets of 100 nm to 500 nm.

We achieved greater success in characterizing water-in-HFO with cryo-SEM than with cryo-TEM. Much of our emulsion study was therefore based on the cryo-SEM technique. The emulsions produced with UIC were visualized using the cryo-SEM technique (see Figure 18(A)), which allows us to image details at a scale of a few nanometers, which is comparable to the size of asphaltene clusters. The asphaltenes in HFOs are a natural surfactant, and they tend to form a solid shell around the water droplets, which stabilizes the emulsions. No surfactant is therefore required to form commercially viable emulsions. Ultrasonically produced emulsions contain smaller water droplets that are more uniformly distributed and very stable. The oil matrix presents holes and bumps, which we suppose are a residue of the water droplets present in the mixture in the form of eutectic ice. The energy-dispersive X-ray analysis technique was used to identify the elements in the obtained images. Figures 18(B) and 18(C) show carbon and oxygen, respectively.

The presence of oxygen confirmed that eutectic ice remained after desublimation. The bumps on the other side correspond to the droplet, which maintained an intact shell surrounding the water. We suppose that the shell that surrounds the water droplets is mainly formed by asphaltenes since no surfactant was added. The asphaltenes have been found to be able to stabilize W/O mixtures (*38, 124, 127–130*).

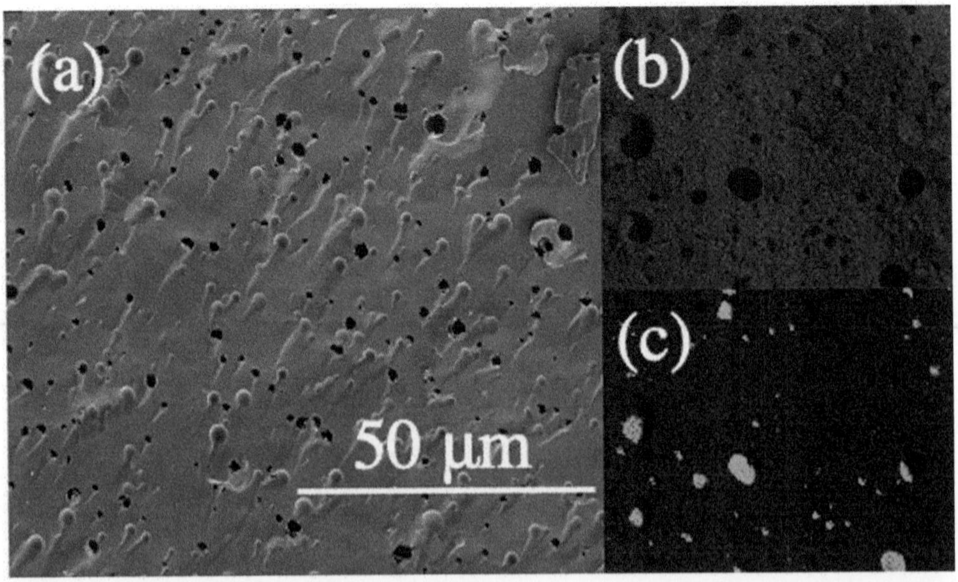

Figure 18. Cryo-SEM results from the emulsified fuel. A raw image is presented in (a), while (b) and (c) show carbon and oxygen energy-dispersive X-ray analyses, respectively.

Figure 19 shows a zoomed-in view of a void from a single sublimated droplet from a UIC water-in-HFO emulsion. Figure 20 presents another view of the void, where the hollowness is clearly evident. Figures 21 and 22 provide images of water droplets obtained through UIC and mechanical mixing. The size of the emulsified droplets was calculated from the SEM images using an in-house algorithm consisting of circle detection coupled with color gradient detection and inspired by the well-known Hough transform algorithm (*131*). Multiple pictures were analyzed, and the sum of the voids was validated against the mass fraction of water used in the emulsion, resulting in a small error. The UIC droplets are much smaller, falling in the nanoscale. Two reasons are proposed to explain this observation. First, UIC guarantees better mixing than mechanical mixing because of the activity of the bubbles, which collapses water emulsions acting in the nanoscale. The second reason for the smaller size may be found in the presence of smaller asphaltenic aggregates.

As mentioned above, asphaltenic aggregates are disrupted by UIC; from the images, the size of the asphaltenic aggregates can be approximately measured (the uncertainty is the resolution of the machine at 2 nm). The size of the layer for mechanically produced samples is between 5 and 10 nm (Figure 2), while the size of the layer for ultrasonically produced samples is about 2–3 nm. A smaller water droplet size is also obtained with less water in the mixture. This can be caused by the fact that the amount of asphaltenes per unit volume becomes larger, and therefore more droplets can be covered on a statistical basis. Figure 23 compares the droplet size distributions of mechanical stirring versus ultrasonically produced emulsions. The implication of this difference in practical applications is enormous. Asphaltenes are the main constituent of cenospheres, and they are responsible for the low carbon efficiency and difficult combustion of HFOs. Asphaltenes will be disrupted when secondary atomization occurs during spray formation, leading to cleaner combustion. This mechanism was described in the past (*132*), and extensive results are reported in the following sections. We have also estimated the droplet size distributions of UIC water-in-HFO emulsions with varied water content (see Figure 24). We believe that with optimized UIC conditions and water content, water-in-HFO emulsions can be designed for various performance criteria.

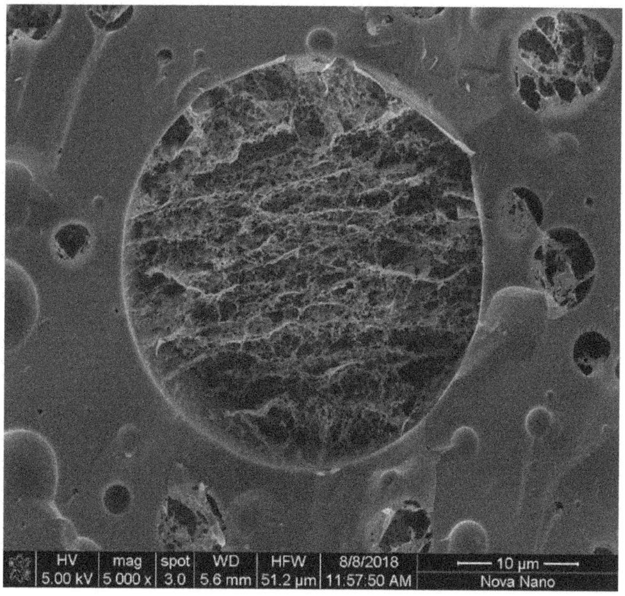

Figure 19. Asphaltenes covering the skeleton of a sublimated water droplet: the hollow sublimated water droplet (void) with oil remnants shows the scaffolding of an asphaltene layer.

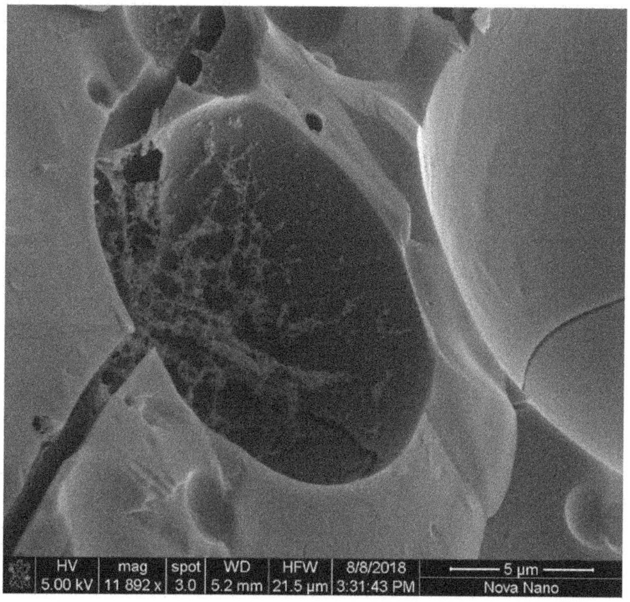

Figure 20. Asphaltene scaffolding as seen from an angled view; the void is more evident in this image.

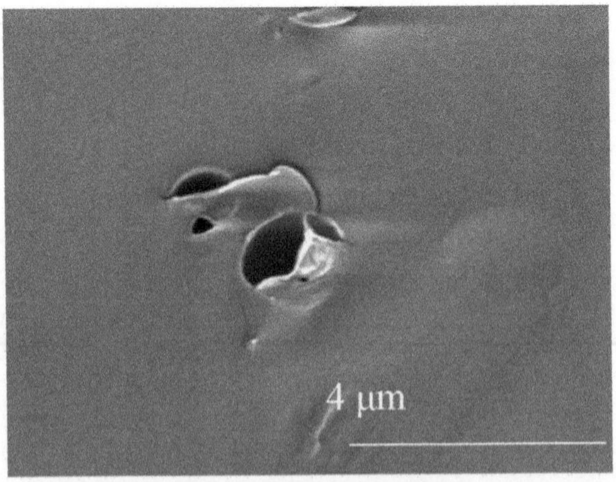

Figure 21. UIC emulsion.

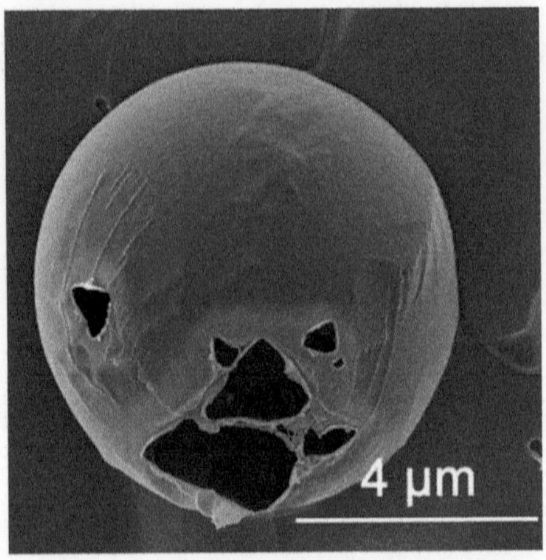

Figure 22. Mechanically produced emulsion.

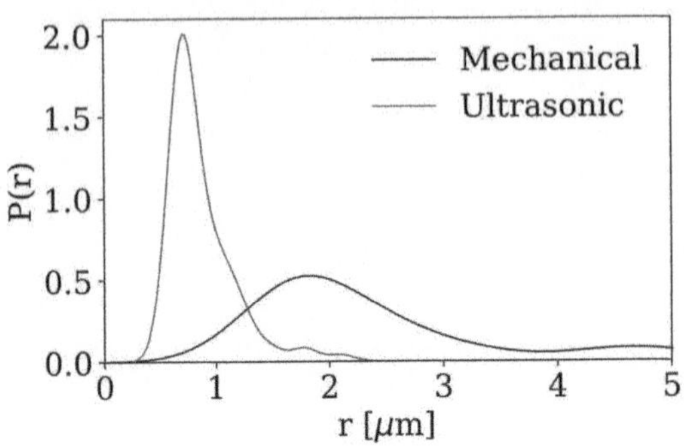

Figure 23. Difference in droplet size obtained through mechanical mixing and UIC. Both tests were performed with a mixture of 20% water in HFO by weight.

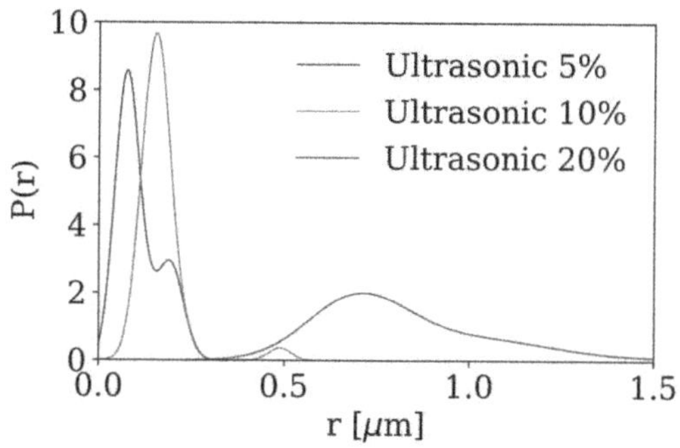

Figure 24. Droplet size distributions of water droplets in emulsified fuel with different water percentages.

High-Speed Visualization

We used high-speed visualization to fundamentally discern how UIC affects the chemical and physical properties of fuel. We characterized the water–fuel emulsification process through sonication. A significant challenge in high-speed imaging of emulsion formation during ultrasonic cavitation is the opaqueness of HFO. This opaqueness is due to the high concentration of asphaltenes and other solidified components in the fuel. A similar problem was noted in our earlier work on imaging of water emulsion droplets inside crude oil samples. Our imaging approach is based on using relevantly thin channels with glass walls, in which the sample is observed, for example, square profile capillaries that have 0.5 × 0.5 mm inner cross sections, and on using near-infrared (NIR) laser illumination, which better penetrates the oil phase than white light illumination can (*133*).

We ran several tests to evaluate the extent of these limitations. In these tests, we used an NIR laser diode of 980 nm, a high-speed monochrome camera (Photron SA-5) with attached micro-objectives that can give up to 30× magnification (~1.5 μm/pixel), and a thin-channel microfluidic cell with internal channel dimensions of 0.1 mm thickness, 8 mm length, and 38 mm width (*134*).

Figure 25 shows some representative examples of an HFO emulsion imaged inside a microfluidic cell. Figure 25(A) is a snapshot of a 2 wt % coarse water droplet emulsion filmed using NIR laser illumination, and Figure 25(B) shows the same sample using regular white-light illumination. The comparison demonstrates the better visibility achieved using NIR illumination. Figure 25(C) shows a thin liquid film with a high water concentration of 20 wt % in HFO with fine water droplets mixed by ultrasonication. At the edge of the film, single water droplets can be discerned. However, once the film is slightly thicker (more than several micrometers), the very high concentration of the water droplets makes the sample opaque (milk effect) to the NIR light, which is absorbed by the water phase. Once the sample is further diluted with HFO to bring the water concentration to below 2 wt %, the droplets' movements inside the thin film are readily observed, as shown in Figure 25(D).

The high-speed camera is capable of imaging sonication bubbles produced by the sonication device (UP400S, Hielscher). As clarified above, visualization inside bulk HFO emulsions is possible only for very thin layers. We used pure water, transparent diesel fuel oil, and diesel fuel with water droplet emulsions as model systems to capture the cavitation bubbles. The sonication frequency was 20 kHz, while the acoustic power density was 105 W/cm^2. Figure 26 shows several high-speed video snapshots of sonication in pure water, demonstrating that imaging under these sonication parameters of individual bubble growth is possible. In the example given, the frame rate is 4000 frames per second (fps) and the shutter speed is 1/40,000 sec (40 kHz), which is comparable with the sonication frequency.

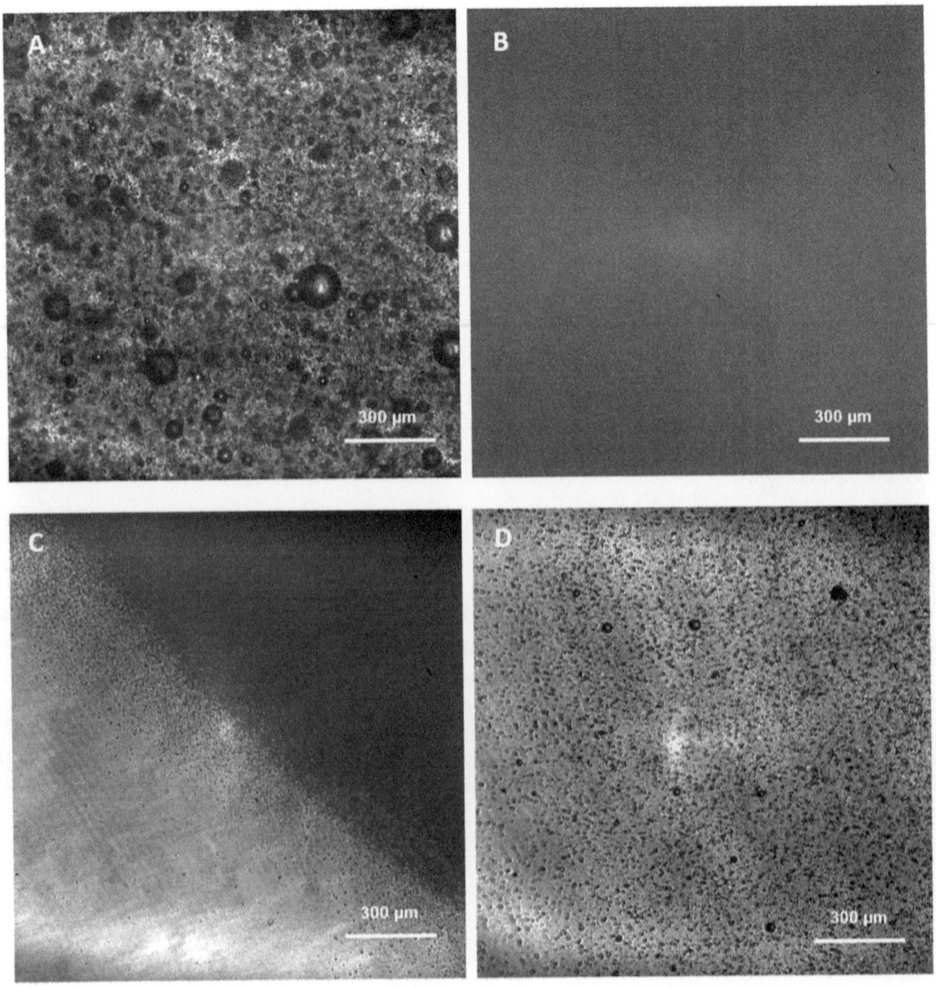

Figure 25. High-speed camera video snapshots of water in an HFO emulsion inside a microfluidic cell channel of 0.1 mm thickness. (A) Coarse emulsion of 2 wt % water content filmed using NIR illumination and (B) the same sample filmed using white-light illumination. (C) NIR illumination of a 20 wt % fine water droplet emulsion film spreading on a glass slide showing the edge of the film. (D) Lower water content (<2 wt %) of fine water droplet emulsion film filmed using NIR illumination.

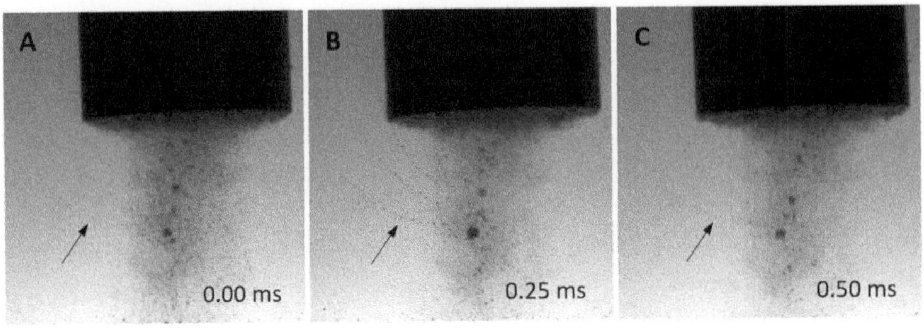

Figure 26. Snapshots from high-speed camera videos showing cavitation bubbles near the sonotrode tip in pure water. Ultrasonication is at 20 kHz. Filming rate 4000 fps, shutter speed 1/40,000 (40 kHz). As marked with the arrows, individual bubbles can be discerned.

Figure 27 demonstrates the capability of our ultra-high-speed Kirana video camera to capture 180 images at up to 5 million images/second at full resolution (924 × 768 px, 180 frames, 5,000,000 fps). The example shows the sonication of pure water. The high filming rates allow for detailed tracking of cavitated bubble progression during the sonic cycle, even capturing bubble burst events (red arrow in Figure 27(D)). Figure 28 shows a cavitation bubble plume at the sonotrode during sonication in pure diesel (Figure 28(A)) or in a diesel–water emulsion (Figure 28(B)). Figure 28(C & D) shows higher magnification images of the area below the sonotrode tip shortly after the sonication is terminated for the respective cases of pure diesel (Figure 28(C)) and diesel–water emulsion (Figure 28(D)) in which finer details can be resolved. In all cases, we can observe the bubble cloud progressing during the application of the ultrasonic pulses with the imaging quality matching literature examples (*112*). These videos were filmed using a filming rate of 1000 fps. The individual cavitation bubble size is expected to be several micrometers below sonication conditions (*135*).

Although the above are preliminary experimental results, they do provide us with valuable information on UIC-based water-in-fuel emulsions. The imaging information can fill a vast gap in the validation of computational models for UIC water-in-fuel emulsions.

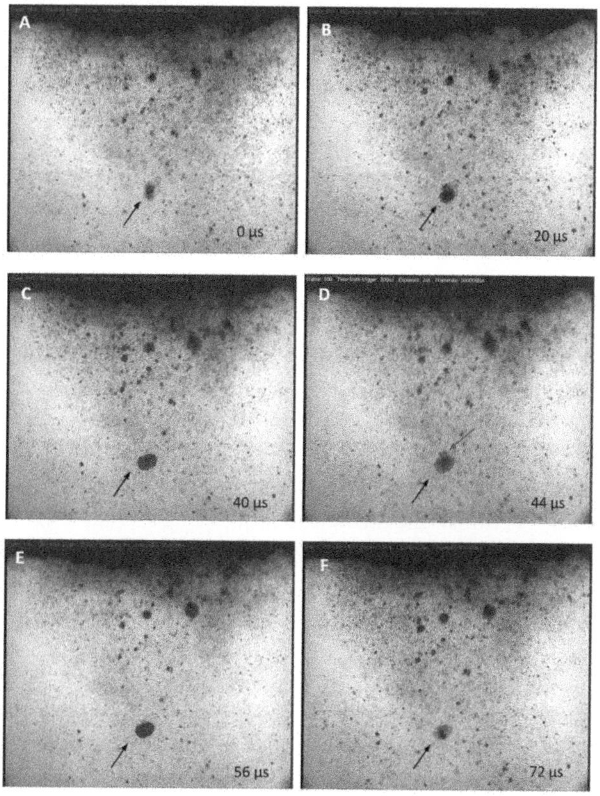

Figure 27. Snapshots from ultra-high-speed camera video showing cavitation bubbles near the sonotrode tip in pure water. Ultrasonication is at 20 kHz, filming rate 500,000 fps, shutter speed 160 ns. As marked with the arrows, individual bubbles pulsate and occasionally burst (D) during the sonication cycle (50 µs) and can be tracked.

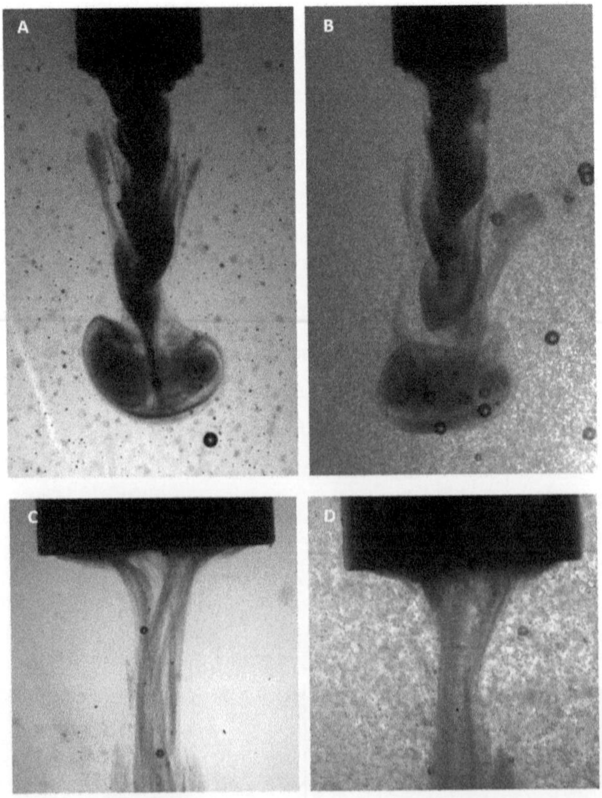

Figure 28. Snapshots from high-speed camera video showing the cavitation bubble structure under the sonotrode tip: (A) general view of the bubble structure in the case of pure diesel fuel and (B) diesel fuel with water droplet emulsion. High-magnification images near the tip shortly after the sonication impulse is stopped for (C) pure diesel or (D) diesel with water droplet emulsion.

4. Oxidative Desulfurization

Brief Overview of ODS

ODS (*136–140*) is a technology for removing sulfur in fuel by oxidizing it to polar sulfoxides or sulfones. The polarity of the formed sulfoxides/sulfones is different from that of the parent sulfur in the fuel, and this enables extraction of the formed sulfones using a suitable extractant. This technique includes two processes: oxidation and the subsequent extraction and separation (*50*, *141*, *142*).

Hydrodesulfurization (HDS) is a widely used, mature commercial method integrated into crude oil refinery processing to remove sulfur compounds from fuel oil and to avoid downstream catalyst poisoning. In HDS, the sulfur compounds are converted into gaseous H_2S with the aid of hydrogen and a catalyst under extreme reaction temperature (300–340 °C) and pressure (20–100 bar). HDS effectively eliminates sulfides/disulfides and thiols but is ineffective on refractory sulfur compounds like thiophenes due to ineffective adsorption of these heterocyclic sulfur compounds on the catalyst's surface (*143*). After the commercial HDS process, the sulfur content in the fuel oil ranges from several dozens to hundreds of parts per million, beyond which the sulfur content does not reduce further and does not meet the sulfur standards. Most of the remaining sulfur compounds are refractory heterocyclic sulfur compounds. If these were to be removed using the HDS technique, then the reactor size would need to be enlarged by five- to 15-fold under much more severe reaction temperature and pressure with more active catalysts. These steps would also result in undesired side reactions, including saturation of olefins and loss of octane number (*143*). Thus, HDS is not economically and technically feasible for total sulfur removal, and alternative methods are necessary.

In recent years, one of the most promising techniques, namely ODS, has been widely studied for deep or ultradeep desulfurization of fuels such as gasoline and diesel (*141*) to meet sulfur emission regulations. Compared with HDS, ODS offers several advantages: (1) ODS is effective against refractory thiophenes and their derivatives under very mild reaction conditions (25–90 °C and ambient pressure), (2) there is no need for expensive hydrogen, and (3) the capital requirement for the ODS process is significantly low.

In recent decades, a number of oxidative systems have been utilized in the ODS process, such as a biphasic hydrogen peroxide system of H_2O_2/organic (formic/acetic) acids (*144–147*), H_2O_2/heteropolyacid (*148–150*), H_2O_2/transition metal salts (*151–153*), as well as single-phase nonhydrogen peroxide systems, O_2, O_3, and N_2O (*154–157*). Among all of these oxidative systems, the hydrogen peroxide system, especially H_2O_2/organic (formic/acetic) acids, is the most well studied and has been demonstrated as the most promising oxidative system because the oxidant, hydrogen peroxide, has the highest percentage of active oxygen (47%) (*143*). In addition, this system is also commercially available, cost efficient, and nonpolluting and produces an environmentally friendly byproduct, namely water (*143, 158*). In this system, the peracids, which have a more powerful oxidation ability and stronger selectivity than H_2O_2, are generated in situ by carboxylic acids reacting with hydrogen peroxide to increase the oxidation ability and selectivity of the process. The main challenge of the hydrogen peroxide system using the ODS process is excessive consumption of the oxidant and the mass-transfer restriction in the biphasic system leading to a slow reaction rate. To overcome mass transfer limitations in the biphasic system, enhanced mass-transfer measures, such as mechanical stirring and sonication (*50*), should be adopted, causing extra energy consumption.

ODS takes advantage of a liquid-phase reaction where sulfur becomes oxidized by an oxygen donor. A variety of oxidants were used in the past for this purpose, such as H_2O_2, molecular oxygen, peracids, ozone, and also air. Three categories of oxidants can be identified depending on the state of reagent and reactant:

- Liquid–liquid reactions, where the oxidizer is water soluble
- Single-phase reactions when the oxidizer is oil soluble
- Gas–liquid reactions when the oxidizer is present as a gas in the mixture

ODS of HFO

Currently, many published ODS studies focus on deep desulfurization of gasoline, diesel, and model fuels.[50] Compared with the number of studies on ODS of gasoline and diesel, studies on ODS of HFO[141] are fewer, and they highlight the complexity of the process. Table 4 summarizes some of the HFO ODS applications. The results in the table indicate that the sulfur-removal efficiency of HFO is generally lower than that of gasoline and diesel. This is due to the high viscosity of HFO, which significantly impedes the biphasic mass transfer, resulting in low oxidation efficiency of sulfur compounds. Several recently developed techniques, however, are suitable for ODS of HFO. Among these techniques, sonication is one of the most promising and will eventually be explored to improve the HFO ODS process.

Table 4. Summary of ODS Applications for HFO

Fuel with Temperature (°C)	*Oxidant*	*Catalyst*	*O/S molar %*	*S Content (wt %)*	*S Removal (wt %)*	*Ref.*
HFO	H_2O_2	Formic acid	~10	2.75	58.5	(*159*)
HFO	H_2O_2	$FeCl_3$	–	3.85	39	(*160*)
Vacuum gas oil	H_2O_2	Acetic acid	100	1.99	>89	(*161*)
HFO	TBHP*	Vanadium(IV) functionalized polymer	–	1.79	50	(*162*)
HFO 1	Potassium superoxide	Acetic acid	2.6	1.7	65	(*163*)
HFO 2	Potassium superoxide	Acetic acid	2.6	2.6	75	–
Bunker C oil	TBHP*	MoO_3	–	3.17	~35	(*164*)
Bitumen	H_2O_2	NaOH	–	4.6	~52	(*165*)

* THBP is tert-Butyl hydroperoxide

In ODS, the reactivity of organosulfur compounds is increased with the increase of electron density on the sulfur atom (*166*). The reactivities of dibenzothiophene (DBT) derivatives are influenced by the electron donation of substituted alkyl groups. Thus, the reactivity decrease in the order 4,6-DMeDBT > 4-MeDBT > DBT, reversing the order of reactivities for HDS (see Figure 29) (*166*). ODS has demonstrated the ability to desulfurize HFO effectively (*141*). However, the above process encounters many difficulties, primarily due to complications imposed by the very high viscosity and molecular structures (asphaltenes) in the fuel. There has consequently not been significant deployment of this technology.

According to our previous study (79), results from FT-ICR-MS of HFO indicate that hydrocarbon compounds containing one sulfur atom (S_1 class) are the most abundant at 52%, followed by 21.4% hydrocarbon compounds that contain two sulfur atoms (S_2 class; see Figure 30). Most of the sulfur species are refractory heterocyclic sulfur compounds, such as benzothiophenes (DBE of 6), DBT (DBE of 9), and benzonaphthothiophene (DBE of 12), and their naphthenic derivatives at DBE of 7–8, 10–11, and 13–15, respectively (see Figure 31). HDS is therefore not suitable for heavy fuels like HFO, whereas ODS has a high oxidation efficiency with these refractory thiophenes and their derivatives under very mild reaction conditions and is potentially the most suitable technique for HFO desulfurization.

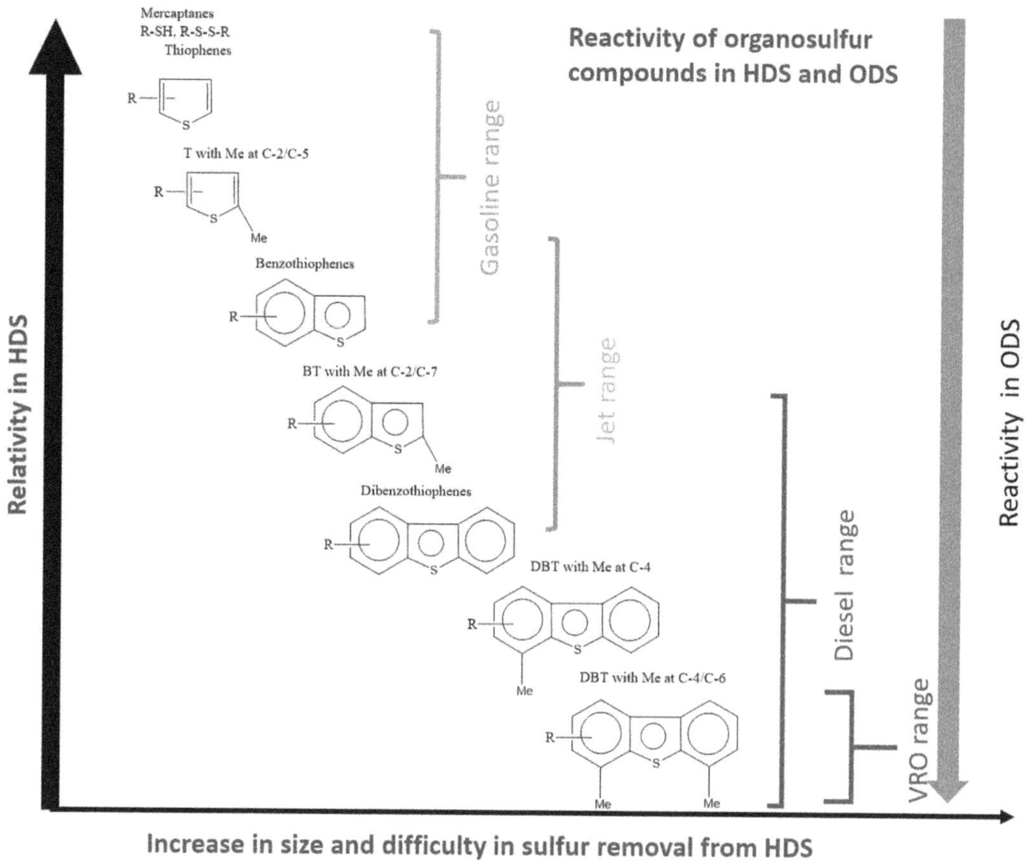

Figure 29. Order of reactivity in HDS and ODS for the different classes of organosulfur compounds. This is a modified version of the illustration from reference (166).

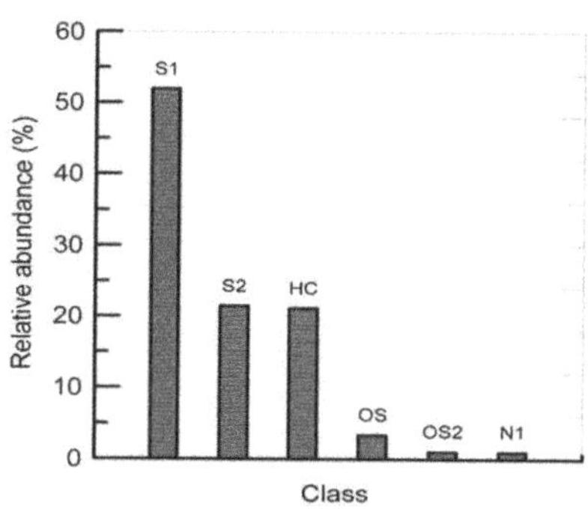

Figure 30. Class distribution plot of the species present in the HFO sample as observed by positive-ion-mode atmospheric pressure chemical ionization Fourier-transform ion cyclotron resonance mass spectrometry (APCI-FT-ICR-MS). Reproduced with permission from (66). Copyright 2015 American Chemical Society.

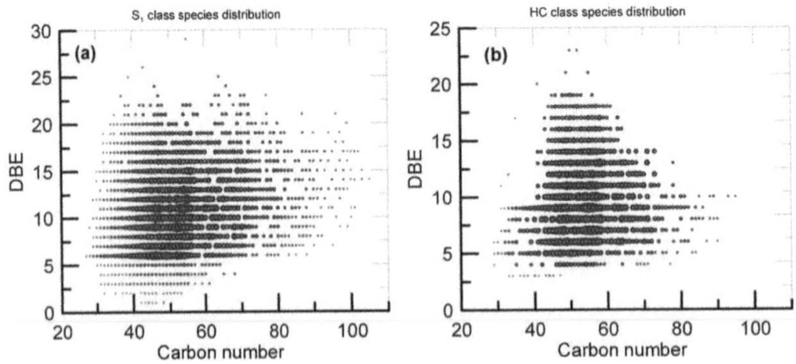

Figure 31. Speciation of sulfur compounds in HFO by FT-ICR-MS67. DBE versus carbon number distribution plots for (a) S_1 and (b) HC class species observed in an HFO sample using positive-ion-mode atmospheric pressure chemical ionization Fourier-transform ion cyclotron resonance mass spectrometry (APCI-FT-ICR-MS). Reproduced with permission from reference (66). Copyright 2015 American Chemical Society.

ODS Without Using Ultrasound Cavitation

Experimental Procedure

Our HFO ODS experiment was conducted in a magnetically stirred water/oil bath reactor. The reaction temperature was precisely controlled up to 400 °C corresponding to water and silicone oil heating, respectively, with the variation within ±0.5 °C. The magnetic stir speed was up to 2600 round/min. Here, we adopted a widely used and cost-efficient H_2O_2/organic acid oxidative system for HFO ODS. Considering the fast decomposition of H_2O_2 under a higher temperature, the temperature of ODS using H_2O_2 oxidant is normally lower than 100 °C (*142, 167*). Thus, the water bath temperature was high enough for the current ODS of HFO.

The detailed experimental procedure was as follows. First, 100 ± 0.5 g HFO (with sulfur content 3.5 wt %) was put into a 500 mL beaker. H_2O_2 oxidant with a specific ratio of H_2O_2/S and organic acid (formic or acetic acid) catalyst with a specific ratio of catalyst/H_2O_2 were separately added into the beaker. A specific amount of extractant (such as water, methanol, or acetonitrile) was also added to the beaker. When the water-bath reactor was heated up to the desired temperature, the beaker containing HFO, oxidative solution, and extractant was put into the bath reactor and magnetically stirred for a specific reaction time for ODS and extraction. After the end of the reaction, the mixture solution was poured into a separation funnel and kept at least for 12 h to ensure the separation of oil in the aqueous phase. The desulfurized HFO was then separately analyzed by ICP-OES and a CHNS analyzer (Thermo Scientific Flash 2000) to obtain sulfur and other elemental compositions. Then the ODS efficiency was calculated.

For sulfur measurement using the ICP-OES method, it should be noted that the HFO sample was first digested by a microwave digester. Then the ICP-OES instrument was used to accurately determine the HFO sulfur content. The HFO digestion procedure was as follows. An accurately weighed HFO sample (less than 200 mg) was added into 8 ml nitric acid (70 wt % concentration) and digested by an ETHOS-E microwave digester (Italy Milestone Corporation). The digestion Teflon ware was heated from room temperature to 220 °C for 20 min and then kept at 220 °C for 15 min. After cooling down to room temperature, it was diluted to 25 mL by distilled water for the ICP-OES analysis.

Results and Discussion

Factors such as reaction temperature, reaction time, oxidant/S ratio, oxidant/catalyst ratio, oxidant catalyst type, extractant type, and extraction time can influence ODS efficiency. In the current preliminary study, we studied the most important factors, oxidant/S ratio, oxidant/catalyst ratio, and extractant type, affecting ODS efficiency. The reaction temperature and time were selected as 60 °C and 60 min, respectively, and the extractant-to-HFO ratio was 1:1 based on the results reported in the literature (*159*, *164*). Based on previous studies (*168–171*), the presence of sulfur-extracting agents like acetonitrile in the ODS process has been proved to enhance ODS efficiency. Thus, we combined the processes of sulfur oxidation and extraction by combining HFO, oxidant/catalyst, and extractant reagents together.

The sulfur and CHN/O contents of the raw and desulfurized HFO are listed in Table 5. To determine the quantity of polar sulfur compounds existing in the raw HFO that can be directly removed by the extractant, HFO was exposed to the extractants, namely, acetonitrile and methanol, without the oxidation process. After extraction by acetonitrile and methanol, the sulfur content in HFO changed from 3.53 to 3.36 and 3.31, respectively. This result indicates that at least 6.2 wt % of the polar sulfur component exists in the raw HFO, and the extraction performance of methanol is better than that of acetonitrile. Then the effect of the O/S ratio and extractant on sulfur removal was investigated. First, the influence of the O/S ratio was studied by varying the catalyst content and keeping the experimental conditions constant. The O/S ratios tested were 1.5, 3, 5, and 10 mol/mol. The quantities used are reported in Table 5. It can be noticed from the atomic composition that no linear correlation is present among the results of the different samples. This is mainly due to the presence of water and acetonitrile residue at the end of the process. This problem indicates that the CHNS analysis is useful in estimating the quality of the separation process but not the efficiency of the ODS reaction in the case of incomplete separation. Comparing the different extractants, namely acetonitrile and methanol, we note that sulfur removal using acetonitrile was better than that using methanol. The sulfur-removal efficiency using methanol was 15.0 and 14.4 wt %, while that using acetonitrile was 22.1 and 21.0 wt % at an H_2O_2/S ratio of 5:1 and 3:1, respectively.

Table 5. Sulfur and CHNO Content of Raw and Desulfurized HFOs

No.	*H_2O_2/S (mole ratio)*	*H_2O_2/Acid (mole ratio)*	*Extractant*	*N (wt %)*	*C (wt %)*	*H (wt %)*	*O (wt %)*	*S (wt %)*
Raw HFO	0	0	0	0.36	85.50	10.84	<0.1	3.53
1	0	0	Acetonitrile	1.01	82.60	10.76	2.27	3.36
2	0	0	Methanol	0.72	83.15	10.84	1.98	3.31
3	5:1	1:1	Acetonitrile	1.17	77.60	10.46	7.77	3.00
4	5:1	1:1	Methanol	0.22	75.14	10.36	11.53	2.75
5	3:1	1:1	Acetonitrile	0.45	75.68	10.04	10.81	3.02
6	3:1	1:1	Methanol	1.03	74.92	10.28	10.98	2.79

Molecular-Level Characterization

Raw HFO and desulfurized HFO samples were ionized by APPI and ESI coupled with FT-ICR-MS, as these have been used extensively to study residual fuel samples. ESI is efficient for ionizing polar molecules (heteroatom-containing $N_nO_oS_s$ species), which are notorious for causing catalyst fouling, corrosion, and deposit formation, whereas APPI is efficient in ionizing species that are not observed by ESI, like benzo- and dibenzothiophenes, polyaromatic hydrocarbons, and cycloalkanes, and it is, therefore, suitable for analyzing the overall composition of petroleum samples.

The mass spectra of HFO and the desulfurized HFO obtained from various O/S ratios using APPI and ESI FT-ICR mass spectrometry are presented in Figures 32 and 33. ESI FT-ICR-MS was employed because it is effective in ionizing heteroatom-containing (N, O, S) species. The mass spectra show the weight distributions of various species in the fuel, which have a significant effect on their physical properties. They also show how the fuel changes on account of the ODS processes. The AMW of HFO calculated from APPI FT-ICR-MS was found to be 549 Da. The AMWs of the desulfurized HFO samples obtained using O/S ratios 1.5, 3, 5, and 10 were calculated as 555, 561, 562, and 566 Da, respectively. The increase in AMW of the desulfurized HFO is due to the addition of the oxygen atoms to the sulfur to form sulfones and sulfoxides.

The abundance of various molecular species in the samples obtained from APPI and ESI FT-ICR-MS is shown in Figures 34 and 35, respectively. HC indicates pure hydrocarbon species. Hydrocarbon molecules containing one, two, and three sulfur atoms are represented with the notation S_1, S_2, and S_3, respectively. Similarly, the notations N, O, and NS indicate hydrocarbon species with one nitrogen atom, one oxygen atom, and one nitrogen and sulfur atom each, respectively. Since ODS resulted in the oxidation of S compounds to form sulfones (SO_2) and sulfoxides (SO) and other species (OS_2), all oxygen- and sulfur-containing heteroatoms are collectively indicated by OS species. This also includes NOS (species containing all three heteroatoms). The quantity (in relative %) of each class is evaluated from the knowledge of the intensity of the specific class divided by the sum total of all measured ion intensities. The S_1, S_2, and S_3 classes in the desulfurized HFO samples were reduced continuously when the O/S ratio was increased as the excess oxidant converted the S species in HFO to sulfones/sulfoxides. The abundance of OS species in HFO is 1.1% as determined using APPI, whereas the abundance of the OS species in the desulfurized HFO sample increased from 8.8% to 18.8% when the O/S ratio was increased from 1.5 to 10. From the ESI FT-ICR-MS, the OS species in HFO increased from 6.7% to 85.3% in the desulfurized sample when the O/S ratio was 10. Similarly, the abundance of the OS species increased in the aqueous phase when the O/S ratio was increased.

The different compound classes and their distributions were then visualized based on their DBE, carbon number, and relative abundance. The molecular class distributions of OS species as a function of DBE and carbon number are presented in Figures 36 and 37. The bubble size indicates the magnitude of the recorded mass peaks. DBE values allow us to understand the molecular structure of the species based on their degrees of unsaturation. The APPI and ESI FT-ICR-MS show that there are very few OS species present in the HFO sample and that the oxidant increases as the OS species increase.

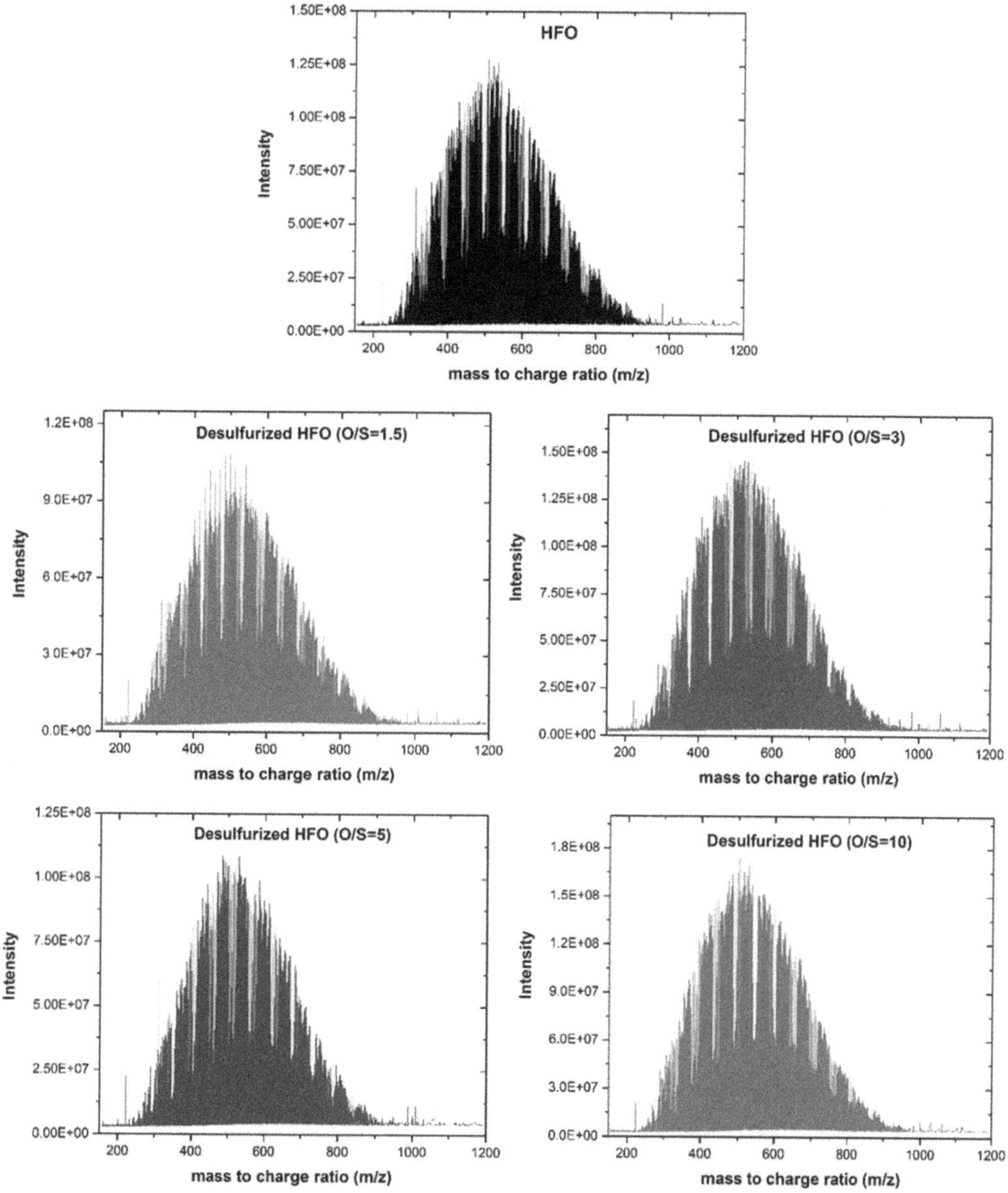

Figure 32. APPI FT-ICR mass spectra of desulfurized HFO samples with varying O/S ratios.

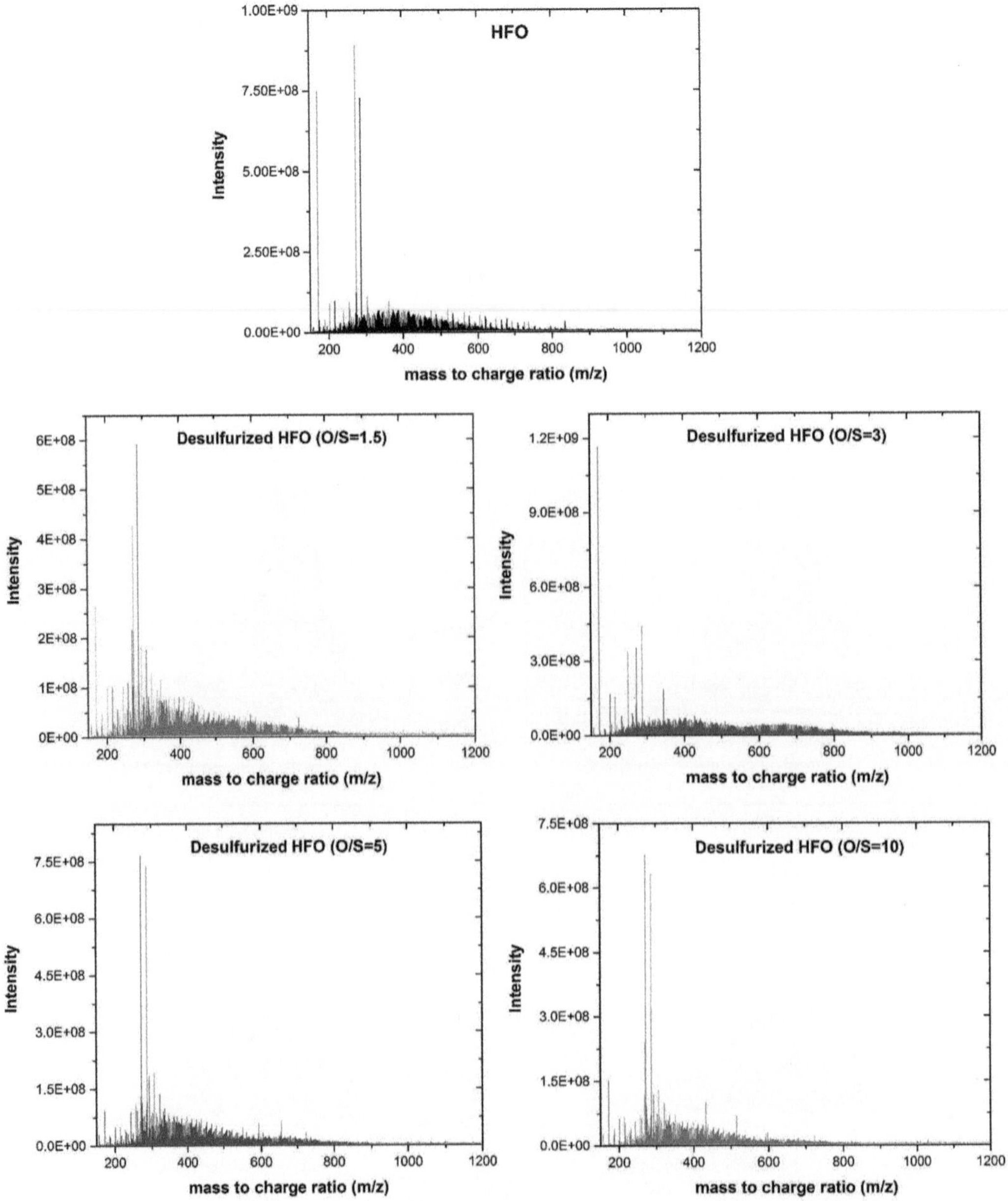

Figure 33. ESI FT-ICR mass spectra of desulfurized HFO samples with varying O/S ratios.

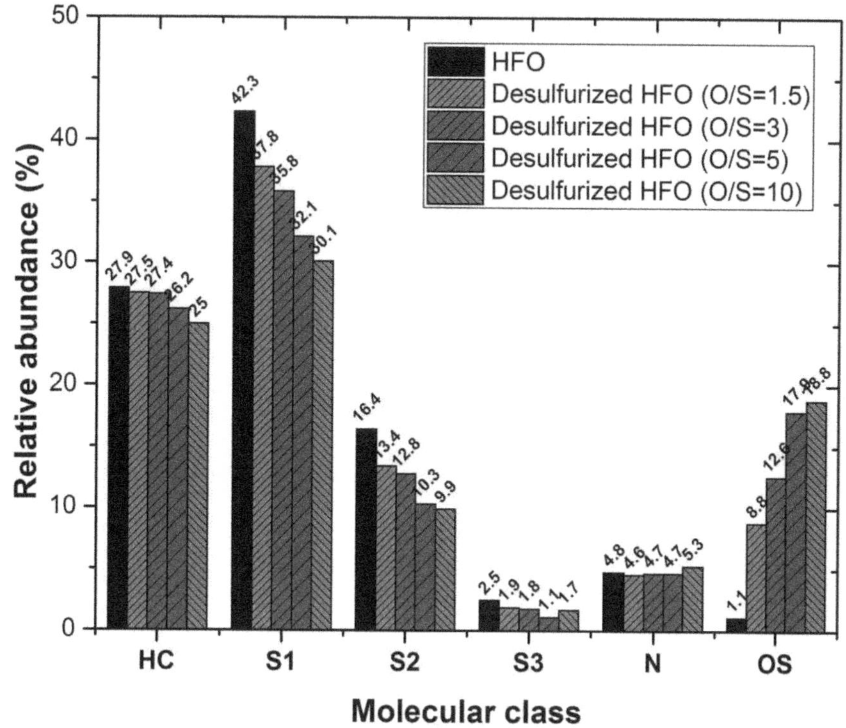

Figure 34. APPI FT-ICR-MS: Relative abundance of molecular classes present in HFO and desulfurized HFO.

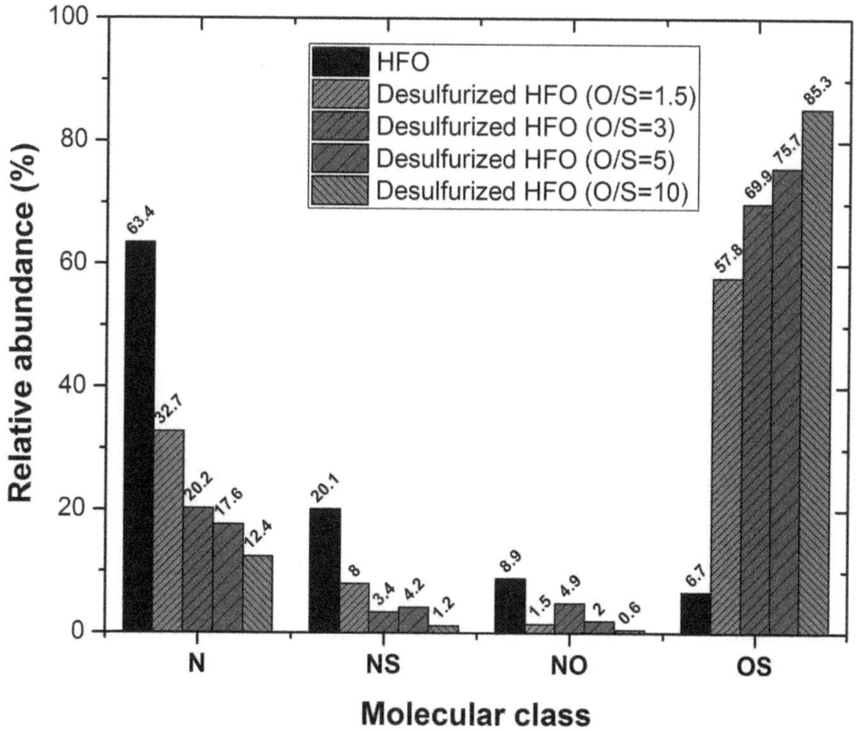

Figure 35. ESI FT-ICR-MS: Relative abundance of molecular classes present in HFO and desulfurized HFO.

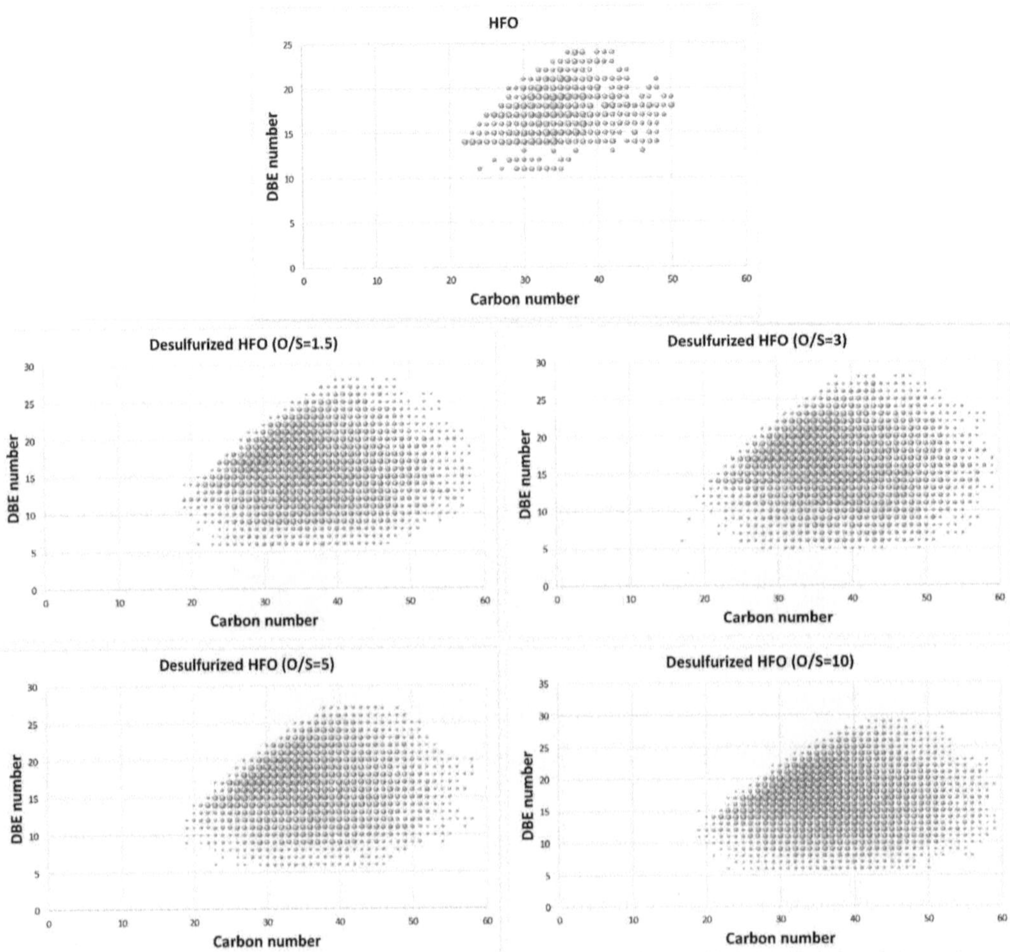

Figure 36. APPI FT-ICR isoabundance plot (carbon number vs. DBE) of HFO and desulfurized HFO of hydrocarbon molecules containing OS heteroatoms.

The ODS experiments presented several challenges, the most important of which was emulsion formation between HFO and the aqueous-phase H_2O_2/catalyst solution. This was observed after the ODS process due to the presence of the emulsifier component (asphaltene) naturally present in HFO. When a high stirring speed (high shear force) was used to enhance the interaction of the sulfur compounds with the oxidant, HFO was well emulsified and the aqueous-phase H_2O_2/catalyst solution became separated from the oil phase, resulting in a decrease in the carbon content of the mixture, whereas the O content increased markedly. Even at lower stirring speed, a small amount of aqueous-phase H_2O_2/catalyst solution was still not completely separated from HFO. These results show that the carbon contents of the desulfurized sample decreased by around 10 wt % after the ODS process, indicating the presence of the aqueous phase in the desulfurized oil. Another challenge was that the HFO became sticky after the ODS process, and its viscosity increased, which is in accordance with the results of some previous studies (*141*, *159*). A reduction in the residence time was expected to be the correct solution to solve problems related to recombination and emulsion formation. A continuous stirring process was the logical solution to obtain better separation and a less viscous product.

We also combined the continuous stirring process with UIC to speed up the oxidation process and achieve good conversion with lower residence time.

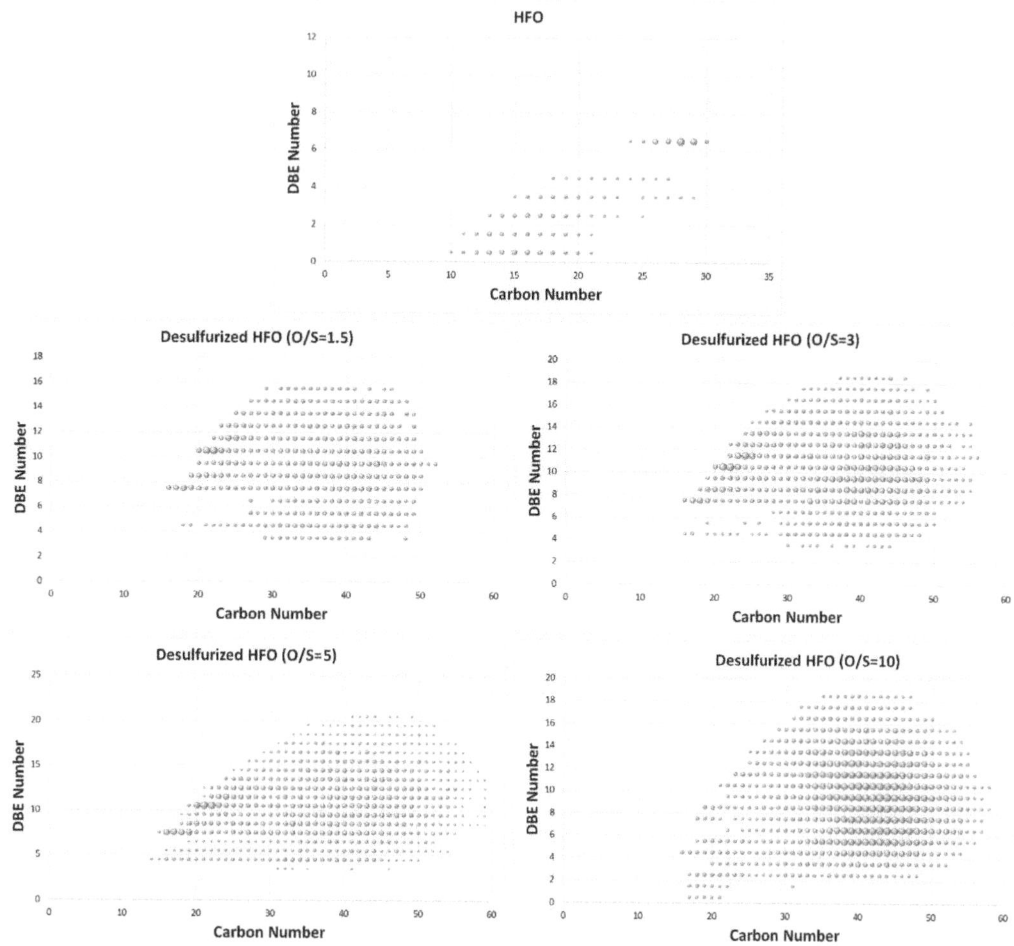

Figure 37. ESI FT-ICR isoabundance plots (carbon number vs. DBE) of HFO and desulfurized HFO of hydrocarbon molecules containing OS heteroatoms.

Ultrasonically Assisted Oxidative Desulfurization (UAOD)

ODS assisted by ultrasonic cavitation has significant efficacy and performance advantages (*50, 164, 165, 172*). Acoustic cavitation introduced by ultrasonic waves (within 20 to 50 kHz for industrial applications) generates extreme temperatures and pressures via the collapse of microbubbles. This phenomenon has been known to cause sonochemical changes leading to the formation of hydroxyl radicals via homolytic cleavage of a water molecule. They are highly reactive with other molecules in the system. A typical product from this system is in situ hydrogen peroxide (H_2O_2). A further advantage of sonication is a reduction in viscosity, achieved by cracking of asphaltenes into smaller, low-boiling-temperature molecules. The sulfur is catalytically oxidized into sulfones, which are easily separated.

The increase in surface area from ultrasonic treatment enhances both the kinetics and the mass transfer in the ODS process as follows. ODS occurs in a biphasic system that has an organic and an aqueous phase. The aqueous phase forms the dispersed phase, and oil as the continuous phase is separated by a layer of asphaltenes and resins in the case of HFO. Upon reaction with a catalyst, the oxidant creates an in situ, highly diffusive oxidative species that diffuses out of the aqueous-phase surface layer of the droplet into the organic phase. This in situ species oxidizes the sulfurous compounds inside the organic phase to their corresponding sulfones. The sulfones, once formed, tend to cover the layer (droplet surface) that is separating the phases and accumulate there. After a while, when this sulfone accumulation is high, further diffusion of the oxidative species from the aqueous phase to the organic phase is hampered, slowing down the ODS process. Thus, mass-transfer limitations are imposed unless sufficiently small droplet sizes are achieved while preparing biphasic emulsions. Therefore, the ODS of liquid fuels can be considerably accelerated by ultrasound treatment, which renders very small (nanoscale) droplets.

The most widely used oxidizing agent in the literature was hydrogen peroxide. Hydrogen peroxide is soluble in oil and commercially available as diluted in water. It contributes to oxidation in a liquid–liquid reaction. Oxidation is expected to take place at the interface between hydrogen peroxide/water and oil. The reaction rate is proportional to the surface area between the two liquids. This makes clear the importance of good mixing to improve the surface area available for the reaction. As discussed in the previous sections, UIC is an effective way to form very fine emulsions. A direct comparison between mechanical and ultrasonically induced cavitation as a mixing method highlights a difference in the obtainable surface area of more than 100 times, which results from the ratio between the surface exposed to an equivalent volume of water in the oil.

Another relevant aspect that makes UIC particularly suitable to improve the efficiency of ODS consists of bubble formation. The bubbles formed by the pressure oscillations of the sonotrode are largely formed by the most volatile species, which in an H_2O_2/HFO/acetic acid system are acetic acid and H_2O_2. The occurrence of cavitation bubbles results in the contribution of another mechanism to the reactivity of the system, the gas–liquid reactions.

UAOD Experiments and Results

UIC-enhanced ODS experiments were performed in a pre-pilot-scale continuous system at KAUST. The UAOD rig commissioning is described in detail in an upcoming publication from our group (*173*). The reactor consisted of a cylindrical vessel in which a sonotrode vibrates at a frequency of 20,000 Hz. The power input of the sonotrode was 2000 W. The fuel flows parallel to the sonotrode. Both oxidizing agent (H_2O_2) and catalyst (CH_3) were mixed before injection into the reactor chamber. The fraction of hydrogen peroxide was kept at 3:1, the stoichiometry of the oxidation reaction, while the amount of acetic acid was kept in a 1:1 molar ratio with the sulfur molecules.

The fuel tested in the facility was a blend of Arabian Extra Light (AXL) and HFO. The choice to dilute HFO in the initial tests was driven by difficulties in operating a continuous process with a highly viscous fuel. The first test consisted of running with AXL. AXL is a relatively low-sulfur fuel (0.7% by mass) compared to HFO, for which the sulfur ranges from 2.7 to 4 wt %. AXL also presents lower viscosity than HFO, which simplifies the operation of the rig. We proceeded by gradually adding HFO to AXL and modifying the process where needed. It is important to note that AXL includes the entire spectrum of organosulfur compounds, and only the heaviest of those are amenable to ODS. In contrast, HFOs include predominantly heavy sulfurous compounds. Intuitively, we expected to see more sulfur reduction in HFO than in AXL from the ODS treatment.

The detailed discussion on the UAOD experiments conducted on the KAUST UAOD rig is presented in a separate publication (*173*). Here, we briefly summarize some key results. We explored three combinations: (1) pure AXL, (2) AXL 10% and HFO 90% by mass, and (3) AXL 50% and HFO 50% by mass. The 50% HFO run with AXL dilution was a sufficient demonstration of the ODS rig's operation capability in handling highly viscous fuels. We managed to run the process smoothly for more than 3 h with no significant problems while achieving a steady operation after approximately 10 min.

The reactor temperature increased with time during the test and reached 75 °C because of the action of the sonotrode. The reactor was operated for two hours with recirculation. Given the size of the reactor, 0.005 m^3, and the flow rate of 10 kg/h, the residence time was about half of the total time spent in the loop. The mixture of AXL/HFO, acetic acid, and hydrogen peroxide was extracted after 30 min, 1 h, and 2 h. The mixture was also washed with an equivalent amount of acetonitrile, and then the sulfones were separated through centrifugation (5 min). The elemental composition of the aqueous and organic phase was measured through ICP-OES to determine the sulfur content and the eventual dilution through mass balance.

The results reported in Figure 38 clearly demonstrate the substantial reduction in sulfur and confirm that ODS preferentially oxidizes heavy molecules, validating literature reports (*166*). In the DBE versus carbon number plot (see Figure 38), the raw fuel includes a wide spectrum of molecules containing sulfur. The carbon atom range goes from 10 to 80 mainly because of the contribution of the vacuum residue fraction of HFO. A large proportion of the sulfurized molecules was removed through the ODS reaction and the subsequent separation that was performed with acetonitrile. The black dots indicate the normalized fraction of sulfurized molecules. It is possible to see that the range of carbon atoms was drastically decreased to between 10 and 40 and that the maximum DBE was reduced compared to the raw fuel. The data confirm the tendency of peroxides to preferentially oxidize large molecules in general and tiophenic groups in particular. This strengthens our premise that HDS and ODS are complementary. As expected, the reduction in sulfur was increased by increasing the amount of HFO in the mixture, as more molecules became available for the oxidation process. Figure 39 reports the results in terms of sulfur percentage reduction by mass. A reduction in the sulfur by nearly 40% by weight was achieved.

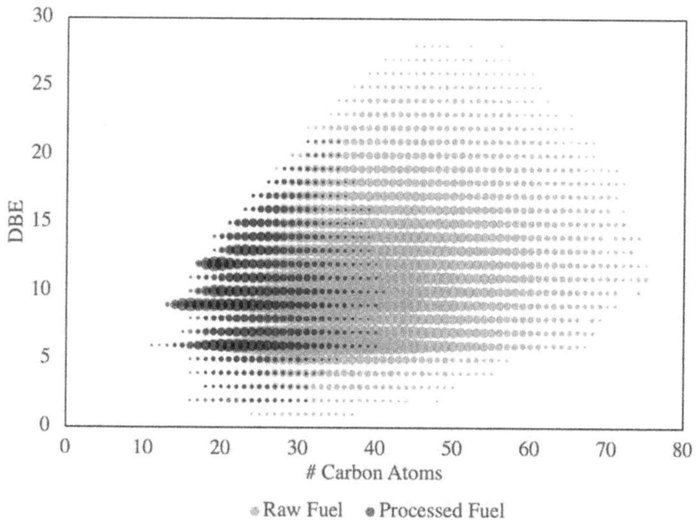

Figure 38. Distribution of sulfur molecules in AXL before and after sonication. The range of sulfurized molecules is substantially reduced after the ODS. The test was performed in the KAUST ODS rig on AXL. Results shown are from FT-ICR-MS performed on the samples. The size of the points reflects the relative amount of the species. DBE stands for double-bond equivalent, a measure of aromaticity.

The separation step was expected to be the limiting step of the ODS process. It is likely that by increasing the amount of HFO, the removal of sulfur will become more substantial because of the higher availability of sulfur species suitable for ODS. The ODS process still requires further optimization with the addition of deasphalting and mild HDS.

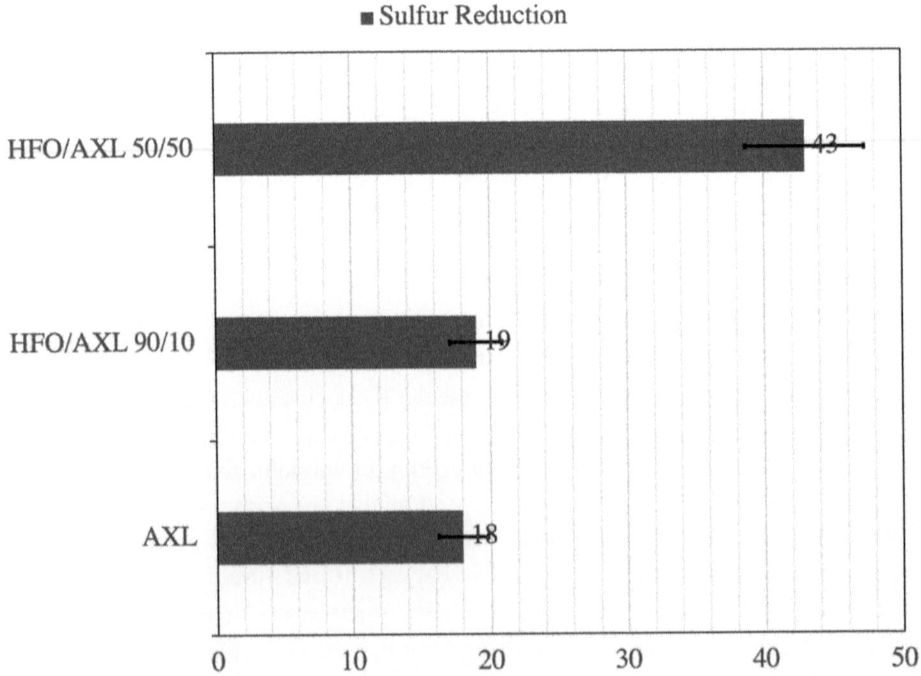

Figure 39. Sulfur reduction by mass % for three samples tested in the KAUST ODS rig facility. The reduction was measured through elemental composition with the ICP-OE technique and cross-validated with the energy-dispersive X-ray technique. Mass balance was accurately calculated to account for dilution, although it was negligible.

5. Conclusions

The utilization of low-grade fuels like HFO for power generation and shipping is an integral part of many oil-rich countries like KSA, and their increasingly diversifying economy. It is necessary to explore the physics behind cavitation-based heavy fuel upgrading, emulsification, desulfurization, multi-phase combustion, and their relationship on HFO combustion optimization. We believe that improving HFO combustion can vastly benefit from enhancements in the upgrading and burning behaviors of asphaltenes. In the current chapter, we have discussed a roadmap for use of "bottom-of-barrel fuel" with high asphaltene content via UIC-based fuel upgrading, desulfurization, and direct use (emulsions). We expect that a strategy of using UIC for asphaltene modification and water-in-HFO-enabled micro-explosions will significantly impact the combustion of HFO. The proposed study has a potential to substantially contribute to the ability to optimize combustion solutions for the clean and efficient burning of HFO in KSA based power plants and shipping. We conclude this chapter with the reiteration of some salient points as follows:

1. We conducted a detailed characterization of the physicochemical properties and molecular-level structural information of HFO, deasphalted oil, and asphaltenes. FT-ICR-MS and ^{1}H NMR analytical methodologies are sufficient to decipher relevant molecular-

level information for heavy viscous fractions of crude oils. Deasphalting of HFO imparts great advantages in terms of transport properties like viscosity and corrosion propensity and induces trace metal removal. However, the reduction of heteroatoms like sulfur and nitrogen is marginal. The proportion of light or less aromatic structural classes increases with deasphalting, which should improve combustion efficiency and reduce PM formation.

2. Sonication is expected to act mostly on the structure of the asphaltenes and, therefore, to influence the physical properties of the mixture. But the fundamental understanding of how the average molecular weight of the asphaltenes changes in time as a result of the sonication process is yet unknown. We delved into the fundamentals of UIC and how it may impact asphaltene structures. The dependence on molecular structure modification on sonication time was explored by extracting samples from the irradiated batch sample at different times. The asphaltene breakup is followed by free-radical recombination reactions to form larger aggregates, and hence could be counterproductive. FT-ICR-MS data acquired from HFO sonication experiments confirmed this point and highlighted the need for an optimized irradiation time.
3. We characterized UIC-based water-in-HFO emulsions using two techniques: high-speed camera visualization and cryo-EM. Capturing the cavitation phenomenon at the timescales of cavitation offered unique information that could be utilized to validate computation models. Cryo-EM imaging provided insights into droplet size distribution and asphaltene's role in emulsion stability. Engineering of optimized emulsions for various combustion applications can be based on information achieved from such characterization techniques.
4. We reported ODS experiments without and with ultrasound (UAOD). ODS is ineffective for lighter sulfur species, which are easily removed by conventional HDS. ODS assisted by ultrasonic cavitation has significant efficacy and performance advantages. Our most recent results convey a 40% reduction of sulfur from AXL/HFO blends following UAOD using the KAUST rig.
5. The biggest identified challenge in UAOD or ODS is the asphaltene content of the HFO. It is responsible for high viscosity, which impedes flow processes and leads to catalyst poisoning and an assortment of undesired heavy PM (sludge-like) formations when oxidized during the ODS process. Hence, deasphalting as a pretreatment of HFO before desulfurization could improve the process by reducing these obstacles and the overall cost.
6. The ODS and HDS processes have mutually reversed reactivity characteristics for the spectrum of organosulfur compounds found in petroleum products. The molecular-level characterization of ODS HFO reveals the extent and kind of structural classes in HFO that are amenable to ODS. A much higher reduction is achievable if mild HDS is used prior to the ODS step. Hence, the ODS and the HDS must be utilized as complementary processes for maximum effectiveness, completeness of sulfur reduction, and overall process optimization.

Overall, we recommend (1) deasphalting, (2) UAOD with mild HDS, and (2) UIC-based water-in-HFO emulsions as some potential solutions for HFO utilization as a fuel in the marine and power-generation sectors.

Acknowledgments

The research reported in this publication was partially supported by the Saudi Electricity Company (SEC) and the Clean Combustion Research Center (CCRC) of King Abdullah University of Science and Technology (KAUST). The authors thank Professor Sigurdur Thoroddsen and Dr. Ivan Uriev Vakarelski for their assistance in conducting high-speed cavitation imaging. We are grateful to Drs. Long Jiang and Aiping Chen for help in the ODS experiments and analytical chemistry work. This research used resources of the KAUST Core Labs. We also thank Drs. Dalaver Anjum, Ali Reza Behzad, Abdel Hamid M. Amwas, Jingyu Liu, and Wen Zhang for cryo-TEM, cryo-SEM, NMR, ICP-OES, and FT-ICR-MS analysis, respectively.

References

1. *World Energy Outlook 2019*; IEA: Paris 2019. https://www.iea.org/reports/world-energy-outlook-2019.
2. Masnadi, M. S.; El-Houjeiri, H. M.; Schunack, D.; Li, Y.; Englander, J. G.; Badahdah, A.; Monfort, J.-C.; Anderson, J. E.; Wallington, T. J.; Bergerson, J. A. Global carbon intensity of crude oil production. *Science* **2018**, *361* (6405), 851–853.
3. Kuhns, R. J.; Shaw, G. H. Peak oil and petroleum energy resources. In *Navigating the Energy Maze*; Springer: Cham, Switzerland, 2018; pp 53−63.
4. Norouzi, N.; Fani, M.; Ziarani, Z. K. The fall of oil age: A scenario planning approach over the last peak oil of human history by 2040. *Journal of Petroleum Science and Engineering* **2020**, *188*, 106827.
5. Ramirez-Corredores, M. M. *The Science and Technology of Unconventional Oils: Finding Refining Opportunities*; Academic Press, Elsevier, 2017.
6. Speight, J. G. Production of Syngas, Synfuel, Bio-Oils, and Biogas from Coal, Biomass, and Opportunity Fuels. In *Fuel Flexible Energy Generation*; Woodhead Publishing, 2016; pp 145−174, ISBN 9781782423782.
7. Dawood, F.; Anda, M.; Shafiullah, G. Hydrogen production for energy: An overview. *International Journal of Hydrogen Energy* **2020**, *45* (7), 3847–3869.
8. Jiang, L.; Elbaz, A. M.; Guida, P.; Al-Noman, S. M.; AlGhamdi, I. A.; Saxena, S.; Roberts, W. L. J. E. Cenosphere Formation during Single-Droplet Combustion of Heavy Fuel Oil. *Fuels* **2019**, 33 (2), 1570–1581.
9. Urban, D. L.; Huey, S. P.; Dryer, F. L. Evaluation of the Coke Formation Potential of Residual Fuel Oils. In *Symposium (International) on Combustion*; Department of Mechanical and Aerospace Engineering, Princeton University, Elsevier: Princeton, NJ, January 1992; Vol. 24, Issue 1, pp 1357−1364.
10. Kurian, V.; Gupta, R. Distribution of vanadium, nickel, and other trace metals in soot and char from asphaltene pyrolysis and gasification. *Energy & Fuels* **2016**, *30* (3), 1605–1615.
11. Kurian, V. *Asphaltene Gasification: Soot Formation and Metal Distribution*. Ph.D. Thesis, University of Alberta, 2016. https://doi.org/10.7939/R3707WT5M.
12. Dunn, K.; Yen, T. A plausible reaction pathway of asphaltene under ultrasound. *Fuel Processing Technology* **2001**, 73, 59–71.

13. Suslick, K. S.; Didenko, Y.; Fang, M. M.; Hyeon, T.; Kolbeck, K. J.; McNamara, W. B.; Mdleleni, M. M.; Wong, M. *Philosophical Transactions of the Royal Society A: Mathematical, Physical and Engineering Science.* **1999**, *357* (1751), 335–353.
14. Suslick, K. S.; Nyborg, W. L. Ultrasound: its chemical, physical and biological effects. *Journal of the Acoustical Society of America* **1990** , February, *87* (2), 919–920.
15. Thompson, L. H.; Doraiswamy, L. Sonochemistry: Science and engineering. *Industrial & Engineering Chemistry Research* **1999**, *38* (4), 1215–1249.
16. Sawarkar, A. N. Cavitation induced upgrading of heavy oil and bottom-of-the-barrel: A review. *Ultrasonics Sonochemistry* **2019**, *58*, 104690.
17. Leong, T.; Ashokkumar, M.; Kentish, S. The fundamentals of power ultrasound-a review. *Springer Singapore Pte Ltd. Acoustics Australia* **2011**, *39*, 54.
18. Avvaru, B.; Venkateswaran, N.; Uppara, P.; Iyengar, S. B.; Katti, S. S. Current knowledge and potential applications of cavitation technologies for the petroleum industry. *Ultrasonics Sonochemistry* **2018**, *42*, 493–507.
19. Kass, M.; Thomas, J.; Sluder, S.; Theiss, T. *A Guide to Fuel, Lubricant, and Engine Concerns Relative to the IMO 2020 Fuel Oil Sulfur Reduction Mandate*; Oak Ridge National Lab (ORNL): Oak Ridge, TN, 2019.
20. Molloy, N. The IMO's 2020 Global Sulfur Cap. What a 2020 Sulfur Constrained World Means for Shipping Lines, Refineries and Bunker Suppliers. *S&P Global Platts*, 2016. https://www.spglobal.com/en/research-insights/articles/what-a-2020-sulfur-constrained-world-means-for-shipping-lines-refineries-and-bunker-suppliers (accessed March 16, 2021).
21. Shabaneh, R.; Al Sadoon, H. M.; Mestneer, R. *Implications of IMO 2020: The Potential for High Sulfur Fuel Oil Penetration in Saudi Arabia*; King Abdullah Petroleum Studies and Research Center (KAPSARC): Saudi Arabia, 2019. https://www.kapsarc.org/research/publications/implications-of-imo-2020-the-potential-for-high-sulfur-fuel-oil-penetration-in-saudi-arabia/
22. Pittalis, E.; Sleiman, T. *Fuel Oil's New Buyers*; S&P Global Platts: London, 2019. https://www.spglobal.com/platts/plattscontent/_assets/_files/en/specialreports/oil/imo-2020-fuel-oil-future-report.pdf (accessed March, 16 2021)
23. Thomson, G. IMO 2020 Is about to Get Real. *Wood Mackenzie*. https://www.woodmac.com/news/opinion/imo-2020-is-about-to-get-real/.
24. Joung, T.-H.; Kang, S.-G.; Lee, J.-K.; Ahn, J. The IMO initial strategy for reducing Greenhouse Gas (GHG) emissions, and its follow-up actions towards 2050. *Journal of International Maritime Safety, Environmental Affairs, and Shipping* **2020**, *4* (1), 1–7.
25. El-Houjeiri, H.; Monfort, J. C.; Bouchard, J.; Przesmitzki, S. Life cycle assessment of greenhouse gas emissions from marine fuels: A case study of Saudi crude oil versus natural gas in different global regions. *Journal of Industrial Ecology* **2019**, *23* (2), 374–388.
26. Jameel, A. G. A.; Alkhateeb, A.; Telalović, S.; Elbaz, A. M.; Roberts, W. L.; Sarathy, S. M. Environmental Challenges and Opportunities in Marine Engine Heavy Fuel Oil Combustion. In *Proceedings of the Fourth International Conference in Ocean Engineering (ICOE2018)*, Springer: Singapore, 2019; pp 1047—1055.
27. Shabaneh, R. . *Data Insight: Fuel Mix for Power Generation and Water Desalination in Saudi Arabia*; KAPSARC: Riyadh, Saudi Arabia, 2020. https://www.kapsarc.org/research/publications/fuel-mix-for-power-generation-and-water-desalination-in-saudi-arabia/ (accessed March 16, 2021).

28. Pei, X.; AbdulJameel, A. G.; Chen, C.; AlGhamdi, I. A.; AlAhmadi, K.; AlBarakati, E.; Saxena, S.; Roberts, W. L. Swirling flame combustion of heavy fuel oil: Effect of fuel sulfur content. *Journal of Energy Resources Technology* **2021**, *143* (8)
29. Ashokkumar, M.; Mason, T. J. Sonochemistry. In *Kirk-Othmer Encyclopedia of Chemical Technology*; John Wiley & Sons, 2001. https://onlinelibrary.wiley.com/doi/full/10.1002/0471238961.1915141519211912.a01.pub2.
30. Suslick, K. S.; Mdleleni, M. M.; Ries, J. T. Chemistry induced by hydrodynamic cavitation. *Journal of the American Chemical Society* **1997**, *119* (39), 9303–9304.
31. Gogate, P. R.; Pandit, A. B. Hydrodynamic cavitation reactors: a state of the art review. *Reviews in Chemical Engineering* **2001**, *17* (1), 1–85.
32. Jang, S. H.; Lee, K. W.; Kim, J. R.; Kim, J. H.; Yoon, S. H.; Cho, I. S.; Choi, J. H. The Combustion Characteristics of Residual Fuel oil Blended with Fuel Additives. *Journal of the Korean Society of Marine Environment & Safety Research* **2016**, August 31, *22* (5), 554–563ISSN 1229-3431(Print)/ISSN 2287-3341(Online).
33. Feldman, N. Control of Residual Fuel Oil Particulate Emissions by Additives. In *Symposium (International) on Combustion*; Elsevier, 1982; Vol. 19, Issue 1.
34. Mulholland, J. A.; Hall, R. E. Fuel oil reburning application for NOx control to firetube package boilers. *J. Eng. Gas Turbines Power.* **1987**, Apr, *109* (2), 207–214.
35. Wang, Z.; Xiong, Y.; Cheng, X.; Liu, M. Experimental study on the flame propagation characteristics of heavy oil oxy-fuel combustion. *Journal of the Energy Institute* **2019**, *92* (6), 1630–1640.
36. Rana, R.; Nanda, S.; Maclennan, A.; Hu, Y.; Kozinski, J. A.; Dalai, A. K. Comparative evaluation for catalytic gasification of petroleum coke and asphaltene in subcritical and supercritical water. *Journal of Energy Chemistry* **2019**, *31*, 107–118.
37. Rydén, M. Chemical Looping Combustion of Liquid Fuels. In *Calcium and Chemical Looping Technology for Power Generation and Carbon Dioxide (CO_2) Capture*; Woodhead Publishing, 2015; pp 287—298.
38. Jacques, M. T., Jordan, J. B., Williams, A., & Hadley-Coates, L. The Combustion of Water-in-Oil Emulsions and the Influence of Asphaltene Content. In *Symposium (International) on Combustion*; Elsevier, January1977; Vol. 16, Ussue 1, pp 307—319.
39. Allouis, C.; L'Insalata, A.; Fortunato, L.; Saponaro, A.; Beretta, F. Study of water–oil emulsion combustion in large pilot power plants for fine particle matter emission reduction. *Experimental thermal and Fluid Science* **2007**, *31* (5), 421–426.
40. Tarlet, D.; Bellettre, J.; Tazerout, M.; Rahmouni, C. Formulation and combustion of emulsified fuel: The changes in emission of carbonaceous residue. *International Journal of Energy Research* **2010**, *34* (8), 688–694.
41. Li, Y. Y.; Hou, S. S.; Sheu, W. J. Investigation on boiler efficiency and pollutant emissions of water/heavy oil emulsions using edge-tone resonant homogenizer. *Fuel* **2014**, *119*, 240–251.
42. Dryer, F. Water Addition to Practical Combustion Systems—Concepts and Applications. In *Symposium (International) on Combustion*; Elsevier, 1977; pp 279—295. https://www.sciencedirect.com/science/article/pii/S0082078477803329.
43. Kadota, T.; Yamasaki, H. Recent advances in the combustion of water fuel emulsion. *Progress in Energy and Combustion Science* **2002**, *28* (5), 385–404.

44. Ballester, J. M.; Fueyo, N.; Dopazo, C. Combustion characteristics of heavy oil-water emulsions. *Fuel* **1996**, *75* (6), 695–705.
45. Ocampo-Barrera, R.; Villasenor, R.; Diego-Marin, A. An experimental study of the effect of water content on combustion of heavy fuel oil/water emulsion droplets. *Combustion and Flame* **2001**, *126* (4), 1845–1855.
46. Kim, W.; Yu, T.; Yoon, W. Atomization characteristics of emulsified fuel oil by instant emulsification. *Journal of Mechanical Science and Technology.* **2012**, Jun, *26* (6), 1781–91.
47. Law, C. K.; Lee, C. H.; Srinivasan, N. Combustion characteristics of water-in-oil emulsion droplets. *Combustion and Flame.* **1980**, Jan 1, *37*, 125–43.
48. Bolszo, C. D.; Narvaez, A. A.; McDonell, V. G.; Dunn-Rankin, D.; Sirignano, W. A. Pressure—Swirl Atomization of Water-in-Oil Emulsions. *Atomization and Sprays* **2010**, *20* (12), 1077.
49. Lin, S.-L.; Lee, W.-J.; Chang, S.-S.; Lee, C.-f.; Lee, L.-F.; Lin, C.-S.; Loong, H. J. E. Energy savings and emission reduction of traditional pollutants, particulate matter, and polycyclic aromatic hydrocarbon using solvent-containing water emulsified heavy fuel oil in boilers. *Fuels* **2011**, *25* (4), 1537–1546.
50. Ja'fari, M.; Ebrahimi, S. L.; Khosravi-Nikou, M. R. Ultrasound-assisted oxidative desulfurization and denitrogenation of liquid hydrocarbon fuels: a critical review. *Ultrasonics Sonochemistry* **2018**, *40*, 955–968.
51. Kozyuk, O. V. *Method and Apparatus for Conducting Sonochemical Reactions and Processes Using Hydrodynamic Cavitation*. U.S. Patent US6,012,492, January 11, 2000.
52. Moncrieff, V. M., Waldron, J. L., Grimes, P., Gunnerman, R. W. *Method for Manufacturing an Emulsified Fuel*. U.S. Patent US7,344,570, March 18, 2008.
53. Waldron J., Grimes P. *Mixing Apparatus and Method for Manufacturing an Emulsified Fuel*. U.S. Patent Application US11/853,548, December 272007.
54. Waldron, J. *APT's Proprietary Emulsion Technology*. https://www.altpetrol.com/emulsion-technology/ (accessed June 2,2021).
55. Cottell, E. W. *Emulsion Fuel System Technology*. http://www.nonoxltd.com/WhitePaper.html (accessed June 2, 2021).
56. Antunes, J. *Enermulsion Inline Ultrasound System*. https://www.tecnoveritas.net/industry/products/enermulsion/ (accessed June 2, 2021).
57. Antunes, J. *A Process for the Reduction of the Sulphur Content of Fuels*. Portugal Patent No. PT110250, November 2, 2019. https://patentscope.wipo.int/search/en/detail.jsf?docId=PT278271147.
58. Colussi, A. J.; Weavers, L. K.; Hoffmann, M. R. Chemical bubble dynamics and quantitative sonochemistry. *Journal of Physical Chemistry A* **1998**, *102* (35), 6927–6934.
59. Hielscher, T. *Ultrasonic Production of Nano-Size Dispersions and Emulsions*; ENS 2005; Paris, France, December 2005; pp 138–143.
60. Garaniya, V.; McWilliam, D.; Goldsworthy, L.; Ghiji, M. J. F. Extensive chemical characterization of a heavy fuel oil. *Fuel* **2018**, *227*, 67–78.
61. Akmaz, S.; Iscan, O.; Gurkaynak, M. A.; Yasar, M. The structural characterization of saturate, aromatic, resin, and asphaltene fractions of Batiraman crude oil. *Petroleum Science and Technology.* **2011**, Jan 1, *29* (2), 160–171.

62. Freitag, N. P.; Exelby, D. R. A SARA-based model for simulating the pyrolysis reactions that occur in high-temperature EOR processes. *Journal of Canadian Petroleum Technology* **2006**, 45 (03)
63. Mullins, O.; Sheu, E.; Hammami, A.; Marshall, A. *Asphaltenes, Heavy Oils, and Petroleomics*; Springer-Verlag: New York, 2007.
64. Peng, P. A.; Morales-Izquierdo, A.; Hogg, A.; Strausz, O. P. J. E. Molecular structure of Athabasca asphaltene: Sulfide, ether, and ester linkages. *Fuels* **1997**, *11* (6), 1171–1187.
65. Vargas, F. M.; Tavakkoli, M. *Asphaltene Deposition: Fundamentals, Prediction, Prevention, and Remediation*; CRC Press: Boca Raton, FL: 2018.
66. Elbaz, A. M.; Gani, A.; Hourani, N.; Emwas, A.-H.; Sarathy, S. M.; Roberts, W. J. E. TG/DTG, FT-ICR mass spectrometry, and NMR spectroscopy study of heavy fuel oil. *Fuel* **2015**, *29* (12), 7825–7835.
67. Barroso, J.; Barreras, F.; Ballester, J. J. F. P. T. Behavior of a high-capacity steam boiler using heavy fuel oil: Part I. High-temperature corrosion. *Fuel Processing Technology* **2004**, *86* (2), 89–105.
68. Barreras, F.; Barroso, J. J. F. P. T. Behavior of a high-capacity steam boiler using heavy fuel oil, Part II: Cold-end corrosion. *Fuel Processing Technology* **2004**, *86* (2), 107–121.
69. Garaniya, V.; McWilliam, D.; Goldsworthy, L. Chemical Characterization of Heavy Fuel Oil for Combustion Modelling. In *World Engineers Convention 2011 (WEC2011)*, 2011; pp 1–11).
70. Abdul Jameel, A. G.; Khateeb, A.; Elbaz, A. M.; Emwas, A. H.; Zhang, W.; Roberts, W. L.; Sarathy, S. M. Characterization of deasphalted heavy fuel oil using APPI (+) FT-ICR mass spectrometry and NMR spectroscopy. *Fuel* **2019**, *253*, 950–963doi:10.1016/j.fuel.2019.05.061.
71. Chaala, A.; Kapoor, M.; Benallal, B.; Gulmissarian, T. Stability and rheological properties of carbon black feedstocks. *Fuel Science & Technology International* **1995**, *13* (1), 71–85.
72. Marshall, A. G.; Rodgers, R. P. Petroleomics: Chemistry of the Underworld. *Proceedings of the National Academy of Sciences* **2008**, *105* (47), 18090–18095.
73. Abdul Jameel, A. G.; Elbaz, A. M.; Emwas, A.-H.; Roberts, W. L.; Sarathy, S. M. J. E. Calculation of average molecular parameters, functional groups, and a surrogate molecule for heavy fuel oils using ^{1}H and ^{13}C nuclear magnetic resonance spectroscopy. *Energy & Fuels* **2016**, *30* (5), 3894–3905.
74. Abdul Jameel, A. G.; Han, Y.; Brignoli, O.; Telalović, S.; Elbaz, A. M.; Im, H. G.; Roberts, W. L.; Sarathy, S. M. Heavy fuel oil pyrolysis and combustion: Kinetics and evolved gases investigated by TGA-FTIR. *J. Anal. Appl. Pyrolysis* **2017**, *127* (August), 183–195doi:10.1016/j.jaap.2017.08.008.
75. Xian, F.; Hendrickson, C. L.; Marshall, A. G. High resolution mass spectrometry. *Analytical Chemistry* **2012**, *84* (2), 708–719.
76. Hur, M.; Ware, R. L.; Park, J.; McKenna, A. M.; Rodgers, R. P.; Nikolau, B. J.; Wurtele, E. S.; Marshall, A. G. Statistically significant differences in composition of petroleum crude oils revealed by volcano plots generated from ultrahigh resolution fourier transform ion cyclotron resonance mass spectra. *Energy & Fuels* **2018**, *32* (2), 1206–1212.
77. McKenna, A. M.; Marshall, A. G.; Rodgers, R. P. Heavy petroleum composition. 4. Asphaltene Compositional space. *Energy & Fuels* **2013**, *27* (3), 1257–1267.

78. Hourani, N.; Muller, H.; Adam, F. M.; Panda, S. K.; Witt, M.; Al-Hajji, A. A.; Sarathy, S. M. Structural level characterization of base oils using advanced analytical techniques. *Energy & Fuels* **2015**, *29* (5), 2962–2970.
79. Elbaz, A. M.; Gani, A.; Hourani, N.; Emwas, A.-H.; Sarathy, S. M.; Roberts, W. L. TG/DTG, FT-ICR mass spectrometry, and NMR spectroscopy study of heavy fuel oil. *Energy & Fuels* **2015**, *29* (12), 7825–7835.
80. Miyabayashi, K.; Naito, Y.; Yamada, M.; Miyake, M.; Ushio, M.; Fuchigami, J.; Kuroda, R.; Ida, T.; Hayashida, K.; Ishihara, H. Compositional analysis of deasphalted oil before and after hydrocracking over zeolite catalyst by Fourier transform ion cyclotron resonance mass spectrometry. *Fuel Processing Technology* **2008**, *89* (4), 397–405.
81. Zhang, T.; Zhang, L.; Zhou, Y.; Wei, Q.; Chung, K. H.; Zhao, S.; Xu, C.; Shi, Q. Transformation of nitrogen compounds in deasphalted oil hydrotreating: characterized by electrospray ionization fourier transform-ion cyclotron resonance mass spectrometry. *Energy & Fuels* **2013**, *27* (6), 2952–2959.
82. Qian, K.; Fredriksen, T. R.; Mennito, A. S.; Zhang, Y.; Harper, M. R.; Merchant, S.; Kushnerick, J. D.; Rytting, B. M.; Kilpatrick, P. K. Evidence of naturally-occurring vanadyl porphyrins containing multiple S and O atoms. *Fuel* **2019**, *239*, 1258–1264.
83. Chen, H.; Nelson, R. K.; Swarthout, R. F.; Shigenaka, G.; de Oliveira, A. H. B.; Reddy, C. M.; McKenna, A. M. Detailed compositional characterization of the 2014 Bangladesh furnace oil released into the world's largest mangrove forest. *Energy & Fuels* **2018**, *32* (3), 3232–3242.
84. Wu, Z.; Rodgers, R. P.; Marshall, A. G. ESI FT-ICR mass spectral analysis of coal liquefaction products. *Fuel* **2005**, *84* (14), 1790–1797.
85. Kuti, O. A.; Yang, S. Y.; Hourani, N.; Naser, N.; Roberts, W. L.; Chung, S. H.; Sarathy, S. M. A fundamental investigation into the relationship between lubricant composition and fuel ignition quality. *Fuel* **2015**, *160*, 605–613.
86. Campuzano, F.; Abdul Jameel, A. G.; Zhang, W.; Emwas, A.-H.; Agudelo, A. F.; Martínez, J. D.; Sarathy, S. M. Fuel and chemical properties of waste tire pyrolysis oil derived from a continuous twin-auger reactor. *Energy & Fuels* **2020**, *34* (10), 12688–12702.
87. Campuzano, F.; Jameel, A. G. A.; Zhang, W.; Emwas, A.-H.; Agudelo, A. F.; Martínez, J. D.; Sarathy, S. M. On the distillation of waste tire pyrolysis oil: A structural characterization of the derived fractions. *Fuel* **2020**, *290*, 120041.
88. Hartman, B. E.; Hatcher, P. G. Hydrothermal liquefaction of isolated cuticle of *Agave americana* and *Capsicum annuum*: Chemical characterization of petroleum-like products. *Fuel* **2015**, *156*, 225–233.
89. Matayeva, A.; Bianchi, D.; Chiaberge, S.; Cavani, F.; Basile, F. Elucidation of reaction pathways of nitrogenous species by hydrothermal liquefaction process of model compounds. *Fuel* **2019**, *240*, 169–178.
90. Giraldo-Dávila, D.; Chacón-Patiño, M. L.; Orrego-Ruiz, J. A.; Blanco-Tirado, C.; Combariza, M. Y. Improving compositional space accessibility in (+) APPI FT-ICR mass spectrometric analysis of crude oils by extrography and column chromatography fractionation. *Fuel* **2016**, *185*, 45–58.

91. Cho, Y.; Witt, M.; Kim, Y. H.; Kim, S. Characterization of crude oils at the molecular level by use of laser desorption ionization Fourier-transform ion cyclotron resonance mass spectrometry. *Analytical Chemistry* **2012**, *84* (20), 8587–8594.
92. Valencia-Dávila, J. A.; Witt, M.; Blanco-Tirado, C.; Combariza, M. Y. Molecular characterization of naphthenic acids from heavy crude oils using MALDI FT-ICR mass spectrometry. *Fuel* **2018**, *231*, 126–133.
93. Pereira, T. M. C.; Vanini, G.; Oliveira, E. C. S.; Cardoso, F. M. R.; Fleming, F. P.; Neto, A. C.; Lacerda, V.; Castro, E. V. R.; Vaz, B. G.; Romão, W. An evaluation of the aromaticity of asphaltenes using atmospheric pressure photoionization Fourier transform ion cyclotron resonance mass spectrometry – APPI(±)FT-ICR MS. *Fuel* **2014**, *118*, 348–357.
94. Klein, G. C.; Kim, S.; Rodgers, R. P.; Marshall, A. G.; Yen, A.; Asomaning, S. Mass spectral analysis of asphaltenes. I. Compositional differences between pressure-drop and solvent-drop asphaltenes determined by electrospray ionization Fourier transform ion cyclotron resonance mass spectrometry. *Energy & Fuels* **2006**, *20* (5), 1965–1972.
95. Klein, G. C.; Kim, S.; Rodgers, R. P.; Marshall, A. G.; Yen, A. Mass spectral analysis of asphaltenes. II. Detailed compositional comparison of asphaltenes deposit to its crude oil counterpart for two geographically different crude oils by ESI FT-ICR MS. *Energy & Fuels* **2006**, *20* (5), 1973–1979.
96. Gaspar, A.; Zellermann, E.; Lababidi, S.; Reece, J.; Schrader, W. Characterization of saturates, aromatics, resins, and asphaltenes heavy crude oil fractions by atmospheric pressure laser ionization Fourier transform ion cyclotron resonance mass spectrometry. *Energy & Fuels* **2012**, *26* (6), 3481–3487.
97. Abdul Jameel, A. G.; Sarathy, S. M. Lube Products: Molecular Characterization of Base Oils. In *Encyclopedia of Analytical Chemistry: Applications, Theory and Instrumentation*, 2006; pp 1–14.
98. Abdul Jameel, A. G.; Naser, N.; Emwas, A.-H.; Dooley, S.; Sarathy, S. M. Predicting fuel ignition quality using ^{1}H NMR spectroscopy and multiple linear regression. *Energy & Fuels* **2016**, *30* (11), 9819–9835.
99. Naser, N.; Jameel, A. G. A.; Emwas, A.-H.; Singh, E.; Chung, S. H.; Sarathy, S. M. The influence of chemical composition on ignition delay times of gasoline fractions. *Combustion and Flame* **2019**, *209*, 418–429.
100. Abdul Jameel, A. G.; Van Oudenhoven, V.; Emwas, A.-H.; Sarathy, S. M. Predicting octane number using nuclear magnetic resonance spectroscopy and artificial neural networks. *Energy & Fuels* **2018**, *32* (5), 6309–6329.
101. Abdul Jameel, A. G.; Naser, N.; Emwas, A.-H.; Sarathy, S. M. Surrogate formulation for diesel and jet fuels using the minimalist functional group (MFG) approach. *Proceedings of the Combustion Institute* **2019**, *37* (4), 4663–4671.
102. Hillier, J. L.; Fletcher, T. H.; Solum, M. S.; Pugmire, R. J. Characterization of macromolecular structure of pyrolysis products from a Colorado Green River oil shale. *Industrial & Engineering Chemistry Research* **2013**, *52* (44), 15522–15532.
103. Fletcher, T. H.; Gillis, R.; Adams, J.; Hall, T.; Mayne, C. L.; Solum, M. S.; Pugmire, R. J. Characterization of macromolecular structure elements from a Green River oil shale, II. Characterization of pyrolysis products by ^{13}C NMR, GC/MS, and FTIR. *Energy & Fuels* **2014**, *28* (5), 2959–2970.

104. Abdul Jameel, A. G.; Elbaz, A. M.; Emwas, A.-H.; Roberts, W. L.; Sarathy, S. M. Calculation of average molecular parameters, functional groups, and a surrogate molecule for heavy fuel oils using ^{1}H and ^{13}C nuclear magnetic resonance spectroscopy. *Energy & Fuels* **2016**, *30* (5), 3894–3905.
105. Menendez, R.; Bermejo, J.; Moinelo, S. R.; Marsh, H. Flash chromatography, extrography and ^{1}H-NMR to characterize coal-derived liquids. *Organic Geochemistry* **1990**, *15* (2), 161–168.
106. Hauser, A.; AlHumaidan, F.; Al-Rabiah, H.; Halabi, M. A. Study on thermal cracking of Kuwaiti heavy oil (vacuum residue) and its SARA fractions by NMR spectroscopy. *Energy & Fuels* **2014**, *28* (7), 4321–4332.
107. Betancourt Cardozo, F.; Avella Moreno, E.; Trujillo, C. A. Structural characterization of unfractionated asphalts by ^{1}H NMR and ^{13}C NMR. *Energy & Fuels* **2016**, *30* (4), 2729–2740.
108. Poveda, J. C.; Molina, D. R. Average molecular parameters of heavy crude oils and their fractions using NMR spectroscopy. *Journal of Petroleum Science and Engineering* **2012**, *84-85*, 1–7.
109. Haqshenas, S.; Ford, I.; Saffari, N. Modelling the effect of acoustic waves on nucleation. *Journal of Chemical Physics* **2016**, *145* (2), 024315.
110. de Andrade, M. O.; Haqshenas, S. R.; Pahk, K. J.; Saffari, N. The effects of ultrasound pressure and temperature fields in millisecond bubble nucleation. *Ultrasonics Sonochemistry* **2019**, *55*, 262–272.
111. Yamamoto, T.; Komarov, S. V. Dynamic behavior of acoustic cavitation bubble originated from heterogeneous nucleation. *Journal of Applied Physics* **2020**, *128* (4), 044702.
112. Tzanakis, I.; Lebon, G.; Eskin, D.; Pericleous, K. Characterizing the cavitation development and acoustic spectrum in various liquids. *Ultrasonics Sonochemistry* **2017**, *34*, 651–662.
113. Hoffmann, M. R.; Hua, I.; Höchemer, R. Application of ultrasonic irradiation for the degradation of chemical contaminants in water. *Ultrasonics Sonochemistry* **1996**, *3* (3), S163–S172.
114. Kishimoto, K.; Kumazawa, Y.; Kuroki, T.; Yamasaki, H.; Okubo, M. Ultrasonically enhanced electrohydraulic discharge for removal of organic compounds. *Journal of Electrostatics* **2020**, *108*, 103502.
115. Rice, F.; Herzfeld, K. The Thermal Decomposition of Organic Compounds from the Standpoint of Free Radicals. VI. The Mechanism of Some Chain Reactions. In *Selected Readings in Chemical Kinetics*; Back, M. H., Laidler, K. J., Eds.; Elsevier, 1967; pp 154–170.
116. Suslick, K. S.; Gawienowski, J. J.; Schubert, P. F.; Wang, H. H. Alkane sonochemistry. *Journal of Physical Chemistry* **1983**, *87* (13), 2299–2301.
117. Mohammadian, E.; Junin, R.; Rahmani, O.; Idris, A. K. Effects of sonication radiation on oil recovery by ultrasonic waves stimulated water-flooding. *Ultrasonics* **2013**, *53* (2), 607–614.
118. Taheri-Shakib, J.; Shekarifard, A.; Naderi, H. Analysis of the asphaltene properties of heavy crude oil under ultrasonic and microwave irradiation. *Journal of Analytical and Applied Pyrolysis* **2018**, *129*, 171–180.
119. Najafi, I.; Amani, M. Asphaltene flocculation inhibition with ultrasonic wave radiation: a detailed experimental study of the governing mechanisms. *Advances in Petroleum Exploration and Development* **2011**, *2* (2), 32–36.

120. Gollapudi, U.; Bang, S.; Islam, M. Ultrasonic Treatment for Removal of Asphaltene Deposits during Petroleum Production. In *SPE Formation Damage Control Symposium*; Society of Petroleum Engineers: Lafayette, Louisiana, 1994. https://onepetro.org/SPEFD/proceedings/94FD/All-94FD/SPE-27377-MS/56795.
121. Hosseini-Dastgerdi, Z.; Tabatabaei-Nejad, S. A.; Khodapanah, E.; Sahraei, E. A comprehensive study on mechanism of formation and techniques to diagnose asphaltene structure; molecular and aggregates: a review. *Asia-Pacific Journal of Chemical Engineering.* **2015**, January, *10* (1), 1–4.
122. Mullins, O. C. J. E. The modified Yen model. *Fuels* **2010**, *24* (4), 2179–2207.
123. *Asphaltenes and Asphalts, 2*; Yen, T. F.; Chilingarian, G. V., Eds.; Elsevier, 2000; Vol. 40B
124. Taylor, P. Ostwald ripening in emulsions. *Advances in Colloid and Interface Science* **1998**, *75* (2), 107–163.
125. Anjum, D. H.; Guida, P.; Behzad, A. R.; Saxena, S.; Roberts, W. L. Optimization of specimen preparation methods for cryo electron microscopy of oil-and-water emulsions. *Microscopy and Microanalysis* **2019**, *25* (S2), 2428–2429.
126. Dryer, F. *Fundamental Combustion Studies of Emulsified Fuels. Final Report*; Department of Mechanical and Aerospace Engineering, Princeton University: New Jersey, 1982.
127. Marrone, N. J.; Kennedy, I. M.; Dryer, F. L. Internal phase size effects on combustion of emulsions. *Combustion Science and Technology* **1983**, *33* (5-6), 299–307.
128. Czarnecki, J.; Tchoukov, P.; Dabros, T. J. E. Possible role of asphaltenes in the stabilization of water-in-crude oil emulsions. *Fuels* **2012**, *26* (9), 5782–5786.
129. McLean, J. D.; Kilpatrick, P. K. Effects of asphaltene aggregation in model heptane–toluene mixtures on stability of water-in-oil emulsions. *Journal of Colloid and Interface Science* **1997**, *196* (1), 23–34.
130. Yarranton, H. W.; Hussein, H.; Masliyah, J. H. Water-in-hydrocarbon emulsions stabilized by asphaltenes at low concentrations. *Journal of Colloid and Interface Science* **2000**, *228* (1), 52–63.
131. Leavers, V. Which Hough transform? *CVGIP: Image Understanding* **1993**, *58* (2), 250–264.
132. Watanabe, H.; Matsushita, Y.; Aoki, H.; Miura, T. J. C. Numerical simulation of emulsified fuel spray combustion with puffing and micro-explosion. *Flame* **2010**, *157* (5), 839–852.
133. Hellesø, S. M.; Atten, P.; Berg, G.; Lundgaard, L. E. Experimental study of electrocoalescence of water drops in crude oil using near-infrared camera. *Experiments in Fluids* **2015**, *6* (6), 1–16.
134. Vakarelski, I. U.; Li, E. Q.; Abdel-Fattah, A. I.; Thoroddsen, S. T. Acoustic separation of oil droplets, colloidal particles and their mixtures in a microfluidic cell. *Colloids and Surfaces A: Physicochemical and Engineering Aspects* **2016**, *506*, 138–147.
135. Brotchie, A.; Grieser, F.; Ashokkumar, M. Effect of power and frequency on bubble-size distributions in acoustic cavitation. *Physical Review Letters* **2009**, *102* (8), 084302.
136. Rajendran, A.; Cui, T.-Y.; Fan, H.-X.; Yang, Z.-F.; Feng, J.; Li, W.-Y. A comprehensive review on oxidative desulfurization catalysts targeting clean energy and environment. *Journal of Materials Chemistry A* **2020**, *8* (5), 2246–2285.
137. Li, J.; Yang, Z.; Li, S.; Jin, Q.; Zhao, J. Review on oxidative desulfurization of fuel by supported heteropolyacid catalysts. *Journal of Industrial and Engineering Chemistry* **2020**, *82*, 1–16.

138. Gieleciak, R.; Litz, K. E.; Rankin, J. P.; De Crisci, A.; Chen, J. Analysis of hydrocarbon compositional changes during oxidative desulfurization of bitumen-derived gas oil. *Energy & Fuels* **2020**, *34* (10), 12088–12102.

139. Vedachalam, S.; Boahene, P.; Dalai, A. K. Oxidative desulfurization of heavy gas oil over a Ti–TUD-1-Supported Keggin-type molybdenum heteropolyacid. *Energy & Fuels* **2020**, *34* (12), 15299–15312.

140. Akopyan, A.; Polikarpova, P.; Gul, O.; Anisimov, A.; Karakhanov, E. Catalysts based on acidic SBA-15 for deep oxidative desulfurization of model fuels. *Energy & Fuels* **2020**, *34* (11), 14611–14619.

141. Houda, S.; Lancelot, C.; Blanchard, P.; Poinel, L.; Lamonier, C. Oxidative desulfurization of heavy oils with high sulfur content: A review. *Catalysts* **2018**, *8* (9), 344.

142. Jiang, Z.; LÜ, H.; Zhang, Y.; Li, C. Oxidative desulfurization of fuel oils. *Chinese Journal of Catalysis* **2011**, *32* (5), 707–715.

143. Bhutto, A. W.; Abro, R.; Gao, S.; Abbas, T.; Chen, X.; Yu, G. Oxidative desulfurization of fuel oils using 18 liquids: A review. *Journal of the Taiwan Institute of Chemical Engineers* **2016**, *62*, 84–97.

144. Jalali, M. R.; Sobati, M. A. Intensification of oxidative desulfurization of gas oil by ultrasound irradiation: Optimization using Box–Behnken design (BBD). *Applied Thermal Engineering* **2017**, *111*, 1158–1170.

145. Parthasarathy, S.; Gomes, R. L.; Manickam, S. Process intensification of anaerobically digested palm oil mill effluent (AAD-POME) treatment using combined chitosan coagulation, hydrogen peroxide (H_2O_2) and Fenton's oxidation. *Clean Technologies and Environmental Policy* **2016**, *18* (1), 219–230.

146. Shiraishi, Y.; Tachibana, K.; Hirai, T.; Komasawa, I. Desulfurization and denitrogenation process for light oils based on chemical oxidation followed by liquid–liquid extraction. *Industrial & Engineering Chemistry Research* **2002**, *41* (17), 4362–4375.

147. Arellano, U.; Shen, J. M.; Wang, J. A.; Timko, M. T.; Chen, L. F.; Vázquez Rodríguez, J. T.; Asomoza, M.; Estrella, A.; González Vargas, O. A.; Llanos, M. E. Dibenzothiophene oxidation in a model diesel fuel using CuO/GC catalysts and H_2O_2 in the presence of acetic acid under acidic condition. *Fuel* **2015**, *149*, 15–25.

148. Choi, A. E. S.; Roces, S.; Dugos, N.; Wan, M.-W. Oxidation by H_2O_2 of bezothiophene and dibenzothiophene over different polyoxometalate catalysts in the frame of ultrasound and mixing assisted oxidative desulfurization. *Fuel* **2016**, *180*, 127–136.

149. de Luna, M. D. G.; Wan, M.-W.; Golosinda, L. R.; Futalan, C. M.; Lu, M.-C. Kinetics of mixing-assisted oxidative desulfurization of dibenzothiophene in toluene using a phosphotungstic acid/hydrogen peroxide system: Effects of operating conditions. *Energy & Fuels* **2017**, *31* (9), 9923–9929.

150. Li, S.-W.; Li, J.-R.; Gao, Y.; Liang, L.-L.; Zhang, R.-L.; Zhao, J.-S. Metal modified heteropolyacid incorporated into porous materials for a highly oxidative desulfurization of DBT under molecular oxygen. *Fuel* **2017**, *197*, 551–561.

151. Xiao, X.; Zhong, H.; Zheng, C.; Lu, M.; Zuo, X.; Nan, J. Deep oxidative desulfurization of dibenzothiophene using a flower-like $WO_3{\cdot}H_2O$ catalyst in an organic biphasic system. *Chemical Engineering Journal* **2016**, *304*, 908–916.

152. Akbari, A.; Omidkhah, M.; Towfighi Darian, J. Facilitated and selective oxidation of thiophenic sulfur compounds using MoO_x/Al_2O_3–H_2O_2 system under ultrasonic irradiation. *Ultrasonics Sonochemistry* **2015**, 23, 231–237.
153. Wang, J.; Zhao, R.; Han, B.; Tang, N.; Li, K. Extractive and oxidative desulfurization of model oil in polyethylene glycol. *RSC Advances* **2016**, *6* (41), 35071–35075.
154. Ding, J.-W.; Wang, R. A new green system of HPW@MOFs catalyzed desulfurization using O_2 as oxidant. *Chinese Chemical Letters* **2016**, *27* (5), 655–658.
155. Wan Mokhtar, W. N. A.; Wan Abu Bakar, W. A.; Ali, R.; Abdul Kadir, A. A. Catalytic oxidative desulfurization of diesel oil by $Co/Mn/Al_2O_3$ catalysts—tert-butyl hydroperoxide (TBHP) system: Preparation, characterization, reaction, and mechanism. *Clean Technologies and Environmental Policy* **2015**, *17* (6), 1487–1497.
156. Abdullah, W. N. W.; Ali, R.; Bakar, W. A. W. A. In depth investigation of Fe/ MoO_3–PO_4/Al_2O_3 catalyst in oxidative desulfurization of Malaysian diesel with TBHP-DMF system. *Journal of the Taiwan Institute of Chemical Engineers* **2016**, *58*, 344–350.
157. Ma, C.; Chen, D.; Liu, F.; Sun, X.; Xiao, F.; Dai, B. Oxidative desulfurization of a model fuel using ozone oxidation generated by dielectric barrier discharge plasma combined with Co_3O_4/γ-Al_2O_3 catalysis. *RSC Advances* **2015**, *5* (117), 96945–96952.
158. Dehkordi, A. M.; Kiaei, Z.; Sobati, M. A. Oxidative desulfurization of simulated light fuel oil and untreated kerosene. *Fuel Processing Technology* **2009**, *90* (3), 435–445.
159. Farshi, A.; Shiralizadeh, P. Sulfur Reduction of Heavy Fuel Oil by Oxidative Desulfurization (ODS) Method. *Petroleum & Coal* **2015**, *57* (3)
160. Flores, R.; Rodas, A.; Chavarria, W. Desulfurization of fuel oils using an advanced oxidation method. *Am. Chem. Soc. Div. Fuel Chem.* **2004**, 9, 341–342.
161. Shiraishi, Y.; Hirai, T. Desulfurization of Vacuum Gas Oil Based on Chemical Oxidation Followed by Liquid–Liquid Extraction. *Energy & Fuels* **2004**, *18* (1), 37–40.
162. Ogunlaja, A. S.; Alade, O. S.; Tshentu, Z. R. Vanadium(IV) catalysed oxidation of organosulfur compounds in heavy fuel oil. *Comptes Rendus Chimie* **2017**, *20* (2), 164–168.
163. Chan, N. Y.; Lin, T.-Y.; Yen, T. F. Superoxides: Alternative oxidants for the oxidative desulfurization process. *Energy & Fuels* **2008**, *22* (5), 3326–3328.
164. Tang, Q.; Lin, S.; Cheng, Y.; Liu, S.; Xiong, J.-R. Ultrasound-assisted oxidative desulfurization of bunker-C oil using tert-butyl hydroperoxide. *Ultrasonics Sonochemistry* **2013**, *20* (5), 1168–1175.
165. Kamal, W. M. I. b. W.; Okawa, H.; Kato, T.; Sugawara, K. Ultrasound-assisted oxidative desulfurization of bitumen. *Japanese Journal of Applied Physics* **2017**, *56* (7S1), 07JE03.
166. Luis Garcia-Gutierrez, J.; Patricia Lozano, I.; Hernández-Pérez, F.; C Laredo, G.; Jimenez-Cruz, F. R & D in oxidative desulfurization of fuels technologies: From chemistry to patents. *Recent Patents on Chemical Engineering* **2012**, *5* (3), 174–196.
167. Julião, D.; Gomes, A. C.; Pillinger, M.; Valença, R.; Ribeiro, J. C.; Gonçalves, I. S.; Balula, S. S. Desulfurization of liquid fuels by extraction and sulfoxidation using H_2O_2 and $[CpMo(CO)_3R]$ as catalysts. *Applied Catalysis B: Environmental* **2018**, *230*, 177–183.

168. Hulea, V.; Fajula, F.; Bousquet, J. Mild oxidation with H_2O_2 over Ti-containing molecular sieves—A very efficient method for removing aromatic sulfur compounds from fuels. *Journal of Catalysis* **2001**, *198* (2), 179–186.
169. Ramírez-Verduzco, L. F.; Torres-García, E.; Gómez-Quintana, R.; González-Peña, V.; Murrieta-Guevara, F. Desulfurization of diesel by oxidation/extraction scheme: Influence of the extraction solvent. *Catalysis Today* **2004**, *98* (1), 289–294.
170. Toghyani, M.; Rahimi, A. UV-irradiation effect on desulfurization of hydrocarbon fuels through an advanced oxidation-extraction process in presence of acid catalyst. *Chemical Engineering and Processing - Process Intensification* **2018**, *133*, 303–311.
171. Yazu, K.; Yamamoto, Y.; Furuya, T.; Miki, K.; Ukegawa, K. Oxidation of dibenzothiophenes in an organic biphasic system and its application to oxidative desulfurization of light oil. *Energy & Fuels* **2001**, *15* (6), 1535–1536.
172. Rahimi, M.; Shahhosseini, S.; Movahedirad, S. Hydrodynamic and mass transfer investigation of oxidative desulfurization of a model fuel using an ultrasound horn reactorch. *Ultrasonics Sonochemistry* **2019**, *52*, 77–87.
173. Hernandez Ponce, C. *Modeling of Sulfur Removal from Heavy Fuel Oil Using Ultrasound-Assisted Oxidative Desulfurization*. Masters Dissertation, King Abdullah University of Science and Technology, Thuwal, Saudi Arabia, 2020. https://repository.kaust.edu.sa/handle/10754/664544.

Chapter 11

Effect of Thermal Treatment on Asphaltene: Conversion Composition and Microstructure Analysis

Yao Zhang, Xiao-Fei Tang,* Ruo-Qian Xu,* Wei Li, Yang-Peng Lin, Ying Liu, Jun-Chen Zhu, and Jun-Lin Zhang

Xi'an Modern Chemistry Research Institute, Xi'an 710065, China

***Emails: tangxiaofei0807@163.com (X.-F. Tang); 546226928@qq.com (R.-Q. Xu).**

The primary objective of the work presented in this chapter was to investigate and quantify how asphaltene molecules change under thermal stress. Nitrogen-protected pyrolysis of the pitch product of solvent deasphalting (SDA) was carried out in an open tube furnace, and the asphaltene samples were extracted from SDA and its thermal cracking residues based on their solubility properties. The asphaltene samples were characterized to further understand the changes that take place in asphaltenes during thermal cracking. With an increase in the operating temperature or reaction time, the asphaltene content of the thermally cracked residue experienced an increasing trend followed by a steady decline that was accompanied by a gradual increase in the yield of the insoluble toluene. The atomic mole ratio n(H)/n(C) decreased significantly from approximately 1.0569 to 0.6618. A comparison between asphaltenes from thermal treatment and SDA was conducted using infrared data analysis to distinguish the aromaticity changes. On the other hand, the aromaticity index ($I_{Ar} = I_{3047}/(I_{3047}+I_{2918})$) increased from the 0.021 to 0.097 with a slight decrease of intensity ratio in the CH_2/CH_3 of the asphaltene molecules, which indicates that thermal treatment promotes the aromaticity of asphaltene. This conclusion was further proved by the weak growth of the D/G ratio (height) during Raman analysis. In addition, the surface of the asphaltene was successfully characterized as a function of the agglomerate particles, numerous porous structures, and smooth surfaces using scanning electron microscopic imaging.

© 2021 American Chemical Society

Introduction

The rapid development of the world economy and the surge in demand for oil products have led to a consistent increase in the consumption of conventional crude oil. In recent years, refineries have paid more attention to exploring emerging and improved technologies, including both carbon rejection methods and catalytic conversion methods. The driving motivation toward these technological explorations is to upgrade the bottom of the barrel with the primary aim of maximizing the liquid yield for more valuable products, ranging from fuels to petrochemical feedstocks (*1–4*).

One of the greatest challenges in upgrading heavy fractions of the crude oil is the substantial amount of asphaltene residues. Asphaltenes are basically the *n*-pentane- or *n*-heptane-insoluble components that concentrate more sulfur, nitrogen, nickel, aluminum, and other impurities than crude oil. Their complex compositions (*5, 6*) have a significant influence on the oil properties and its processing, often causing problems such as coking, fouling, rapid catalyst deactivation, and a reduced reaction rate (*7–10*). Asphaltenes can also cause serious problems in the transportation and storage of petroleum products (*11–17*), thus making the understanding of how asphaltenes respond to thermal treatment imperative. Most of the earlier published studies have been primarily focused on the coking phenomenon under thermal stress (*7–9, 18–20*) and kinetic model development (*21*). Studies related to asphaltene behavior during thermal stress loading were conducted mostly under hydrogenation conditions. For example, Merdrignac et al. (*22*) studied the evolution of the asphaltene structure during hydroconversion conditions with an ebullated bed and concluded that: (1) the asphaltene unit size decreases when conversion increases, and (2) the aromaticity of asphaltenes increases due to dealkylation. They also found that the remaining asphaltenes were dissociated in smaller aggregates, an aspect that was considered related to dissociation mechanisms (*23*). Some researchers have examined the changes in the asphaltene properties during hydrotreatment of the heavy crudes. Trejo et al. (*15*) observed different asphaltene morphologies that were related to the mixture of the solvents used for fractionating asphaltenes and the corresponding hydroprocessed reaction conditions. Ancheyta et al. (*24*) found that the structural properties changed more drastically when the temperature was higher with an increase in the aromaticity of the asphaltenes. In their study, Camacho-Bragado et al. (*17*) observed that the precipitated asphaltene molecular aggregates had a structure that resembled a fullerene onion observed from high resolution electron microscopy and energy dispersive spectrometers.

Although research on asphaltene has been widely reported in the literature, studies focused on understanding the noncatalytic thermal conversion and molecular structure changes of asphaltenes are still insufficient (*25–28*). Solmaz et al. (*25*) reported that Raman asphaltenes have heteroaromatic-fused aromatic rings with alkyl and naphthalene substituents located along their peripherals, which apparently represents the first step in asphaltene conversion. AlHumaidan et al. (*26–28*) draw considerable attention to changes in the asphaltene structure during thermal cracking of residual oils. In their study, it was observed from FTIR testing that thermal cracking of the residual oils mainly affected the paraffinic side chains and had a relatively limited effect on the aromatic core. Subsequent XRD testing further indicated that cracking severity had not substantially affected the layer distance between the aromatic sheets (d_m), the distance between the aliphatic chains and naphthenic sheets (d_c), and the average diameter of the aromatic sheet (L_a). However, a decrease in the cluster diameter (L_c) was observed. As the crack severity increased, the cluster diameter (L_c) decreased due to the loss in the aliphatic carbon and a decrease in the number of aromatic sheets per stack (M). The NMR investigations on the pitch products were also found to be consistent with the previous conclusion. In this study, the pitch product of the solvent deasphalting (SDA) was chosen as the feedstock to investigate how the asphaltene molecules changed under thermal stress loading. The characteristic peaks of infrared (IR) and Raman not only prove the existence of functional groups, but the intensity ratio of functional groups was calculated to reveal change during the thermal cracking reaction. This change in functional groups provides us with new ideas for understanding the composition of asphaltenes.

Experimental Section

Materials

n-heptane (C_7H_{16}, 99.9%) was purchased from Tianjin Fuyu Fine Chemical Limited Company (Tianjin, China). Toluene (C_7H_{14}, 99.9%) was purchased from Chengdu Kelong Chemical Limited Company (Chengdu, Sichuan, China). SDA was provided by a Petrochemical Company; the specific properties are shown in Table 1.

All the chemicals were analytical grade and used as received without further purification.

Table 1. Properties of SDA Used in the Thermal Cracking Process

Properties of SDA	*Value*
Density, $g{\cdot}cm^{-3}$ (20°C)	1.03
Carbon residue, wt %	33.61
C, wt %	87.91
H, wt %	9.91
S, wt %	0.52
N, wppm	0.89
Atomic molar ratio n(H)/n(C)	1.35
Saturate, wt %	23.71
Aromatics, wt %	30.28
Resin, wt %	27.09
C_7-asphaltenes, wt %	18.92

Thermal Reaction Experiments

The tube reactor, which was inserted in a tubular furnace, was first purged with nitrogen and then heated to the preset temperatures with 5 mL/min of nitrogen gas. When the predetermined reaction temperatures (namely 450, 500, 520 and 550 °C, respectively) were reached, a crucible with 20 g of SDA was charged into the tube reactor for the desired reaction time. The final thermal cracking products were collected and subjected to various testing techniques for characterization. The effects of the reaction temperature and reaction time on the thermal cracking of asphaltenes were also evaluated and discussed in the subsequent text. When analyzing the reaction temperature, the reaction time was held constant at 8 min. To evaluate the reaction time, the reaction temperature was similarly held constant at 500 °C.

Precipitation of Asphaltenes

The "Determination of Petroleum Asphalt Four Components" analysis method is widely used to separate asphaltenes based on the nature of asphaltenes, which are essentially insoluble in low boiling-point alkane (e.g., *n*-pentane or *n*-heptane) and soluble in toluene (which is more polar). The asphaltene was obtained by extracting 1 g of the sample in 50 mL of *n*-heptane. Thus, asphaltene was collected as *n*-heptane insoluble. The *n*-heptane insoluble on the filter paper was washed with *n*-heptane using a Soxhlet extractor until it became colorless. Then, the *n*-heptane insoluble was extracted with 70 mL of toluene for at least 2 h or until the drop was colorless. Thereafter, the solvent was evaporated, dried, and weighed to obtain the asphaltene. The toluene insolubles on the filter paper were also collected for subsequent use in the study. The dried solid asphaltene was used in the subsequent material characterization tests.

The parent asphaltene sample (Asp-0) was extracted from the SDA. The thermally treated asphaltene samples were obtained from the residues after thermal cracking. Based on the reaction temperature ranging from 450 to 550 °C, the asphaltenes were extracted and designated as follows: Asp-450, Asp-500, and Asp-550, respectively. And, based on the 0 to 8 min reaction time, the asphaltenes extracted from the reaction residue were denoted as follows: Asp-3, Asp-3.5, Asp-4, Asp-5, Asp-6, Asp-7, and Asp-8, respectively.

Characterization

The asphaltene contents of C, H, N, and S were determined by a VARIOEL-III type element analyzer (EA, Elementar cario EL CUBE) at room temperature. Analysis of chemical bonds in the samples were carried out using an IR spectrum (IR, BRUKER TENSOR 27), with a wavelength accuracy of 0.01 cm^{-1} and approximately 16 times the cumulative scanning capability. The Raman spectrum were obtained using a multifunctional laser microscopy Raman spectroscopy (Raman, HORIBA HR Evolution) with an equal wavelength accuracy of 0.01 cm^{-1} at two times the cumulative scan capability. The surface morphology of C7-asphaltene was observed using a scanning electron microscope (SEM, FEI Quanta-600F) at an acceleration voltage of 20.00 kV. The SEM digital images of the asphaltene particles were taken at 500, 5000, and 10,000 times, respectively.

Results and Discussion

Effects of Temperature and Time on Asphaltene Conversion

Variation of the asphaltene content is one of the main considerations during thermal cracking of SDA. The relationship between reaction temperature or time and the percentage of residue remaining is shown in Figure 1. The amount of remaining solid residue shows a decreasing trend with an increase in the reaction time or temperature. When the temperature range is 450 to 550 °C, the corresponding residue range is 53.34% to 48.76%, which shows a slight difference. The remains (residue) tend to be stable in the temperature range of 500–550°C. At this time and within this temperature range, temperature changes (i.e., increments) have little effect on the progression of the thermal reaction. By contrast, the influence of time on the cracking residue was observed to be more prominent. The reaction is relatively rapid before 6 min, as can be seen in the slope of the graphical curve in Figure 1. After 6 min of reaction, the residual solid content is 42.3%, with most of the SDA having reacted.

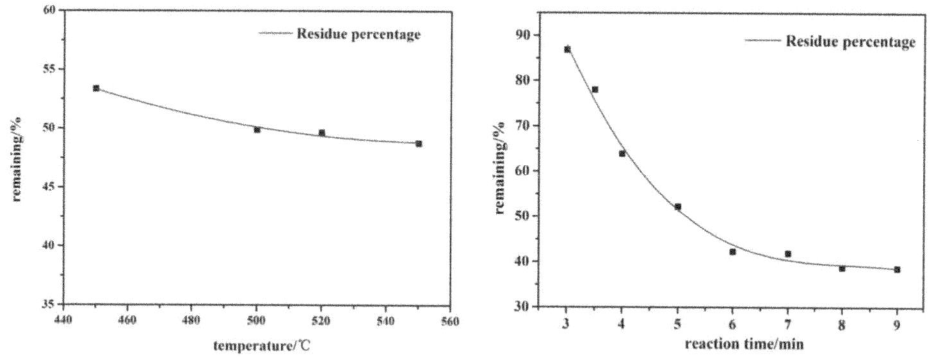

Figure 1. Residue remaining percentage under different temperature and time.

The composition distribution of the cracked residue at different reaction conditions is shown in Figure 2. When the reaction temperature was raised to 450°C (despite the reaction time remaining constant at 8 min), the asphaltene content extracted from the reaction residue is the highest (23.91%). As the temperature continued to rise, the asphaltene content decreased to only 11.26% at 550 °C. Similarly, when the reaction temperature remained constant at 500 °C, the asphaltene content under a different reaction time reached the maximum value at 5 min, which is 27.94% in Figure 2. After this (i.e., 5 min), the asphaltene content begins to decrease until reaching 9 min. Overall, the asphaltene content exhibits an increasing response trend followed by a downward decay, both as a function of reaction time and temperature. This change in composition distribution is largely attributed to the open reaction system. The open reaction system is the light petroleum products produced in the pyrolysis reaction, such as paraffin and aromatic hydrocarbons, that continue to overflow. As a result, the content of the three components continues to decrease. Similarly, the heavy components also continue to accumulate alongside the occurrence of aggregation and coking. Thus, the net result is that toluene insoluble also continues to increase. At the start of the reaction, the content of the asphaltene increases when there is less coking. However, as the coking progresses become more intense, the asphaltenes are converted into toluene insoluble with larger molecular weights and more complex structures. In this state (i.e., toluene insoluble), the content of asphaltene will accordingly decrease. Thus, the option reaction temperature should be around 500 °C with a required reaction time of approximately 6 min to reduce coking.

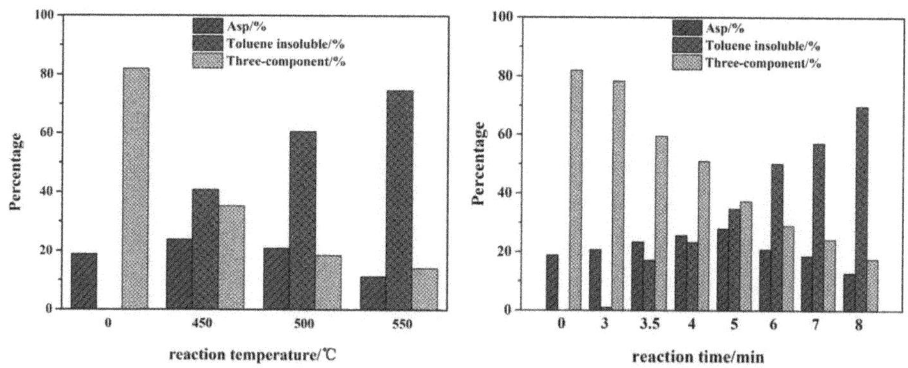

Figure 2. Residue distribution at different reaction temperatures and times (Asp represents asphaltene content and Three-component represents the addition of saturates, aromatics, and resins content).

Elemental Analysis

As listed in Table 2, the elemental analysis results of asphaltenes show that the degree of cracking affects the redistribution of elements. In particular, C and H are greatly affected by the degree of reaction while S and N are insignificantly affected. The C and H content of the asphaltenes in residues have obvious changes compared with the parent asphaltene (i.e., C = 85.72 and H = 7.55, respectively). The atomic molar ratio, namely n(H) / n(C), decreased significantly from 1.0569 before the reaction to 0.6618 (i.e., lowest) at the end of the reaction. This decay in the n(H)-to-n(C) ratio is caused by the scission and rearrangement of the C-C bonds in the alkyl side chain of the asphaltene during thermal cracking reaction, which ultimately results in a decrease in the aliphatic hydrogen content and an increase in the aromatic hydrogen content. On the contrary, changes in the sulfur and nitrogen contents of corresponding asphaltenes were marginal. This is mainly due to the presence of sulfur and nitrogen in the asphaltenes as heterocyclic compounds, which have strong aromaticity and are not prone to breakage to desulfurize and denitrify.

Table 2. Elemental Analysis of Asphaltenes Extracted from SDA and Its Residues

Sample	*Elemental, wt %*				*n(H)/n(C)*	*n(N)/n(C)*	*n(S)/n(C)*	*n(H)/n(S)*
	C	H	N	S				
Asp-0	85.72	7.55	1.21	0.52	1.06	1.21×10^{-2}	2.27×10^{-3}	2.82
Asp-450	89.58	6.34	1.61	0.63	0.85	1.54×10^{-2}	2.64×10^{-3}	2.27
Asp-500	89.79	5.39	1.74	0.63	0.72	1.66×10^{-2}	2.63×10^{-3}	1.92
Asp-550	90.35	4.98	1.78	0.69	0.66	1.69×10^{-2}	2.86×10^{-3}	1.76
Asp-3	89.65	6.83	1.54	0.64	0.91	1.47×10^{-2}	2.68×10^{-3}	2.44
Asp-3.5	89.71	6.42	1.59	0.68	0.86	1.52×10^{-2}	2.84×10^{-3}	2.29
Asp-4	89.76	6.09	1.66	0.76	0.81	1.59×10^{-2}	3.18×10^{-3}	2.17
Asp-5	89.87	6.01	1.66	0.70	0.80	1.58×10^{-2}	2.92×10^{-3}	2.14
Asp-6	89.92	6.00	1.62	0.76	0.80	1.54×10^{-2}	3.17×10^{-3}	2.13
Asp-7	90.15	5.95	1.56	0.73	0.79	1.48×10^{-2}	3.04×10^{-3}	2.11
Asp-8	89.79	5.39	1.74	0.63	0.72	1.66×10^{-2}	2.63×10^{-3}	1.92

IR Spectroscopy Analysis

Owing to the structural diversity of asphaltene molecules, the main peak area of the IR spectrum can be divided into three wavelength regions: (1) aromatic and aliphatic C-H stretch mode regions from 3200 to 2750 cm^{-1}; (2) CC and CM (M is sulfur, nitrogen, and other heteroatoms) stretching as well as CH and MH deformation mode regions from 1800 to 1000 cm^{-1}; and (3) aromatic C-H out-of-plane deformation as well as the skeletal mode regions below 1000 cm^{-1}.

As demonstrated in the IR curve (denoted as Asp-0 in Figures 3 and 4), the parent asphaltene from SDA basically shows characteristic peaks (*11*, *26*) that correspond to modes of the functional groups in the asphaltene spectrum. The IR spectra of the asphaltene samples of the corresponding reaction conditions were analyzed and compared to their parent asphaltene to characterize the composition changes in the asphaltene structure during the thermal cracking process. Overall, the

IR spectrum showed and proved that the extension of the residence and the increase in the cracking temperature has a profound effect on the asphaltene composition (i.e., the functional group distribution). The spectra of the heat-treated asphaltene samples exhibited similar characteristics to the parent asphaltene, with some regular changes in the peak intensity. Looking at Figures 3 and 4, it can be inferred that the following components are present in the mixture:

(1) Long paraffin molecules or saturated naphthenes (2918 and 2851 cm^{-1} -ν_{CH}; 1454, 1374, and 720 cm^{-1} -$\delta_{CCH\ CH2}$, as well as many C-C stretching at the 950–1100 cm^{-1} range). Obviously, there are strong peaks observed at 2918 and 2851cm^{-1} (Figure 3a) that are attributed to the saturated CH stretching vibrations as well as 1454 and 1374 cm^{-1} (Figure 3b), which corresponds to the CH bending vibration of the -CH_2 and -CH_3 structures.[26] The change in paraffin before and after the reaction is the most obvious, which indicates that during the thermal reaction process the chain hydrocarbons present in the asphaltene may undergo breakage of the C-H bonds or isomerization transformation.

(2) Aromatic hydrocarbons (1593 and 1588 cm $^{-1}$ -$\nu_{C=C}$; 3047cm^{-1} -ν_{CH}; 650–900 cm^{-1} -δ_{CCH}). This includes the aromatic ring C-H with stretching modes ranging from 3000 to 3100 cm $^{-1}$(Figure 3a), stretching modes around 1590 cm^{-1} (Figure 3b), and the C-H out-of-plane bending modes at 900–650 cm^{-1} (Figure 3c). This indicates a significant decrease in the C-H stretching mode and a slight increase in the C-H out-of-plane bending modes during thermal cracking, which became more prominent as the reaction intensity increased. The multiple peaks of the C-H out-of-plane bending vibration indicate that there are multiple aromatic rings (monosubstituted and polysubstituted aromatic compounds) in the asphaltene structure. The bending vibration absorption peak shows the diversity and complexity of the aromatic ring substituents in asphaltenes (*11*, *29*, *30*).

(3) Carbonyl compounds (1700–1750 cm^{-1} range) and ethers (1000–1200 cm^{-1} range). The small number or magnitude of the weak peaks observed at 1725 cm^{-1} and 1000–1200 cm^{-1}, respectively, were attributed to the stretching vibrations of the carbonyl compounds (C=O) and ethers (O-C-O).

The change in the structural response trend as a function of time was the same as the temperature. In general, the thermally treated asphaltene spectrum not only has most of the characteristics of the parent asphaltene, but also has more aromatic structures that can be observed from the spectrum at 900–650 cm^{-1}.

Most importantly, the intensity ratios of the main peaks in the IR spectrum for the exact analysis of the asphaltene groups were compared before and after thermal cracking. The effects of heat treatment on the aliphatic part of the asphaltene molecule were evaluated using the intensity ratio of methylene to methyl [nCH_2/mCH_3], according to the different intensity ratio calculation methods proposed in the literature (*26*, *31*, *32*). Figure 5 and 6 illustrate how the aliphatic chain of thermally cracking asphaltene molecules change with temperature and time using the intensity ratio of the CH_2/CH_3 stretching vibration (I_{2918}/I_{2951}) and bending vibration (I_{1454}/I_{1373}), respectively. Under different reaction temperatures, the three intensity (I_{2918}/I_{2951}, I_{1454}/I_{1373}, $[I_{2918}+I_{2851}]/I_{2951}$) ratios range from 2.49 to 2.14, 1.40 to 0.97, and 4.58 to 3.99, respectively, with all experiencing a downward response trend. Again, the changes in the response trends were similar to that of the reaction time. This indicates that the long chains were converted to the shorter chains during the thermal cracking process with the asphaltene molecules losing more of their CH_2 than CH_3 components.

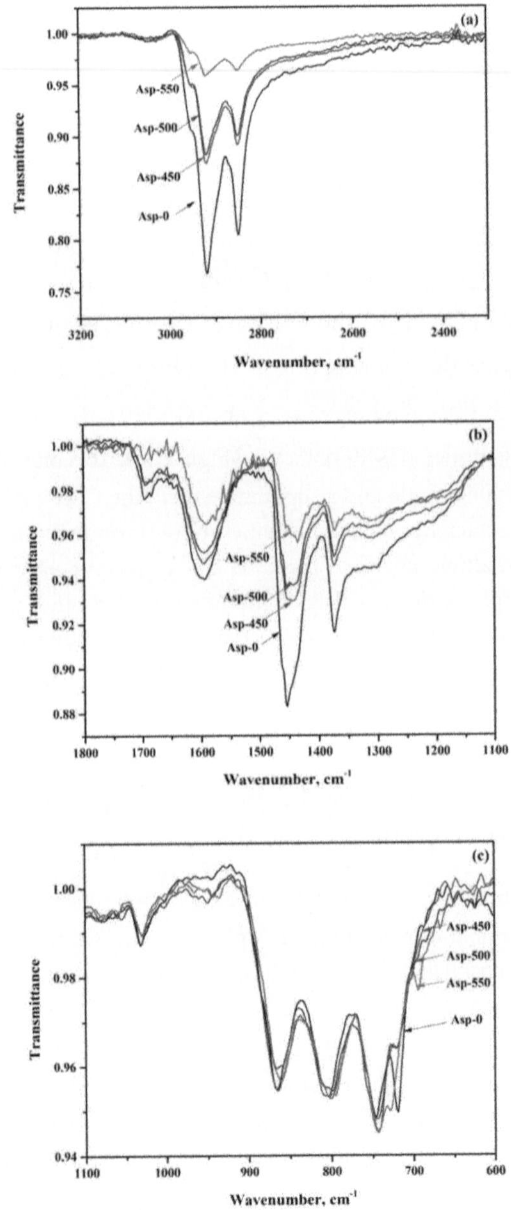

Figure 3. Section plots of IR spectra of asphaltenes separated from SDA (black) and its residues after cracking over 8 min at 450 °C (red), 500 °C (blue), and 550 °C (purple).

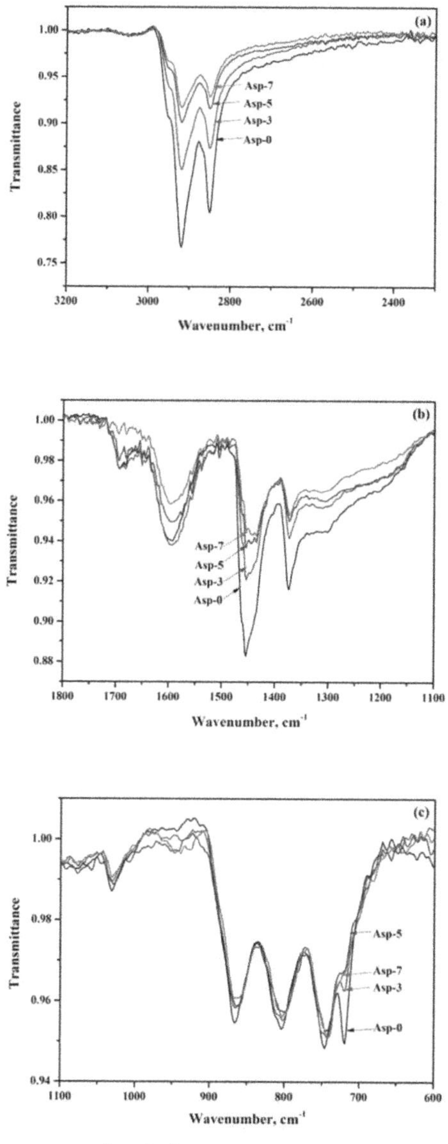

Figure 4. Section plots of IR spectra of asphaltenes separated from SDA (black) and its residues after cracking at 500 °C over 3 min (red), 5 min (blue), and 7 min (purple).

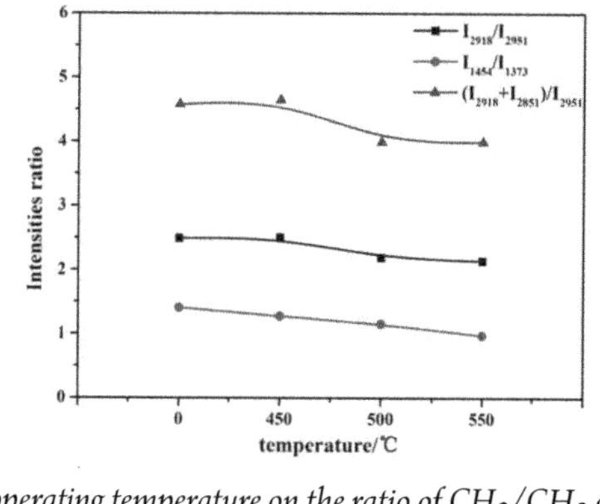

Figure 5. Effect of operating temperature on the ratio of CH_2/CH_3 group of asphaltene.

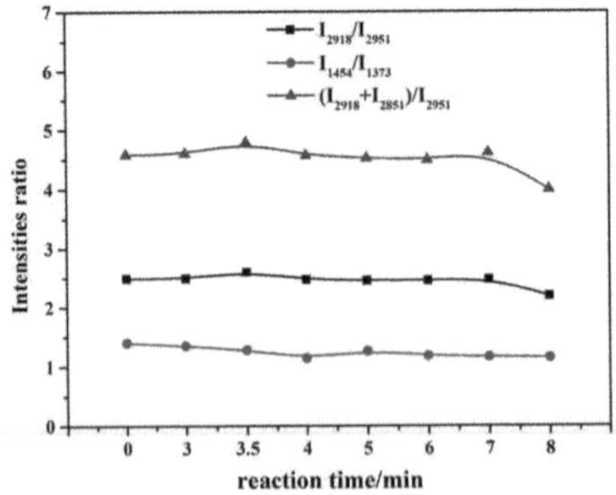

Figure 6. Effect of reaction time on the ratio of CH_2/CH_3 group of asphaltene.

Similarly, the aromaticity of asphaltenes from SDA and the residues were monitored through changes in the concentration of aromatic and aliphatic C-H bonds. This was accomplished by using the intensity ratio of 2918 and 3047 cm^{-1} to evaluate the amount of aliphatic and aromatic hydrogen as well as the calculation of the hydrogen aromaticity index ($I_{Ar} = I_{3047} / [I_{3047} + I_{2918}]$) (33). On the other hand, the intensity ratio between the C=C aromatic stretch and the aliphatic CH_2 (I_{1593}/I_{2918}) also revealed and substantiated the aromaticity of the asphaltenes (34). Figures 7 and 8 show the calculation results of $I_{Ar} = I_{3047} / (I_{3047} + I_{2918})$ and I_{1593}/I_{2918}. The figures clearly demonstrate how the aromaticity of thermally cracking asphaltene molecules change with temperature and time. The aliphatic-to-aromatic C-H stretching ratio of Asp-0 is the lowest (0.26) compared to another asphaltene sample. Particularly, at the higher reaction temperature of 550 °C, the aromaticity index of Asp-550 reached 3.23 times that of Asp-0 based on $I_{Ar} = I_{3047} / (I_{3047} + I_{2918})$ computations. The enhancement of aromaticity was also proved and substantiated by using the intensity ratio between the C=C aromatic stretch at 1593 cm^{-1} and the aliphatic CH_2 at 2918 cm^{-1} (I_{1593}/I_{2918}).

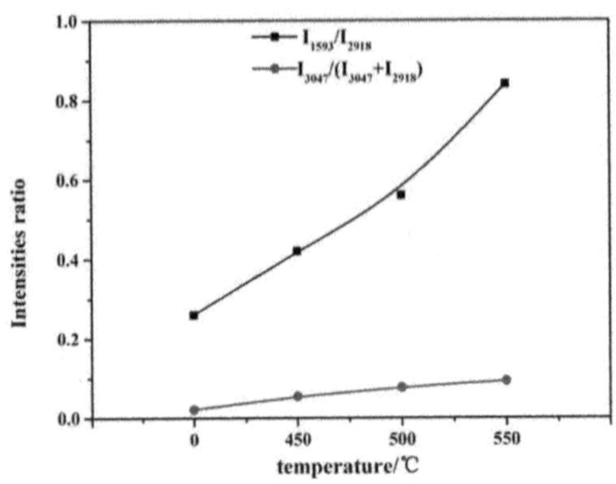

Figure 7. Effect of operating temperature on asphaltene aromaticity.

The results presented in Figures 6–8 confirm that thermal treatment mainly affected the scission of the long alkyl side chains in asphaltenes. However, the aromaticity of the asphaltenes was also enhanced. Furthermore, as the cracking temperature increased, the asphaltene molecules become more difficult to decompose.

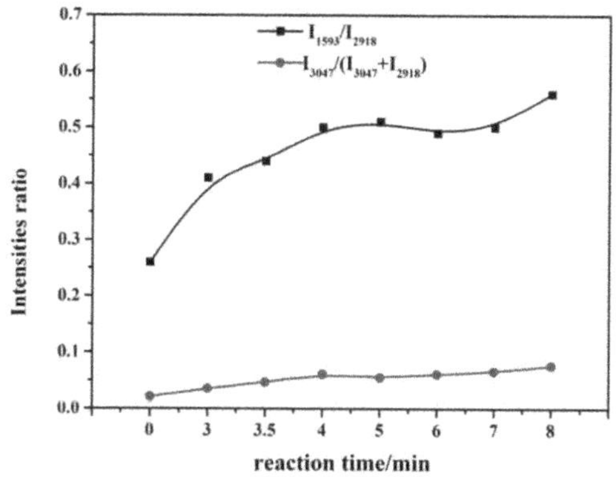

Figure 8. Effect of reaction time on asphaltene aromaticity.

Raman Spectroscopy Analysis

The asphaltene samples from thermal cracking under different operating temperatures and reaction times were further characterized using Raman spectroscopy analysis. Figures 9 and 10 show two gentle Raman ridges for the asphaltene samples as a function of temperature and time effects. Because asphaltenes are mainly composed of fused ring aromatic hydrocarbons, the fluorescent effect of the aromatic rings makes the asphalt in the qualitative Raman spectrum only have gentle Raman ridges without sharp Raman peaks. As reported in the literature, this is the Raman spectrum characteristic associated with most asphaltenes (35–37).

The Raman peak around 1350 cm^{-1} reflects the disordered structure of the organic matter and the defects between the structural units, which contains a defect band called the D peak. The D peak is strictly related to the presence of aromatic rings with defects, (such as bond angle or bond length defects) or the presence of substituent groups. The other sharp and intense Raman peaks within the range of 1580–1610 cm^{-1} belong to the vibration of the C=C bonds, which contains a graphitic band called the G peak. The G peak occurs in aromatic ring and olefinic molecular units.

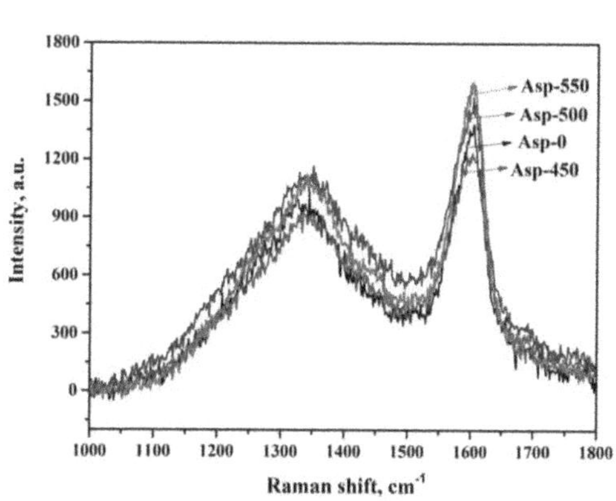

Figure 9. Raman spectra of asphaltenes at different operating temperatures.

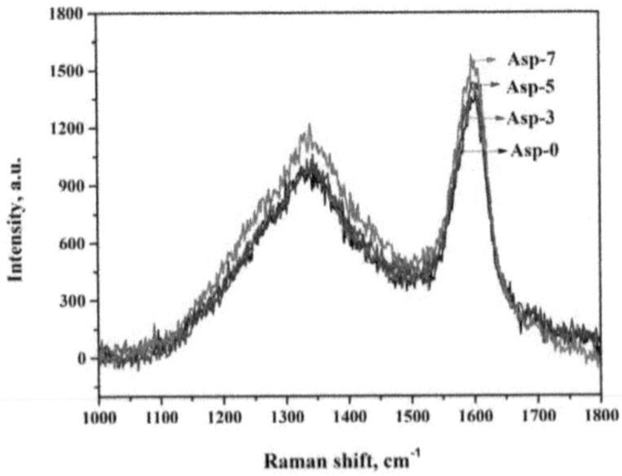

Figure 10. Raman spectra of asphaltenes on different reaction time.

From the Raman spectra of the asphaltenes in the reactant and residues, the pair of Raman peaks (namely 1344 and 1604 cm^{-1}) are consistent. These two peaks are very stable. In particular, the 1604-cm^{-1} G peak is very prominent, which is the Raman characteristic of high-carbon carbonized asphalt. As the degree of reaction intensifies, the G peak becomes sharper and sharper. As shown in Table 3, the changes in the D/G ratio (height) generally have an increasing response trend as a function of temperature and time, respectively.

Table 3. D/G Ratio (Height) of Raman Spectra Obtained from Asphaltenes

Sample	*Asp-0*	*Asp-450*	*Asp-500*	*Asp-550*	*Asp-3*	*Asp-5*	*Asp-7*
Condition		8 min, 450 °C	8 min, 500 °C	8 min, 550 °C	3 min, 500 °C	5 min, 500 °C	7 min, 500 °C
D/G (height)	0.68	0.71	0.70	0.72	0.72	0.73	0.77

SEM Characterization Analysis

Figure 11 shows SEM images of the asphaltenes before and after thermal cracking. Based on a comparison of the SEM photographs at a magnification of 500 times after full grinding (Figure 11a), it can be seen that the Asp-0 extracted from the SDA has a relatively simple microscopic morphology after being physically broken with most of it existing as irregular particles of different sizes and a few pore structures with a smooth and soft surface, respectively. Compared with the Asp-0 in the SDA, the Asp-8 extracted from the SDA particles is smaller in terms of the thermal cracking residue from the SEM photographs at 500 times (Figure 11b). This indicates that the asphaltene is more likely to break during physical grinding. The microscopic morphology is relatively rich with an irregular porous structure of a loose surface, loose layer, smooth and soft surface structure, or an irregular pore structured surface form (see Figures 11c–f). Trejo et al (*15, 38, 39*) have reported similar results. Furthermore, it was determined from energy dispersive spectrometer (i.e., characterize element types and content analysis) that these morphologies had a similar elemental composition comprising mostly of C, S, and O, respectively, and slightly V, Ni, and Si, respectively (*15*).

It is speculated that the difference in the microscopic morphology before and after the reaction is mostly due to the reduction in the alkyl chains and the strong superposition of the aromatic nuclei after the smooth surface is thermally reacted, which is also consistent with the IR data. The asphaltene surface after the reaction is porous and loose, which may be related to the accepted theory of colloidal structure for the asphaltenes (*14*). There is a balance of dissolution between the asphaltene and the colloid, with the porous structure being formed through the dissolution of the colloid during the precipitation of the asphaltene. This is a result of more colloids being left inside or around the asphaltene during the thermal reaction, thus causing the distance between the asphaltene particles to increase and the pores to correspondingly increase. Additionally, the different components of the asphaltene as a complex blend show different dissolubility properties in the solvent and different styles of precipitation, which results in an inhomogeneous material matrix.

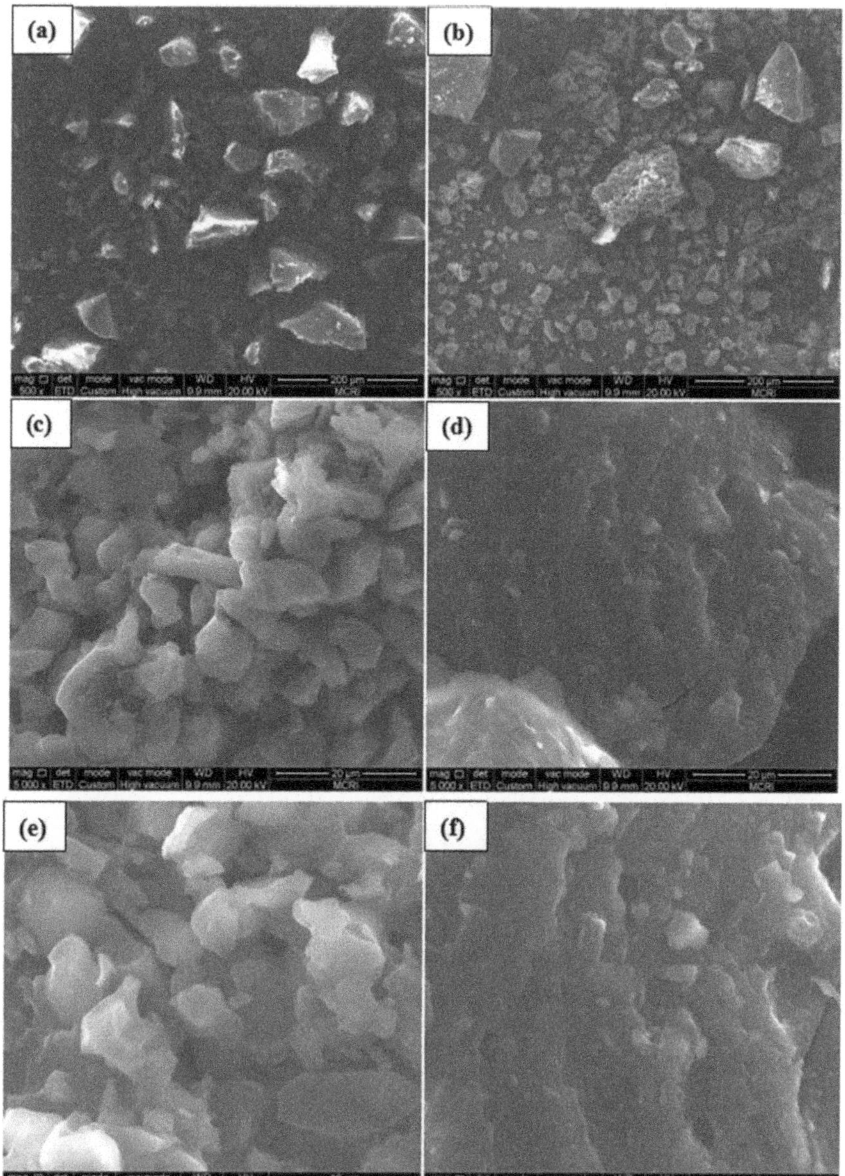

Figure 11. SEM image of Asp-0 and Asp-8; (a), (b) SEM images of Asp-0 and Asp-8 respectively at 500 times; (c), (d) SEM images of Asp-8 at 5000 times; (e), (f) SEM images of Asp-8 at 10,000 times (Asp-0, unreacted; Asp-8, reacted 8 min under 500°C).

Conclusions

The main scientific contribution of this work is characterizing, revealing, and quantifying the conversion, composition, and microstructure of asphaltenes in an open reaction system protected by nitrogen. An increase in the operating temperature or reaction time causes the asphaltene H/C ratio to significantly decrease from 1.0569 before the reaction to 0.6618 (lowest) near the end of the reaction. The aromaticity index ($I_{Ar} = I_{3047}/(I_{3047}+I_{2918})$) increased from the lowest 0.021 to 0.097 after the reaction, which confirmed that asphaltene aromaticity was enhanced with thermal treatment. This conclusion was further substantiated through the intensification of a G peak intensity in the Raman analysis. Furthermore, SEM imaging analysis showed that the surface of the thermally cracked asphaltene was looser and the pore structure increased in comparison to the parent asphaltene. Therefore, it is recommended that more attention should be paid to the changes in the asphaltene with respect to other heavy oil or residual oil during thermal reactions. Evidently, this will beneficially provide a feasible way for evaluating and quantifying the subsequent changes that may occur in the process of refining petroleum and its by-products.

References

1. Long, J.; Wang, Z.; Huang, W.; Zu, D. Application of Heavy Solvent Deasphalting in Sulfur-Containing Residuum Processing. *Pet. Process. Petrochem.* **2004**, 35, 1–5.
2. Joshi, J. B.; Pandit, A. B.; Kataria, K. L.; Kulkarni, R. P.; Sawarkar, A. N.; Tandon, D.; Tandon, D.; Ram, Y.; Kumar, M. M. Petroleum Residue Upgradation via Visbreaking: A Review. *Ind. Eng. Chem. Res.* **2008**, 47, 8960–8988.
3. Rodríguez-Reinoso, F.; Santana, P.; Palazon, E. R.; Diezc, M. A.; Marsh, H. Delayed Coking: Industrial and Laboratory Aspects. *Carbon* **1998**, *36*, 105–116.
4. Sawarkar, A. N.; Pandit, A. B.; Samant, S. D.; Joshi, J. B. Petroleum Residue Upgrading via Delayed Coking: a Review. *Can. J. Chem. Eng.* **2007**, *85*, 1–24.
5. Luo, P.; Wang, X.; Gu, Y. Characterization of Asphaltenes Precipitated with Three Light Alkanes under Different Experimental Conditions. *Fluid Phase Equilib.* **2010**, *291*, 103–110.
6. Duyck, C.; Miekeley, N.; Silveira, C. L. P. D.; Aucélio, R. Q.; Campos Patrícia Grinberg, R. C.; Brandão, G. P. The Determination of Trace Elements in Crude Oil and its Heavy Fractions by Atomic Spectrometry. *Spectrochim. Acta* **2007**, *62*, 939–951.
7. Meng, X.; Xu, C.; Gao, J. Coking Behavior and Catalyst Deactivation for Catalytic Pyrolysis of Heavy Oil. *Fuel* **2007**, *86*, 1720–1726.
8. Yang, L.; Meng, X.; Xu, C.; Gao, J. Influencing Factors of Coking for Heavy Oil Catalytic Pyrolysis. *J. Petrochem. Univ.* **2009**, 22, 22–25.
9. Lin, X.; Fan, Y.; Shi, G.; Liu, H.; Bao, X. Coking and Deactivation Behavior of HZSM-5 Zeolite-Based FCC Gasoline Hydro-Upgrading Catalyst. *Energy Fuels* **2007**, *21*, 2517–2524.
10. Absi-Halabi, M.; Stanislaus, A. Effect of Process Conditions and Catalyst Properties on Catalyst Deactivation in Residue Hydroprocessing. In *Deactivation and Testing of Hydrocarbon-Processing Catalysts*; ACS Symposium Series 634; American Chemical Society: Washington, DC, 1996; pp 229–237.
11. Ilyin, S.; Arinina, M.; Polyakova, M.; Bondarenko, G.; Konstantinov, I.; Kulichikhin, V.; Malkin, A. Asphaltenes in Heavy Crude Oil: Designation, Precipitation, Solutions, and Effects on Viscosity. *J. Pet. Sci. Eng.* **2016**, *147*, 211–217.

12. Ghannam, M. T.; Hasan, S. W.; Abu-Jdayil, B.; Esmail, N. Rheological Properties of Heavy & Light Crude Oil Mixtures for Improving Flowability. *J. Pet. Sci. Eng.* **2012**, *81*, 122–128.
13. Fakher, S.; Ahdaya, M.; Elturki, M.; Imqam, A. Critical Review of Asphaltene Properties and Factors Impacting its Stability in Crude Oil. *J. Pet. Explor. Prod. Technol.* **2020**, *10*, 1183–1200.
14. Liu, Y.-J. Microstructure Characterization of Asphaltenes from Atmospheric Residue Before and After Hydroprocessing. *J. Fuel Chem. Technol.* **2012**, *40*, 1086–1091.
15. Trejo, F.; Ancheyta, J.; Rana, M. S. Structural Characterization of Asphaltenes Obtained from Hydroprocessed Crude Oils by SEM and TEM. *Energy Fuels* **2009**, *23*, 429–439.
16. Ibrahim, Y. A.; Abdulhameed, M. A.; Al-Sahhaf, T. A.; Fahim, M. A. Structural Characterization of Different Asphaltenes of Kuwaiti Origin. *Pet. Sci. Technol.* **2003**, *21*, 825–837.
17. Camacho-Bragado, G. A.; Santiago, P.; Marin-Almazo, M.; Espinosa, M.; Romero, E. T.; Murgich, J.; Rodriguez-Lugo, V.; Lozada-Cassou, M.; Jose-Yacaman, M. Fullerenic Structures Derived from Oil Asphaltenes. *Carbon* **2002**, *40*, 2761–2766.
18. Rudnick, L. R.; Galya, L. G. Structural and Chemical Changes During Coking of Two Petroleum Residua: Hydrogen and Carbon Distribution of Liquid Products. *Energy Fuels* **1991**, *5*, 733–738.
19. Verkoczy, B. Factors Affecting Coking in Heavy Oil Cores, Oils and SARA Fractions Under Thermal Stress. *J. Can. Petrol. Technol.* **1993**, *32*, 25–33.
20. Zhao, Y.-X.; Wei, F.; Yu, Y. Effects of Reaction Time and Temperature on Carbonization in Asphaltene Pyrolysis. *J. Pet. Sci. Eng.* **2010**, *74*, 20–25.
21. Banerjee, D. K.; Laidler, K. J.; Nandi, B. N.; Patmore, D. J. Kinetic Studies of Coke Formation in Hydrocarbon Fractions of Heavy Crudes. *Fuel* **1986**, *65*, 480–484.
22. Merdrignac, I.; Quoineaud, A. A.; Gauthier, T. Evolution of Asphaltene Structure During Hydroconversion Conditions. *Energy Fuels* **2006**, *20*, 2028–2036.
23. Gauthier, T.; Danial-Fortain, P.; Merdrignac, I.; Guibard, I.; Quoineaud, A. A. Studies on the Evolution of Asphaltene Structure during Hydroconversion of Petroleum Residues. *Catalysis Today* **2008**, *130*, 429–438.
24. Ancheyta, J.; Centeno, G.; Trejo, F.; Marroquín, G. Changes in Asphaltene Properties During Hydrotreating of Heavy Crudes. *Energy Fuels* **2003**, *17*, 1233–1238.
25. Solmaz, A.; Mehmet, A. G.; Muzaffer, Y. The Effect of Temperature on the Molecular Structure of Raman Asphaltenes during Pyrolysis. *J. Anal. Appl. Pyrolysis.* **2012**, *96*, 139–145.
26. Alhumaidan, F. S.; Hauser, A.; Rana, M. S.; Lababidi, H. M. S. Impact of Thermal Treatment on Asphaltene Functional Groups. *Energy Fuels* **2016**, *30*, 2892–2903.
27. Alhumaidan, F. S.; Hauser, A.; Rana, M. S.; Lababidi, H. M. S.; Behbehani, M. Changes in Asphaltene Structure During Thermal Cracking of Residual Oils: XRD Study. *Fuel.* **2015**, *150*, 558–564.
28. Alhumaidan, F. S.; Hauser, A.; Rana, M. S.; Lababidi, H. M. S. NMR Characterization of Asphaltene Derived from Residual Oils and Their Thermal Decomposition. *Energy Fuels* **2017**, *31*, 3812–3820.
29. Wu, L.-L.; Deng, W.-N.; Li, C.; Zhang, H.-Y.; Wang, X.-J. Analysis of Coal Tar Heavy Component Asphaltene Properties and Speculation on the Effect of Hydrocracking Green Coke. *J. Fuel Chem. Technol.* **2014**, *42*, 938–944.

30. Hong, Kun.; Ma, F.-Y.; Zhong, M.; Liu, J.-M.; Mo, W. L. Structural Analysis of Residual Heavy Asphaltenes and its Effect on Coke Generation in the Thermal Reaction Process of Hydrogen. *J. Fuel Chem. Technol.* **2016**, *44*, 357–365.
31. Coelho, R. R.; Hovell, I.; de Mello Monte, M. B.; Middea, A.; Lopes de Souza, A. Characterization of Aliphatic Chains in Vacuum Residues (VRs) of Asphaltenes and Resins Using Molecular Modeling and FTIR Techniques. *Fuel Process. Technol.* **2006**, *87*, 325–333.
32. Fossen, M.; Kallevik, H.; Knudsen, K.; Sjöblom, J. Asphaltenes Precipitated by a Two-Step Precipitation Procedure. 2. Physical and Chemical Characteristics. *Energy Fuels* **2011**, *25*, 3552–3567.
33. Guillen, M. D.; Iglesias, M. J.; Dominguez, A.; Blanco, C. G. Fourier Transform Infrared Study of Coal Tar Pitches. *Fuel* **1995**, *74*, 1595–1598.
34. Poveda, J.; Molina, D.; Martínez, H.; Florez, O.; Campillo, B. Molecular Changes in Asphaltenes within H_2 Plasma. *Energy Fuels* **2014**, *28*, 735–744.
35. Liu, M.-J.; Wei, F.; Yang, X.-M.; Dong, S.-A.; Li, Y.-J.; He, X.-J. Synthesis of Porous Graphene-Like Carbon Materials for High-Performance Supercapacitors from Petroleum Pitch Using Nano-$CaCO_3$ as a Template. *New Carbon Mater.* **2018**, *33*, 316–323.
36. Zhang, N.; Tian, Z.-J.; Mao, G.-J.; Wu, S.-H.; Liu, J.-X.; Tuo, Q. Raman Spectroscopic Characteristics of Bitumen Inclusions. *Geochemica* **2009**, *38*, 174–179.
37. Fang, C.-C.; Xiong, Y.-Q.; Li, Y.; Liang, Q.-Y.; Chen, Y. Raman Spectra Characteristics of Solid Bitumens Generated During Oil Cracking. *Geochemica* **2015**, *44*, 196–204.
38. Perez-Hernandez, R.; Mendoza-Anaya, D.; Mondrago-Galicia, G.; Espinosa, M. E.; Rodriguez-Lugo, V.; Lozada, M.; Arenas-Alatorre, J. Microstructural Study of Asphaltene Precipitated with Methylene Chloride and n-hexane. *Fuel* **2003**, *82*, 977–982.
39. Trejo, F.; Ancheyta, J.; Morgan, T. J.; Herod, A. A.; Kandiyoti, R. Characterization of Asphaltenes from Hydrotreated Products by SEC, LDMS, MALDI, NMR, and XRD. *Energy Fuels* **2007**, *21*, 2121–2128.

Chapter 12

Low-Pressure Hydrocracking of Wax over $Pt/SiO_2-Al_2O_3$ to Produce Kerosene for Synthetic Jet Fuel

Felix Link, Nuvaid Ahad, and Arno de Klerk*

Department of Chemical and Materials Engineering, University of Alberta, Edmonton, Alberta T6G 1H9, Canada

***Email: deklerk@ualberta.ca**

Hydrocracking was reviewed with a focus on wax hydrocracking over platinum-promoted catalysts with the specific intent of producing kerosene suitable for synthetic jet fuel. Examples were provided of new experimental work that highlighted aspects of platinum catalysis at low pressure and of kerosene characterization from wax hydrocracking at 2 MPa over 0.5 wt % Pt on amorphous silica-alumina. Several observations were made that have implications for the nature of products that will be obtained from wax hydrocracking over $Pt/SiO_2-Al_2O_3$ at low pressure. Dehydrogenation over Pt is not of equal probability along the carbon chain, leading to an increased probability of forming 2- and 3-branched species as primary products. The hydrogenation rate over Pt is affected by steric constraints around the double bond alkenes. This increases the relative probability of acid catalysis in relation to hydrogenation of branched alkenes when compared to linear alkenes. The general sequence in the hydrocracking mechanism should therefore be interpreted as: dehydrogenation → protonation → $j \times$ (acid catalyzed conversion steps) → deprotonation → hydrogenation, where $j \geq 1$. During low-pressure hydrocracking, the contribution of Pt catalyzed skeletal isomerization and dehydrocyclization reactions increases in importance. Cycloalkanes in kerosene produced by wax hydrocracking is valuable because it increases the composition window that enables blending of synthetic jet fuel that meets specification.

Introduction

The production of liquid transport fuels from renewable and waste materials is one potential strategy to reduce the future use of fossil fuels. Judging by commercial use, the most successful approaches thus far were to produce alcohols for blending in motor gasoline, and to hydroprocess plant oils and waste animal fats for blending in diesel fuel (*1*, *2*). A large volume of renewable material for blending in jet fuel is absent from the portfolio.

© 2021 American Chemical Society

When the prospect for producing a jet fuel blending material from renewable and waste materials was reviewed, it was concluded that, "...only HEFA [Hydroprocessed Esters and Fatty Acids] fuels are already being produced at industrial scale...but may not be sufficiently scalable due to its reliance on waste fats and vegetable oil..." and as alternatives "... F–T [Fischer–Tropsch] fuels are closest to commercialization..." (*3*).

Fischer–Tropsch synthesis (FTS) is a process that converts synthesis gas (CO and H_2) into mixture of heavier hydrocarbons and oxygenates (*4*). The synthesis gas can be produced from a wide variety of renewable and waste materials by gasification (*5*), making it scalable to produce large volumes without being technically constrained by the availability of raw material as feed. The product from FTS cannot be used directly for blending in jet fuel.

One production pathway is gasification of renewable raw material, synthesis gas cleaning and conversion by low-temperature FTS to a product rich in wax, followed by hydrocracking of the wax (Figure 1). Wax hydrocracking is the primary, and often the only, refining process employed, to produce a kerosene product for blending into jet fuel. The literature includes several studies reporting on Fischer–Tropsch wax hydrocracking as a source of kerosene-range blend material for jet fuel (*6*).

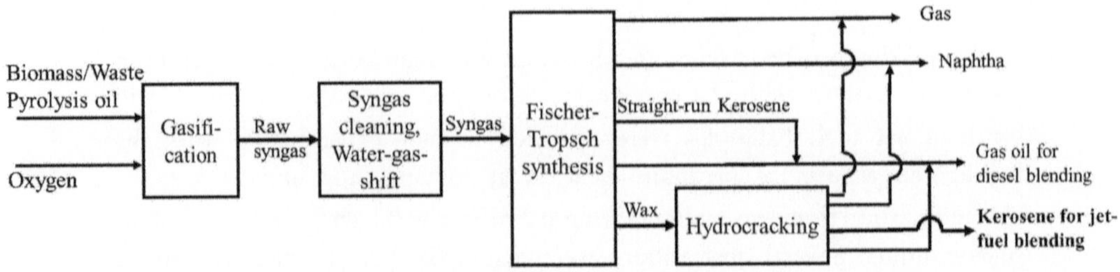

Figure 1. Example for a block flow scheme of kerosene-range blend material production from renewable resources via FTS and hydrocracking. The pathways leading to kerosene material are bold.

From a product perspective, this work explored low-pressure hydrocracking of wax to produce kerosene for synthetic jet fuel. A critical review of the literature is presented with a focus on the use of platinum for hydrogenation–dehydrogenation, as well as the sequence and types of reactions relevant to hydrocracking at lower pressure. The review is supplemented with new experimental work. In the context of the review, the products from wax hydrocracking over $Pt/SiO_2–Al_2O_3$ at 2 MPa were interpreted and the observations were related to the compound-class requirements for producing synthetic jet fuel (*7, 8*).

Wax Hydrocracking in Broader Context

Hydrocracking Process

The demand for liquid transportation fuels requires refiners to use techniques that increase the yield of naphtha, kerosene, and gas oil product from crude oils. Hydrocracking is a key technology to upgrade low metal content, high boiling point, and high molecular weight feedstocks to transportation fuels. Fundamentally, the molecular weight of the feedstock is decreased by breaking carbon–carbon bonds. The formed carbon–carbon double bond is then saturated with hydrogen, leading to a paraffinic product. Hydrocracking is always accompanied by hydroisomerization, which increases the degree of branching of the alkanes in the product (*9*). Hydroisomerization is valuable because it decreases the freezing point of the product and it decreases the low-temperature viscosity of the product. Collectively these are referred to as the cold flow properties and are important specification related properties of jet fuels and diesel fuels.

It is useful to briefly explain the differences between conventional petroleum hydrocracking and wax hydrocracking processes.

Conventional hydrocracking is used for a variety of feedstocks, including gas oils, vacuum gas oils, deasphalted oils and products from thermal cracking, and fluid catalytic cracking (*10*). Crude oil-based feedstocks can contain a significant fraction of aromatics and compounds with heteroatoms like oxygen, sulfur, and nitrogen. Therefore, hydrogenation of aromatics, as well as the removal of heteroatoms, are important side reactions for these feedstocks (*11*). Hydrotreating as a first reaction step, and hydrotreating occurring concomitantly with hydrocracking, make petroleum hydrocracking an exothermic process. The unconverted hydrogen after product knock-out contains H_2S and NH_3 produced from hydrodesulfurization and hydrodenitrogenation. The formation of H_2S during the process also means that the hydrocracking catalyst will be sulfided and remain in a sulfided state during operation. Therefore, the design of petroleum hydrocrackers resembles the design of petroleum hydrotreaters and both are operated at high pressure (*12–14*).

Typically, waxes consist mainly of alkanes, specifically *n*-alkanes. This is also the case for wax derived from low-temperature Co- and Fe-based FTS (*15*, *16*). These materials are sulfur-free and hydrocracking is near isothermal, since the endothermic cracking is balanced with the exothermic hydrogenation of the alkenes produced by cracking. There are no other exothermic hydrotreating reactions occurring as found in petroleum hydrocracking. There are also no other H_2 consuming reactions and there is no need for high-pressure operation beyond that required to suppress deactivation by coke-forming reactions. The absence of H_2S also implies that reduced metal hydrocracking catalysts can be used. When sulfided metal hydrocracking catalysts are used, sulfur-containing compounds or H_2S must be co-fed to the hydrocracker to maintain the catalyst in a sulfided state (*8*).

Hydrocracking Catalyst

Hydrocracking catalysts have been reviewed in the literature (*17*, *18*). Hydrocracking requires bifunctional catalysts that provide metal sites for hydrogenation and dehydrogenation, as well as acid sites for isomerization and cracking reactions. Providing sufficiently strong hydrogenation activity to balance acid-catalyzed cracking activity is challenging for sulfided base metal promoted catalysts, whereas this is a constraint encountered with reduced platinum only when acid support materials have a high activity and high concentration (*19*).

Further discussion this chapter is limited to platinum promoted amorphous silica–alumina (Pt/SiO_2–Al_2O_3) hydrocracking catalysts. Reference to different metals and support materials will be made only when pertinent to illustrate specific concepts.

Within the general class of Pt/SiO_2–Al_2O_3 catalysts, the ratio of SiO_2 to Al_2O_3 can be varied during support synthesis and which influences the acid site strength and density. Amorphous SiO_2–Al_2O_3 contains Brønsted and Lewis acid sites. The total acid site density increases with increasing Al_2O_3 content, while the ratio of Brønsted to Lewis acid sites increases (*20*). The increase is not monotonous over the whole concentration range and it typically passes through a maximum, as illustrated by the change in acid sites over the concentration range from pure silica to pure alumina (*21*).

The amount of platinum impregnated on the support material can also be varied during catalyst synthesis, along with the method of metal impregnation, which influenced the metal surface area per unit catalyst surface area. Thus, the ratio between metal activity and acid activity of the catalyst can be adjusted. To a certain extent, the ratio between hydrogenation and dehydrogenation activity and acid activity can be used to manipulate the activity for hydroisomerization compared to hydrocracking of the catalyst (*22*).

The textural properties of silica–alumina supports strongly depend on preparation and calcination conditions (*17*). Nevertheless, silica–alumina as acid support does not impose the same level of geometric constraints as found in zeolite-based silica–alumina support materials. The geometric constraint experienced during hydrocracking influences the product distribution to some extent and the isomer distribution resulting from hydroisomerization was proposed as method to evaluate pore constraints (*23*). Geometric constraints in zeolite-based supports are also exploited for selective hydroisomerization in catalytic dewaxing (*24*). Generally, zeolite catalyst supports exhibit higher acid site strength and density compared to amorphous support materials (*17*). These differences are pointed out to provide context for the observations in this study compared to platinum promoted zeolite hydrocracking catalysts.

Using the metal platinum also differs in aspects of its behavior from other reduced metals employed for hydrocracking catalysts. Among the most notable differences of platinum compared to other reduced metals used for hydrocracking are with respect to hydrogenolysis, double-bond isomerization, and dehydrocyclization. Platinum is less prone to methanolysis than reduced nickel and palladium (*25, 26*). Platinum is also less prone to double-bond isomerization of alkenes compared to reduced nickel and palladium (*27*). The latter becomes relevant when discussing the mechanism of hydrocracking in more detail, which relies on the formation of alkenes. Platinum also exhibits a propensity for dehydrocyclization reactions (*28*). As with the support materials, these differences are noted for comparison with hydrocracking catalysts that employ different reduced metals than platinum.

Industrial Application of Pt/SiO_2–Al_2O_3 for Wax Hydrocracking

Pt/SiO_2–Al_2O_3 is frequently reported as suitable a catalyst for hydrocracking of FT waxes (*22, 29, 30*). The comparably low acid site density and strength, combined with the highly active hydrogenation function, leads to a high selectivity for middle distillates (i.e., kerosene and atmospheric gas oil), which are often cited as the preferred product (*17, 22, 31*).

The largest industrial applications of noble metal wax hydrocracking are Shell Middle Distillate Synthesis processes used for the gas-to-liquids facility in Bintulu, Malaysia and the Pearl gas-to-liquids facility in Ras Laffan, Qatar. The wax is produced in a Co-based FT process and one of the downstream process steps is a wax hydrocracking process called "heavy paraffin cracking". Based on the mild operating conditions reported for this process, 300–350 °C, 3–5 MPa (*32*), the "proprietary catalyst" employed is likely of the general type Pt/SiO_2–Al_2O_3.

At a smaller scale, a Pt/SiO_2–Al_2O_3 catalyst was also used for wax hydrocracking by Eni/IFP–Axens in the Sannazzaro GTL pilot/demonstration-scale plant (*33*), although it was not explicitly disclosed as such. This chapter will discuss other smaller scale applications and laboratory studies.

Alkane Hydrocracking Mechanism

The mechanism of hydrocracking of *n*-alkanes is described in terms of a sequence of reaction steps that take place on different sites of the catalyst. Little variation in the general description of the hydrocracking steps have occurred over the past the decades, illustrated in Figure 2. (*9*, *17*, *34*).

A) Dehydrogenation of the alkane at a metal site to form an alkene.
B) Desorption of the alkene from the metal site and diffusion to an acid site.
C) Protonation of the alkene at the acid site to form a carbocation.
D) Hydride shifts and alkyl shifts that lead to an isomerized carbocation.
E) Breaking of a C–C bond in the carbocation (β-scission) to form a shorter carbocation and a shorter alkene.
F) Deprotonation of the carbocation to form an alkene.
G) Diffusion of alkene to metal site.
H) Hydrogenation of alkene at the metal site to form an alkane.

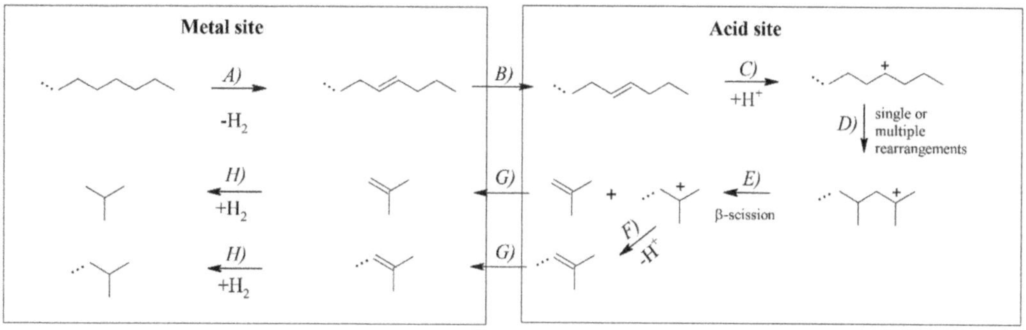

Figure 2. Exemplary pathway for hydrocracking over a bifunctional catalyst. The steps of hydrocracking named previously appear in italics.

For completeness, mention can be made of attempts to explain the hydrocracking steps using hydrogen spillover. However, convincing arguments indicate that observations that were attributed to hydrogen spillover, were more likely explained by bifunctional catalysis (*35*).

Dehydrogenation and hydrogenation reactions are reviewed in more detail further in the chapter. If a noble metal (e.g., platinum) is used for the metal sites in hydrocracking, the reaction rate of hydrogenation and dehydrogenation is typically high. Therefore, it is often assumed that hydrogenation and dehydrogenation are in equilibrium and the surface concentration of alkenes is high enough that the rate determining steps are occurring at the acid sites (*36*). It should be realized that this is an assumption that effectively disguises the complexity of hydrogenation–dehydrogenation equilibria of individual species.

Like hydrogenation and dehydrogenation, protonation and deprotonation at the acid site are nominally viewed as reversible reactions. The carbocation formed from protonation of an alkene can undergo hydride and alkyl shifts as well as cracking. Although cracking and dimerization can be seen as forward and reverse reactions, cracking is assumed to be an irreversible reaction during hydrocracking. This is just an assumption, which ignores the possibility that alkenes produced by cracking may indeed react with carbocations on the catalyst. If one were to consider "in cage" reactions, analogous to what is described for free-radical cracking, then there is a possibility the alkene will react with the same carbocation species that may or may not have rearranged on the catalyst by hydride shift.

The nature of the alkenes and how they participate in the hydrocracking reaction, increase in importance with a decrease in hydrogen partial pressure. At lower pressure, the "equilibrium" concentration of alkenes increases, and this is therefore an aspect that is emphasized in this chapter.

Hydride Shifts

Hydride shifts alter the position of the positive charge in the carbocation. Hydride shifts are reported to occur very rapidly compared to alkyl shifts and cracking reactions (*37*). Therefore, this reaction is not considered to be rate determining. This assumption implies that the position of the double bond is immaterial, because the rate of hydride shift is so rapid. If this is a valid assumption, then all alkenes of the same skeletal structure will be subject to rapid hydride shift and would be indistinguishable during subsequent acid-catalyzed reactions since the position or distribution of positions of the carbocations would be similar.

The position of the positive charge at a specific methyl group is energetically favored if this methyl group has many methyl substituents (i.e., tertiary carbocations are more stable than secondary carbocations and secondary carbocations are more stable than primary carbocations). Thus, the thermodynamic equilibrium favors tertiary carbocations in higher concentrations than secondary carbocations. Primary carbocations are energetically unfavored (*38*). There is the tacit assumption, due to the high rate of hydride shifts, that "equilibration" will take place, but in practice it is just a way to express the thermodynamic driving force. It is difficult to imagine how such a localized phenomenon would lead to bulk thermodynamic equilibration.

Isomerization Reactions

Two types of alkyl shifts occur in carbocations of branched species. Carbocation rearrangements that do not alter the degree of branching are called A-type isomerization. The position of the branch in a branched carbocation is changed in an A-type isomerization step (*39, 40*). In branched species, repeated A-type isomerization reactions can shift the position of the branch toward a restricted equilibrium distribution (*39*). The equilibrium is restricted because A-type isomerization, by definition, does not enable access to all the skeletal isomers of that carbon number.

Carbocation rearrangements that alter the degree of branching in the carbocation are called B-type isomerization. The degree of branching of carbocations is shifted toward the equilibrium of branched and linear carbocations, which is an equilibration of all skeletal isomers of that carbon number. Typically, B-type isomerization is reported to be slower than A-type isomerization (by a factor of approximately 50 times for decanes) (*41*).

It is generally stated that this isomerization proceeds via protonated cyclopropane (PCP^+) intermediate, as explained by Sie (*42, 43*). Also suggested is the possibility that larger cyclic structures are involved in such rearrangements, which would explain the formation of isomers with longer side-chains (*44*). The proposal that explains isomerization as a monomolecular process via a larger cyclic structure is assuming that no additional reactions take place. This type of isomerization can also be explained by consecutive monomolecular skeletal isomerization reactions, or by cracking, hydride shift of the carbocation, and in cage readdition of the alkene formed by cracking, which is a bimolecular process.

Despite the customary depictions of the PCP^+ intermediate (Figure 3), it was suggested on theoretical grounds that the actual intermediate might not be a formal closed-ring structure, but that it is actually a hybrid structure (*45*). The structure that was proposed (rightmost in Figure 3) is a hybrid that captures some of the features of both the corner- and edge-protonated structures and are both examples having three center two electron (3c–2e) bonds. This is a bond type that will be encountered again in this chapter.

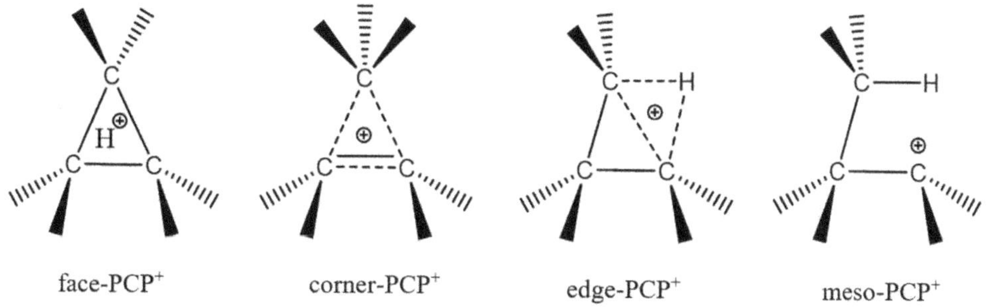

Figure 3. Different PCP+ intermediate structures that were proposed to represent the intermediate state in skeletal isomerization.

Although skeletal isomerization is usually discussed only in terms of acid catalysis, platinum is also capable of skeletal isomerization (*46–49*).

Cracking via β-Scission

The rupture of a C–C bond in a carbocation leads to the elimination of an alkene. The second product is a shorter carbocation. The rate of β-scission is determined by the structure of the parent carbocation and the formed carbocation (Figure 4).

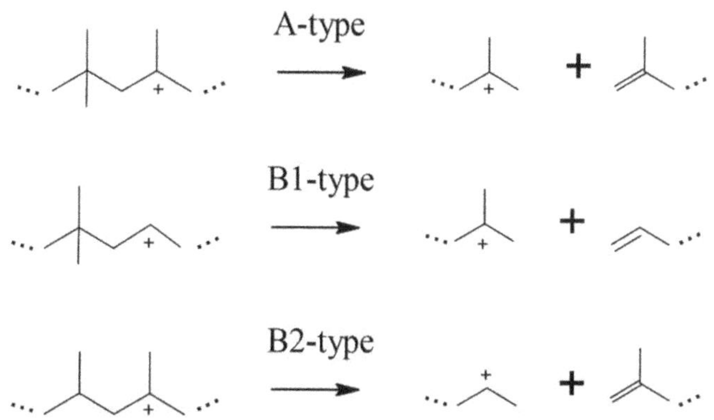

Figure 4. Carbocation rearrangements during A- and B-type β-scission.

The β-scission involving two tertiary carbocations (A-type β-scission) proceeds more rapidly than β-scission involving a tertiary and a secondary carbocation (B-type β-scission). The reason is that tertiary carbocations are more favored energetically compared to secondary carbocations. Scission rearrangements that involve two secondary carbocations (C-type β-scission) are considerably slower and the rate of these reactions is negligible for short chain feedstocks under typical hydrocracking conditions if the catalyst does not exhibit shape-selectivity (*50*).

Both A- and B-type β-scission rearrangements require branching in the carbocations. A-type β-scission proceeds via trimethyl-branched carbocations and B-type β-scission proceeds via dimethyl-branched carbocations. Therefore, several B-type isomerization steps are necessary before cracking of a linear carbocation. In addition, several A-type isomerization steps might be required to obtain a carbocation configuration that is suitable for β-scission.

If the deprotonation of the carbocation, desorption of the alkene, and hydrogenation of the formed alkene are rapid, alkanes are formed without further secondary acid-catalyzed reactions (e.g., cracking).

Lacking the occurrence of secondary cracking reactions leads to a product carbon number distribution that represents the distribution of primary cracking products. If a single *n*-alkane is used as feedstock, then the product carbon number distribution will be symmetrical with the center at half of the carbon number of the feed molecule if cracking is of equal probability along the carbon chain, except for the two end-most C–C bonds. This distribution is a feature of so-called "ideal hydrocracking", which is observed over specific catalysts at low-severity hydrocracking conditions (e.g., Pt/Ca–Y and Pt/SiO_2–Al_2O_3) (*9*).

Hydrogenolysis and Protolysis

The classic carbocation mechanism described in this chapter indicates that the formation of C_2 products is energetically unfavorable and that it is not possible to form a C_1 product. One would therefore not anticipate any methane and only little ethane as cracking products. Both methane and ethane are typically formed in low amounts during hydrocracking. Although the formation of C_2 products can be explained by carbocation β-scission, a different explanation for the formation of methane must be sought.

There are two pathways that can explain the formation of methane during hydrocracking: hydrogenolysis and protolysis. Neither of these reactions were shown in the general hydrocracking mechanism in Figure 2.

Hydrogenolysis reactions take place in presence of metal catalysts and is a reaction where rupture of a given bond is accompanied by addition of hydrogen across the broken bond. Hydrogenolysis of waxes involves the rupture of a carbon–carbon bond, formation of carbon–hydrogen bonds and it leads to the formation of lower molecular mass products (*51*, *52*). Temperature needed for these reactions generally exceed 100 °C (*53*).

The sequence of steps leading to hydrogenolysis are (*52*): (1) adsorption of the alkane on the metal, (2) dehydrogenation of the alkane to form a hydrogen-depleted adsorbed species, (3) C–C scission of the hydrogen-depleted adsorbed species to form two separate hydrogen-depleted adsorbed species, (4) hydrogenation of the hydrogen-depleted species, and (5) desorption of the alkane products.

Hydrogenolysis occurs at the same conditions as carbon deposition and similar intermediates are formed in both. A sufficient temperature is required to provide the activation energy of C–C bond rupture. With an increase in temperature, the rate of dehydrogenation increases and the rate of hydrogenolysis increases (*54*). Hydrogen pressure typically has an inverse effect on the rate of hydrogenolysis (*52*). Although the term hydrogenolysis implies scission by hydrogen, it misrepresents the reaction, because the scission follows on dehydrogenation and hydrogenation plays no role in the scission, only in removal of the products after scission. Hydrogenolysis is therefore anticipated to increase in importance when operating hydrocracking at lower pressure.

Protolysis is the result of acid catalysis. When an alkane is protonated by strong Brønsted acid, it results in the formation of a three center two electron (3c–2e) bond in the intermediate (*55*). This intermediate can decompose by α-scission, as illustrated in Figure 5.

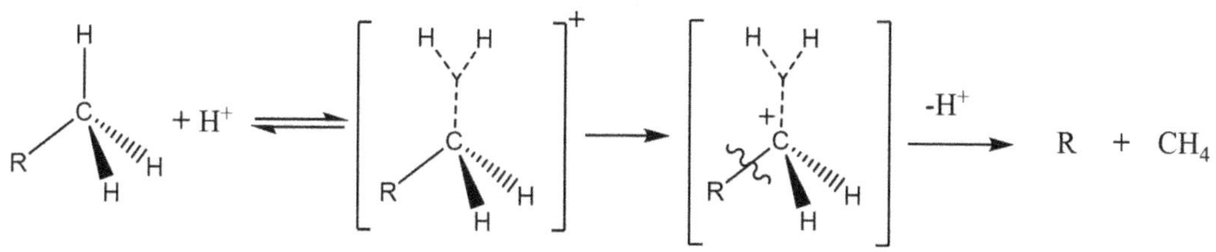

Figure 5. Pathway for protolysis via α-scission.

In presence of both alkanes and alkenes, the rate of alkene protonation is much higher than alkane protonation. Cracking by α-scission after formation of a 2e–3c bond will be minor compared to cracking by carbocation β-scission, but protolysis is a pathway that cannot be ruled out completely.

Dehydrocyclization

Dehydrocyclization results in removal of hydrogen (which is also a dehydrogenation reaction) and the formation of new C–C bond (*53*). To some extent, it is the opposite reaction of hydrogenolysis, because dehydrogenation does not lead to bond scission, but to bond formation. In fact, the two reactions are closely related (*49*).

As discussed, Pt has a propensity for dehydrocyclization reactions (*28*), although this type of reaction is not restricted to Pt. Seen in the context of dehydrocyclization being an opposite reaction to hydrogenolysis, the tendency of Pt to perform hydrodecyclization is a matter of selectivity. The bond forming nature of platinum can also be seen from its ability to perform skeletal isomerization, which according to the Anderson–Avery mechanism, proceeds through an adsorbed cyclic intermediate (*47, 48*). The size of the cyclic intermediate is affected by the carbon chain length of the alkane, with longer alkanes being prone to cyclopentane and cyclohexane formation. Thus, there is a link between the skeletal rearrangement and dehydrocyclization catalysis.

An alkane on a Pt catalyst can be dehydrocyclized to cycloalkanes or alkanes can directly be aromatized to aromatic products (*34, 56*). Over Pt catalysts, cycloalkanes and aromatics have been observed as byproducts during hydrocracking of *n*-alkanes (*34*) and *n*-alkane rich feedstocks (*57*).

However, most studies on hydrocracking of *n*-alkanes do not mention or discuss the formation of cyclic products. The possible formation routes of cyclic and aromatic hydrocarbons over Pt are discussed in this chapter.

Hydrocracking of Model *n*-Alkanes and Waxes

Reactions using model *n*-alkane compounds are less relevant for industrial application than waxes spanning several carbon numbers. Nevertheless, studies employing model *n*-alkanes offer fundamental understanding about the reactions occurring during hydrocracking and the influence of operating conditions. Accordingly, it is useful to first review hydrocracking with model *n*-alkanes before discussing more applied studies that employed waxes.

Hydrocracking with Model *n*-Alkanes

Dependence of Conversion on Operating Conditions

The operating conditions of hydrocracking can be manipulated to achieve desired yields of targeted products. The influence of these conditions on conversion and selectivity are summarized in the following using *n*-alkane feedstocks as example.

During hydrocracking of *n*-alkanes, the *n*-alkane conversion increases with increasing temperature (Figure 6). The observed relationship is due to an increase of the reaction rate of the kinetically controlled isomerization and hydrocracking reactions (*58*). At lower temperatures, hydroisomerization is the dominant reaction with little to no hydrocracked product formation. Thus, under these mild conditions the product predominantly contains unconverted and isomerized feed molecules. Upon increasing temperature, the yield of isomerized feed molecules goes through a maximum (Figure 6). As discussed, hydroisomerization precedes hydrocracking; consequently, isomerized molecules are consumed by hydrocracking reactions. At high temperatures, complete conversion of *n*-alkane molecules can be achieved with little to no isomerized feed molecules remaining in the product. Similar trends as shown in Figure 6 have also been observed for longer chain model *n*-alkane feedstocks up to *n*-C_{44} (*58–60*).

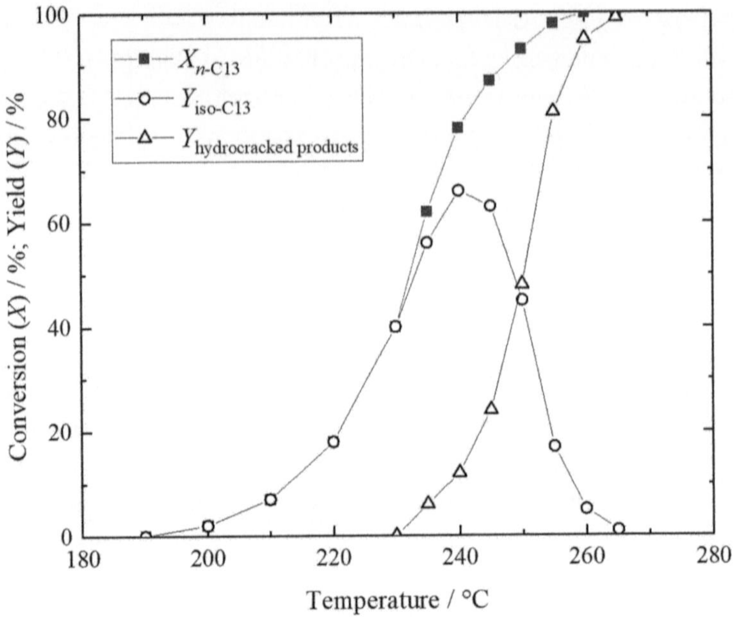

Figure 6. Conversion of n-C_{13} over 0.5 wt % Pt–CaY and yield of isomerized C_{13} and hydrocracked products as a function of the reactor temperature. Conditions: p_{n}–C_{13} = 13 kPa, p_{H2} = 4.0 MPa, m_{cat}/n_{n}–C_{13} = 750 g h mol^{-1}. Drawn using the data reported in ref (9).

It has been stated that for a given *n*-alkane, the isomer yield compared to cracking yield (Figure 6) is a unique function of the *n*-alkane conversion (*58*). Thus, any manipulation of the operating conditions that affects the conversion will in turn also affect the selectivity for cracked and isomerized products.

At otherwise constant conditions, the weight hourly space velocity (WHSV) can be varied to influence the conversion during hydrocracking (*61*). At lower WHSV, higher contact time between feed molecules and catalyst lead to increasing conversion and vice versa. In contrast to a change of the temperature, a change in WHSV does not affect the vapor–liquid equilibrium during trickle phase operation of hydrocracking.

The *n*-alkane conversion increases with a decrease in pressure (*58*). This can be explained by higher amounts of alkenes on the catalyst surface since lower pressure favors dehydrogenation. When the temperature is sufficiently high, cyclic and aromatic species start appearing in the hydrocracked product (*34, 62*). Thermodynamically, the driving force to produce cyclic and aromatic species increase at constant temperature with a decrease in pressure.

An increase in the ratio of H_2 to *n*-alkane feed had a positive impact on conversion for *n*-hexadecane (*n*-C_{16}), whereas it was negligible for *n*-octacosane (*n*-C_{28}) (*58*). This was explained in terms of the vapor–liquid equilibrium in combination with the ratio of vapor to liquid phase. The increase in H_2 to *n*-alkane does not affect the vapor–liquid equilibrium, but it affects the amount of the hydrocarbon species present in each phase. The products of hydrocracking are more volatile than the *n*-alkane feed (i.e., the partial pressure of the products of hydrocracking is higher than the partial pressure of the *n*-alkane feed). Vapor–liquid equilibrium causes the relative concentration of the more volatile products to increase in the vapor phase, but the H_2 to *n*-alkane ratio determines the total amount of material that can be accommodated in the vapor phase and thereby removed from the liquid phase. When comparing *n*-C_{16} and *n*-C_{28}, hydrocracking of *n*-C_{16} is likely to produce products that have a high vapor pressure, which would cause more effective removal of hydrocracked products from the liquid phase than for the hydrocracked products from *n*-C_{28}. Thus, heavier reactive feed molecules are concentrated in the liquid phase, leading to a higher conversion.

One implication of these observations is that the relationship between conversion of *n*-alkanes and temperature is more complex. Temperature affects both reaction kinetics and vapor–liquid equilibrium.

Dependence of Reactivity on the Chain Length

In wax hydrocracking, the feedstock contains a mixture of hydrocarbons with different chain lengths. Knowledge about the reactivity of *n*-alkanes as a function of their chain length offers an indication about the reaction rates of these alkanes in the mixture.

For *n*-alkanes from C_6 to C_{11}, Weitkamp (Table 1) observed a significant increase of the initial *n*-alkane reaction rate with increasing carbon chain length. This was explained by stronger adsorption of the heavier compounds on the catalyst surface compared to the lighter compounds.

To support the interpretation that the difference in initial reaction rate was due to a difference in adsorption strength, the Henry adsorption coefficients were determined for different *n*-alkanes (Table 2) (*64*). The resulting higher alkene concentrations of heavier compounds on the catalyst surface leads to a faster desorption of cracked products from acid sites, which increases the cracking reaction rate, but can also affect the product distribution. Secondary cracking reactions are suppressed by a rapid desorption of cracked products from the acid sites due to their lower adsorption strength. Kinetic analysis showed that the intrinsic reaction rates (not considering adsorption and desorption) also increase with increasing chain length. However, the adsorption constants increase more drastically than the intrinsic reaction rates with increase of chain length (*64*).

Table 1. Initial Reaction Rate During *n*-Alkane Hydrocracking over 0.5 wt % Pt/CaY as a Function of Feed *n*-Alkane Chain Length. Data from ref (*63*).

Feed	*Initial reaction rate / mol g⁻¹ h⁻¹*
n-hexane	0.13×10^{-2}
n-heptane	0.38×10^{-2}
n-octane	0.53×10^{-2}
n-nonane	1.1×10^{-2}
n-undecane	1.3×10^{-2}

Conditions: $T = 275$ °C, $p = 3.9$ MPa, $n_{H2}/n_{n\text{-alkane}} = 17{:}1$.

Table 2. Henry Adsorption Constants for n-Alkanes on Na–Y Zeolite at 233 °C. Data from ref (*64*).

Component	*Henry adsorption constant / mol kg⁻¹ Pa⁻¹*
n-hexane	7.88×10^{-5}
n-heptane	21.5×10^{-5}
n-octane	56.4×10^{-5}
n-nonane	148×10^{-5}

The impact of vapor–liquid equilibrium was not explicitly considered by the aforementioned studies. Unless the conditions were such that hydrocracking was performed as a purely vapor phase process for all the model *n*-alkane feed materials, vapor–liquid equilibrium would affect the observed reaction rates.

The increase of the reaction rate with increasing chain length has also been observed for longer chain *n*-alkanes up to *n*-C_{28}. However, for *n*-C_{36} and *n*-C_{44}, a lower reaction rate than for *n*-C_{28} was observed, with a decreasing trend with increasing chain length (*60*). It was speculated that geometric constraints in the catalyst pores could be responsible for the decrease of reactivity of the *n*-C_{36} and *n*-C_{44}, compared to *n*-C_{28}, as well as aspects related to relative adsorption strength and steric effects.

There are multiple possible explanations, and it is worthwhile remembering that conversion, irrespective of whether it is hydroisomerization or hydrocracking, requires reaction to take place at two different sites. Any attribute of the *n*-alkane or catalyst that affects transport, may contribute to the observations that there is a monotonic increase in conversion rate for *n*-C_6 to *n*-C_{11} (*63*), but a decrease for *n*-C_{28} to *n*-C_{44} (*60*). In fact, when the acid support is a zeolite, there may even be other effects, such as the so-called "window"-effect, that could alter transport (*65*).

Product Distribution

In their early work on hydrocracking, Coonrad and Garwood used *n*-hexadecane to study the product distribution of hydrocracking over a Pt/SiO_2–Al_2O_3 catalyst (*34*). At conversion levels of 53%, they observed a symmetrical carbon number distribution in the product centered around C_8 as shown in Figure 7 (*34*). This indicates what was called "ideal hydrocracking" behavior, where each

C–C bond in the parent alkane has a similar cracking probability except for the two C–C bonds at either end of the carbon chain. Similar symmetrical distributions were observed by other groups using different *n*-alkanes under mild conditions (*9, 66*).

For higher conversions, the carbon number distributions from hydrocracking shifts toward shorter chain products (Figure 7) (*34, 66*). This has been explained by the occurrence of secondary cracking reactions. Cracked products can undergo isomerization and the isomerized products can then crack again by the same cracking mechanism as the parent molecule.

The cracked products consist mainly of branched alkanes (Table 3) (*9, 61, 63, 66*).

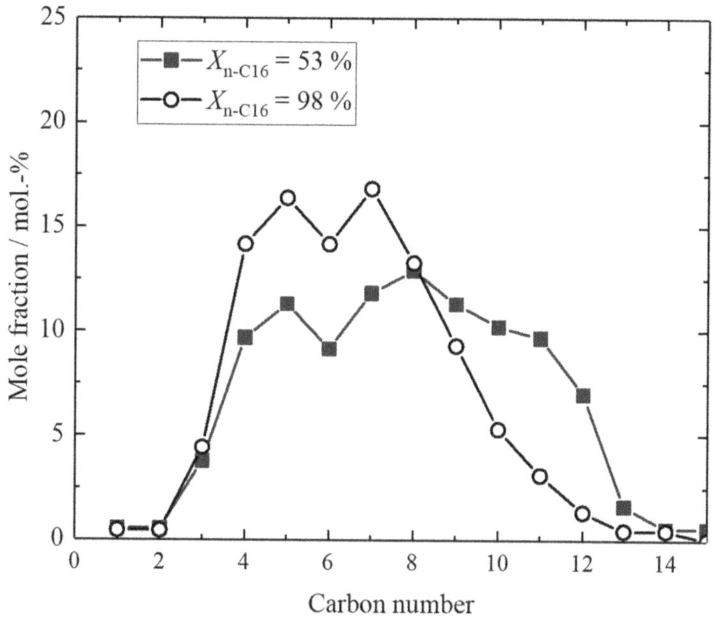

Figure 7. Molar alkane carbon number distribution of the cracked product from hydrocracking of n-hexadecane at different conversions. Naphthenic products are omitted. Drawn using the data reported in ref (34).

Table 3. Mole Fraction of Isoalkanes in the Cracked Product with Different Carbon Numbers Obtained from Hydrocracking of Different Feed Alkanes over Pt/CaY at 50% Cracking Conversion. Data from ref (63).

Feed	*Fraction of isoalkanes in the product with the indicated carbon number / mol %*					
	C_4	C_5	C_6	C_7	C_8	C_9
n-heptane	87.5	56.0	71.0			
n-octane	63.5	91.5	62.0	76.0		
n-nonane	63.7	75.8	91.0	71.0	79.0	
n-decane	65.7	73.0	78.0	91.0	73.0	
n-undecane	67.3	75.7	78.1	83.1	92.0	
n-dodecane	65.6	74.1	78.0	82.2	85.5	92.0

The observed extent of isomerization (Table 3) could be explained by the predominance of A-type cracking, which is the fastest cracking reaction route. The products of A-type cracking are two monomethyl-branched cracked products (Figure 4) and would therefore lead to isomerized products. However, such an explanation is quite onerous in terms of the degree of prior hydroisomerization that should be achieved. Secondary hydroisomerization of the cracked products could also contribute to an increase in branching. At the same time, such an explanation could not explain the high isobutane fraction, which exceeded the thermodynamic equilibrium at the test conditions. When the extent of branching exceeds the thermodynamic equilibrium concentration, secondary hydroisomerization will contribute to the conversion of branched products to linear products. The direction that secondary hydroisomerization will change the product depends on the extent of branching introduced into the primary cracked product.

Except monomethyl-branched species it approximates for the branching of hydroisomerized feed molecules do not reach the thermodynamic equilibrium distribution (*67*).

Two additional observations can be made with respect to the example in Table 3. Cracked products that have a chain length of n-3 when n is the carbon chain length of the feed had an isoalkane content of around 90 mol % and was higher than the average isoalkane content. Linear cracked products of chain length n-3 can be formed via C-type β-scission (Figure 8), which is reported to be very slow since two secondary carbocations are involved in the rearrangement (*39*), or the even slower cracking from a linear carbocation. Cracking to produce a branched C_{n}-$_3$ and C_3 product mostly took place from a dibranched C_n species, since cracking from a monobranched or linear species was reportedly slow. The second observation is that for non-C_{n}-$_3$ products, the fraction of isoalkanes increased slightly with an increase in the carbon number of the fragment.

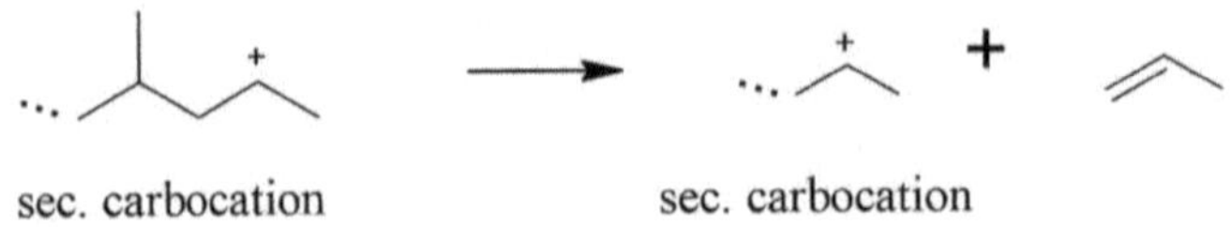

Figure 8. C-type β-scission rearrangement.

These observations should be put into perspective. The high isoalkane content of C_{n}-$_3$ products is not consistently observed (*68*). The increase in isoalkane content with increasing number of carbon is also limited. Limitations related to the number of isomerization possibilities diminishes with an increasing carbon number and the ratio of isoalkanes to total alkanes in the hydrocracked product reaches a value that becomes practically independent of carbon number for $>C_{10}$ (*68*). This can be seen from the distribution shown in Figure 9 (*68*).

Hydrocracking of Waxes

Waxes are mixtures of heavy hydrocarbons, typically with a high proportion of *n*-alkanes. One can anticipate that most of the observations that were made with respect to model *n*-alkane feed materials will apply equally to waxes. This was to some extent illustrated by Figure 9, which shows product characterization from hydrocracking of model *n*-alkane feeds, mixtures of model *n*-alkane feeds, and wax.

Literature on wax hydrocracking includes studies on sulfided base metal, as well as noble metal promoted catalysts on various acidic support materials. The use of Pt-promoted hydrocracking catalysts for wax hydrocracking commonly seeks to obtain the highest selectivity to material in the distillate range for diesel fuel blending, or to hydroisomerize the wax to produce lubricant base oils (*60, 68–81*). Some studies can also be found using Pt-promoted hydrocracking catalysts and polyethylene as feed (*82–84*). Polyethylene is essentially a wax with higher carbon numbers than found in FT or petroleum-derived waxes.

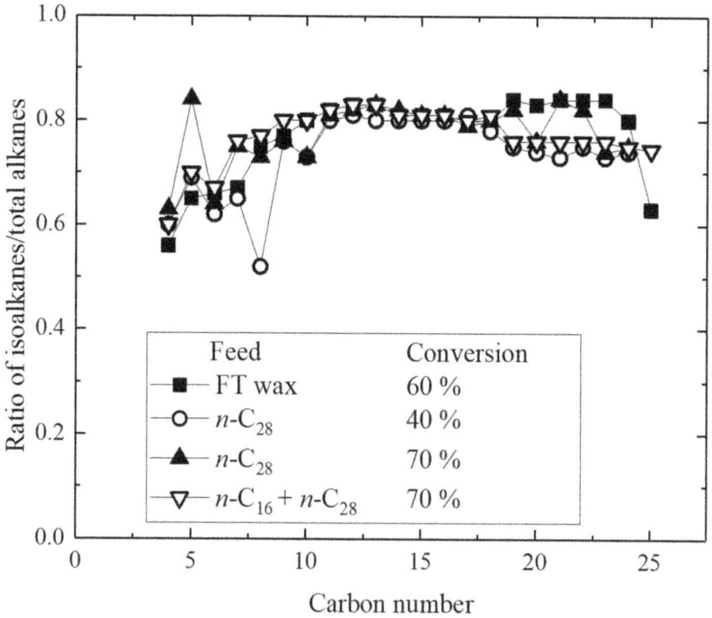

Figure 9. Ratio of isoalkanes to total alkanes in the hydrocracking product of different feedstocks at different conversion levels at 3.5 MPa as a function of the carbon number. Drawn using the data reported in ref (68).

There are two aspects of the hydrocracked product distribution that are important to achieve high quality refined products: the degree of isomerization and the boiling point distribution. The degree of isomerization refers to the number of branches on the carbon chain, which mainly affects the flow behavior at low temperature and at what temperature the material freezes to produce a solid phase. This is important for the blending of lubricant base oil, diesel fuel (especially in colder climates), and jet fuel. The boiling point distribution affects the use of the material, since different types of refined products have different boiling point ranges. Producing products with a boiling point <140 °C is typically less desirable, since it is difficult to blend high octane number motor gasoline with hydrocracked naphtha.

Similar selectivities for middle distillates have been observed experimentally during wax hydrocracking over sulfided NiMo/SiO_2–Al_2O_3 and Pt/SiO_2–Al_2O_3 catalysts and for moderate conversions (Table 4). The observed selectivities were similar for light and heavy FT waxes, as well as refined paraffin wax. However, at conversions above 80%, Leckel (*78*) observed that the selectivity for naphtha over middle distillate increases more strongly for sulfided base metal catalysts than for Pt/SiO_2–Al_2O_3. At the same time, different degrees of branching may be obtained using different catalysts.

FT wax hydrocracking has also been conducted using Pt supported on crystalline zeolites not represented in Table 4 (*69*). A lower selectivity for middle distillates is typically found for zeolite catalysts compared to the amorphous silica–alumina supported catalysts listed in Table 4.

Other parameters (i.e., pressure and H_2 to wax ratio) have little effect on the carbon number selectivity, although conversion increases with a decrease in pressure as long as catalyst deactivation is limited (*77, 80*).

Table 4. Literature-Reported Values of the Selectivity for C_{10}–C_{22} Products from Hydrocracking at Different Pressures, Temperatures, and Using Different Catalysts

Feed	*Catalyst*	*Pressure MPa*	*Temperature °C*	*Conversion of C_{23+} %*	*Selectivity for $C_{10}C_{22}$ %*	*Ref.*
FT wax	Ni–Mo/SiO_2–Al_2O_3	3.5	350	34	82	(*80*)
FT wax	Ni–Mo/SiO_2–Al_2O_3	7	365	37	84	(*80*)
FT wax	Pt/SiO_2–Al_2O_3	3.5–6.5	338–360	20–60	80–90	(*73*)
FT wax	Pt/SiO_2–Al_2O_3	7	340–360	40–90	80	(*78*)
Paraffin wax	Pt/SiO_2–Al_2O_3	3.5	300–330	10–60	70–80	(*70*)

The base metal catalysts are used in the sulfided state.

Pressure has a noticeable effect on the isoalkane content of the lighter fractions, which increases with a decrease in pressure (*77*). This appears to be a general trend, with the isoalkane content increasing as the pressure is decreased (*74*).

Temperature also has an effect on the isoalkane to *n*-alkane ratio in the cracked products. When all other conditions are kept constant, the amount of isoalkanes increases with an increase in temperature and the associated increase in conversion (*74*). The opposite is found for the feed (Figure 6). At the same time, if the impact of temperature is studied as a function of conversion by varying the space time, the isoalkane to *n*-alkane ratio in the cracked products increases as the temperature is decreased for a specific conversion (*59*).

A progressive decrease of the mass fraction of long-chain *n*-alkanes is observed with increasing conversion (*73*). The impact of vapor–liquid equilibrium to enrich the liquid phase in heavier hydrocarbons was also observed for waxes and was confirmed by vapor–liquid equilibrium calculations (*85*).

The data in Table 4 suggests that the nature of the metal does not affect the carbon number distribution of the product. However, differences in the carbon number distribution of products at near similar conversion were reported for wax hydrocracking over Pt and Pd hydrocracking catalysts that used the same amorphous Al_2O_3–SiO_2 support material (*70*). It is not clear whether the observed differences can be ascribed to a difference in metal-to-acid site activity ratio instead of the type of metal. It is also emphasized that the base metal catalysts used in Table 4 are in the sulfided state, with correspondingly lower hydrogenation–dehydrogenation activity (*19*).

Unsulfided (reduced) base metal catalysts exhibit a high activity for hydrogenolysis over hydroisomerization and hydrocracking. Methane selectivities of up to 50% were observed during hydrocracking of *n*-C_{14} over reduced CoMo/SiO_2–Al_2O_3 and gaseous product selectivities of around 20% was found during hydrocracking of wax over reduced Ni/zeolite catalysts (*86, 87*). Sulfiding of the catalyst suppresses hydrogenolysis. However, continuous addition of sulfur species to the otherwise sulfur-free wax feedstock is required to maintain the catalyst in a sulfided state.

A topic that is not often reported or commented on, is the formation of cyclic products during wax hydrocracking. For hydrocracking of FT wax over PtW/SiO_2–Al_2O_3 at 7 MPa and 360 °C, Leckel identified 1.5 wt % aromatics in the distillate fraction of the product (*57*). The aromatics consisted mainly of short chain alkyl-substituted monoaromatics. As pressure is decreased the extent to which cyclic and aromatic products are formed can be increased (*88*).

Overview of Pt Catalysis Relevant to Wax Hydrocracking

This chapter deals with platinum promoted hydrocracking catalysis and it is worthwhile to look at metal catalysis by platinum in more detail to provide a useful discussion of low-pressure hydrocracking over Pt/SiO_2–Al_2O_3.

Dehydrogenation over Pt

Dehydrogenation is the first step in hydrocracking (Figure 2). Dehydrogenation of alkanes over platinum involves three major steps: (1) homolytic cleavage of C–H bond and formation of an alkyl species which is chemisorbed on the catalyst (i.e., forming a C–Pt bond), (2) cleavage of another C–H bond and formation of an alkene, and (3) desorption of the alkene (*89*).

Dehydrogenation and hydrogenation are reversible and equilibrium limited reactions. To achieve higher conversion of alkanes to alkenes, dehydrogenation can be operated at higher temperature or lower pressure. The equilibrium constant for conversion of alkanes to alkenes at a given set of operating conditions increases with an increase in carbon number for light *n*-alkanes (*53, 90*). This effect is significant only for light, normally gaseous *n*-alkanes. For heavier material, as the *n*-alkanes typically found in the feed to a wax hydrocracker, there is little difference in the equilibrium constant for dehydrogenation of different carbon number *n*-alkanes. Kinetically, the rate of dehydrogenation of heavy alkanes to alkenes is higher than that of light alkanes (*91*).

An assumption made in the description of dehydrogenation is that the C–H bonds in alkanes that are equivalent have an equal chance of being dehydrogenated (*92*). This implies that the primary dehydrogenation products of an *n*-alkane will result in a nearly equal probability of the different *n*-alkene double-bond isomers being formed. In the description of the hydrocracking mechanism, it is further assumed that the dehydrogenation–hydrogenation reaction is equilibrated (*36*). This implies that the alkene distribution will resemble the thermodynamic equilibrium following from repeated hydrogenation–dehydrogenation. To achieve such equilibrium it is necessary that over platinum's dehydrogenation–hydrogenation reactions should be fast compared to other reactions.

The assumption that dehydrogenation of *n*-alkanes to *n*-alkenes is equilibrated over Pt was experimentally investigated and compared to calculated equilibrium values (Table 5) (*93, 94*).

Table 5 shows a difference between the equilibrium and experimental *n*-hexene distribution; therefore, equilibrium was not reached. The relative amount of 1-hexene formed was higher compared to the amount anticipated at equilibrium, while 3-hexene isomers were less than anticipated at equilibrium.

Platinum exhibits a low activity for double-bond isomerization through a partially dehydrogenated species, compared to other metals such as Pd and Ni. The order of double-bond migration activity for over metal catalysts is palladium > nickel >> rhodium ~ ruthenium > osmium ~ iridium ~ platinum and the order is unchanged by support or solvent (*95*). The isomer distribution shown in Table 5 indicates a preference for adsorption of specific carbons on Pt. Although it is not clear whether this observation is also valid for longer chain *n*-alkanes, it nevertheless challenges the notion that dehydrogenation is an equilibrated reaction over a Pt-promoted catalyst.

In a study conducted by Paál et al. using 3-methylpentane as feed, the initial selectivity for dehydrogenation over Pt favored *cis*- and *trans*-3-methyl-2-pentenes (*96*). A pulsed reactor system was employed, and it was found that within a short period the dehydrogenation, selectivity changed to favor the dehydrogenation to 3-methyl-1-pentene, which should have been less abundant if the dehydrogenation was equilibrated. The higher selectivity to 3-methyl-1-pentene remained for the duration of the tests, even after catalyst rejuvenation with hydrogen. Dehydrogenation of the branched alkane did not result in an equilibrated mixture of branched alkenes, analogous to the observation made in Table 5.

Table 5. Distribution on *n*-Alkene Isomers Obtained After Dehydrogenation of *n*-Hexane over Pt/Al_2O_3 at $T = 397\ °C, p = 0.1$ MPa and WHSV = 27.7 h^{-1}

Description	*Selectivity / %*				
	1-hexene	*trans-2-hexene*	*cis-2-hexene*	*trans-3-hexene*	*cis-3-hexene*
experimental	28	33	26	8	5
equilibrium	12	24	31	20	13

For comparison, the restricted equilibrium composition of *n*-hexene isomers is given that were obtained from ref (*94*).

Dehydrogenation of alkanes to alkenes is not the only reaction that takes place over Pt. In addition to the formation of dienes, the dehydrogenation of the alkane on Pt is an intermediate step to several other reactions catalyzed by Pt (mentioned previously), namely, hydrogenolysis, skeletal isomerization, and dehydrocyclization. Preference for dehydrogenation at specific carbon positions would influence the type of primary products obtained over Pt.

Hydrogenolysis over Pt

Pt is not prone to hydrogenolysis except under extreme conditions. When hydrogenolysis occurs, the products include methane, but it is not predominantly formed via methanolysis (*97, 98*). The order of activity of noble metals for hydrogenolysis follows: Ru > Ir > Rh >> Pt > Pd (*99*).

Because the first step in hydrogenolysis is dehydrogenation, then the low hydrogenolysis activity is a matter of inherent selectivity, with reactions such as alkene desorption, skeletal isomerization and dehydrocyclization being competing reaction pathways with higher selectivity than hydrogenolysis. Over Pt, the selectivity for isomerization compared to hydrogenolysis is orders of magnitude higher than for any of the other noble metals, including Pd, despite the lower hydrogenolysis activity of Pd compared to Pt (*46*).

Skeletal Isomerization and Dehydrocyclization over Pt

The Anderson–Avery (*47, 48*) description of skeletal isomerization over Pt is such that skeletal isomerization and dehydrocyclization are two different outcomes of the same reaction, rather than two different pathways. An alkane reacting over a Pt catalyst can follow different pathways for isomerization, as shown in Figure 10.

The first type of skeletal isomerization is thought to proceed through a cyclopropane-type intermediate. The surface intermediate leading to this type of isomerization follows from 1,3-

adsorption of the alkane on Pt according to the Anderson–Avery description. The cyclopropane structure that is formed is not protonated, but is otherwise analogous to the intermediate for skeletal isomerization over an acid catalyst as shown in Figure 3. Cycloalkanes can be isomerized following on 1,3-adsorption. Therefore, it is possible for alkylcycloalkanes to be isomerized to ring-opened products, as shown in Figure 11.

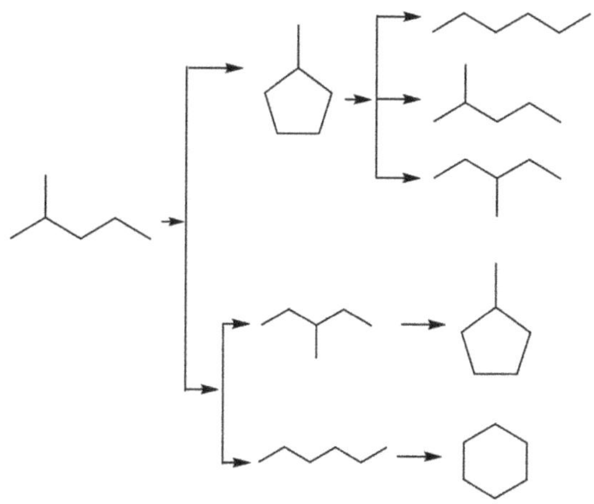

Figure 10. Isomerization and dehydrocyclization over platinum.

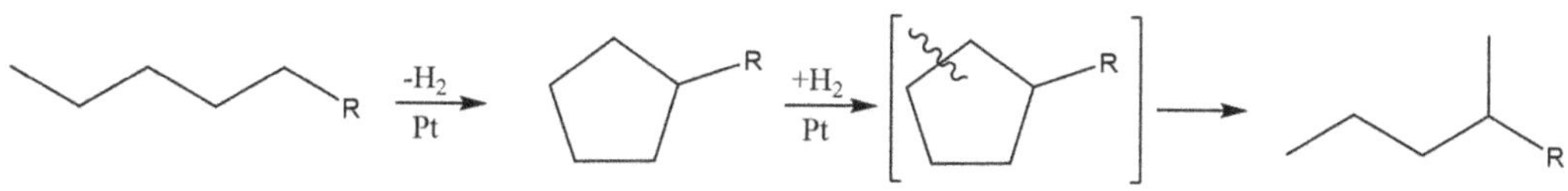

Figure 11. Example of isomerization by dehydrocyclization and hydrogenolysis on platinum.

The second type of skeletal isomerization is ring closure that follows from 1,5- or 1,6-adsorption of the alkane on Pt. This reaction can also be described as a dehydrocyclization reaction. If the dehydrocyclization of 1,6-adsorption is followed by further dehydrogenation, it is possible to form aromatic hydrocarbons. Over Pt the direct dehydrocyclization of *n*-alkanes to aromatics is possible (*28*). Thus, it is possible to form benzene directly from *n*-hexane and from cyclohexane.

It is evident that the nature of the initial adsorption of the alkane on Pt affects the pathway that follows dehydrogenation. It is therefore not surprising that the relative rate of these reactions exhibit structure sensitivity, as demonstrated by differences in reaction rates over different single-crystal surfaces (*100*). At low hydrogen partial pressure, the initial chemisorption on Pt is rate controlling and involves the scission of C–H bonds (*28*). The possibility of reaction by different pathways would then depend on which carbons of the alkane were chemisorbed on the Pt surface.

(1) 1,2-Adsorption can lead to dehydrogenation of the *n*-alkane to an *n*-alkene.
(2) 1,3-Adsorption can lead to skeletal isomerization of the *n*-alkane to an iso-alkane.
(3) 1,5-Adsorption can lead to dehydrocyclization of the *n*-alkane to cyclopentane-species.
(4) 1,6-Adsorption can lead to dehydrocyclization of the *n*-alkane to a compound that contains a cyclohexane or benzene ring.

The hydrogen partial pressure, as well as the nature of the Pt surface, is anticipated to affect the outcome because some of these pathways also involve hydrogenolysis. The effect of the hydrogen partial pressure and the temperature on the selectivity of these reactions was reported in several studies (*100–104*).

Our work found that all of the pathways were active under typical dehydrogenation conditions, as illustrated by the primary C_6 products from the conversion of *n*-hexane over Pt/Al_2O_3 shown in Table 6 (*93*).

Table 6. Product Distribution of Different Isomers Obtained After Dehydrogenation of *n*-Hexane over Pt/Al_2O_3 at $T = 397$ °C, $p = 0.1$ MPa and WHSV = 27.7 h^{-1}

Product	*Selectivity to C_6 products / wt%*[a]
hexenes	49
iso-alkanes	19
cycloalkanes	18
C_6H_{10} species [b]	6
C_6H_8 species [b]	3
benzene	6

[a] Hydrogenolysis leading to lighter than C_6 products was minor but was observed. [b] Multiple species that potentially included unsaturated acyclic and cyclic compounds.

Hydrogenation over Pt

Platinum is an active metal for alkene hydrogenation, but it is not the most active compared to other reduced metals, with the sequence being Rh > Ru > Pd > Pt > Ir ≈ Ni > Co > Fe (*105*). The chain length and the structure of the alkenes affect their rate of hydrogenation. Similar effects of chain length and structure are observed over Pt catalysts and hydrotreating catalysts employing other metals.

During the liquid phase hydrogenation of 1-alkenes over Pt-, Pd-, and Rh-based catalysts, it was (broadly speaking) observed that the rate of hydrogenation decreased with increasing chain length for alkenes from C_6 to C_{17} (*106*). In some instances, two alkenes would have near similar hydrogenation rates. A similar trend was observed for C_6 to C_8 cycloalkenes, but the rate of hydrogenation of cyclopentene was slower than cyclohexene and cycloheptene (*107*). During vapor phase hydrogenation of over reduced (not sulfided) $CoMo/Al_2O_3$ it was similarly observed, broadly speaking, the rate of hydrogenation decreased with increasing chain length for alkenes from C_2 to C_8 (*108*). When naphtha and distillate was hydrotreated as combined feed over a sulfided $NiMo/Al_2O_3$ catalyst, the conversion of alkenes in the distillate was lower compared to that in the naphtha fraction (*109*).

Simultaneous hydrogenation of 1-hexene with 1-tridecene and 1-undecane over Pt was used to determine the relative adsorption coefficients for the alkene-pairs. These results suggest that the length of the substituent of a 1-alkene has no effect on its adsorption coefficient and the differences in reaction rate come from a difference of reaction rate at the catalyst surface (*106*).

The skeletal structure of alkenes has a considerable influence on the rate of hydrogenation. Linear alkenes are more reactive than branched alkenes, the reason for this being that the reactivity of alkenes is affected by the alkyl (or other) groups bonded to the C=C. The presence of a substituent other than hydrogen in the alkene affects the apparent hydrogenation rate constant and the adsorption of alkenes on the catalyst. The rate of alkene hydrogenation decreases with an increase in the number of substituents at the double bond (Figure 12) (*106, 110*).

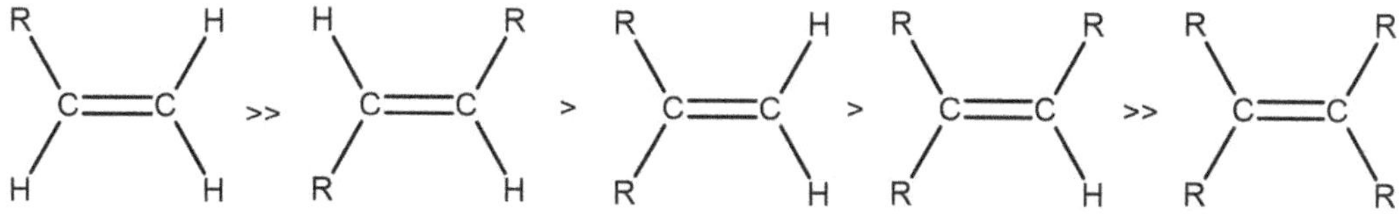

Figure 12. Qualitative dependence of reactivity of substituted alkenes on the degree and position of substitution.

However, not only the number of, but also the structure of the substituent, affects alkene hydrogenation. Bulkier substituents at the double bonded carbon led to lower reaction rates (*108*). Linear free energy relationships were used to find correlations to the dependence of the reaction rate of substituted alkenes on the degree of substitution and the steric effects of the substituents (*111*). Strong correlations between the initial hydrogenation reaction rate and the steric substituent constant used in the Taft-equation were observed (Figure 13). This leads to the conclusion that steric hindrance is responsible for the difference in observed reaction rates for substituted alkenes (*112*).

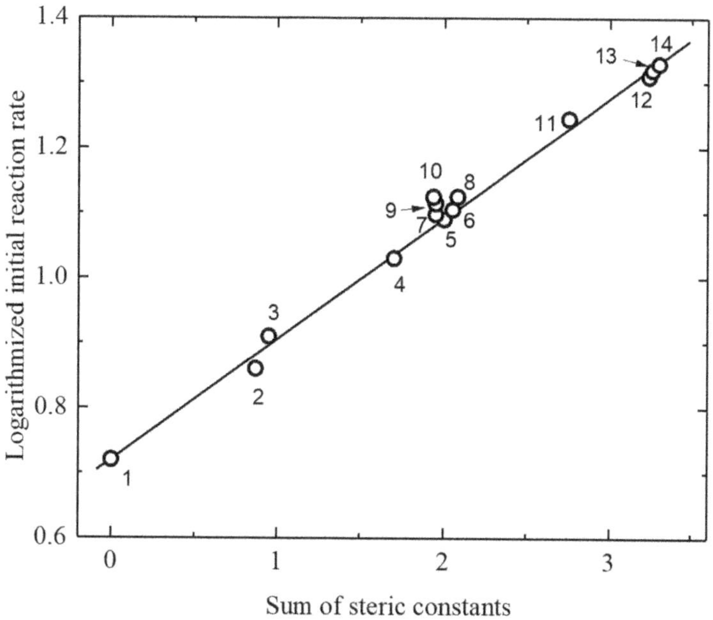

Figure 13. Correlation of relative rates of alkene hydrogenation over a Pt/SiO2 catalyst at T = 293 K against the sum of Taft steric constants of substituents on the double bond. Alkenes according to numbering: (1) 2,3-dimethyl-2-butene, (2) 2,4-dimethyl-2-pentene, (3) 2-methyl-2-hexene, (4) 4-octene, (5) 3-hexene, (6) 2-hexene, (7) 2,3-dimethyl-3-butene, (8) 3,3-dimethyl-1-butene, (9) 4-methyl-2-pentene, (10) 4,4-dimethyl-l-pentene, (11) 4-methyl-1-pentene, (12) 1-octene, (13) 1-heptene, (14) 1-hexene. Drawn using the data from ref (112).

The same principle can be used to explain reactivity differences in linear alkanes. The rate of hydrogenation of internal alkenes is slower than that of 1-alkenes of the same carbon chain length (*113*). It is likewise observed for the hydrogenation of branched alkenes, where the number of substituents surrounding the C=C group affects the hydrogenation rate. For example, the rate of hydrogenation of 2,4,4-trimethyl-1-pentene is faster than that of 2,4,4-trimethyl-2-pentene over Pt/Al_2O_3 and Co/SiO_2 catalysts (*114, 115*).

The impact of steric effects is also seen in the hydrogenation of dienes. In dienes there is a preference for the hydrogenation of the alkene group with the least steric hinderance. For example, Okuhara et al. studied the selectivity of isoprene (2-methyl-1,3-butadiene) hydrogenation over MoS_2 catalyst (*116*). They observed higher selectivity toward formation of 2-methyl-1-butene (77.0%) than 3-methyl-1-butene (20.6%), which indicated that the alkene group with the least steric hinderance was preferentially hydrogenated.

In conclusion, it appears that the impact of steric hinderance on alkene hydrogenation is a general phenomenon observed over Pt-based catalysts as well as other reduced and sulfided hydrotreating catalysts.

Products Observed during Pt/SiO_2–Al_2O_3 Wax Hydrocracking at 2 MPa

The review of hydrocracking with an emphasis on the use of Pt as metal and hydrocracking at lower pressure highlighted a few aspects to be aware of that could be different to conventional high-pressure hydrocracking over sulfided base metal catalysts. These aspects were:

(1) Alkane conversions over Pt proceed through alkane chemisorption and dehydrogenation. The relative position of the carbon atoms involved in the adsorption affected the likely outcome. The initial dehydrogenation of the alkane on adsorption can lead to desorption of an alkene, but it can also lead to the desorption of a skeletally isomerized alkane or dehydrocyclization to a cycloalkane or an aromatic. This was also experimentally demonstrated by reaction over Pt/Al_2O_3 (Table 6).

(2) Platinum has low activity for hydrogenolysis leading to products with shorter carbon chain lengths than the feed and the formation of such products during hydrocracking is likely to be the result of acid-catalyzed cracking.

(3) Dehydrogenation leading to the desorption of an alkene may not be an equilibrated reaction (Table 5). The evidence presented indicated a preference for dehydrogenation closer to the extremities of the *n*-alkane chain, although no corroboration was if this was generally true. However, if *n*-alkane chemisorption that occurs closer to the extremities of the chain was preferred, it would suggest that Pt catalyzed skeletal isomerization would tend to more readily form 2- and 3-methyl-branched isomers.

(4) The fact that "ideal hydrocracking" could be observed over Pt-based catalysts suggests that acid-catalyzed isomerization reactions may be more prone to equilibration than dehydrogenation.

(5) The hydrogenation rate of alkenes is affected by steric hinderance in the proximity of the double bond. Under hydrocracking conditions, the concentration of alkenes with tertiary carbons is therefore likely to be higher than that of alkenes with secondary and primary carbons. This in turn affects the probability of protonation and acid-catalyzed reactions such as cracking.

The products from wax hydrocracking over a Pt/SiO_2–Al_2O_3 catalyst at 2 MPa were fractionated, analyzed, and interpreted using these observations.

Product Distribution of Hydrocracking at 2 MPa

Wax hydrocracking using a medium wax (melting point 54 °C) was performed in a packed bed reactor loaded with 0.5 wt % Pt on a commercial Siralox-30 amorphous silica–alumina support (30 wt % SiO_2). The hydrocracker was operated at a liquid hourly space velocity of 2 h^{-1}, hydrogen-to-wax feed ratio of 600 m^3/m^3, and pressure of 2 MPa. A continuous run was performed lasting 113 h to obtain around 10 kg of product. An atmospheric residue conversion between 23 and 30% was maintained over most of the duration by controlling the temperature in the range 310–345 °C. The product was fractionated in a distillation column to obtain different narrow-cut distillation fractions, which were characterized. Table 7 shows the results of the analysis of four narrow distillation cuts representing different distillate products of a synthetic fuels refinery (*117*).

Not all the compounds in each fraction could be identified. Of relevance to synthetic jet fuel production is that all the *n*-alkanes could be identified, which means that the unassigned compounds in Table 7 are not *n*-alkanes. In addition, all monomethyl-branched alkanes were unambiguously identified in all product cuts. The unassigned compound in the light kerosene and medium kerosene cut likely consist mainly of dimethyl- and multimethyl-branched alkanes.

Table 7. Mass Fraction of Hydrocarbon Types in Hydrocracking Product Distillation Cuts Derived via Gas Chromatography Analysis

Hydrocarbon compound class	*Mass fraction of hydrocarbon type in cut / wt %*			
	Naphtha [a] *<140 °C*	*Light kerosene* *140–145 °C*	*Medium kerosene* *200–205 °C*	*Heavy kerosene* *260–265 °C*
Aromatic	3	1	3	-[e]
Cycloalkanes	11	10	5	-[e]
Linear alkanes	20	18	14	7
Monomethyl-branched alkanes	40	43	29	18
Monoethyl-branched alkanes	1	6	5	75[c]
Dimethyl-branched alkanes	8	18	35[b]	75[c]
Multibranched alkanes	-[d]	2	35[b]	75[c]
Unassigned compounds	15	3	9	75[c]
Average carbon number	8.0	9.1	11.6	14.9
Degree of branching[f]	1.2	1.3	1.8	~ 2.1

[a] Products that elute before *n*-pentane were not identified or included in the analysis. [b] Sum of dimethyl-branched and multibranched alkanes. [c] Sum of monoethyl-branched, dimethyl-branched, multibranched alkanes and unassigned compounds. [d] Could not be unambiguously identified for quantification. [e] Could not be accurately determined by one dimensional gas chromatography due to chromatographic resolution. [f] This is an estimate based only on compound identified as branched. When dimethyl and multibranched (assumed trimethyl) compounds could not be clearly separated, the assumption was made that they were equally represented.

Hydrocarbon Compound Classes in the Hydrocracking Product

Formation of n-Alkanes

The number of *n*-alkanes in the cracked product will be highlighted, because it affects the quality of fuel products that can be produced. For the utilization of the hydrocracking product as jet fuel blending material, a low-mass fraction of *n*-alkanes is favorable as low-mass fractions of *n*-alkanes lead to low freezing points.

The mass fraction of *n*-alkanes decreases with increasing boiling point of the fractions from 20 wt % in the naphtha cut to 7 wt % in the heavy kerosene cut. From primary hydrocracking, *n*-alkanes can be formed via B-type β-scission (Figure 4) or C-type β-scission (Figure 8). The latter was expected to be contributing less based on the reported differences in cracking rate, with the rate of B-type β-scission > C-type β-scission (*50*).

The light hydrocracking products in the naphtha likely reflect the composition of primary cracking products, as they are less prone to secondary isomerization by being present mainly in the vapor phase at the hydrocracking conditions employed ($T > 310$ °C, $p = 2$ MPa). Therefore, the mass fraction of 20 wt % *n*-alkanes in the naphtha cut emphasizes the importance of B-type β-scission and to a lesser extent C-type β-scission. This has some interesting implications.

Of the three possibilities mentioned, B1-, B2-, and C-type cracking, B2-type is the only cracking that follows from a tertiary carbocation. However, B2-type cracking has the specific additional requirement of dibranching separated by one carbon (i.e., having a $-CH(CH_3)-CH_2-CH(CH_3)-$ configuration). In *n*-heptane hydroisomerization, isomerization to yield 2,4-dimethylpentane is not particularly favored or disfavored, yet, cracking to produce C_3+C_4 was overwhelmingly favored compared to other combinations of cracking products (*118*). Furthermore, the increase in *n*-heptane hydrocracking conversion as temperature was increased was not preceded by an increase in the formation of 2,4-dimethylpentane. Unless the rate of hydroisomerization to 2,4-dimethylpentane increased to match the rate of cracking to maintain the amount of 2,4-dimethylpentane and supply precursors for cracking, it appears unlikely that a $j,(j+2)$-dibranched alkane, which is required for B2-type cracking, is the main precursor species for cracking. At the same time, it could be argued that the limited amount of $j,(j+2)$-dibranched alkanes is indicative of the high rate of cracking of the protonated intermediate, which would limit the amount of $j,(j+2)$-dibranched alkenes desorbing to be hydrogenated to alkanes.

B1-type cracking requires the prior formation of an alkane that has dibranching on the same carbon. This is an onerous requirement. It was shown that dibranching on the same carbon follows from skeletal isomerization of dibranching on adjacent carbons (*119*). This means that it is slow to form a dibranched species with both branches being the same carbon, because it requires more isomerization steps than are required to form dibranched isomers that have branching on different carbons. Similarly to the observations about 2,4-dimethylpentane, there was no increase in 2,2-dimethylpentane concentration that preceded heptane hydrocracking to mainly C_3+C_4 products (*118*).

C-type cracking requires only one methyl branch in the precursor carbocation. However, it proceeds via two secondary carbocation intermediates and is therefore slow compared to B-type cracking. For *n*-heptane hydrocracking, C-type cracking is the only cracking mode that can yield *n*-butane. The virtual absence of *n*-butane in the cracking product of *n*-heptane could be employed as argument that even though B-type cracking requires one more isomerization step in the precursor cation, it is the strongly preferred cracking route (*50*). It is difficult to see how C-type cracking would meaningfully contribute to cracking if double-bond isomerization is fast. It is stated that the rate of double-bond isomerization is much faster than that of cracking (*50*), which means that the carbocation should migrate to the tertiary carbon in the precursor species before there is much chance of cracking. A more plausible explanation is that dehydrogenation and protonation form the precursor species for C-type cracking, but that isomerization takes place to produce the 2,4-dimethyl pentane-2-carbocation, which is the precursor for B1-type cracking, which then takes place as second acid-catalyzed step.

Viewed collectively, there is a conundrum, because the precondition for dehydrogenation and protonation of the indicated alkane-precursor species for each of B1-, B2-, and C-type cracking challenges one or more observations in literature.

In the literature, many studies that result in conversion versus isomerization relationships (as shown in Figure 6) use an increase in temperature to increase conversion. The common perception following on the relationships and relative cracking rates for types A to C (Figure 4 and Figure 8) is that hydrocracking proceeds through hydroisomerization to branched species and then the branched species are eventually cracked.

A few studies made use of space velocity at constant temperature, which provides a measure of the initial products during hydrocracking. At constant low conversion, the ratio of cracking to isomerization products that was obtained over catalysts with different Pt loading and the same acid support, and the same Pt loading and acid support with different silica-to-alumina ratios, were not constant (*120*). At first glance this is anticipated, because it is known that the metal-to-acid activity ratio affects hydrocracking catalysis. Highlighting this shows that the number of cracking products does not have a direct causal relationship to the amount of hydroisomerization products. This does not imply that skeletal isomerization is not important in facilitating cracking, but rather that skeletal isomerization on an acid site is not necessarily followed by hydrogenation on a metal site. Isoalkanes are not the only precursor species for cracking. The amount of isomerized species in the hydrocracked product increased at the expense of cracked products as the probability of encountering a metal site to convert alkenes to alkanes and alkanes to alkenes increased (i.e., high metal site to acid site ratio) (*67*).

It is worthwhile to reiterate that hydrogenating branched alkenes is slower than hydrogenating linear alkenes. For a branched alkene to survive running the gauntlet and not be subject to acid-catalyzed conversion again depends on the probability of being hydrogenated relative to being protonated. This does not call the sequence of events shown in Figure 2 into question, but rather shows the importance of the number of acid-catalyzed steps occurring before hydrogenation of the product(s). The hydrocracking mechanism that illustrates only a single acid-catalyzed transformation between dehydrogenation and hydrogenation is therefore misleading because multiple acid-catalyzed steps may take place prior to alkene desorption.

Using rank-analysis to distinguish primary from nonprimary reaction products, Girgis and Tsao found that monobranched products were clearly primary products and multibranched products were clearly not primary products of hydroisomerization (*121*). The authors used poorly fitting curves to show that cracking products could be disqualified as primary products, but in their discussion of the results it was clear that they were ambivalent about it. Girgis and Tsao noted that "...the possibility of some of the cracking products being formed directly from *n*-hexadecane cannot be ruled out..." (*121*). Based on the results, judgment about cracking products being primary or nonprimary is uncertain. There is a probability that for some of the reacting species, the sequence of acid-catalyzed steps leading to cracking may well occur before the alkene has had a chance to be hydrogenated. Again, it does not call the sequence of events shown in Figure 2 into question. Instead, this suggests that repeated reaction at acid sites prior to hydrogenation cannot be ruled out. Differently put, the body of experimental evidence is more easily explained by the sequence dehydrogenation → protonation → $j \times$ (acid-catalyzed conversion steps) → deprotonation → hydrogenation. This does not mean that j cannot be 1, but neither does it require j to be 1.

In conclusion, the *n*-alkane content of product cuts in Table 7 can be explained superficially, but the detailed description of the sequence of reactions are not fully convincing and it appears that multiple acid-catalyzed reaction steps may take place between each dehydrogenation and hydrogenation event.

Formation of Isoalkanes

The isoalkane content in the kerosene fraction is particularly important to achieve low onset of freezing, and is required for blending to jet fuel. The degree of branching is increasingly important as the boiling point of the cut increases, since monobranched species generally have a higher freezing point than dibranched species. The observed increase in the amount of material that is dibranched or multibranched with an increase in boiling point (Table 7) is desirable for jet fuel blending.

The kerosene fractions in Table 7 had a high isoalkane content. The ratio of isoalkanes to total alkanes is 0.79 and 0.83 for the light and medium kerosene cut, respectively. These values are comparable with values of around 0.8 for all compounds from C_9–C_{22} (*68*). For the heavy kerosene fraction, if it is assumed that all unassigned compounds are multibranched alkane isomers, the isoalkanes to total alkanes ratio is higher than 0.9. This is slightly higher than reported values from hydrocracking at higher pressure and at similar conversions (*68*, *70*, *122*).

An aspect that is not clear is to what extent the observed branching was a consequence of hydroisomerization *before* cracking or *after* cracking. If branching was introduced mainly before cracking, then the degree of branching of the isomerized wax should be very high, since some branching is lost on cracking (e.g., Figure 4). When a hydroisomerized FT wax was solvent separated to recover only the more branched fraction, the average degree of branching found ranged from 1.8–2.9 (*123*). The highly branched material that can be solvent separated is only a fraction of the total amount of hydroisomerized wax, with the rest comprising unconverted and monobranched (*60*, *123*). To explain the extent of branching in the cracked product, it would require hydroisomerization to a higher degree of branching to take place at the same rate as hydrocracking to lighter products. This is an onerous requirement to meet to explain all of the branching in the cracked product. These observations indicate that the high degree of branching that was reported for the medium and heavy kerosene cuts in Table 7 includes a contribution of hydroisomerization *after* cracking.

There are two contributing factors that would reduce the degree of branching that can be achieved by hydroisomerization *after* cracking. First, at a hydrocracking pressure of 2 MPa, less of the light material will be present in the liquid phase. The vapor pressure of C_{10} alkenes is of the order 2 MPa at 330–340 °C. Second, the shorter the carbon chain length becomes, the less space there is for isomerization. This is particularly evident in the limitations encountered in the development of an effective C_7 hydroisomerization process (*118*).

There are some previous observations that are pertinent. It was pointed out that dehydrogenation had position preferences and that dehydrogenation of branched species may be structure sensitive. Hydrogenation of alkenes was structure sensitive. The more substituents exist at the double bond and the bulkier these substituents are, the less prone that alkene is for hydrogenation over Pt. Therefore, compared to branched alkenes, unbranched alkenes are hydrogenated faster. Branched alkenes remain alkenes longer and have a higher chance of participating in further acid catalysis, potentially leading to cracking, causing a loss of branched material by cracking.

When considering the secondary dehydrogenation of an alkane, the position of dehydrogenation is not equilibrated over Pt. The primary position of the positive charge of the carbocation formed upon protonation of the alkene is therefore possibly not in equilibrium. The probability of dehydrogenating an alkane with a high degree of branching might be lower and, when it takes place the position of dehydrogenation will tend not to be at the position of branching.

Cycloalkanes and Aromatics

The primary value of cycloalkanes in kerosene from wax hydrocracking is that the cycloalkanes have a higher density than *n*- and isoalkanes, which makes it easier to blend a jet fuel that meets the lower limit of the density specification. The value of aromatics in kerosene is to improve both density and elastomer compatibility. Both aspects will be discussed in more detail. The compounds are also important in the naphtha fraction because heavy naphtha is usually employed as feed to the catalytic naphtha reformer. The cycloalkane (N = naphthene) and aromatic (A) content of the feed affect the N+2A value, which is a measure of the ease of conversion to high octane number gasoline, with higher numbers being preferable (*124*).

The mass fraction of cyclic compounds ranges from 8–14 wt % in the naphtha, heavy naphtha, and light kerosene cuts. A more detailed composition of the cyclic compounds identified in the naphtha cut is shown in Table 8 (*117*).

Table 8. Composition of Cyclic Hydrocarbons in the Naphtha Cut (Boiling Point < 140 °C) Derived from Hydrocracking of Paraffin Wax at 2 MPa over $Pt/SiO_2–Al_2O_3$

Compound	*Mass fraction of cyclic compounds / wt %*
Benzene	1
Toluene	24
Methylcyclopentane	3
Ethylcyclopentane	4
Dimethylcyclopentanes	11
Trimethylcyclopentanes	3
Cyclohexane	2
Methylcyclohexane	22
Other alkylcyclohexanes	31

Methylcyclopentane has the highest mass fraction among the C_6 cyclic compounds. The thermodynamic equilibrium favors methylcyclopentane under the studied conditions. The equilibrium ratio for methylcyclopentane/cyclohexane at reaction conditions was between 5 and 7 (*125*), whereas the ratio observed in the hydrocracking product was 1.5. At 2 MPa and 320 °C under similar H_2/hydrocarbon ratio as used for the hydrocracking experiment, the calculated thermodynamic equilibrium ratio of methylcyclohexane/toluene was 7.7, whereas the observed ratio in the product from hydrocracking was 0.9. These products are formed by direct dehydrocyclization over Pt. Secondary isomerization and dehydrogenation are possible, but the cyclic products are not an equilibrated mixture.

The complexity of the cyclic compounds increases in the heavier cuts. Table 9 illustrates the composition of identified cyclic compounds in the light kerosene cut. The aromatic compounds that were identified in the light kerosene cut consist of not only monoaromatic compounds, but about two thirds of these compounds are bicyclic naphtheno–aromatic compounds (i.e., indane and tetralin derivatives). Similarly, *cis*-and *trans*-decalin were identified as saturated bicyclic nonaromatic compounds in addition to alkylcyclohexanes.

Table 9. Composition of the Cyclic Hydrocarbons Identified in the Light Kerosene Cut Obtained from Hydrocracking of Paraffin Wax at 2 MPa in Terms of Compound Class

Compound	*Mass fraction of cyclic compounds / wt %*
Alkylbenzenes	10
Alkylindanes	12
Tetralin and alkyltetralins	9
Alkylcyclohexanes	43
Decalins	26

The formation of bicyclic compounds from *n*-alkylbenzenes with a side chain $>C_4$ has been observed in the literature under hydrocracking conditions (*126*, *127*). The reaction is proposed to proceed via cyclization of the side chain. Direct ring closure over Pt on a nonacid support has also been observed (e.g., the direct ring closure of *o*-ethyl toluene to indane and of *n*-propyl benzene to indane) (*128*). These reactions represent ring closure between two aliphatic carbons and between an aliphatic and an aromatic carbon. No previous report of the direct route for the formation of bicyclic saturated compounds like *cis*- and *trans*-decalin was found. It is nevertheless evident that ring closure of alkylcyclohexanes could lead to decalin formation in the same way as any of the other cyclic compounds.

To summarize, hydrocracking at 2 MPa leads to a highly isomerized cracked product with higher amounts of aromatics and cycloalkanes formed that usually reported for hydrocracking at higher pressures. The benefit of this type of operation for kerosene as blend material for jet fuel was pointed out (*87*). Both cycloalkanes and monoaromatics are useful compounds in jet fuel blends.

Kerosene from Low-Pressure Wax Hydrocracking as Jet Fuel Blend Material

Kerosene-range hydrocarbons can be broadly classified into five groups: *n*-alkanes, isoalkanes, cycloalkanes, alkenes, and aromatics. The specifications outlined in ASTM D7566 directly and indirectly set limits for specific component groups in semisynthetic jet fuel blends (*129*). Some of the key specification properties and the classes that are relevant to each are tabulated in Table 10.

Due to the origin of hydroprocessed FT material (e.g., obtained by wax hydrocracking) inherently has a high heat of combustion. The heat of combustion is not a listed specification requirement for the blend material, although it is for the final blended jet fuel, which must have a heat of combustion exceeding 42.8 MJ kg^{-1} (*129*).

Table 10. Specifications of Selected Properties for Blended Semisynthetic Jet Fuel Outlined in ASTM D7566 (*129*) and the Hydrocarbon Class Critical to Meet the Specification

Property	*Jet A-1 specification*	*Critical compound class for the property*
Aromatic content	8–25 vol %	Aromatics content directly regulated
Smoke point	Min. 25 mm	Maximum bicyclic aromatic content
Density at 15 °C	775–840 kg m^{-3}	Lower density (*n*-alkanes, isoalkanes, alkenes) in balance with higher density (cycloalkanes, aromatics)
Existent gum	7 mg (100 ml)	Maximum alkenes content, which includes *n*-alkenes, isoalkenes and cycloalkenes
Freezing point	Max. −47 °C	Maximum *n*-alkanes content and to a lesser extent isoalkanes with low degree of branching
Viscosity at −40 °C	Max. 12 mm^2 s^{-1}	All classes, but relationship between viscosity and composition not clear

In most cases the type of compounds affecting the property is understood, but there are two notable exceptions. The relationship between the structure of the hydrocarbons and their cryogenic viscosity and freezing point contribution in the mixture is not clear (*130*). What is empirically known is that kerosene blend components obtained from wax hydrocracking tend to result in blends with higher cryogenic viscosity. It is also known that including material with a high freezing point temperature, such as *n*-alkanes, increases the freezing temperature of the mixture. It is therefore possible to indicate some broad relationship between composition and the type of species that might affect freezing point.

The influence of the composition of the products from low-pressure wax hydrocracking on kerosene blend material is discussed to note beneficial or detrimental directional changes.

Impact of Compound Classes on Properties Relevant for Blending of Jet Fuel

n-Alkanes and Isoalkanes

Noncyclic alkanes have the most favorable turbine combustion characteristics of all compound classes in kerosene (*129*). Therefore, a large mass fraction of these compounds is desirable. They consist of *n*-alkanes and isoalkanes, which mainly differ in their freezing points. The melting points (equivalent to the freezing point for a pure substance) of compounds with the same carbon number can differ drastically as shown for selected C_{12}-alkanes in Table 11.

Table 11. Melting Point and Density of Selected Noncyclic C_{12} Isomers

Compound	*Melting point / °C*	*Density at 20 °C / kg m^{-3}*
n-dodecane	−12	748
2-methylundecane	−46	746
4-methylundecane	−69	750–751
2,4-dimehtyldecane	−90	743–752

The position and the number of branches in an isoalkane affects how much the freezing point decreases compared to the *n*-alkane. Since the freezing point specification only sets a maximum limit for the freezing point, a high fraction of isoalkanes, preferably dibranched and multibranched molecules, is desirable. The pressure at which hydrocracking is conducted affects the alkene concentration and vapor–liquid equilibrium. Directionally, a decrease in pressure increases the alkene concentration and thereby probability of skeletal isomerization, which is beneficial. At the same time a decrease in pressure increases the amount of kerosene-range molecules that will be in the vapor phase. Material in the vapor phase is less likely to be converted over the wax hydrocracking catalyst with liquid filled pores. If the kerosene material in the vapor phase has a low degree of branching, the decreased probability of hydroisomerization could be detrimental to the freezing point of the final product.

Cycloalkanes

Cycloalkanes exhibit similarly favorable combustion characteristics as *n*- and isoalkanes. Compared to *n*-alkanes with the same carbon number, cycloalkanes exhibit lower melting points, somewhat like isoalkanes, as illustrated for selected C_{10}-alkanes in Table 12. Density is the main difference between isoalkanes and cycloalkanes relevant to jet fuel specifications, with the density of monocyclic cycloalkanes being around 70–85 kg m^{-3} higher (Table 12) and bicyclic cycloalkanes exhibit an even higher density.

Table 12. Melting Point and Density of Selected C_{10} Hydrocarbons

Compound	*Formula*	*Melting point / °C*	*Density at 20 °C / kg m^{-3}*
n-decane	$C_{10}H_{22}$	−30	730
2-methylnonane	$C_{10}H_{22}$	−74	727
n-butylcyclohexane	$C_{10}H_{20}$	−75	800
tert-butylcyclohexane	$C_{10}H_{20}$	−41	813
cis-decalin	$C_{10}H_{18}$	−43	887
n-butylbenzene	$C_{10}H_{14}$	−88	860
tert-butylbenzene	$C_{10}H_{14}$	−58	867
tetralin	$C_{10}H_{12}$	−36	969
naphthalene	$C_{10}H_8$	+80	978[a]

[a] Reported at 82 °C.

The presence of some cycloalkanes is desirable in kerosene used for jet fuel because it is the only hydrocarbon compound class together with aromatics that provides higher density. Over a Pt-promoted hydrocracking catalyst cyclic compounds are formed mainly by dehydrocyclization over Pt. A decrease in pressure directionally favors dehydrocyclization to produce cycloalkanes and is therefore beneficial.

Alkenes

Alkenes exhibit a low sooting propensity, which makes them generally desirable for combustion in turbines. However, their poor thermal stability leading to formation of gums limits their inclusion in kerosene used for jet fuel to less than about 1% (*129*). Products from hydrocracking of wax contain virtually no alkenes due to the equilibrium of hydrogenation/dehydrogenation that favors hydrogenation. Directionally, a decrease in pressure increases the equilibrium alkene concentration, but the analysis of the kerosene cuts from wax hydrocracking showed that even under the relatively low pressure of 2 MPa, no alkenes were identified (*117*). However, as the pressure is decreased further the alkene content becomes significant, as illustrated by the results in Table 6.

Aromatics

The less favorable combustion properties of aromatic hydrocarbons compared to aliphatic hydrocarbon in turbine engines is the reason for a maximum aromatic content specification of 25 vol % in jet fuel (Table 10). Bicyclic aromatic compounds have a high tendency for sooting and their concentration is limited more narrowly by the smoke point specification. The minimum aromatic content specification of 8 vol % in jet fuel is to prevent shrinkage of aged elastomer seals in the fuel system of an aircraft. Wax hydrocracking produces little aromatics; however, the number of aromatics in the product is increased directionally by a decrease in pressure. In addition to meeting the minimum aromatic content specification, the secondary benefit of an increase in aromatics content is the increase in the density (Table 12).

Impact of Hydrocarbon Composition on Jet Fuel Specification Requirements

An excellent overview of the composition of kerosene cuts in the boiling range of 190–230 °C with respect to jet fuel specifications was prepared in the work of Cookson et al. (*7*). The mass fractions of three compound groups, *n*-alkanes, aromatics, and the lump of iso- and cycloalkanes, were analyzed in several samples, as well as properties relevant for jet fuel specification. Via statistical methods, empirical linear equations were derived that predict the properties (e.g., density, freezing point, etc.), as a function of the mass fraction of each of the compound classes. A generalized version of these equations is the following:

$$\text{Property} = a_1 \cdot \omega_{n\text{-alkanes}} + a_2 \cdot \omega_{\text{iso- and cycloalkanes}} + a_3 \cdot \omega_{\text{aromatics}} \quad (1)$$

a_i ... empirical coefficient for compound class i
ω_i ... mass fraction of compound class i

From these relationships, the compositional requirements of kerosene cuts for the use as jet fuel can be plotted as lines in a ternary diagram. This gives an excellent indication about whether petroleum-derived and coal tar-derived kerosene with a known composition will meet the specifications. Since the methodology was to relate compositional information to product properties, the same methodology can be applied to kerosene that was derived from other sources than those that were considered by Cookson et al. (*7*). From a composition point of view, it is shows that for a kerosene cut to be suitable for jet fuel, it should contain a high percentage of isoalkanes and cycloalkanes and that only a limited percentage of *n*-alkanes can be tolerated. The number of aromatics is directly regulated (Table 10) but affects some of the other requirements.

Although the methodology to relate kerosene composition to jet fuel properties using Equation 1 may be generally applied, the empirical coefficients that were reported for petroleum and coal tar-derived kerosene (7), do not necessarily describe the relationship for all fuels. This is due to the lumping of iso- and cycloalkanes. Two kerosenes that have different ratios of iso- to cycloalkanes, even if it has the same total amount of iso- and cycloalkanes, will exhibit a different relationship between this lumped compositional group and the fuel properties. It is therefore anticipated that the empirical constants for petroleum and coal tar in (7) are not directly applicable to kerosene cuts with different iso- to cycloalkane ratio (e.g., straight run FT kerosene or kerosene from wax hydrocracking).

Isoalkanes have a lower density compared to cycloalkanes, as illustrated by Table 12. Kerosene derived from wax hydrocracking has a higher proportion of isoalkanes in relation to cycloalkanes compared to most petroleum and coal tar-derived products. The low amount of cycloalkanes in the isoalkane-rich synthetic kerosene derived from hydrocracking of FT wax often leads to densities below the specification for jet fuels (*131*). Thus, the minimum density requirement could be limiting the composition for the wax hydrocracking-derived kerosene if the iso- to cycloalkane ratio is not considered.

To show the impact of the mass fraction of cycloalkanes, the empirical coefficients for the density originally published by Cookson et al. (7) were modified for material with high low-mass fractions of cycloalkanes and high iso- to cycloalkane ratio. Two kerosene materials, Sasol isoparaffinic kerosene (IPK) with boiling range of 175–230 °C (*6*), and the kerosene cut with a boiling point of 180–240 °C of the product from wax hydrocracking at 2 MPa (*117*), were used as basis for the modification. These materials were selected because they exhibit similar initial and final boiling points as the kerosene cuts characterized for the original data by Cookson et al. (7). The main difference between the samples from the original paper and the two kerosene materials is the number of cycloalkanes in the lumped iso- and cycloalkane fraction. Thus, only the empirical coefficient describing the effect of the mass fraction of iso- and cycloalkanes on the inverse specific gravity, a_2 in Equation 1, was modified. This was done by calculating the abovementioned coefficient that correctly predicts the experimentally determined density of the two kerosene cuts with low-mass fractions of cycloalkanes. The results are shown in Table 13.

Table 13. Empirical Coefficients to Determine the Specific Gravity Based on the Hydrocarbon Group Composition According to Equation 1[a]

Description of kerosene	*Density / kg m⁻³*	*Empirical coefficients for inverse specific gravity / m3 kg⁻¹*		
		a_1	a_2	a_3
Kerosene from Cookson et al. (7)	-	1.4503	1.1821	1.0300
Sasol IPK (from (6))	769	1.4503	1.2880	1.0300
Kerosene from wax hydrocracking at 2 MPa	766	1.4503	1.2977	1.0300
Average for both synthetic kerosenes with cycloalkane mass fraction 5–10 wt %	-	1.4503	1.2929	1.0300

[a] Original data by Cookson et al. (7) and modified coefficients for synthetic IPK cuts.

For both kerosenes with low cycloalkane mass fractions, empirical coefficients for the prediction of the inverse specific gravity based on the mass fraction of iso- and cycloalkanes is significantly higher than the one derived for petroleum-based kerosene cuts. This reflects the impact of cycloalkanes on the density once again. The nominal values for the two synthetic kerosenes are quite similar, at least when comparing their difference with the difference from the one derived for petroleum-based kerosene. Thus, the average values of the coefficients can be used to approximately describe the density of synthetic kerosene cuts with a cycloalkanes content.

The coefficients have been used to calculate the composition requirements for density of synthetic kerosene cuts with a low-mass fraction of cycloalkanes. This is shown in Figure 14, together with the original data obtained by Cookson et al. (7). The range shown for the synthetic kerosene cuts is only approximately correct because it is only based on two samples, which do not exactly match the boiling point cuts used by Cookson et al. (7). However, it illustrates that a cycloalkane content in kerosene can greatly reduce the composition window that will enable blending of on-specification jet fuel.

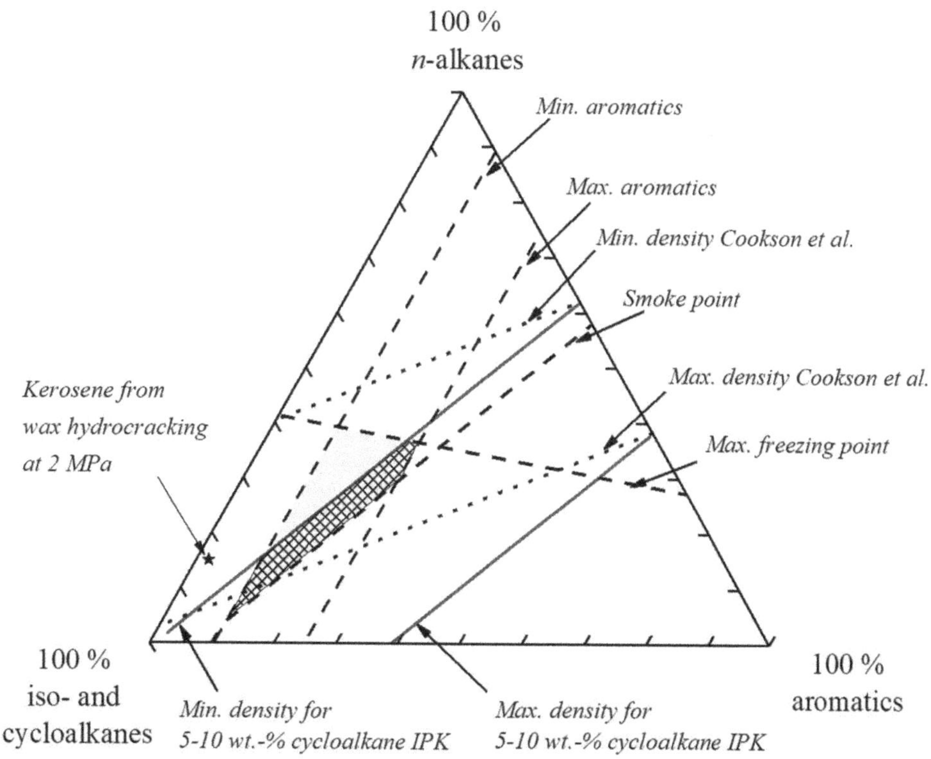

Figure 14. Compositional requirements to satisfy jet fuel specifications in (129). The density limits from Cookson et al. (7) and those calculated for IPK with a low amount of cycloalkanes are both indicated. The compositional space to obtain Jet A-1 from petroleum and coal tar is highlighted in gray and the checkered area is for IPK with low cycloalkane content. The composition of the kerosene cut (180–240 °C) obtained from wax hydrocracking at 2 MPa is marked with an asterisk.

Kerosene from Wax Hydrocracking at 2 MPa for Semisynthetic Jet Fuel Blending

ASTM D7566 specifies requirements for the blending of semisynthetic jet fuel, which is a mixture of synthetic kerosene and petroleum-derived kerosene. Kerosene from oligomerization followed by hydrotreating is called isoparaffinic kerosene (IPK) and kerosene from wax hydrocracking is called synthesized paraffinic kerosene (SPK). For semisynthetic jet fuel, the volume fraction of IPK or SPK is limited to maximum 50 vol % with the other 50 vol % (or more) being petroleum-derived kerosene.

The hydrocarbon group composition and selected properties of several FT derived IPK and SPK materials are listed in Table 14 (*6*), and compared to the 180–240 °C kerosene cut derived from wax hydrocracking at 2 MPa (*117*).

Table 14. Composition and Properties of Different Synthetic Jet Fuel Blend Materials (*6*) Compared to the 180–240 °C Kerosene from Wax Hydrocracking at 2 MPa in this Work

Description		*Sasol IPK*	*Syntroleum S-8*	*Shell GTL*	*Sasol GTL-1*	*Sasol GTL-2*	*This work*
Composition							
n-Alkanes	wt %	2	18	42	72	26	15
Isoalkanes	wt %	88.5	82	57	26	72	70–75
Cycloalkanes	wt %	9.5	0	2	2	2	5–8[a]
Aromatics	wt %	<1	0	<1	0	0	1–2[a]
Properties							
IBP[c]	°C	175	144	156	144	178	180
FBP[d]	°C	230	276	198	208	266	240
Freezing point	°C	<−65	−51	−54	−49	−60	<−70[b]
Density	kg m^{-3}	769	755	736	733	762	766

[a] Estimated from gas chromatography of narrow cuts. [b] From differential scanning calorimetry. [c] IBP = initial boiling point. [d] FBP = final boiling point.

The composition, freezing point, and density of the kerosene obtained by wax hydrocracking at 2 MPa are well within the range of the other FT derived kerosenes. The mass fraction of aromatics is slightly higher, which is likely due to the hydrocracking pressure of 2 MPa, which promotes aromatics formation.

At present ASTM D7566 (*129*) is listing only SPK blend materials that were qualified and that contained less than 0.5 wt % aromatics before blending, which therefore places a specification limit on these materials. There is no reason for this amount of aromatics in SPK to be an absolute limit, given that the blended kerosene requires the aromatics content to meet specification (Table 10). However, no SPK materials with a higher aromatics content have been qualified as semisynthetic jet fuel blending materials yet.

Conclusions

Low-pressure wax hydrocracking over Pt/SiO_2–Al_2O_3 to produce kerosene for blending in synthetic jet fuel was investigated utilizing the literature and supplementing literature observations with experimental observations. Conclusions drawn from this work are:

(1) Dehydrogenation over Pt is not of equal probability along the carbon chain. The alkene distribution from alkane dehydrogenation over Pt is not equilibrated. Dehydrogenation

close to the end of the hydrocarbon chain appears to be favored. In hydrocracking this implies that it may lead to a higher probability of forming 2- and 3-branched species as primary products from hydroisomerization.

(2) Hydrogenation rate is affected by steric constraints around the C=C double bond. Alkene groups with more steric crowding have a lower rate of hydrogenation. In hydrocracking, the implication is that branched alkenes formed by skeletal isomerization over acid sites are slower to be hydrogenated. Branched alkenes with the C=C on a tertiary carbon have a higher chance of not being hydrogenated but being protonated again compared to linear alkanes or branched alkenes where the C=C is further away from branching position.

(3) This study's critical evaluation of literature and experimental observations suggests that there may be more than one consecutive reaction taking place at acid sites prior to hydrogenation. The general sequence in the hydrocracking mechanism should be interpreted as: dehydrogenation → protonation → $j \times$ (acid-catalyzed conversion steps) → deprotonation → hydrogenation, where $j \geq 1$.

(4) Catalysis of Pt proceeds through alkane chemisorption and dehydrogenation, where the position of the carbon atoms involved in the adsorbed intermediate affects the probability of the type of subsequent reaction. On Pt, alkanes can be dehydrogenated to alkenes, skeletally isomerized to isoalkanes, or dehydrocyclized to cycloalkanes and aromatics. The contribution of Pt catalyzed reactions other than dehydrogenation increases in importance during low-pressure hydrocracking.

(5) The degree of branching in kerosene produced by hydrocracking is at least partly determined by hydroisomerization after cracking. The degree of branching in the kerosene affects the freezing point, which is important for jet fuel. As hydrocracking pressure is decreased the rate of kerosene hydroisomerization may increase, but only if vapor–liquid equilibrium does not cause much of the kerosene to be in the vapor phase.

(6) The cycloalkane content in kerosene produced by wax hydrocracking is important and low-pressure hydrocracking over Pt promotes the formation of cycloalkanes. A low cycloalkane content in kerosene can greatly reduce the composition window that will enable blending of synthetic jet fuel that meets specification.

References

1. Demirbas, A. Biofuels sources, biofuel policy, biofuel economy and global biofuel projections. *Energy Convers. Manag.* **2008**, *49*, 2106–2116.
2. Baladincz, P.; Hancsók, J. Fuel from waste animal fats. *Chem. Eng. J.* **2015**, *282*, 152–160.
3. Schäfer, A. W. The Prospects for Biofuels in Aviation. In *Biofuels for Aviation: Feedstocks, Technology and Implementation*; Chuck, C. J., Ed.; Academic Press: Amsterdam, 2016; pp 3–16.
4. Maitlis, P. M. What is Fischer–Tropsch? In *Greener Fischer–Tropsch Processes for Fuels and Feedstocks*; Maitlis, P. M., De Klerk, A., Eds.; Wiley-VCH: Weinheim, 2013; pp 1–15.
5. Basu, P. *Biomass Gasification and Pyrolysis: Practical Design and Theory*; Academic Press: Amsterdam, 2010.
6. Moses, C. A. *Comparison of Semi-Synthetic Jet Fuels Final Report*; CRC Project No. AV-2-04a; Coordinating Research Council, Inc., Universal Technology Corporation, 2008.
7. Cookson, D. J.; Lloyd, C. P.; Smith, B. E. Investigation of the chemical basis of kerosene (jet fuel) specification properties. *Energy Fuels* **1987**, *1*, 438–447.

8. De Klerk, A. *Fischer–Tropsch Refining*; Wiley-VCH: Weinheim, 2011.
9. Weitkamp, J. Catalytic hydrocracking mechanisms and versatility of the process. *ChemCatChem* **2012**, *4*, 292–306.
10. Bricker, M.; Thakkar, V.; Petri, J. Hydrocracking in Petroleum Processing In *Handbook of Petroleum Processing*; Treese, S., Pujadó, P., Jones, D., Eds.; Springer: Cham, 2015; pp 317–359.
11. Minderhoud, J. K.; Van Veen, J. A. R. First-stage hydrocracking: Process and catalytic aspects. *Fuel Process. Technol.* **1993**, *35*, 87–110.
12. *Handbook of Petroleum Refining Processes*, 4 ed.; Meyers, R. A., Ed.; McGraw-Hill: New York, 2016.
13. Robinson, P. R.; Dolbear, G. E. Hydrocracking. In *Springer Handbook of Petroleum Technology*; Hsu, C. S., Robinson, P. R., Eds.; Springer: Heidelberg, 2017; pp 713–776.
14. Kaiser, M. J.; De Klerk, A.; Gary, J. H.; Handwerk, G. E. *Petroleum Refining: Technology, Economics, and Markets*, 6 ed.; CRC Press: Boca Raton, 2020.
15. Cookson, D. J.; Smith, B. E. Determination of the structures and abundances of alkanes and olefins in Fischer–Tropsch products using ^{13}C and ^{1}H N.M.R. methods. *Fuel* **1989**, *68*, 776–781.
16. *Fischer–Tropsch Waxes*; Le Roux, J. H., Oranje, S., Eds.; Sasol: Sasolburg, 1984.
17. Scherzer, J.; Gruia, A. *Hydrocracking Science and Technology*; CRC Press: Boca Raton, 1996.
18. Saab, R.; Polychronopoulou, K.; Zheng, L.; Kumar, S.; Schiffer, A. Synthesis and performance evaluation of hydrocracking catalysts: A review. *J. Ind. Eng. Chem.* **2020**, *89*, 83–103.
19. Pirngruber, G. D.; Maury, S.; Daudin, A.; Alspektor, P. Y.; Bouchy, C.; Guillon, E. Balance between (de)hydrogenation and acid sites: Comparison between sulfide-based and Pt-based bifunctional hydrocracking catalysts. *Ind. Eng. Chem. Res.* **2020**, *59*, 12686–12695.
20. Tanabe, K. *Solid Acids and Bases*; Academic Press: New York, 1970; pp 45–101.
21. Daniell, W.; Schubert, U.; Glöckler, R.; Meyer, A.; Noweck, K.; Knözinger, H. Enhanced surface acidity in mixed alumina-silicas: A low-temperature FTIR study. *Appl. Catal. A* **2000**, *196*, 247–260.
22. Bouchy, C.; Hastoy, G.; Guillon, E.; Martens, J. A. Fischer–Tropsch waxes upgrading via hydrocracking and selective hydroisomerization. *Oil Gas Sci. Technol.* **2009**, *64*, 91–112.
23. Ernst, S.; Kumar, R.; Neuber, M.; Weitkamp, J. Characterization of the effective pore width of molecular sieve solids by means of catalytic test reactions. *Stud. Surf. Sci. Catal.* **1988**, *39*, 531–540.
24. Van Veen, R.; Minderhoud, J.; Huve, L.; Stork, W. Hydrocracking and Catalytic Dewaxing. In *Handbook of Heterogeneous Catalysis*; Ertl, G., Knözinger, H., Weitkamp, J., Eds.; Wiley-VCH: Weinheim, 1997; pp 2017–2037.
25. Matsumoto, H.; Saito, Y.; Yoneda, Y. Contrast between nickel and platinum catalysts in hydrogenolysis of saturated hydrocarbons. *J. Catal.* **1970**, *19*, 101–112.
26. Matsumoto, H.; Saito, Y.; Yoneda, Y. The classification of metal catalysts in hydrogenolysis of hexane isomers. *J. Catal.* **1971**, *22*, 182–192.
27. Bond, G. C.; Webb, G.; Wells, P. B.; Winterbottom, J. M. Patterns of behavior in catalysis by metals. *J. Catal.* **1962**, *1*, 74–84.
28. Davis, B. H. Alkane dehydrocyclization mechanism. *Catal. Today* **1999**, *53*, 443–516.

29. Calemma, V.; Correra, S.; Perego, C.; Pollesel, P.; Pellegrini, L. Hydroconversion of Fischer–Tropsch waxes: Assessment of the operating conditions effect by factorial design experiments. *Catal. Today* **2005**, *106*, 282–287.

30. De Klerk, A.; Furimsky E. *Catalysis in the Refining of Fischer–Tropsch Syncrude*; Royal Society of Chemistry: Cambridge, 2010; pp 165–182.

31. Busca, G. Silica-alumina catalytic materials: A critical review. *Catal. Today* **2019**, *357*, 621–629.

32. Schrauwen, F. J. M. Shell Middle Distillate Synthesis (SMDS) Process. In *Handbook of Petroleum Refining Processes*, 3ed.; Meyers, R. A., Ed.; McGraw-Hill: New York, 2004; pp 15.25–15.40.

33. Perego, C.; Bortolo, R.; Zennaro, R. Gas to liquids technologies for natural gas reserves valorization: The Eni experience. *Catal. Today* **2009**, *142*, 9–16.

34. Coonradt, H. L.; Garwood, W. E. Mechanism of hydrocracking. Reactions of paraffins and olefins. *Ind. Eng. Chem. Proc. Des. Dev.* **1964**, *3*, 38–45.

35. Prins, R. Hydrogen spillover. Facts and fiction. *Chem. Rev.* **2012**, *112*, 2714–2738.

36. Derabandere, B.; Froment, G. Influence of the hydrocarbon chain length on the kinetics of the hydroisomerization and hydrocracking of *n*-paraffins. *Stud. Surf. Sci Catal.* **1997**, *106*, 379–389.

37. Froment, G. F. Kinetics of the hydroisomerization and hydrocracking of paraffins on a platinum containing bifunctional Y-zeolite. *Catal. Today* **1987**, *1*, 455–473.

38. Brouwer, D. M.; Hogeveen, H. Electrophilic Substitutions at Alkanes and in Alkylcarbonium Ions. In *Progress in Physical Organic Chemistry*, Vol. 9; John Wiley & Sons, Inc: Hoboken, NJ, 1972; pp 179–240.

39. Martens, J. A.; Tielen, M.; Jacobs, P. A. Attempts to rationalize the distribution of hydrocracked products. III. Mechanistic aspects of isomerization and hydrocracking of branched alkanes on ideal bifunctional large-pore zeolite catalysts. *Catal. Today* **1987**, *1*, 435–453.

40. Jacobs, P. A.; Martens, J. A. Introduction to acid catalysis with zeolites in hydrocarbon reactions. *Stud. Surf. Sci. Catal.* **1991**, *58*, 445–496.

41. Alvarez, F.; Ribeiro, F. R.; Perot, G.; Thomazeau, C.; Guisnet, M. Hydroisomerization and hydrocracking of alkanes: 7. Influence of the balance between acid and hydrogenating functions on the transformation of n-decane on PtHY catalysts. *J. Catal.* **1996**, *162*, 179–189.

42. Sie, S. T. Acid-catalyzed cracking of paraffinic hydrocarbons. 2. Evidence for the protonated cyclopropane mechanism from catalytic cracking experiments. *Ind. Eng. Chem. Res.* **1993**, *32*, 397–402.

43. Sie, S. T. Acid-catalyzed cracking of paraffinic hydrocarbons. 3. Evidence for the protonated cyclopropane mechanism from hydrocracking/hydroisomerization experiments. *Ind. Eng. Chem. Res.* **1993**, *32*, 403–408.

44. Martens, J. A.; Jacobs, P. A. Evidence for branching of long-chain *n*-alkanes via protonated cycloalkanes larger than cyclopropane. *J. Catal.* **1990**, *124*, 357–366.

45. Sandbeck, D. J. S.; Markewich, D. J.; East, A. L. L. The carbocation rearrangement mechanism, clarified. *J. Org. Chem.* **2016**, *81*, 1410–1415.

46. Maire, G. L. C.; Garin, F. G. Metal Catalysed Skeletal Reactions of Hydrocarbons. In *Catalysis Science and Technology*, Vol. 6; Anderson, J. R., Boudart, M., Eds.; Springer-Verlag: Berlin, 1984; pp 161–226.
47. Anderson, J. R. Metal catalyzed skeletal reactions of hydrocarbons. *Adv. Catal.* **1973**, 23, 1–90.
48. Anderson, J. R.; Avery, N. R. The isomerization of aliphatic hydrocarbons over evaporated films of platinum and palladium. *J. Catal.* **1966**, 5, 446–463.
49. Barron, Y.; Maire, G.; Muller, J. M.; Gault, F. G. The mechanisms of hydrogenolysis and isomerization of hydrocarbons on metals II. Mechanisms of isomerization of hexanes on platinum catalysts. *J. Catal.* **1966**, 5, 428–446.
50. Martens, J. A.; Jacobs, P. A.; Weitkamp, J. Attempts to rationalize the distribution of hydrocracked products. I. Qualitative description of the primary hydrocracking modes of long chain paraffins in open zeolites. *Appl. Catal.* **1986**, *20*, 239–281.
51. Sinfelt, J. H. Catalytic hydrogenolysis on metals. *Catal. Lett.* **1991**, 9, 159–172.
52. Sinfelt, J. H. Specificity in catalytic hydrogenolysis by metals. *Adv. Catal.* **1973**, 23, 91–119.
53. Bond, G. C. *Metal-Catalyzed Reactions of Hydrocarbons*; Springer: New York, 2005.
54. Blakely, D. W.; Somorjai, G. A. The dehydrogenation and hydrogenolysis of cyclohexane and cyclohexene on stepped (high Miller index) platinum surfaces. *J. Catal.* **1976**, *42*, 181–196.
55. Olah, G. A.; Prakash, G. K. S.; Williams, R. E.; Field, L. D.; Wade, K. *Hypercarbon Chemistry*; Wiley-Interscience: New York, 1987.
56. Chaar, M. A.; Butt, J. B. Conversion of n-heptane over Pt- and Pd/SAPO-11 catalysts. *Appl. Catal. A* **1994**, *114*, 287–293.
57. Leckel, D. A closer look at the intrinsic low aromaticity in diesel hydrocracked from low-temperature Fischer–Tropsch wax. *Energy Fuels* **2009**, 23, 32–37.
58. Rossetti, I.; Gambaro, C.; Calemma, V. Hydrocracking of long chain linear paraffins. *Chem. Eng. J.* **2009**, *154*, 295–301.
59. Calemma, V.; Peratello, S.; Perego, C. Hydroisomerization and hydrocracking of long chain *n*-alkanes on Pt/amorphous SiO_2-Al_2O_3 catalyst. *Appl. Catal. A* **2000**, *190*, 207–218.
60. Calemma, V.; Peratello, S.; Stroppa, F.; Giardino, R.; Perego, C. Hydrocracking and hydroisomerization of long-chain *n*-paraffins. Reactivity and reaction pathway for base oil formation. *Ind. Eng. Chem. Res.* **2004**, *43*, 934–940.
61. Regali, F.; Boutonnet, M.; Järås, S. Hydrocracking of n-hexadecane on noble metal/silica–alumina catalysts. *Catal. Today* **2013**, *214*, 12–18.
62. Schulz, H. F.; Weitkamp, J. H. Zeolite catalysts. Hydrocracking and hydroisomerization of *n*-dodecane. *Ind. Eng. Chem. Prod. Res. Dev.* **1972**, *11*, 46–53.
63. Weitkamp, J. The Influence of Chain Length in Hydrocracking and Hydroisomerization of *n*-Alkanes. In *Hydrocracking and Hydrotreating*; Ward, J. W., Qader, S. A., Eds.; ACS Symposium Series 20; Amercan Chemical Society: Washington, DC, 1975; pp 1–27.
64. Denayer, J. F.; Baron, G. V.; Souverijns, W.; Marte,ns, J. A.; Jacobs, P. A. Hydrocracking of *n*-alkane mixtures on Pt/H–Y zeolite: Chain length dependence of the adsorption and the kinetic constants. *Ind. Eng. Chem. Res.* **1997**, *36*, 3242–3247.
65. Wei, J. Nonlinear phenomena in zeolite diffusion and reaction. *Ind. Eng. Chem. Res.* **1994**, 33, 2467–2472.

66. Steijns, M.; Froment, G.; Jacobs, P.; Uytterhaeven, J.; Weitkamp, J. Hydroisomerization and hydrocracking. 2. Product distributions from *n*-decane and *n*-dodecane. *Ind. Eng. Chem. Prod. Res. Dev.* **1981**, *20*, 654–660.

67. Alvarez, F.; Ribeiro, F. R.; Perot, G.; Thomazeau, C.; Guisnet, M. Hydroisomerization and hydrocracking of alkanes. 7. Influence of the balance between acid and hydrogenating functions on the transformation of *n*-decane on PtHY catalysts. *J. Catal.* **1996**, *162*, 179–189.

68. Kang, J.; Ma, W.; Keogh, R. A.; Shafer, W. D.; Jacobs, G.; Davis, B. H. Hydrocracking and hydroisomerization of *n*-hexadecane, *n*-cctacosane and Fischer–Tropsch wax over a Pt/SiO_2–Al_2O_3 Catalyst. *Catal. Lett* **2012**, *142*, 1295–1305.

69. Tomasek, S.; Lonyi, F.; Valyon, J.; Wollmann, A.; Hancsók, J. Hydrocracking of Fischer−Tropsch paraffin mixtures over strong acid bifunctional catalysts to engine fuels. *ACS Omega* **2020**, *5*, 26413–26420.

70. Suárez París, R.; L'Abbate, M. E.; Liotta, L. F.; Montes, V.; Barrientos, J.; Regali, F.; Aho, A.; Boutonnet, M.; Järås, S. Hydroconversion of paraffinic wax over platinum and palladium catalysts supported on silica-alumina. *Catal. Today* **2016**, *275*, 141–148.

71. Seo, M.; Lee, D-W.; Lee, K-Y.; Moon, D. J. Pt/Al-SBA-15 catalysts for hydrocracking of C_{21}-C_{34} *n*-paraffin mixture into gasoline and diesel fractions. *Fuel* **2015**, *143*, 63–71.

72. Pölczmann, G.; Valyon, J.; Szegedi, Á.; Mihályi, R. M.; Hancsók, J. Hydroisomerization of Fischer–Tropsch wax on Pt/AlSBA-15 and Pt/SAPO-11 catalysts. *Top. Catal.* **2011**, *54*, 1079–1083.

73. Calemma, V.; Gambaro, C. Effect of feed distribution on hydrocracking of Fischer–Tropsch Wax. In *Synthetic Liquids Production and Refining*; De Klerk, A., King, D. L., Eds.; ACS Symposium Series 1084; American Chemical Society: Washington, DC, 2011; pp 239–253.

74. Gamba, S.; Pellegrini, L. A.; Calemma, V.; Gambaro, C. Liquid fuels from Fischer–Tropsch wax hydrocracking: Isomer distribution. *Catal. Today* **2010**, *156*, 58–64.

75. Kobayashi, M.; Saitoh, M.; Togawa, S.; Ishida, K. Branching structure of diesel and lubricant base oils prepared by isomerization/hydrocracking of Fischer–Tropsch waxes and alpha-olefins. *Energy Fuels* **2009**, *23*, 513–518.

76. Liu, Y.; Murata, K.; Okabe, K.; Inaba, M.; Takahara, I.; Hanaoka, T.; Sakanishi, K. Selective hydrocracking of Fischer–Tropsch waxes to high-quality diesel fuel over Pt-promoted polyoxocation-pillared montmorillonites. *Top. Catal.* **2009**, *52*, 597–608.

77. Leckel, D. O. Low-pressure hydrocracking of coal-derived Fischer–Tropsch waxes to diesel. *Energy Fuels* **2007**, *21*, 1425–1431.

78. Leckel, D. O. Noble metal wax hydrocracking catalysts supported on high-siliceous alumina. *Ind. Eng. Chem. Res.* **2007**, *46*, 3505–3512.

79. Leckel, D.; Liwanga-Ehumbu, M. Diesel-selective hydrocracking of an iron-based Fischer−Tropsch wax fraction (C_{15}−C_{45}) using a MoO_3-modified noble metal catalyst. *Energy Fuels* **2006**, *20*, 2330–2336.

80. Leckel, D. Hydrocracking of iron-catalyzed Fischer−Tropsch waxes. *Energy Fuels* **2005**, *19*, 1795–1803.

81. Zhang, S.; Zhang, Y.; Tierney, J. W.; Wender, I. Anion-modified zirconia: Effect of metal promotion and hydrogen reduction on hydroisomerization of *n*-hexadecane and Fischer–Tropsch waxes. *Fuel Process. Technol.* **2001**, *69*, 59–71.

82. Bin Jumah, A.; Anbumuthu, V.; Tedstone, A. A.; Garforth, A. A. Catalyzing the hydrocracking of low density polyethylene. *Ind. Eng. Chem. Res.* **2019**, *58*, 20601–20609.
83. Miller, S. J.; Shah, N.; Huffman, G. P. Conversion of waste plastic to lubricating base oil. *Energy Fuels* **2005**, *19*, 1580–1586.
84. Ding, W.; Liang, J.; Anderson, L. L. Depolymerization of waste plastics with coal over metal-loaded silica-alumina catalysts. *Fuel Process. Technol.* **1996**, *49*, 49–63.
85. Correa, S.; Calemma, V.; Pellegrini, L.; Bonomi, S. Waxy streams characterisation to perform VLE calculations in hydrocracking process. *Chem. Eng. Trans.* **2005**, *6*, 849–854.
86. Böhringer, W.; Kotsiopoulos, A.; de Boer, M.; Knottenbelt, C.; Fletcher, J. C. Q. Selective Fischer-Tropsch wax hydrocracking–opportunity for improvement of overall gas-to-liquids processing. *Stud. Surf. Sci. Catal.* **2007**, *163*, 345–365.
87. Hodala, J. L.; Jung, J.; Yang, E.; Hong, G. H.; Noh, Y. S.; Moon, D. J. Hydrocracking of FT-wax to fuels over non-noble metal catalysts. *Fuel* **2016**, *185*, 339–347.
88. Miller, S. J.; O'Rear, D. J. Highly Paraffinic, Moderately Aromatic Distillate Fuel Blend Stocks Prepared by Low Pressure Hydroprocessing of Fischer–Tropsch Products. U.S. Patent Application US2007/0187291, 2007.
89. Sanfilippo, D. Dehydrogenation of paraffins. Key technology for petrochemicals and fuels. *CatTech* **2000**, *4*, 56–72.
90. Bhasin, M. M.; McCain, J. H.; Vora, B. V.; Imai, T.; Pujadó, P. R. Dehydrogenation and oxydehydrogenation of paraffins to olefins. *Appl. Catal. A.* **2001**, *221*, 397–419.
91. Jiang, H.; Ren, S.; Zhou, L.; Wang, Y.; Cao, J. Kinetic model of heavy paraffins dehydrogenation. *Petrol. Sci. Technol.* **2015**, 33, 1305–1313.
92. Germain, J. E. *Catalytic Conversion of Hydrocarbons*; Academic Press: London, 1969; p 91.
93. Ahad, N., unpublished results.
94. Stull, D. R.; Westrum, E. F. Jr.; Sinker, G. C. *The Chemical Thermodynamics of Organic Compounds*; Wiley: New York, 1969.
95. Rylander, P. *Catalytic Hydrogenation in Organic Syntheses*; Academic Press, Inc: New York, 1979.
96. Paál, Z.; Dobrovolszky, M.; Tétényi, P. Transformations of 3-methylpentane and 3-methylpentenes over platinum black catalyst. *J. Catal.* **1976**, *45*, 189–197.
97. Matsumoto, H.; Saito, Y.; Yoneda, Y. Contrast between nickel and platinum catalysts in hydrogenolysis of saturated hydrocarbons. *J. Catal.* **1970**, *19*, 101–112.
98. Matsumoto, H. The classification of metal catalysts in hydrogenolysis of hexane isomers. *J. Catal.* **1971**, *22*, 182–192.
99. Carter, J. L.; Cusumano, J. A.; Sinfelt, J. H. Hydrogenolysis of *n*-heptane over unsupported metals. *J. Catal.* **1971**, *20*, 223–229.
100. Davis, S. M.; Zaera, F.; Somorjai, G. A. Surface structure and temperature dependence of light-alkane skeletal rearrangement reactions catalyzed over platinum single-crystal surfaces. *J. Am. Chem. Soc.* **1982**, *104*, 7453–7461.
101. Offer'eva, T. G.; Krasavin, S. A.; Bragin, O. V. Effect of hydrogen partial pressure on catalytic transformations of C_6 alkanes and methylcyclopentane on Pt/C. *Bull. Acad. Sci. USSR, Div. Chem. Sci.* **1981**, *30*, 447–452.

102. Krasavin, S. A.; Bragin, O. V. Effect of hydrogen of conversions of C_6 hydrocarbons in the presence of alumina-platinum catalysts. *Bull. Acad. Sci. USSR, Div. Chem. Sci.* **1982**, *31*, 1171–1176.
103. Zimmer, H.; Dobrovolszky, M.; Tétényi, P.; Paál, Z. Hydrogen control of platinum-catalyzed skeletal reactions of alkanes: selectivity and surface species. *J. Phys. Chem.* **1986**, *90*, 4758–4764.
104. Rohrer, J. C.; Hurwitz, H.; Sinfelt, J. H. Kinetics of the catalytic dehydrocyclization of n-heptane. *J. Phys. Chem.* **1961**, *65*, 1458–1460.
105. Fabrizio, C.; Trifiro, F. Classification of industrial catalysts and catalysis for the petrochemical industry. *Catal. Today.* **1997**, *34*, 269–279.
106. Růžička, V.; Červený, L. Untersuchung der hydrierung in der flüssigen phase. IV. Zusammenhang zwischen der struktur einiger olefine und ihrer hydrierungsgeschwindigkeit. (Engl. Transl. "Liquid phase hydrogenation. IV. Correlation of the hydrogenation rate of several olefins with their structure"). *J. Prakt. Chem.* **1969**, *311*, 135–140.
107. Jardine, I.; Mcquillin, F. J. Mechanisms of hydrogenation. Part V. The hydrogenation of olefinic and acetylenic substances. The influence of stereochemistry and substitution on chemisorption and on chromatography. *J. Chem. Soc.* **1966**, 458–462.
108. Uchytil, J.; Jakubickova, J.; Kraus, M. Hydrogenation of alkenes over a cobalt-molybdenum-alumina catalyst. *J. Catal.* **1980**, *64*, 143–149.
109. De Klerk, A.; Strauss, M. J. Opportunities for efficiency improvement in high temperature Fischer–Tropsch hydroprocessing units. *Prepr. Pap.-Am. Chem. Soc., Div. Fuel. Chem.* **2008**, *53*, 313–314.
110. Lebedev, S. V.; Kobliansky, G. G.; Yakubchik, A. O. The relative rates of catalytic hydrogenation of different types of unsaturated compounds. Part I. Aliphatic ethylenic derivatives. *J. Chem. Soc. Trans* **1925**, *127*, 417–440.
111. Kraus, M. Linear correlations of substrate reactivity in heterogeneous catalytic reactions. *Adv. Catal.* **1967**, *17*, 75–102.
112. Kraus, M. Substituent Effects. In *Handbook of Heterogeneous Catalysis*; Ertl. G., Knözinger, H., Weitkamp, J., Eds.; Wiley-VCH: Weinheim, 1997; pp 1051–1064.
113. Ghosh, P.; Andrews, A. T.; Quann, R. J.; Halbert, T. R. Detailed kinetic model for the hydrodesulfurization of FCC Naphtha. *Energy Fuels* **2009**, *23*, 5743–5759.
114. Lylykangas, M. S.; Rautanen, P. A.; Krause, A. O. I. Hydrogenation and deactivation kinetics in the liquid-phase hydrogenation of isooctenes on Pt/Al_2O_3. *Ind. Eng. Chem. Res.* **2004**, *43*, 1641–1648.
115. Lylykangas, M. S.; Rautanen, P. A.; Krause, A. O. I. Liquid-phase hydrogenation kinetics of isooctenes on Co/SiO_2. *Appl. Catal. A* **2004**, *259*, 73–81.
116. Okuhara, T.; Itoh, H.; Miyahara, K.; Tanaka, K. Hydrogenation of dienes and the selectivity for partial hydrogenation on a molybdenum disulfide catalyst. *J. Phys. Chem.* **1978**, *82*, 678–682.
117. Link, F., unpublished results.
118. Chica, A.; Corma, A. Hydroisomerization of pentane, hexane, and heptane for improving the octane number of gasoline. *J. Catal.* **1999**, *187*, 167–176.

119. Condon, F. E. Catalytic Isomerization of Hydrocarbons (Mechanisms, Kinetics, and Thermodynamics). In *Catalysis VI: Alkylation, Isomerization, Polymerization, Cracking and Hydroreforming*; Emmett, P. H., Ed.; Reinhold: New York, 1958; pp 43–189.
120. Batalha, N.; Pinard, L.; Bouchy, C.; Guillon, E.; Guisnet, M. *n*-Hexadecane hydroisomerization over Pt-HBEA catalysts. Quantification and effect of the intimacy between metal and protonic sites. *J. Catal.* **2013**, *307*, 122–131.
121. Girgis, M. J.; Tsao, Y. P. Impact of catalyst metal-acid balance in *n*-hexadecane hydroisomerization and hydrocracking. *Ind. Eng. Chem. Res.* **1996**, *35*, 386–396.
122. Calemma, V.; Gambaro, C.; Parker, W. O.; Carbone, R.; Giardino, R.; Scorletti, P. Middle Distillates from Hydrocracking of FT Waxes: Composition, Characteristics and Emission Properties. *Catal. Today* **2010**, *149*, 40–46.
123. Kobayashi, M.; Saitoh, M.; Ishida, K.; Yachi, H. Viscosity properties and molecular structure of lube base oil prepared from Fischer–Tropsch waxes. *J. Jpn. Petrol. Inst.* **2005**, *48*, 365–372.
124. *Catalytic Naphtha Reforming*, 2 ed.; Antos, G. J.; Aitani, A. M., Eds.; Marcel Dekker: New York, 2004.
125. Ciapetta, F. Cycloalkanes. *Ind. Eng. Chem.* **1953**, *45*, 159–162.
126. Sullivan, R. F.; Egan, C. J.; Langlois, G. E. Hydrocracking of alkylbenzenes and polycyclic aromatic hydrocarbons on acidic catalysts. Evidence for cyclization of the side chains. *J. Catal.* **1964**, 3, 183–195.
127. Russell, C. L.; Klein, M. T.; Quann, R. J.; Trewella, J. Catalytic Hydrocracking Reaction Pathways, Kinetics, and Mechanisms of n-Alkylbenzenes. *Energy Fuels* **1994**, *8*, 1394–1400.
128. Davis, B. H.; Venuto, P. B. Paraffin dehydrocyclization. Distribution of aromatic products obtained with "nonacidic" supported Pt catalysts. *J. Catal.* **1969**, *15*, 363–372.
129. *Standard Specification for Aviation Turbine Fuel Containing Synthesized Hydrocarbons*; ASTM D7655-20a; ASTM International: West Conshohocken, PA, 2015.
130. Coetzer, R. L. J.; Joubert, T. S.; Viljoen, C. L.; Nel, R. J. J.; Strydom, C. A. Response surface models for synthetic jet fuel properties. *Appl. Petrochem. Res.* **2018**, *8*, 39–53.
131. De Klerk, A. Aviation Turbine Fuels through the Fischer–Tropsch Process. In *Biofuels for Aviation Feedstocks, Technology and Implementation*; Chuck, C. J., Ed.; Academic Press: Amsterdam, 2016; pp 241–259.

Editors' Biographies

Ajay K. Dalai

Dr. Ajay K. Dalai obtained his B.Sc. Tech. (Petro-Chem) from LIT, Nagpur, India, and M.Tech. in Chemical Engineering from IIT, Kanpur, India, and obtained his Ph.D. in Chemical Engineering with Professors N.N. Bakhshi and N.M. Esmail from University of Saskatchewan. He is working as a Canada Research Chair (Tier 1) and Distinguished Professor in the Department of Chemical and Biological Engineering at the University of Saskatchewan. He has served as Head of Chemical Engineering, as Associate Dean, Research and Partnership in the College of Engineering, and as Special Adviser to VP Research Signature Areas of Energy and Mineral Resources. Dr. Dalai's research focus is the novel catalyst development for gas to liquid technologies, biodiesel production, hydrogen/syngas production, hydroprocessing, value-added products from biomass, and pollution control. Dr. Dalai has published over 550 research papers, mostly in heterogeneous catalysis and catalytic processes in international journals and conference proceedings. He has submitted several patent applications. His ground-breaking research has led to over 28,000 citations of his work and H-index of 80. He serves in the Editorial Board of several prestigious journals. Professor Dalai has received fellowship from different societies such as Royal Society of Canada, Royal Society of Chemistry in UK, International Association of Advanced Materials, American Institute of Chemical Engineering, Canadian Academy of Engineering, Chemical Institute of Canada, Engineering Institute of Canada, and Indian Institute of Chemical Engineers. He is a Fulbright Fellow from USA, DAAD Fellow from Germany and JSPS Fellow from Japan. Professor Dalai has received several national and international awards. Recently, he received the Indian Institute of Chemical Engineers DRC Life Time Achievement Award, Canadian Society of Chemical Engineering R.S. Jane Memorial Award for outstanding contributions to Chemical Engineering Profession, Royal Society of Canada's Miraslow Romanoswki Medal for Outstanding Contribution to Environmental Science, and University of Saskatchewan's George Ivany Internationalization Award.

Dady B. Dadyburjor

Prof. Dadyburjor received his undergraduate degree from the Indian Institute of Technology, Bombay, and graduate degrees from the University of Delaware, working with Jon H. Olson and Stanley I. Sandler. He carried out post-doctoral research with Eli Ruckenstein at SUNY Buffalo. Dr. Dadyburjor's research interests are primarily in heterogeneous catalysis, clean fuels and chemicals, deactivation and regeneration of catalysts and the design of novel catalysts specific for selective reactions. He is a member of the Editorial Board of the Journal of Fuel Chemistry and Technology and a director of the Division of Energy and Fuels of ACS. He is a Fellow of ACS and a Fellow of the American Institute of Chemical Engineers (AIChE).

Ying Zheng

Dr. Ying Zheng is a Professor in the Department of Chemical and Biochemical Engineering at Western University, Canada. Dr. Zheng joined as Assistant Professor in the Department of Chemical Engineering at the University of New Brunswick in 1999 where she was promoted to Associate Professor with Tenure in 2004 and to full professor in 2008. She was awarded the Canada Research Chair (Tier II) in Chemical Process & Catalysis in 2012. In 2015, she joined the University of Edinburgh as Chair Professor of Chemical Engineering/Reaction Engineering. She is a well-known researcher in the field of catalysis and reaction engineering. She is an elected fellow of Canadian

© 2021 American Chemical Society

Academic of Engineering, the Chemical Institute of Canada, Institute of Chemical Engineer and member of the College of New Scholars of the Royal Society of Canada. She is also awarded with the CSChE Syncrude Canada Innovation Award, Imperial Oil Research Award and Humboldt Fellowships.

Aijun Duan

Dr. Duan is a Professor of Petrochemical Processing and Catalysis Engineering in the College of Chemical Engineering and Environment at the China University of Petroleum, Beijing, China. She received her Ph.D. degree in Chemical Engineering from the China University of Petroleum, Beijing (2002); M.Tech. in Chemical Engineering from the China University of Petroleum, EastChina (1993); B.Sc. Tech. in Chemical Engineering from the China University of Petroleum, EastChina (1991). Her research areas focus on the synthesis of novel materials and catalysts for ultra-clean fuel production technologies, reaction kinetics for hydrogenation of transportation fuels, efficient green transformation and optimal utilization of fossil energy. Dr. Duan has published over 200 research papers in several international journals and conference proceedings with citations over 5,700 and an H-index of 45.

William L. Roberts

Dr. Roberts is a Professor of Mechanical Engineering at the King Abdullah University of Science and Technology (KAUST) in Saudi Arabia. His research interests include experimental combustion, propulsion, and laser-based optical diagnostics for reacting flows. Of fundamental interest is the complex interaction between the various length and velocity scales in turbulent flows and the chemical kinetics associated with combustion. His focus is on understanding these interactions in canonical flames, using advanced techniques to measure scalar and vector quantities of interest. He is currently building a unique high-pressure combustion capability at KAUST, which will be used to understand combustion phenomena, particularly the formation of pollutants such as soot, occurring in practical combustion hardware such as gas turbines and internal combustion engines.

Sonil Nanda

Dr. Sonil Nanda is a Research Associate at the University of Saskatchewan in Saskatoon, Saskatchewan, Canada. He received his Ph.D. degree in Biology from York University, Canada; M.Sc. degree in Applied Microbiology from Vellore Institute of Technology (VIT University), India; and B.Sc. degree in Microbiology from Orissa University of Agriculture and Technology, India. Dr. Nanda's research areas are related to the production of advanced biofuels and biochemical through thermochemical and biological conversion technologies such as gasification, pyrolysis, carbonization, torrefaction and fermentation as well as agronomic applications of biochar, phytoremediation of heavy metal contaminated soils and carbon sequestration. He is the editor of books entitled "*New Dimensions in Production and Utilization of Hydrogen*" (Elsevier), "*Recent Advancements in Biofuels and Bioenergy Utilization*" (Springer Nature), "*Biorefinery of Alternative Resources: Targeting Green Fuels and Platform Chemicals*" (Springer Nature), "*Fuel Processing and Energy Utilization*" (CRC Press), "*Bioprocessing of Biofuels*" (CRC Press) and "*Biotechnology for Sustainable Energy and Products*" (I.K. International Publishing House Pvt. Ltd.).

Indexes

Indexes

Author Index

Subject Index

A

B

D

F

U

W

Z

Printed in the USA/Agawam, MA
January 20, 2022

788083.002